Pa Bell

A. Jean de Grandpré

& THE METEORIC RISE OF BELL CANADA ENTERPRISES

LAWRENCE SURTEES

Random House
Toronto

Published in Canada in 1992 by Random House of Canada Limited, Toronto.

Canadian Cataloguing in Publication Data
Surtees, Lawrence, 1956-
    Pa Bell

Includes index
ISBN 0-394-22142-7

1. Grandpré, A. Jean de. 2. Bell Canada Enterprises – History.
3. Businessmen – Canada – Biography. I. Title

HE8870.B45S87 1992   384.6'065'71   C91-093990-X

Jacket design: Falcom/Design & Communication Inc.
Author photo: Jo. E Jones, St. Catharines, Ont.

Printed and Bound in Canada

10 9 8 7 6 5 4 3 2 1

Dedicated to Nancy Gale, for her love,
to Jessica, so she may learn,
and to the fond memory of Walter J. Surtees, for all he taught

Table of Contents

Acknowledgments

The idea for this book originated during a noon-time walk in the early spring of 1989 when a former colleague asked if I had any ideas for a book. "Just one," I replied, as I thought about the then pending retirement of A. Jean de Grandpré and the fact that neither a full account of the man nor the conglomerate he built had yet been told.

I wish to express my appreciation to Mr. de Grandpré for agreeing to be interviewed for this project and giving generously of his time. He and his family jealously guard their privacy, a shield I could not pierce, but I am indebted to his brother, Louis-Philippe de Grandpré, for sharing his family records and his recollections, and his son, Jean-François de Grandpré.

My deep thanks to all those people who agreed to be interviewed. This book relies on interviews with almost one hundred people who have been connected with Mr. de Grandpré's life and career. Those who consented are identified in a list accompanying the Bibliography, and I would also like to thank those who wished to remain anonymous, as well as the countless sources who, while not mentioned in this book, have informed my understanding of the companies and events described over the last decade of my reporting career. I am particularly grateful to the following for their generosity and hospitality: Robert Carleton Scrivener, Dr. H. Rocke Robertson, Louise Brais Vaillancourt, Edmund Fitzgerald, Walter Light, Jim Thackray, Francis Fox, Dr. John Meisel, and Arthur Campeau.

For research help, I wish to thank Stephanie Sykes, chief librarian and director of Bell Canada's corporate archives in Montreal, for obtaining hard-to-find records and answering many queries. I am grateful to the staff of the National Library and Public Archives of Canada in Ottawa, and I would like to acknowledge my gratitude to Jacline Labelle of the public examination room of the CRTC in Hull, Quebec, for her patient assistance in meeting my requests over the past ten years and Celia Donnelly, librarian at *The Globe and Mail* in Toronto, for her assistance and sense of humor. And I would like to thank Father Louis Bertrand Raymond, director of alumni affairs at Collège Jean-de-Brébeuf in Montreal, for providing a copy of the 1940 class list.

Various public relations officers went out of their way to arrange interviews. I am appreciative for the support shown by Monic Houde, Linda Gervais, Rod Doney and Doug Peck at Bell Canada; Stephen Bowen, Gerald Levitch and John Lawlor at Northern Telecom; and Anne Kelly at Bell Canada International Inc.

I owe a special thanks to Stephanie MacKendrick in Toronto for starting me down the journalistic path that led to this book. When I set out in Ottawa to begin covering the regulated communications sector, I was privileged to be tutored by five exceptional lawyers who have taught me how the industry, and Ottawa, really work. I wish to thank Stuart M. Robertson for his friendship and support, T. Gregory Kane, Charles Dalfen, Gordon Kaiser and Allan R. O'Brien.

My wife, Nancy Gale, and daughter, Jessica, suffered through every word of this book. I dedicate this work to them for their support. Nancy has patiently read every word of the many drafts and suggested many beneficial changes, proving again that she is my best editor. I also dedicate this work to the memory of my late father, whose work as an electronics scientist exposed me in childhood to the technologies that are so central to the businesses described in this book.

I am indebted to Andrew MacIntosh, now at the Montreal *Gazette*, for that inspirational noon-time walk and for sharing some of his documents cited in Chapter Three. Nor would this book have been written without the coaxing and enthusiastic support of Helen Heller, my agent; Ed Carson, former publisher of Random House of Canada Ltd.; and Doug Pepper, executive editor.

Introduction

On May 3, 1988, A. Jean de Grandpré strode up to the podium in the Congress Hall of the Montreal Convention Centre and stunned shareholders at the BCE Inc. annual meeting by announcing he was stepping down as chief executive officer of the giant holding company. The news ended over two years of speculation and rumor about when de Grandpré would retire from the conglomerate he had created in 1983. Although sixty-six years old at the time, he had once joked there were no retirement rules at BCE.

As he began a retrospective look back on his career, a twenty-year-old photographic portrait of BCE's founder and chairman flashed onto a giant screen behind him. The dark hair was now gone, but his fit, powerful build gave the impression that he could go on leading the corporation for many more years. In projecting the larger-than-life likeness, however, the meeting's organizers unwittingly captured the perception that many people had of de Grandpré — an aristocratic and imposing titan of the Canadian business establishment. The more than eight hundred shareholders in attendance seemed to agree and responded by giving him a standing ovation after he completed his address.[1]

Yet de Grandpré was not prepared to retire completely. He retained his post as chairman, which allowed him to be the titular head of BCE. A year later he relinquished that position at the 1989 annual meeting. On that occasion, though, there were no ovations. His successor had already begun dismantling the diversified empire which had been tarnished by the acquisition of a number of troubled enterprises.

His business career is a story of both change and conflict.

Change was the hallmark of de Grandpré's tenure at Bell Canada and BCE — a period that was the most exciting era in the telephone industry's history. Bell Canada's name is not normally associated with either change or excitement. Steeped in the century-old traditions associated with providing what the industry calls "POTS" — the nickname for "Plain Old Telephone Service" — and encouraged to be fiscally conservative by its hundreds of thousands of shareholders, it is little wonder that Bell has been perceived as staid and boring. Its aggressive opposition to competition made it the corporate personification of "monopoly"; its conservative investment strategy led it to be touted as Canada's preeminent "widow-and-orphan" stock; and its doctrinaire adherence to the principles of the Bell system did not foster imaginative management or daring vision. Radical technological change in the postwar era turned the industry on its head. The technologies of the Information Revolution led to major market innovations and spawned new businesses that clamored for regulatory and policy changes to end the telephone company's cherished monopoly status.

One of the most important changes of this period was the transformation of Bell Canada's manufacturing arm, Northern Electric Company (now Northern Telecom Limited), into Canada's first global manufacturer since Massey-Ferguson. De Grandpré not only inherited the vision to build Northern Telecom into an independent global manufacturer, but was also handed the task of managing the many conflicts wrought by those changes.

Guided by a global outlook, he consciously set out to make the utility a world-class competitor. "I am an internationalist," de Grandpré once said regarding his upbringing by his French-Canadian mother and American-born father. Well traveled, he boasted of his companies doing business in ninety-six countries, and he maintained apartments in New York and Toronto, in addition to his home in Montreal. Yet he was not an adventurer. There are pictures of him shaking hands with Arab potentates, but no one ever heard of de Grandpré kicking off his shoes and dancing with a Saudi sheik in a Bedouin tent village. His is a safe, sanitized world of chauffeur-driven limousines, corporate jets and elevator rides to secure high-rise office suites and condominium apartments.

Yet there was a paradox underlying de Grandpré's bid to go global. While seeking an entrée to foreign markets, he was loathe to reciprocate at home, fighting every regulatory decision that sought to make any of Bell's markets competitive. Given a choice, he would have liked to have competed abroad, but not at home.

Bell's greatest success came in Saudi Arabia, where, in 1978, the utility won one of three parts of the largest telecommunications infrastructure contract ever awarded. But the Saudi contract also marked a watershed in Bell's corporate evolution by setting de Grandpré and the Canadian Radio-television and Telecommunications Commission on a confrontational path that led ultimately to the corporate reorganization which made the regulated telephone utility a subsidiary of Bell Canada Enterprises Inc., now named BCE Inc.

In terms of its scope and impact on the corporate landscape, the formation of BCE was one of the most important reorganizations of large regulated business since the early twentieth century. Although BCE is only nine years old, it is one of the largest corporations in Canada. By most measures, it is one of the wealthiest — in 1986, it was *the* wealthiest. BCE was the first domestic corporation to earn more than $1 billion in profit after taxes, a feat it achieved less than three years after the 1983 reorganization was consummated. Within that period, de Grandpré transformed Bell Canada from a sleepy monopoly utility into a giant management holding company at the pinnacle of Canada's corporate establishment. Revenue almost tripled in the past decade, to almost $17 billion when he stepped down as chairman in 1989 from $6 billion in 1980. Its firms do business on five continents and its products and services, including telephone service, telecommunications equipment and financial services, touch Canadians in all corners of the country. BCE's position as the most widely owned corporation in Canada also means that its enormous profit is shared with the largest number of shareholders in the country.

BCE isn't just big — its corporate clout has an enormous impact on the economic well-being of Canada:

- Its largest subsidiary, Bell Canada, has more than seven million subscribers in Ontario, Quebec and the Northwest Territories;
- It has almost 280,000 shareholders and its stock trades on fourteen stock exchanges in North America, Europe and Asia;
- BCE and its almost one hundred corporate affiliates and subsidiaries employ 119,000 people;
- It owned almost $42 billion in assets at the end of 1990 and ranked among the top three Canadian corporations by revenue and profit;

- BCE can raise over $3 billion a year in debt and equity and, in a good year, pays out more than $1 billion of dividends to its shareholders.

De Grandpré's power and that of the corporation he headed grew hand-in-hand. By the early 1980s, his achievements and stature eclipsed most of his peers in corporate Canada, unseating Paul Desmarais, head of Montreal-based Power Corporation, as the "most important francophone in Canadian industry."[2] His board appointments to several of the top *Fortune 500* corporations and connections throughout the international business community also gave him respect and influence abroad that few Canadian businesspeople have attained. What most distinguished the feisty and sometimes mercurial de Grandpré from his peers, however, was his burning desire to have his way over equally influential and rival politicians, mandarins, competitors and public interest groups. And it was the use of his power that attracted so much attention.

His response to competitors often reflected the adage, "Don't get mad — get even." He sometimes used the corporate jet to unexpectedly confront and intimidate an adversary, ranging from cabinet ministers to newspaper publishers. Many have been taken aback by the intensity of his argument, interpreting his passion as a personal attack. He particularly recalls an instance when he vigorously disagreed with former Governor-General Jeanne Sauvé when she was minister of communications. Although they had grown up in the same neighborhood, he appeared to be so angry with her that she telephoned his wife at home and asked her, "'Is it because I'm a woman, or why? I've never had such a difficult conversation in my life [as] with your husband today.'" Hélène de Grandpré replied that her husband "'probably just wanted to tell you what he wanted and it will be forgotten tomorrow.'"[3]

Francis Fox, former federal minister of communications, marveled at de Grandpré's complex personality: "He was a hard-working, hard-driving, quick-witted individual. He was also very friendly, but sort of an overpowering steamroller. Jean de Grandpré . . . has an . . . extremely friendly and warm personality, but there is a lot of steel in his personality at the same time."[4] The one word Fox says that describes de Grandpré is "tough": "Toughness, I think, is a fundamental trait of Jean de Grandpré. The difficulty, I suppose, is that people who are tough have a tendency to be inflexible." [5]

But the prevalent impression of de Grandpré is of a monopolist's monopolist, reflected in his nickname, Pa Bell. Given the scale and scope of BCE's interests, it is little wonder that de Grandpré has often equated his wishes with the national interest: What's good for BCE or Bell is good for Canada. In the wake of de Grandpré's retirement and BCE's "restructuring" — Bay Street parlance for shucking money-losing or unproductive assets — some investors have asked whether BCE's shareholders and, indeed, whether Canada, is better off economically because of the corporation's creation. However, de Grandpré's transformation of Canada's largest monopoly utility, Bell Canada, into a management holding company which became the largest corporation in Canada transcends BCE's tattered acquisition record.

Chapter 1
The Making of Pa Bell

In all societies . . . two classes of people appear — a class that rules and a class that is ruled.

— GAETANO MOSCA, ELEMENTI DI SCIENZA POLITICA (1896)[1]

A. Jean de Grandpré set out for his first day at school on a sunny day in September 1927. It was a short walk, past rows of Montreal's famous walk-up apartments to the private elementary school, Jardin d'Enfants. Operated by the Sisters of Providence, the school was located next to the Institute for the Deaf and Dumb on Rue St-Denis. But Jean had a more eventful first day than most children. Although suitably dressed for the warm day, his short pants were prohibited attire at the Roman Catholic school. The offended nuns sent the boy home to change into more suitable clothing, telling him not to return unless he conformed to their rule to wear either stockings or breeches.

"That's ridiculous, it's too hot," his mother told him when he unexpectedly appeared at home. Her son would defy the school's rigid dress code. She sent him back to school with a note for the nuns that threatened to enroll him in the neighboring school run by the Jewish community. "I can't understand why my son cannot show his knees when your own John the Baptist is only covered with a lamb's skin!" Jean's mother told the nuns in her letter.[2] Anxious not to lose the tuition, the sisters capitulated and allowed the boy to take his seat in his shorts.

Bespectacled, skinny and often sickly in his early youth, Jean had been ridiculed by youngsters in his neighborhood before that morning.

But the experience, which was orchestrated by his mother as much for his own benefit as anyone else's, made the boy a *cause célèbre* among his classmates. And the lesson of that first day at school was not lost on the six-year-old: it is preferable to lead than to be led.

Albert Jean de Grandpré's name and background has occasionally misled observers to believe that he straddles two cultures. He is the son of an American-born Montreal insurance company executive, Roland de Grandpré, and a Quebec-born mother, Aline (neé) Magnan. However, although his father was born in Fall River, Massachusetts, on Independence Day, July 4, 1891, his ancestry is entirely French Canadian. The family is descended from Charles Duteau, a Frenchman who arrived near Batiscan in 1662, making the family one of the oldest in Quebec. After his marriage to Jeanne Rivard a decade later, the couple settled at Champlain, near Trois-Rivières. Their son, Alexis Duteau, was born in 1698 and married Marie Charlotte Brisset, the daughter of a seigneur, in 1722. At the time, the only way to hold land in New France was by way of a concession from a seigneur, who in turn held the land by a grant from the King of France. Alexis was granted land on the island of Dupart near Soreil on the St. Lawrence. "They were the first to settle there, which is understandable, since there were only 3,000 settlers in all of Canada in the late 1600s," says de Grandpré's older brother, Louis-Philippe, who has traced the family's lineage.[3]

Alexis's son, Jacques, was born in 1724. For reasons not recorded, he both changed the spelling of Duteau to Dutaut and added the surname "de Grandpré," which means "of the big meadow." One of his sons moved off the island to a new parish near Saint-Cuthbert, where his family farmed the land for the next century. It was there that de Grandpré's grandfather, also named Louis-Philippe, was born in 1862.

After studying at classical colleges in L'Assomption and Joliette, de Grandpré's grandfather graduated from medical school in 1885 and set up practice in Manville, Rhode Island, later that year. De Grandpré says his grandfather emigrated to southern New England as a young man because of its milder climes. More likely, like many francophones of those years, he wanted to escape the depressed economic climate of Quebec in the 1880s. Fall River was a natural choice because it had a large francophone community of ex-patriot Quebeckers. Although now a depressed city whose stone grist mills house discount factory outlets, Fall River was then a prosperous textile center located on the east shore of the Sakonnet River next to the Rhode Island-Massachusetts state line, about thirty-two kilometers from the summer mansions of America's aristocracy in Newport.

The young doctor's practice did well and he married a well-established physician's daughter, Marie Louise (neé Beaudry), in 1887. Five years later, she died of a sudden illness when de Grandpré's father was just a year old. When his own health began to deteriorate soon after, Louis-Philippe left his son and daughter in the care of a nanny in Fall River and went to California, thinking the warmer climate would help his condition. He then traveled to France to do post-doctoral studies in the new field of infectious diseases and public health. Returning from Paris in 1896, he moved his children first to Montreal and five years later back to New England, to Worcester, Massachusetts.

De Grandpré's father, Roland, received most of his schooling at the Grammar and Classical High School in Worcester, but decided to move to Montreal alone at the age of fifteen in 1906 to complete his high-school education at the Commercial and Technical High School. "He became fluently bilingual at the time. He studied French by himself by [using] the books of grammar. He had the professor's books, and not the student's books, and he could correct whatever he was working on by himself,"[4] de Grandpré recalls being told.

About two years later, Louis-Philippe remarried and joined his son and daughter in Montreal, where he was appointed medical inspector in the city's infectious diseases department and was responsible for public health in local schools.[5]

Roland landed his first job the following year as a junior at Evans & Johnson, a small Montreal insurance company. He worked at several insurance companies in Montreal, rising through the ranks with steady promotions, despite his lack of formal actuarial or business training. As with French, he taught himself what he needed to know. "He was a self-made man in the true sense of the word," de Grandpré says. "He was known to be probably the most expert technician of insurance in Canada — that was his reputation. My father was the only person I know who has read a dictionary of insurance!"[6]

His aunt Anita was equally accomplished and driven to succeed. A nun at the large convent run by the Sisters of the Holy Name on the north slope of Mount Royal in Outremont, she held a Ph.D. in mathematics and was the director of studies at the convent school.

Roland married Aline Magnan on May 10, 1916, and the following year became the Continental and Fidelity-Phenix's examiner in charge of eastern Canada. Their first child, Louis-Philippe, was born on February 6, 1917. Roland was a dashing figure, not at all stern looking, with dark, straight hair parted to one side, a trim mustache and small, round wire-rimmed glasses. Yet he was a short, rather frail man who always spoke very softly.[7] By contrast, his wife was a tall, erect, strong-willed woman who came from a family of twelve children.

Aline's father, Arthur Magnan, was a Montreal tax collector. A former hardware merchant, he had been hired by the federal government as director of the customs and excise office to clean up a scandal in the hardware export-import business. De Grandpré's maternal grandmother was descended from a family of distinguished lawyers headed by Charles C. de Lorimier, a former judge in the Quebec Superior Court. Judge de Lorimier was widely regarded in his time for writing one of the most important treatises on Quebec civil law, a thirteen-volume annotated compendium of the Quebec Civil Code. He was born in Dubuque, Iowa, where his father had taken up residence after being exiled from Lower Canada for his role in supporting the 1837-38 rebellions against British rule. His brother was less fortunate.

François Marie Thomas Chevalier de Lorimier, de Grandpré's great-great-great uncle, was one of the dozen *"les patriotes"* hanged by the British for their role in the rebellions led by Louis-Joseph Papineau. De Lorimier was hanged at the gallows outside the Montreal jail on February 15, 1839. He remains a martyr for the cause of Quebec independence to this day because of an extraordinary political testament he wrote to his compatriots from his cell on the eve of his death, a passionate document that remains the manifesto of Quebec independence. In his letter, de Lorimier expressed the hope that his country would one day be liberated from British rule. His emotional appeal to his fellow *patriotes* ended with the cry,

> As for you, my compatriots, may my execution and that of my companions on the scaffold be of use to you. . . . For them do I die on the gibbet the infamous death of the murderer . . . . and for them do I die exclaiming: "Long live liberty! Long live independence!" (. . . et pour eux je meurs en m'écriant: "Vive la liberté, vive l'indépendance!")[8]

De Grandpré was born in Montreal on September 14, 1921, the middle child in a family of three sons. He grew up in the era that historian E. H. Carr termed the "twenty years' crisis,"[9] the period of unbridled political corruption and excessive, over-heated economic growth between the two world wars, which spawned the Great Depression and the advent of regulated monopolies. Roland was promoted Quebec Superintendent of Continental and Fidelity-Phenix early in 1921, and the family enjoyed the prosperity of the Roaring Twenties through his hard work. They would have been considered middle class at that time, living in modest comfort in a walk-up flat on Rue St-Hubert, a busy north-south thoroughfare that parallels Rue St-Denis and St-Urbaine. De Grandpré's brother remembers that the street teemed with children. "I think we had more fun than children do today because there were so many more of us and because there wasn't as much money around. We had to make our own fun then. We all lived in flats, with the stairs outside, so there were many more families nearby than in the suburbs."[10] He remembers one neighbor who lived a couple of flats away who belonged to a notorious gang of bank robbers. After the murder of the driver of an armored car during a heist in the early Twenties, the neighbor was hanged. For the most part, though, the neighborhood was populated by young professionals.

Nineteen twenty-one was an important year for de Grandpré's thirty-year-old father for another reason. He was appointed Canadian manager, or chief agent, of two U.S.-based insurance firms, the New Hampshire Fire Insurance Company of Manchester, New Hampshire, and Granite State Fire Insurance Company, a position he held for the next thirty-five years until his premature death in 1956 from liver cancer. The respect with which he was held by his peers was shown when he was elected head of the two leading organizations of the industry; he served as president of both the Canadian Underwriters Association and the Dominion Board of Insurance Underwriters during 1954-55.

The Montreal of de Grandpré's boyhood was the mercantile and financial capital of Canada. His father's offices were located in the Insurance Exchange Building at 276 St. James Street, which is now known by its original name, Rue St-Jacques. The building stood next to the stately Molson's Bank, on the south side of what was then the most important street in Montreal's financial district. Directly across the street was the headquarters of the Canadian Imperial Bank of Commerce; to the east, about a block away, the tomb-like head office of the Bank of Montreal; and, just a few doors to the west, the stately façade of the Royal Bank's headquarters, allowing the street to lay claim to being the hub of Canadian capitalism.

His new appointment admitted Roland de Grandpré to the ranks of the Montreal Establishment. He joined the Montreal Reform Club, the now defunct haven for Liberal party organizers, supporters and politicians, rather than the more popular Club Saint-Denis. Although Roland was not as politically active as his father, he had a number of political connections and did some electoral work at the provincial level. He was a good friend of Hector Perrier, former secretary of the province of Quebec, and helped organize his campaigns. "My dad was a Liberal, and he was a Democrat because he was born in the United States. He was never a nationalist in the narrow sense; he was a nationalist from a Canadian standpoint. He loved the country and tried to do what was best for it. He thought that the world would belong to whoever wanted to move forward," de Grandpré says, echoing the theme of Sir Wilfrid Laurier's Liberalism. "It is not surprising that I inherited that same vision of life. I cannot understand a narrow nationalism, not in this country, and the view that we should concentrate on the Canadian scene."[11] While proud of his francophone roots, as evident in his early choice of his French middle name over his English first name, de Grandpré eschewed Quebec nationalism and often recited Sir Wilfrid's famous line that the twentieth century would belong to Canada.

He attributes his father's and his own political views to his grandfather's influence. A close friend of both Sir Lomer Gouin, Liberal premier of Quebec from 1905 to 1920, and Joseph-Israel Tarte, minister of public works from 1896 to 1902 in Sir Wilfrid's first government,[12] the elder de Grandpré was very active in the Liberal party and often took to the campaign hustings to give speeches for his friends.

In addition to politics, all three generations shared a similar view of religion and the Church. "We were always religious in the sense that we were going to church at least once a week. We were very respectful of the Church as an institution, but we were never subservient to the Church people, in the sense that we were not prepared to accept everything without question," de Grandpré says. "We were looking at our faith with open eyes, instead of being blinded. There is an expression in French, '*la foi du charbonnier*,' which means to accept everything the way it is presented to you, never applying your own judgment. That is not the way we were brought up." Although de Grandpré received most of his schooling from nuns and priests, "we were never prepared to accept the words of a priest just because he was saying it."[13]

He also perceived his father's business career as a challenge to Catholic orthodoxy and another example of how the family refused to blindly accept the views of the Church on matters of commerce and career paths. As de Grandpré explains it, the teachings of the Church led to the economic domination of Quebec's francophones. "At the time, money was dirty. For a long, long time, we were told by the priests, 'money is for the Protestants,' and not to go into mercantile operations. It stinks, but that's the way it was. That's why the WASPs [White Anglo-Saxon Protestants] dominated. They were in the business world; they were retailers, marketers and traders. It's not by accident, it's part of the traditions."[14]

Although his view of the Church's role in Quebec is rejected by most historians, what matters is de Grandpré's life-long perception that his family's regard for commerce went against the cultural norm of the francophone world. To the young, impressionable de Grandpré, his father's career proved that a francophone could succeed in the anglophone business world. The perceived questioning of the Catholic orthodoxy's conservative assumptions by de Grandpré's parents remains one of the greatest influences on his intellectual and personal development. And the respect for the profit motive that both parents instilled in their sons also helped prepare the boys for their subsequent careers in the business world as corporate lawyers and directors. In fact, de Grandpré's older brother wanted to study commerce after graduating from law school, but could not afford to continue his studies because of the Depression.

De Grandpré remembers their home as being quiet, apart from visits by family and school friends. Most of all, he relished his grandfather's weekly visits every Sunday morning after mass. "He would always bring us chocolate bars."[15] In fact, the only vice de Grandpré recalls his father having was an addiction to chocolate; he would consume about four to five pounds a week. Roland was introverted and rarely socialized, except to play the occasional game of bridge or to meet with his political cronies at the Reform Club. "My father had no hobbies. He never played any sport; he never played outdoor hockey. He was a worker. That's all he was. . . . And that's the kind of environment in which the three children were brought up."[16] Reading was the closest thing to a hobby that his father had. He read every book and magazine he could get his hands on. That thirst for knowledge and love of books rubbed off on his sons, and the family often spent the evenings together reading.

His parents also passed on their view that with knowledge and discipline came advancement. De Grandpré's father differed from the well-heeled members of the Establishment because he did not represent "old money." He assumed his place in the executive ranks through his own travails and discipline, rather than nepotism or an inherited position. Nor did he rest once he got there. Propelled by a desire to advance his family's social position further, he often took on extra work.

Success and ambition permeated the household. Discipline was the cardinal virtue and freedom came after discipline, a view that reflected not only the preachings of the Roman Catholic Church, but also the values of two parents who strived to better their own lot in life and who believed that their children, through better schooling and the same hard work, could do far better than they. The motto of the household was hung in a picture frame on the wall and constantly reminded de Grandpré that, "Everything that needs to be done has to be done right."[17] Yet he denies that it was a pressured or stressful environment. "It was a very relaxed environment, but we had to produce the goods; as long as we were producing, everyone was happy."[18]

Aline de Grandpré was delegated the role of disciplinarian, a role that suited her more assertive nature and that she was forced to fill due to the very long hours that her husband worked each day and most weekends. Louis-Philippe recalls their father often worked twelve hours a day, seven days a week during the Twenties: "My mother had a theory of fathers, about how much they contribute to creating children: about 5 per cent."[19] But, de Grandpré says, "If we had gone wrong, I'm sure that my father would have taken over."[20]

Both parents shared the same ambitions for their children. Aline, however, had a burning desire for her sons to succeed and to be the absolute best. And not just the best in popular fields of the day, such as law, medicine or the priesthood. She wanted her sons to be *leaders*. She succinctly expressed that sentiment in a story de Grandpré is fond of recounting. She once said half-jokingly, "I have three sons, and I want one to be the pope, the second one to be a judge of the Supreme Court of Canada and the third one to be the prime minister."[21] All three brothers, in fact, became lawyers, and Louis-Philippe was appointed to the Supreme Court of Canada by Prime Minister Pierre Elliott Trudeau in 1974.

Aline had a sense of humor, though, and she often surprised her sons' friends with the jokes she told at the supper table. "She had three boys, so she had to bring them up the way you bring up boys. You don't always tell them very demure jokes."[22] Although they weren't overly risqué, they "were not the kind of jokes that were expected from a mother who appeared to be very proper." Roger Dussault, one of de Grandpré's boyhood friends, recalls that, as well as being a friendly woman, ". . . she was a leader. She didn't mince words."[23] Her sons inherited her outspokenness.

A Bell Canada employee who accompanied de Grandpré and his mother to the Montreal airport many years later was struck by the affection that mother and son showed toward each other and remembers the tremendous respect and devotion that he showed to her. De Grandpré acknowledges that he owes much to his mother, particularly for his education. She wanted her sons to receive the best schooling possible. "We were not that extraordinary, it's just that we had a mother that could teach us things," de Grandpré says, referring to his mother's decision to keep her sons at home during what is now called kindergarten. "I remember vividly how I learned what the equator was and what the solar system was . . . she used a rubber band, a round orange and the axis was a knitting needle." After his morning lesson, he had the rest of the day to play. He remembers her saying, "You won't play in the snow when you're forty years old."[24]

De Grandpré has fond recollections of his childhood. His older brother says he was a "normal child" who liked to "play on the street like everybody else. He was no more or less playful or studious than the norm."[25] For the first three years of his life, though, he was in poor health. Aline had to feed him special food, and his eyeglasses caused him to be the brunt of jokes from other children and regular teasing from his brothers. Apart from the teasing, the three boys were close and enjoyed playing together.

The summers of his early childhood were spent on a farm at Isle Castort on Lac St-Pierre, a large widening of the St. Lawrence River near Berthierville, Quebec, which is about eighty kilometers northeast of Montreal. His father rented a house for the summer on an aunt's farm. It was a large, prosperous farm with about fifteen horses and at least forty head of cattle. Even on their holiday, however, the family's work ethic prevailed and de Grandpré helped out with the chores. His relatives installed a small seat on the front of a horse-drawn hay wagon so that the six-year-old boy could help lead the horses while the older men worked the thresher and baler. The family also spent many Christmas holidays at the farm, exposing de Grandpré to the traditions of the rural Québécois. Along with midnight Christmas feasts, he remembers getting in trouble one winter when a horse got stuck in a snow fort he had built.

In 1931 as the Great Depression struck, the family moved to a new house in Outremont. Today, the three-story, cube-shaped red brick house at 31 Beloeil Avenue is squeezed between more recently built sprawling bungalows. At that time, though, it stood on a large lot, which had been purchased from the Sisters of the Holy Name Convent, where Anita taught. Starved for cash, the convent had subdivided its orchard on the north slope of Mount Royal. Located two blocks west of the Outremont city hall, Beloeil is a short street that runs south of Côte Ste-Catherine and up the steep slope of Mount Royal until stopping abruptly at Maplewood. The house was certainly not grand or elegant by Mount Royal standards, but its location meant Roland de Grandpré had arrived.

De Grandpré forged his closest, and life-long, friendships after the move to Beloeil Avenue. He met Roger Dussault on a neighborhood tennis court when he was ten years old, and the two have been friends for almost sixty years. He lost another very close friend from those years, Jean Dadépatie, in a tragic automobile accident in 1957.

Asked if de Grandpré possessed any qualities that distinguished him from his childhood peers, Dussault says "his leadership. He was always *the* leader," but quickly adds that de Grandpré was "not the type who imposed his view on others."[26] They were both active in sports. In addition to tennis, they played baseball and hockey together. De Grandpré and his brothers organized hockey games on their backyard skating rink in the winter and, in the summer, ran ping-pong tournaments in their basement with as many as fifteen friends. Their mother provided food and drink. "We knew that she was . . . exacting and demanding. On the other hand, she was prepared to make life very pleasant for us. If you had not produced at school, the ping-pong tournaments and the hockey games would come to an end,"[27] de Grandpré recalls.

The francophone district of Outremont on the north side of the mountain, privileged as it was, has nevertheless been described as an isolated enclave, encircled by anglophones who dominated Montreal's society at the time. De Grandpré was raised in a household where French was the spoken language, and he received a rigorous, traditional classical education at private French-speaking schools. Although he says he was bilingual when he left Jardin d'Enfants, he was not exposed fully to English culture or institutions until he entered law school at McGill University on the other side of Mount Royal.

Beloeil Avenue was also far removed from the economic hardships that engulfed most of Canada during those years. Although his father took a one-third cut in pay and "was lucky to keep his job," de Grandpré admits that he and his brothers "never suffered."[28] In addition to the move into a more expensive house, the family owned a car, had hired a maid and spent summers in the country.[29] Their father also enrolled his three sons in exclusive private schools at the height of the Depression.

It was a brisk twenty-minute walk along the wide Côte Ste-Catherine to Collège Jean-de-Brébeuf, which de Grandpré attended for six of the eight years between 1932 and 1940. Although he won a bursary to attend a rival school, de Grandpré was enrolled in the prestigious Collège Brébeuf after he completed his primary schooling. During his first year, the school changed its policy and compelled students to board on campus, even if they lived a short distance away, so de Grandpré was transferred for his second and third years to Collège Sainte-Marie. In his fourth year, he moved back to Collège Brébeuf after it removed its boarding requirement.

Collège Brébeuf played a crucial role in forging de Grandpré's character and personality. It combined high school with a four-year undergraduate degree and, like every student, he not only walked through the doors of its imposing pillared façade six days a week, Monday through Saturday, but was also obligated to return to the cream-colored buildings on Sunday morning to celebrate mass.

Founded in 1928, Brébeuf was named after Jean de Brébeuf, the Jesuit missionary priest who arrived in Canada in 1625 and was martyred after being killed by Indians in 1649. When de Grandpré attended the school, it was regarded as one of the finest academic institutions in Canada and offered one of the best liberal arts educations available. But it was more than the Montreal equivalent to Toronto's privileged Upper Canada College; it was the flagship of the Church-run *collèges classiques*. From the early 1900s on, these schools were assailed by their critics for a reliance on a traditional core curriculum steeped in the humanities, which gave short shrift to mathematics and science. Critics argued that the Catholic curriculum did not teach its students to search critically for the truth and that the teaching was from summaries, rather than the original works of literature and philosophy.[30] While students of Brébeuf may have agreed with that assessment of the Catholic education system in general later in life, they did not share that view about their own instruction at Brébeuf, which was noted for its rigorous discipline, long hours and demanding classical curriculum that stressed rationality and Cartesian logic. The curriculum included entire courses on rhetoric and debate, syntax, composition, belles-lettres, several philosophy courses, four languages (French, English, Latin and Greek), mathematics, algebra, physics, chemistry and biology.

Brébeuf stood out from the typical *collège classique* in one other important way. It was elitist. Nationalists and other critics of the classical education system pointed out that the high cost of tuition made education at Brébeuf the prerogative of the sons of the upper class. In 1954, for example, the average annual fee of $700 for tuition and room and board required a family income of over $3,000, placing education at the college out of reach of more than 80 per cent of French-Canadian families.[31] Brébeuf could not list a single working class student among its class lists of more than six hundred students enrolled in the 1953 academic year.[32] That disparity, which bred the heightened sense of privilege attached to education at the college, was possible as long as the Church maintained its monopoly over the educational system in Quebec. Even Rev. Rodolphe Tremblay, a former principal at the school, recognized its elitism. "I don't like to say it, but it had a snob reputation," Father Tremblay told the Canadian Press in an interview in December 1986, a month after the Jesuits gave up the college because their religious order no longer had enough priests to run it.[33] The privileged status of Brébeuf during de Grandpré's attendance contrasted starkly with the deprivation of the Depression: as his classmates handed around the works of Locke, de Tocqueville and Thoreau, the soup kitchens handed out food to the swelling ranks of the destitute and unemployed.

For those able to afford it, education at a private school is, as McMaster University sociologist Wallace Clement termed it in his 1975 analysis of the corporate elite, a "life-long asset,"[34] rewarding its recipient not only with the profit of superior teaching, but also the added dividend of friendships with members of the future elite. The list of de Grandpré's classmates in the 1940 graduating class is studded with the names of recent political and professional leaders: Pierre Elliott Trudeau; Gilles Lamontagne, Quebec lieutenant-governor and former minister of national defence; Guy Viau, former curator of Quebec's provincial museum and former head of Canada House in Paris; Marcel Cazavan, former president of Quebec's giant pension fund, the Caisse de dépôt et placement du Québec; Guy Monty, the former president of Hydro Québec International; and Roger Dussault, co-founder of the largest school textbook publishing company in Quebec.[35]

Unlike de Grandpré's father, they were not rags-to-riches, self-made men. De Grandpré and his peers came from ambitious, privileged homes and were molded purposely by their Jesuit teachers for greater things. The teachers at Brébeuf shared the same ambition for their charges as de Grandpré's mother held for her sons: to make men to lead others. That purpose, viewed as the mission of the Jesuits' educational order in Canada, challenges the widely popular view which holds that leaders make themselves.

Students were educated about the roles and responsibilities of leadership. That aspect of their instruction was steeped in the values of Catholicism, which "taught the difficult and negative aspects of those roles and . . . the responsibilities of having greater rights," says a former student. "We were taught to follow very strict standards, to desire to be the best. And we were told, don't expect to be thanked for doing something good; that is expected."[36]

The success the priests at Brébeuf had in attaining their goal of producing future leaders can be measured by the high number of degrees they have conferred on future political leaders, such as Trudeau, Quebec premier Robert Bourassa, and former premier Pierre Marc Johnson, as well as scores of cabinet ministers and business and professional leaders, including former federal communications minister Francis Fox and stockbroker and former chairman of the Montreal Chamber of Commerce, Jean Ostiguy. "The Jesuits were extraordinary teachers," de Grandpré says. "They motivated us properly and we absorbed it like sponges." He viewed them as highly successful in their mission to "build personalities, as well as leaders."[37]

Looking back at his years at Brébeuf, de Grandpré still marvels at his classmates, not so much because of their success later in life, but because of how well they did as a class. "It was probably the best class that went through Brébeuf," de Grandpré says, echoing a view shared by the priests who are still there. "Would you believe that, once or twice, five of us had an average of 90 per cent in every single exam, and 90 per cent overall? This has never been repeated nor had it been seen before," he boasts, proud of his good fortune in having been among a group of ambitious, bright young men who were driven to learn.[38] De Grandpré does not attribute the unprecedented standing of his class only to coincidence, adding that their voracious appetites for knowledge were fed by a coterie of brilliant teachers.

De Grandpré says he had no single mentor at Brébeuf, but he mentions several teachers who most inspired him: A. Brossard, who taught rhetoric, M. Buist and M. Lamarche, who taught philosophy; and Robert Bernier. Father Bernier, whom Trudeau has called his mentor, taught de Grandpré and Trudeau humanities, or *belles-lettres*. He told Edith Iglauer in a 1969 interview for *The New Yorker* magazine, "There were about fifty in my class, and the idea was to train a boy's mind and personality so that he would be ready for anything and would learn rapidly. It was quite a special group. . . ." At the time, Father Bernier was a young priest from Manitoba, "discovering the world of culture and America at the same time they did."[39] In addition to teaching his students French, Greek and Latin literature, he exposed them to Cézanne, Picasso and other contemporary artists, to modern architects like Le Corbusier and to classical musicians through concerts at the Montreal Symphony. In Father Bernier's teachings, the Jesuit's pursuit of the rational was combined with an appreciation of beauty.

People who have met both de Grandpré and Trudeau notice certain qualities of their intellects and leadership abilities, which they attribute to their common instruction at Brébeuf. "I think there was something that he [de Grandpré] had from his education that Trudeau also had: an ability to synthesize information and to think of problems very logically, like mathematics,"[40] says Robert Bandeen, former president of Canadian National Railway, who dealt with both men in the 1970s.

Propelled by his ambition and intellect, de Grandpré excelled at Brébeuf, in intramural and extracurricular activities as well as his studies. Not surprisingly, given his family background, debating was one of his favorite pastimes. Known as the Rhétoricien, he sharpened his rhetorical skills in Brébeuf's debating society and developed his Churchillian sense for the comeback line by debating against Trudeau. "He was a good debater and made a great speech," Dussault says of his classmate. "He was actually a lot like Trudeau. He doesn't leave much to chance."[41]

In addition to the debating society, he and Trudeau were among the co-founders of a new body, Académie sciences-arts de Brébeuf, set up to promote the intellectual development of its members through discussion and to encourage their further cultural development. They were intellectual bons vivants who referred to themselves as "les snobs." De Grandpré was the first president of the Academy and Trudeau was its secretary. Reflecting their select status, they gave their body the Latin motto, *Viam veritatis elegi* ("I am chosen to see the truth").

De Grandpré was also involved in student politics and distinguished himself further among his peers by serving as class president. Trudeau served as his vice-president and Guy Viau as secretary. Much has been made about a perceived rivalry between de Grandpré and Trudeau during those years. Says de Grandpré, "He was a very private person and I was not what you would call a close friend. We were classmates and we respected each other. But I would never ring his doorbell." De Grandpré's future wife, Hélène Choquet, on the other hand, was a member of Trudeau's circle of friends and went out with him a few times during their school years.[42] De Grandpré did develop a close friendship during his student years with Trudeau's sister, Suzette.

Although de Grandpré dismisses the role of class president as "not that significant," a student from a later class says, "the really big guy on campus was the class president. At Brébeuf, to be class president is the ultimate honor. It is also a very great responsibility." Essentially, it involved representing the class before the priests and acting as class spokesman. In a sense, the class president practiced what the Jesuits preached, applying their lessons and values of leadership under the priests' watchful eyes. In later years, those who sought the position campaigned early, usually several years before they reached their graduating year, but the post was less political and more honorific when de Grandpré held it.

Politically, Brébeuf had a tide of its own that ebbed and flowed in its embrace or rejection of Quebec nationalism. "It flirted with nationalism," says a former student who later helped organize the federalists in the 1980 referendum on Quebec sovereignty. The nationalist tide was very much out at Brébeuf during de Grandpré's time there, unlike, for example, at Collège Sainte-Marie, where the rector had highly politicized the school in the late 1930s. Quebec's nationalists were later challenged by two great anti-nationalists who were classmates of de Grandpré — Pierre Vadeboncoeur and Pierre Trudeau, who co-edited an intellectual journal named *Cité libre*. It was at Brébeuf, under the tutelage of teachers like Father Bernier, that Trudeau and Vadeboncoeur shaped their own criticism of traditional French-Canadian nationalism. A hallmark of the Citélibristes was their rejection of nationalism as conservative, isolationist and xenophobic, believing instead that liberalism devoid of ideological sentiment held the key to the modernization of Quebec society along with democratic and socioeconomic reforms. Although de Grandpré was more moderate and conservative, he shared a similar political outlook.

His more moderate views likely appealed to the Jesuit administration at Brébeuf, who placed a premium on the ability to lead by consensus. "I believe de Grandpré would have been the preferred candidate of the

Jesuits, who perceived themselves as the army of God," says the former student. "I could see de Grandpré on his knees praying in allegiance to them more than I could see Trudeau."

The yearbook of Brébeuf's tenth conventum, or graduating class, declared that de Grandpré represented the spirit and aspirations of the class and was "the right man."[43] His classmates believed he was "predestined" to be the president of his class, and described him as a commanding and stately figure who was "very convincing" and "of restrained eloquence." They further praised his "great common sense" and said he fulfilled his duties with a sense of style. Rating his performance as "excellent," they added that, in accordance with the Jesuit teachings to lead by consensus, de Grandpré requested permission from the class to perform his duties. Nor was any deed "too little" for his attention.[44] This paean to their class president ends by saying he was also very stylish ("chic"), considerate, amiable and always smiling.

After graduating from Brébeuf with a *baccalauréat* summa cum laude, de Grandpré intended to follow in his grandfather's footsteps and become a medical doctor. His science marks were good enough and, aided by a strong letter of recommendation from Father Donat Boutin, prefect of studies at Brébeuf, he was accepted into medicine at McGill University. He took extra courses in the summer of 1940 so that he could skip the pre-med curriculum and enroll directly into first-year medicine. But, soon after, he changed his mind and opted to follow his older brother into law. He credits the education he received at Brébeuf for enabling him to make such a radical switch with ease.

After rejecting medicine as a career, de Grandpré left the comfort of the francophone world that he grew up in to study law at the anglophone-dominated McGill University. Despite the profound changes he faced because of that choice, he was determined to conquer whatever hurdles lay in his path. He expropriated the slogan of his class at Brébeuf, *Etre quelq'un* ("To be someone"), as his personal motto at McGill. "Be someone," declared the quote above his entry in the university's 1943 yearbook.[45] "It gave me a target and set a goal — to realize yourself, and to be above the crowd."[46] It also reflected the ambition of the middle son not only to succeed, but to find his own identity.

He entered the law faculty at McGill fascinated by foreign policy and international affairs and bent on pursuing a career as a diplomat in the foreign service at what is now the Department of External Affairs.

But between his second and third years, in 1942, "the government appointed three or four ambassadors to very important posts, and they were all given to political appointees. I said to myself, 'I am going to waste my time, my career, doing the daily work in the embassies and waste my younger years in the department. And when there is a very important appointment, in Tokyo or London or Paris or Washington, it will be given to a political appointee. The hell with that,' and I decided to stay in law."[47]

On the advice of his father he chose McGill over the Université de Montréal so that he could become bilingual. "He wanted me to be perfectly at ease in either language because, he said, 'you'll be practicing in North America.' I had to earn a living with my confreres from across the country and one big asset that I had was to be perfectly bilingual — my father was perfectly bilingual — and there was no choice."[48] De Grandpré was also influenced by his older brother, Louis-Philippe, who had graduated from McGill's law faculty two years earlier.

Making the switch from Brébeuf, where he was well known and popular on campus, to McGill, with its alien WASP culture and where he was unknown was not without its difficulties. In his first year, he wrote most of his exams and papers in French, while he set out to refine his skills in English. The greatest difficulty, he found, was in Roman law, "because the pronunciation in Latin is so different in English than in French."[49]

But his McGill classmates not only learned quickly who he was, because of his keen involvement in student affairs, but also that de Grandpré was *someone* because of his connection with one of Canada's leading lawyers at that time, the Hon. F. Philippe Brais. The instruction of law was markedly different from today in one crucial respect. During the 1940s, a young student articled for a law firm at the same time he or she began formal instruction in a law faculty. Before classes had even started, de Grandpré began work on August 4, 1940, in the firm of Brais Campbell alongside his older brother, who had worked at the small firm since graduating in 1938. "I had never worked in a law firm. I had never seen a law book," de Grandpré says. He remembers his first case well because it was assigned to him by his mentor-to-be at 8:30 that morning. Brais called the young man into his office and said, "'There's an inquest at the coroner's court at ten o'clock. Could you represent the company?' I didn't ask him, where is it?, but I said, fine, I'll be there, and then I walked out and went into my brother's office and I said, where the

hell is the coroner's court?" De Grandpré still remembers the name of
the company — it was Apex Insurance. The inquest was investigating
the death of an Apex client, a man who had been electrocuted by a
faulty light while delivering coal to the basement of a building. "I
didn't ask any legal questions, but I asked a lot of questions about the
laws of physics. So that was my introduction to law. From that point
on, it never stopped."[50]

After he graduated, de Grandpré says one of his law professors told
him, "'You're a living example of how useless our course is because
you've not attended 50 per cent of the lectures.'"[51] Lectures were held
in the morning between 8:00 and 10:00 and again in the afternoon
between 4:00 and 6:00. Students worked at a law firm in between.
However, as de Grandpré's brother says, attending classes at those
times was often difficult because "at the time, a young lawyer was supposed to be in the office before the seniors in the morning and he was
supposed to leave the office after them at night."[52]

Brais was one of the most powerful denizens of Montreal's
Establishment and of the Canadian corporate elite from the 1930s to
the 1950s. Born in Montreal in 1894, he was educated in law at McGill
University, called to the bar in 1917 and started his own practice in
1923 after working for the late Senator George G. Foster and former
Quebec chief justice J. E. Martin. Although he specialized in insurance
law, he also served as Crown prosecutor of Montreal from 1922 to
1930, which explains his practice of both corporate law and criminal
litigation. In the next two decades, he became a senior member of the
Montreal bar and served as president of the Canadian Bar Association
from 1944-45. His corporate directorships gave him a great deal of
influence in some of the leading boardrooms of the day: Canadian
Pacific Railway, Sun Life Assurance Company, Rediffusion
Incorporated, Montreal Trust Company, Banque Canadienne
Nationale and Quebec Airways.

Brais's political connections were equally formidable at both the
provincial and federal level. He was appointed by Prime Minister
Mackenzie King as co-chairman of the National War Finance
Committee alongside Harry Ratcliffe, the former president of the
Toronto investment firm of McLeod Young Weir & Company, from
1941 to 1945. As the committee's name suggests, their job was to direct
and manage the financing of the war effort and administer the war
bonds program. Brais also served as co-chairman of the Quebec war
bond campaign with Ian MacNaught of Sun Life. Louis-Philippe de

Grandpré remembers the many hours his boss had to spend signing war bonds and associated documents for the committee. "I had to learn how to sign his name to help him get through the stack of bonds."[53]

Like de Grandpré's father, Brais was a member of the Liberal-dominated Reform Club. He often took his young students for lunch to the club where, Louis-Phillipe recalls, the powerful mentor would "open the world to us." Brais was very involved in Quebec provincial affairs and was appointed in February 1940 both government leader of the Legislative Council, or upper house, of the Quebec National Assembly and minister without portfolio in the cabinet of Premier Adélard Godbout. Newspapers at the time reported that his appointment was resented by some members of the anglophone community, who wanted an Irish Catholic to be appointed. But Brais quickly developed a reputation that gained him widespread respect for his ability to approach political problems with a legal mind and to eschew political expediency or partisanship in favor of moderation.[54] He was a staunch opponent, however, of Quebec nationalists and defended the Godbout government from the attacks of former premier Maurice Duplessis and the Bloc Populaire. After the war, Brais continued to sit in the Quebec upper house, where he remained an opponent of Duplessis's Union Nationale party, which defeated Godbout's government in the 1944 election.

Brais never ran for federal office, although he had been touted as a possible candidate by the King government in 1941 to replace Ernest Lapointe, the terminally ill minister of justice, as Quebec leader in the federal cabinet. However, according to historian Dale Thomson, Brais was too closely identified with the English-Canadian financial community. That factor also precluded Brais from being a candidate for the national leadership of the Liberal party and running against Louis St. Laurent after the war.[55]

De Grandpré first came to know of Brais through his father. The two men worked in the same industry, knew each other socially at the Reform Club and worked near each other on St. James Street. Roland hired Brais in 1931 to represent him in a dispute with a contractor involved in building their new house in Outremont. De Grandpré recalls the contractor had done a poor job putting in the floors — there were not enough nails to hold the floorboards down. "Our society, if you go back fifty years, was very closely knit," says Brais's daughter, Louise Brais Vaillancourt, whom de Grandpré appointed to the board of directors of Bell Canada in 1975. "Francophones who were in the business world at that time all knew each other."[56]

Brais knew just about everyone. What made him famous throughout the country, however, was his work as a defense counsel for the Montreal-based Bronfman family in a sensational trial in the 1930s. The distillers who owned Seagram's were charged with smuggling and conspiracy to ship bootleg liquor to the United States during Prohibition. Brais was one member of a high-powered army of lawyers retained by Samuel Bronfman's family. He and his colleagues won the case. Judge Jules Desmarais, who was later appointed head of the Quebec Liquor Commission, threw it out of court on June 15, 1935, after ruling that the Crown had failed to establish any proof of the alleged conspiracy.[57]

Although de Grandpré sought out his articling position without intercession by either his father or brother, Brais was reluctant to have two brothers in the same firm. De Grandpré speculates that he was finally given the position because it was only a junior articling job. Whatever the case, the young student relished the fame that no doubt rubbed off on him from being associated with such a powerful figure. Even today, he boasts, "I was the envy of my class."[58] He recalls that he was "given a long leash" by his mentor and allowed to assume far more responsibility on cases than many of his peers. Louis-Philippe has a similar recollection of his experience with Brais: "He was a wonderful teacher. He gave me a lot of rope and I was lucky enough not to hang myself with it. I'm sure he did believe that anyone who wanted to succeed in life had to swim, and he helped me and, eventually, Jean, to swim — and to swim fast."[59]

Brais's daughter says her father "would have been very difficult to work with, I'm sure. He had a reputation for winning cases and he did not suffer fools gladly."[60] The steely eyed, ruggedly built Brais even looked intimidating. In court, he was described as being "cold as ice" with a caustic wit. He also had a restless, if not boundless, energy that made him an exceptionally tough taskmaster. He started each day by walking the first mile to work, down Outremont through the fashionable Square Mile district, before being picked up by his driver. As a young man, he used to canoe from Montreal to Toronto via the Rideau Canal system and the Great Lakes and once canoed from Montreal to New York City via the Richelieu and Hudson rivers.

Brais was meticulously thorough. His lessons to his articling students were steeped in a work philosophy that required young lawyers "to always be sure of your facts, to check and to double check and to also be true to yourself," recalls his daughter. In retrospect, many "high-powered francophone lawyers today say that everything they learned in law, they learned from him."[61] Echoing that sentiment, de Grandpré says, "Everything I know in law or about business I learned from him."[62] As an articling student, he says Brais taught him a number of values that are critical to leadership — loyalty, research, discipline and hard work. "And a certain degree of brinkmanship." Brais "was not a man who spent long hours discussing commas," de Grandpré's brother adds. "He went to the substance of any problem. . . . He was able to brush away anything that was not of use. And he was a wonderful pleader. He was very exacting." Yet he was "charitable in his own gruff way."[63]

De Grandpré recalls Brais's generosity in giving him access to his Ford convertible in the mornings. The young student would begin his day at Brais's house in Outremont at about 7:30, where he would be given instructions for the morning's work or asked to deliver a stack of war bonds. The older lawyer would then set off on foot and de Grandpré would drive the car to nearby McGill to make his first class at 8:00 and then to the law office at about 9:00. "It was quite a perk to have a car during the war, especially with the special gas ration coupons." The perks and experience certainly made up for the low pay an articling student received. De Grandpré's salary for his first year was $3 a week, which was doubled to $6 in 1941. Despite the meager earnings, he says Brais treated him well. "He was very fair, and he told you if something was wrong, and he gave praise if it was right. And he never took credit for something I or someone else did."[64]

Insurance law was not as boring a field as it might seem. For a young articling student, Brais's practice provided a training ground unrivalled by almost any other area of law for its depth. Insurance cases exposed a practitioner to numerous legal fields, including corporate and commercial law, civil liability and litigation. Before the advent of no-fault automobile insurance in Quebec, even minor accidents were usually pleaded before the courts. "That was a wonderful training ground because we were in court day in and day out," Louis-Philippe says. "We learned how to question a witness, we learned how to present a case to the court, and we learned not to lose sight of the substance of a case, or to spend useless time. In the old days, a case that lasted three days was a very long case."[65]

De Grandpré remembers one extraordinary case he worked on as a student in his last year of law school in 1943. It was a civil action that arose from a crash of a Quebec Airways plane in northern Labrador. The survivors of the crash died of starvation because the radio in the plane had not been strong enough to send a message to rescuers. The diary that MacFarlane, one of the survivors, kept after the crash detailed the anguish of hearing the failed efforts of the search planes to locate the wreck. "MacFarlane took pictures that we recovered when the bodies were discovered. And, in the diary, we could read that they were starving to death," de Grandpré says. "It was the most poignant thing you could ever read. From an ego point of view, it was not a difficult or noteworthy case. But the human side of it was something that left an indelible mark."[66]

The rigorous training and broad exposure to the law that Brais's practice provided allowed de Grandpré to excel in his course work at McGill. Although he may not have spent as much time in class as he did at the office, he had some very able teachers, including: former cabinet minister Edouard-Fabre Surveyer for civil law; Frank R. Scott, one of Canada's leading experts on constitutional law and civil liberties, for contract law; Judge O. S. Tyndale for civil procedure; John P. Humphreys, former president of the United Nations Human Rights organization, for Roman law; and former Supreme Court of Canada justice Gerald Fauteux, for criminal law. De Grandpré says he had no academic mentor at McGill, although he was close to Fauteux, who was one of Brais's neighbors.

Somehow, de Grandpré also found time to help run the affairs of the law students' society. He was president of his class from 1941 to 1943, secretary-treasurer of the law students' society in 1941-42, president of that society in 1942-43, and served on the students' executive council.[67] Mandatory military training in the militia at McGill was fit into his schedule as well. De Grandpré did not have to face the emotionally charged prospect of conscription for service in the armed forces because he did not complete his legal training until 1944 — "We were admitted to the bar one year retroactively to when we graduated, and by that time the war in Europe was winding down."[68]

Those who failed were conscripted, which gave de Grandpré and his classmates a compelling incentive to excel. While most of the francophone community in Quebec opposed conscription, academic deferment was seen as a great privilege by the teachers and the university administration at anglophone-dominated McGill. In the eyes of the university leadership, de Grandpré and his peers owed a great debt to society and to those who served in the war because of their privileged deferment. Said Sir Edward Beatty, chancellor of McGill and president of the Canadian Pacific Railway, "The only question which you have to ask yourselves is whether your opportunities to share in the building of a better world will be so used as to justify you before that great multitude who now offer and give their lives that you may have freedom."[69]

De Grandpré first made headlines in Montreal in 1941 for scoring the highest marks in his class on the first-year final exams. Placing first in his class in civil law and Roman law, second in criminal law and third in international law earned him McGill's Adolphe Meilhiot Prize. He was the first francophone to win the award, which resulted in widespread newspaper coverage and glowing stories in *La Presse*, *Le Devoir* and *La Patrie*. The following year he repeated the accomplishment, placing first on the second-year law exams and winning the Alexander Morris Exhibition prize and more headlines.

On May 26, 1943, de Grandpré received his degree in civil law at the age of twenty-one, graduating summa cum laude. He won the *Elizabeth Torrance gold* medal for placing first out of his class of five on his final year exams, a prize from the Association of Young Barristers for distinction in his finals on civil procedure and the award from the Bar Association of Montreal for the highest standing in commercial law.[70] One award he decided not to accept, the Macdonald traveling scholarship, carried a $1,200 stipend — a substantial sum in 1943 — which allowed the student to study anywhere he or she chose. However, the options were narrow that year. It was virtually impossible to study in Europe, and there was some risk in studying law in the United States because of their draft rules. De Grandpré agonized over whether to accept the award, and sought the advice of his father, Brais and several instructors, including Judge Tyndale, Judge Fauteux and the head of the Bar Association of Montreal, John L. O'Brien. Their consensus was that he should turn it down because of the war. "I would probably waste a year and lose some of the opportunities that were coming my way at that time because of the publicity surrounding the honors that I had received on my graduation." His father was also "dead against" accepting the scholarship because he argued that his son had a chance to "step into a very great firm with a very high profile."[71]

However, when de Grandpré asked Brais shortly after his graduation whether there was an opportunity to join the firm as a lawyer, Brais told him, "No, I don't want to have two brothers in the firm. I'd be too dependent on a family connection within the firm."

He then asked, "What will you do if your brother asks you to do something? Are you going to tell him to go to hell? Are you going to do it?"

De Grandpré says he told Brais that "in the office, he's not my brother . . . I'm not going to treat him like my brother. I'm going to treat him as a senior lawyer who is asking me to do something and I'll do it."

"Well, I'm not inclined to have you in my firm," Brais responded, "but I saw the way you performed and I'll take you on."[72]

As it happened, Brais's concerns over any potential conflicts between the two brothers were eliminated not long after de Grandpré was called to the bar. His older brother gave Brais notice that he was leaving to join another law practice. Louis-Philippe joined a larger firm that would give him more opportunity to practice litigation. De Grandpré thus had a valuable opportunity, as he says, to "move into my brother's shoes."

De Grandpré did just that. "I think Dad was very proud of him. Jean had a very ethical frame of mind and took to heart the business of the firm,"[73] says Louise Brais Vaillancourt, adding that her father was impressed by the younger man's intellect and judgment and "really took a liking to him." The two, in fact, developed an almost father-and-son relationship, particularly as de Grandpré assumed greater responsibilities at the firm in the late Forties. "You have to understand that Philippe Brais had one son, and he decided he was not going to be a lawyer — he was going to be an engineer," de Grandpré says. "And I think this was a disappointment, that his son was not going to inherit his firm. I really became the *de facto* son. That's why I was so close to him over the years." He felt that Brais had so much confidence in him that if he stayed there long enough, he would inherit the firm.[74]

As de Grandpré assumed more of the case load at the three-man firm, Brais took on increased responsibilities in the corporate and professional arenas. His partner, Alexander John "Jack" Campbell, left in 1946 after his appointment as a judge to the Supreme Court of Quebec. The years from 1945 to 1949 were "a very difficult period" for the young lawyer as Brais tried but failed to find another senior partner to share the load. That failure placed an inordinate strain on de Grandpré, who at this point was only in his mid-twenties. Finally, Brais hired André Montpetit as a partner, "but André Montpetit could not work with Philippe Brais . . . the chemistry was not there," de Grandpré says. After eighteen months, Montpetit left and the firm's moniker was changed to Brais & de Grandpré.[75]

The term "the Cold War" had not yet been coined when the tensions that spawned it gave rise to two sensational legal cases in Canada, one of espionage and the other of diplomatic intrigue. After Brais's services were retained in both, de Grandpré was involved, albeit as a bit player.

The roots of the anti-communist hysteria and civil service purges that characterized the Cold War period can be traced to Ottawa on the night of September 5, 1945, when a twenty-six-year-old military cipher clerk named Igor Gouzenko defected from the Soviet Embassy. With the diminutive, blond-haired man, his wife and two-year-old child came a small mountain of secret documents, which he had spirited out of the embassy's highly secure code room continuously for a month before his defection. Those decoded ciphers documented an extensive Soviet espionage network run out of Ottawa by Gouzenko's boss, Colonel Nikolai Zabotin.

The Gouzenko defection led to the arrest and trial of Fred Rose, the first member of Parliament to stand trial for spying. The case riveted the nation for three months in 1946. "I don't want to leave you with the impression that I was in the heart of the Gouzenko case — far from it,"[76] de Grandpré says of his work for Brais on the case. But Brais played one of the most critical roles at the trial, and it was certainly the most famous case de Grandpré was ever connected with.

Gouzenko's defection was kept secret for six months, during which time it was not even revealed by Prime Minister Mackenzie King to his full cabinet.[77] As investigators pored over the revelations, RCMP officers hoped to capture the web of bureaucrats and sympathizers who were implicated. To that end, the King government invoked the draconian powers of the War Measures Act in October 1945.[78] The Act, which gave the RCMP the powers to search or detain any suspect without having to lay a formal charge, kept the case secret. E. K. Williams, the president of the Canadian Bar Association, was retained to examine the Gouzenko documents, and he advised the Corby committee, the top-level body struck to manage the affair, that criminal proceedings were premature and, that with the exception of four people, "no prosecution with the evidence now available could succeed."[79] He also recommended that two judges be appointed to a royal commission, which would sit in secret, would not be bound by ordinary rules of evidence and would not allow those summoned before it to be represented by counsel.[80]

The Canadian government was forced to act, however, when the U.S. journalist Drew Pearson stated in a broadcast on February 4, 1946, that King had informed U.S. president Harry Truman about Soviet spy activities in Canada. The following day, King informed the full cabinet of the Gouzenko revelations[81] and obtained approval for an order-in-council creating a royal commission under Supreme Court judges Roy Kellock and Robert Taschereau to investigate and report on the espionage activities.[82] The commissioners appointed Gerald Fauteux and D. W. Mundell, both of the Department of Justice, and Williams as their counsel. The hearings began on February 15, 1946, with Gouzenko as the first witness and, on the same day, King announced the arrest of several individuals.[83] Under the War Measures Act order, eleven men and two women were detained.[84]

King's statement provoked a sensation throughout the Western world. But in Canada the most controversial part of the affair had yet to begin. In early March, shortly after the Taschereau-Kellock commission filed its first interim report with the government,[85] Dr. Raymond Boyer, an assistant professor of chemistry at McGill University employed in secret explosives research and one of those arrested, implicated a member of Parliament in the spy ring. The MP, who used the cover name "Debouz," was identified as Rose, the MP for Cartier-Montreal and the first and only Communist party candidate to ever hold a seat in Parliament.[86]

When the Taschereau-Kellock commission reported its findings on Rose to the government in early March 1946, the Department of Justice was not sure how to proceed. It faced the unprecedented question, at least in Canada, of whether parliamentary privilege precluded Rose's arrest once the new session of Parliament opened the following week. Justice Minister Louis St. Laurent referred two important questions to the commission's lawyers:

(a) whether a federal member of Parliament who would have committed a crime under the Official Secrets Act can be arrested either during the coming session of Parliament or within the few days which remain before its opening?
and
(b) whether . . . it would be advisable or inadvisable to obtain the issue of a warrant of arrest against this member at the same time as that being issued against the individual who divulges to this member verbatum important war secrets for the benefit of a foreign power?[87]

Those questions found their way onto Brais's desk on March 9, 1946. He was already involved in the espionage affair, having been appointed by St. Laurent as prosecutor of the Boyer case, and was well known to Williams, who had succeeded him as president of the Canadian Bar Association. In his answer dated March 12, 1946, Brais told St. Laurent that Parliamentary immunity does not apply to any criminal offense and that "a member who has committed an indictable offense is therefore liable to arrest at any time and any place except on the floor of the house when it is sitting."[88] On the second question, Brais concluded that Boyer and Rose should be charged separately, but at the same time, and that each should also be charged with conspiracy.[89]

While members listened to the Governor-General deliver the Speech from the Throne that opened the second session of the Twentieth Parliament on March 14, 1946, Staff Sergeant René Noel swore a warrant for Rose's arrest for violating the Official Secrets Act.[90] Rose was arrested at his home that night shortly after he returned from Parliament.[91]

Brais was named special federal prosecutor for the Rose trial and de Grandpré helped him research and prepare his case. He remembers a maze of code names; the list of the sixty-three cover names published in the Taschereau-Kellock report fills three pages. Says de Grandpré: "I was doing the research, the leg work, and I was trying to understand the labyrinth of all . . . the cover names that were used by the spies, trying to identify who they were. . . . Anyway, it was a complex web of names, and I was involved in making sure that we were tracking these individuals properly."[92] Brais was also assisted by intelligence officers from the RCMP who were assigned to investigate the case. In court, his team included the chief Crown prosecutor of the Montreal district, as well as two other special prosecutors and lawyers from the Quebec attorney-general's office assigned to the case.

Brais established a *prima facie* case against Rose at his preliminary hearing on March 26, 1946, when Boyer admitted that he gave the MP information to be passed to Moscow on the secret RDX explosive.[93] Rose was arraigned on April 15, pled not guilty to unlawfully communicating secret information to the Soviet Union and conspiracy, and was committed to trial on May 20. The trial ended on June 13 after a three-week hearing,[94] and Rose was convicted shortly before midnight on June 16 after a twelve-member jury deliberated for only twenty-seven minutes.[95] Sentenced to six years in prison, he was released after serving four years of his term and moved to Warsaw, where he died at the age of 76 in 1983.

De Grandpré was also involved in an intriguing diplomatic feud that began in 1946 between the new communist government of Poland and the former Free Polish government-in-exile for control of Poland's state treasures, which had been rescued during Hitler's blitzkrieg in September 1939 and sent to Canada for safe keeping during the war. The most prized treasure was the eleventh-century Polish coronation sword, Szezerbiec, the battle sword of Poland's first king, Boleslaw I the Valiant. The thirty-four steamer trunks also included: swords, armor and paintings from the twelfth century; other coronation regalia of Polish kings from the fourteenth century; 140 gold and silver sixteenth-century tapestry curtains; manuscripts and prayer books from the twelfth, thirteenth and fourteenth centuries; a two-volume edition of the Gutenberg Bible; and the original manuscripts of composer Frederic Chopin.[96]

The saga of the treasures began when they were assembled and smuggled out of the Wawel Royal Castle at Krakow by Dr. Stanislaus Swierz-Zalewski, curator of the castle's museum.[97] They arrived in Canada aboard the Polish ship *Batorv* on July 16, 1940, after Swierz-Zalewski eluded Hitler's armies and air force in Romania, France and then England.[98] The federal government stored the trunks in a fourth-floor office in the records storage building at the Central Experimental Farm in Ottawa. Access to the room was only given to Swierz-Zalewski, who remained curator of the treasures, and to Joseph Polkowski, an employee of the Polish government-in-exile's legation in Ottawa. But just before the communists took control in 1944 (Canada recognized the new government on July 7, 1945[99]), the Polish exiles arranged for Wenceslas Babinski, a Polish diplomat in Canada, to hide the treasures. He dispersed them widely that spring, placing two crates of the most valuable articles in a Bank of Montreal vault on Sparks Street in Ottawa, eight trunks in the Convent of the Precious Blood in Ottawa and twenty-four trunks at the Monastery of the Redemptorist Fathers at Sainte-Anne-de-Beaupré outside Quebec City.

After discovering that the storeroom at the Experimental Farm was almost empty, the representatives of the Polish communist government traced the whereabouts of the treasures. When they heard that the artifacts were about to be moved again, they asked the federal government on May 16, 1946, for help to prevent their removal. Once more, they were too late. The religious orders told External Affairs the treasures had been removed only days before.

Although Louis St. Laurent, then secretary of state for External Affairs, agreed to help the communists, he said the government could not secure possession of any object stored outside of government buildings. The disappearance of the treasures was made public that November, leading to reports from anti-communist Poles in Montreal that the treasures were sacred relics and belonged to the Roman Catholic Church. In late December, R. G. Riddell, head of the First Political Division at External Affairs, met with Z. R. Bielski, the Polish chargé d'affaires, and recommended that the legation obtain legal advice. He said the Polish government could proceed with a court action "for the recovery of material which they claimed belonged to the Polish State, if they could discover where the material was located."[100]

Enter Brais and de Grandpré. Shortly after Christmas, the Polish legation retained Brais as their counsel. "Mr. Brais and myself had a number of meetings with them and the RCMP," de Grandpré says.[101] The Polish legation wanted the lawyers to persuade Ottawa to hand the matter over to the police. Only if the RCMP were brought in could the Polish government track down the treasures, which would then enable them to start a legal action for their recovery. Brais and de Grandpré mediated those secret negotiations throughout the summer of 1947 and, by late September, Brais convinced St. Laurent to bring in the RCMP. A couple of days after the police were instructed to begin a search for the treasures, the Polish legation revealed that it had submitted a diplomatic request for police assistance and declared its intention to commence a legal action to recover the treasures.[102] De Grandpré and Brais then turned to planning a legal case that would be handled in either the Ontario or Quebec provincial court.

The RCMP did not publicly confirm the order to find the treasures until December 22, 1947. RCMP Commissioner S. T. Wood revealed his instructions ordered him only to keep the treasures under surveillance. A few weeks later, in late January 1948, RCMP Corporal J. R. R. Carriere tracked the trunks to the stone caves in the basement of the Hotel Dieu hospital and convent in the lower town of Quebec City. Lester B. Pearson, then under-secretary of External Affairs, informed the Polish legation of the results of the investigation on February 13, 1948: "As we agreed with your solicitors, we are communicating the information to you for any further action you and your solicitors may wish to take."[103]

Once the RCMP gave de Grandpré's client the information on the whereabouts of the treasures, the Polish chargé d'affaires delivered an

ultimatum to the convent's Mother Superior on February 21, 1948. His
demand that the artifacts be relinquished within four days prompted her
to appeal to the Quebec government for assistance.[104] Acting under the
pretext that the RCMP had violated the sanctity of the Hotel Dieu convent and conducted an illegal search, Quebec's premier Maurice
Duplessis, a rabid anti-communist, impounded the objects and secretly
moved them to the steel-encased vaults of the Quebec provincial museum. Duplessis charged on March 4 in the Quebec National Assembly
that Quebec ministers in the feberal cabinet were "collaborators of Stalin
and his Polish government" and that the treasures were the property of
the Roman Catholic Bishop of Krakow.[105] Although the federal government angrily denounced the seizure, it refused to intrude on Quebec's
prerogatives, leaving the Polish treasures hostage to the vagaries of
Canadian federalism, and in Duplessis's custody, until after his death.
They were finally released to the Polish government on January 3, 1961
— fifteen years after the Polish legation first tried to get them back — by
Liberal premier Jean Lesage. The trunks set sail from Boston aboard the
Polish freighter, *Krynica,* the next day.

"I was only twenty-six and I had been involved in two of the most
international cases of the day," de Grandpré boasted in an interview.
Although he correctly pointed out that the Polish treasures case never
reached the courts, his claim that he and Brais "negotiated a settlement"[106] for the return of the artifacts does not stand up to scrutiny. In
fact, he had left Brais's firm before the matter was finally resolved by
the death of Duplessis.

Jean de Grandpré married Hélène Choquet on September 27, 1947, in
a traditional Roman Catholic ceremony at the imposing Notre Dame
cathedral located in the heart of Old Montreal. The grand Catholic
church, renowned for its gothic design and gilded ceiling, is one of
the largest cathedrals in North America and the most important
church in Montreal.

Tall and elegant, Hélène is the daughter of Azarie Choquet, a Montreal
police department official. She grew up on Mansfield Avenue in the same
Outrement neighborhood that her husband lived in. Although they had
many mutual friends and acquaintances, including Jeanne Sauvé, the former governor-general of Canada, the two did not start dating until de
Grandpré graduated from Collége Jean-de-Brébeuf. Hélène was a kindergarten teacher and established her own school before her marriage.

De Grandpré does not come across as the incurably romantic type, yet he is deeply devoted to his wife, and a framed black-and-white photograph of Hélène graces one of the walls of his Montreal office. Although he spoke with great affection for her, he seemed awkward when asked how she had won his heart. His answer reflected the restrained rationality instilled by his Jesuit training, rather than the passionate reminiscences of a young man in love: "My wife is an outstanding lady. She has a good sense of humor. She loves children and loves a quiet home, which I've not been able to give her because she's been busy following me. She's a very devoted wife and mother." He paused for a moment, and then added, "She's a very great asset."[107]

Their courtship started as a summer romance while he was vacationing with his family in July 1940. During his teens, de Grandpré spent the summers in the rustic countryside of the Gaspé Peninsula where the St. Lawrence River widens on its way to the mouth of the Gulf of St. Lawrence. His seven-year courtship of his bride-to-be began on the tennis court of the grand hotel at Notre Dame du Portage, near Rivière-du-Loup, about 440 kilometers northeast of Montreal. It continued throughout his law school years and into his first three years as a lawyer working for Brais. Because they each lived with their parents until they were married, it was a traditional, old-fashioned courtship, with meetings in their families' parlor rooms before a date.

The couple often appear to be a study in contrasts. Outgoing and charming, Hélène does not exude the same forcefulness or emotion as her husband. Close friend Edmund Fitzgerald, the former chairman and chief executive officer of Northern Telecom, describes Hélène as "a very unique, very pleasant and very easy-going person. . . . She keeps him from getting too serious."[108] Keeping de Grandpré from being consumed by his work, on the other hand, is often a futile endeavor, as Hélène discovered at the end of their honeymoon.

They spent the three-and-a-half weeks of their honeymoon driving along the Atlantic seaboard of the United States before returning to Montreal on the evening of Friday, October 24, 1947. The next morning, Brais telephoned to say he needed the answer to a difficult legal question for a case that was going to be argued before the Supreme Court of Canada at 10:30 on Monday. "And I said, 'Yes sir'. . . . I didn't tell him I was just coming back from my holidays and I can't do it. So I did it," de Grandpré says, adding he provided his boss with the answer that he needed "and we won the case."[109] He still recalls the

legal issues and names of the parties involved. The case was Marie
Thériault v. H. Huctwith et al., an appeal of a Quebec Court of
Appeal judgment involving a 1941 Montreal traffic accident. It
received wide attention because the accident subsequently claimed the
life of a well-known Quebec portrait artist, Alphonse Jongers, who
was a passenger in the car driven by Léontine Thériault.

Jongers was severely injured and sued her and the owner and driver
of the truck that hit Thériault's car. The lower court ruled that the
accident was the fault of both drivers and ordered them to pay Jongers
$8,500 plus costs. The owner and driver of the truck appealed the rul-
ing, but Jongers died as a result of his injuries before the case could be
heard. The case took a byzantine twist when Jongers's will appointed
Thériault as executrix of his estate. That automatically made her the
plaintiff in the appeal to the Supreme Court of Canada, even though
she was already a defendant in the same case.[110]

The case is famous in law schools because it raised the extraordinary question, Can you sue yourself? — the question that Brais wanted answered by de Grandpré on the last weekend of his honeymoon.

In working the problem out, he decided Thériault was legally obligated to appeal the case on behalf of the estate. Because she had paid
the amount assessed of all defendants in the original trial, she now had
the right to sue the other co-defendants. In doing so, she wasn't suing
herself, despite the oddity of being both a plaintiff and a defendant.
The Supreme Court agreed and found that Thériault's right to sue the
other defendants for amounts due to her had been ignored by the
Court of Appeal.

Not long after their first anniversary, the first of de Grandpré's four
children, Jean-François, was born on November 12, 1948. He was followed by Lilianne in 1950, Suzanne in 1952 and Louise in 1957. Like
their father, the children had a warm, but disciplined home life. Not
only did they attend the same school in Outremont that their father
had — Collége Jean-de-Brébeuf became co-educational in the Sixties
— but they later moved into de Grandpré's childhood home on
Beloeil Avenue after their grandfather died. Despite de Grandpré's
many absences, both parents were very involved in their children's
upbringing, recalls a friend of Jean-François'.

That upbringing instilled the same strong work ethic in his children
that de Grandpré grew up with and likely contributed to their future
success: Jean-François, who is married to Michelle Décary, now practices law at his father's law firm, Lavery O'Brien; Lily, who married

David James, a partner at the management consulting firm of McKinsey & Company, studied journalism at the University of Western Ontario in London, Ontario, and works for the Bank of Montreal; Suzanne, who married Claude Baillargeon, a Montreal lawyer who is a partner with her brother at Lavery O'Brien, graduated in nursing from McGill; and Louise, who lives in Toronto with her husband, Michael Merrithew, an executive at Northern Telecom, is active with McGill's alumni organization in Toronto.

"We all work too hard,"[111] Jean-François says. While he says his father's success served as an inspiration, he attributes his own and his sisters' achievements to his mother, because de Grandpré was "so much caught up in his work." Although in recent years, Hélène can usually be found at her husband's side at major functions and travels regularly with him when business takes him away from Montreal, that wasn't the case in the early years of their marriage. When they started their family, business frequently took de Grandpré away on long trips, occasionally for weeks at a time, leaving Hélène to tend the hearth as her mother-in-law had before her.

After his marriage and the birth of his son, de Grandpré found it increasingly difficult to cope with the workload and sixteen-hour days he was required to put in at Brais's firm. His frustration was exacerbated by Brais's refusal to hire another lawyer, despite his pleas for help. At one point in early 1949, he talked with Brais about the prospect of his older brother rejoining the firm, but, for reasons he still does not understand, those discussions failed to bring any results. He lost weight from the strain and felt that he was going to either ruin his health or ruin his reputation by doing a poor job.[112]

In June 1949, de Grandpré decided to leave Brais's firm. "He was very hurt when I left. He never thought I would do that to him," de Grandpré says, recalling that Brais later asked him, "Why did you do that to me? Did I do something wrong? What could I have done to keep you with me?"[113] As it turned out, Brais's previous partner, Alexander Campbell, resigned from the Superior Court bench a month after de Grandpré's resignation and resumed his partnership with Brais. The workload was further eased with the hiring of an articling student, John J. Pepper, who would remain with the firm for the next two decades.

De Grandpré didn't strike out fully on his own. At the age of twenty-seven, he started a legal practice with his older brother, Louis-Philippe, who had decided to leave the firm of Letourneau Tansey Monk de Grandpré Lippe and Tremblay, where he had practiced law since leaving Brais's firm in 1944. Louis-Philippe brought another Brais protegé with him, Harold Tansey, with whom he had worked since 1935 when they had articled together.[114] The three men hung out their shingle on July 5, 1949, as Tansey de Grandpré & de Grandpré in a small office in the Insurance Exchange Building at 276 St. James Street — the same building where de Grandpré's father worked.

The firm was set up to specialize in insurance law and related litigation and grew slowly in its first decade. By 1957 there were seven lawyers on staff. The successor firm, Lavery O'Brien, now has 110, many brought on through mergers with other small Montreal firms over the past twenty years.

The two brothers never considered asking their younger brother to join them at their new firm. Although Pierre had been called to the Quebec bar the year before they created their partnership, de Grandpré says taking him on would have been ill-advised for two reasons: "One, I think that Mr. Tansey would have been slightly invaded by the family and he would have felt uncomfortable. It is all right to be in partnership with two brothers, but in partnership with three is just a bit too much. The second one is, we were not quite sure our partnership would succeed. And we always took the view that we could not put all the family's eggs in the same basket."[115]

Practicing law together for seventeen years forged a strong bond between the two brothers, which was reinforced by the close friendship that developed between their wives and children. "My daughter who lives in Montreal is very, very close to Philippe's daughter who lives just a few blocks away." And de Grandpré says, "The chemistry of the couples makes a big difference." Pierre, in contrast, is separated from his wife. Nevertheless, see each other regularly. "He's an outstanding lawyer,"[116] de Grandpré says of his younger brother, who is head of the Montreal law firm of de Grandpré Godin.

With their extensive contacts throughout the insurance business and expertise in courtroom litigation, the firm soon developed a small but select client list that included some of the largest and best-known insurance companies in Canada: Lloyd's of Canada, which was the domestic arm of the British-based underwriting syndicate; Sun Alliance; Chubb; Prudential; All-State; Continental; and Commercial Union. In law, says de Grandpré, it is the client that specializes the lawyer. "You get a number of cases in one discipline and you become the expert in that discipline. It didn't take very long [after] we formed our partnership that we became the leading insurance and tort negligence law firm in the province of Quebec and maybe in Canada for that matter."[117]

Their reputation in the insurance industry was aided as well by the name of their father, their previous work for Brais and by another powerful denizen of the Montreal legal community, John L. O'Brien, who also had an extensive insurance law practice. In addition to serving as Pierre's mentor in the early years of his articling, O'Brien boosted Louis-Philippe's career in 1952 by introducing him to the executive of the Canadian Bar Association (CBA) and recommending him for a position on the CBA's insurance committee.[118] Its reputation for being dull and secretive disguised the insurance industry's tremendous influence and power in the corporate world, a position enhanced by its extraordinary financial wealth. In that period, corporate financing required the participation of the large insurance companies. Only in the middle of the postwar boom was the industry's clout eclipsed by the growth of the chartered banks. The elder de Grandpré's appointment to the committee that lobbied for legal change and that set standards and policies for specialists practicing in the field of insurance law helped cement Tansey de Grandpré's status among the industry elite as a leading firm.

The partners developed very close personal relationships typified by their open-door office policy. The outgoing and assertive temperament and style of the de Grandpré brothers, who both thrived on litigation work, was complemented by the soft-spoken, quiet Tansey. "He was a very effective trial lawyer because he was so low key," says J. Vincent O'Donnell, who joined the firm in 1957. Eventually, Tansey chose to litigate less and spend a greater portion of his time behind the scenes working on legal opinions for the firm's corporate clients. De Grandpré, on the other hand, was the kind of lawyer "who could step into any problem and field," O'Donnell says. "He had the kind of mind that grasped things very easily and had an ability to take complex things and make them appear simple."[119]

As time went on, the partners branched out into other areas of the law. "It is very seldom that you see a lawyer who is starting a firm or leading a firm who will stay forever in the discipline that launched him in his career," explains de Grandpré. "In fact, I can identify . . . three periods in my practice of law in the firm: the first period was negligence and insurance; the second period was labor relations; and the third one . . . was administrative law."[120] Although those three legal specialities bore little relation to each other, the diverse experience

provided by them gave de Grandpré a superb background for his subsequent career at Bell Canada's corporate head office.

His first major move away from insurance law came through a
political appointment. In December 1955, he was selected to serve for a
year-and-a-half term as co-counsel with Ottawa lawyer James M. (Jack)
Coyne to the Royal Commission on Broadcasting. Known as the
Fowler Commission — after its chairman, Robert MacLaren Fowler —
the inquiry was struck by Prime Minister Louis St. Laurent to consider
the respective roles of both public and private broadcasters and to
examine the role of the CBC (Canadian Broadcasting Corporation) in
regulating the domestic broadcasting industry in the wake of the
advent of television.[121] In addition to Fowler, who was a lawyer and
head of the Canadian Pulp and Paper Association, the government
appointed two other commissioners: Edmond Turcotte, editor-in-chief
of a prominent French-language Liberal newspaper, *Le Canada,* and
former ambassador to Colombia; and James Stewart, an officer of the
Canadian Imperial Bank of Commerce. They were "all people who
had close links with C. D. Howe and with Louis St. Laurent," de
Grandpré says.[122] Although his work didn't leave him much time for
political activities, he had his own ties to St. Laurent through his past
work with Brais, as an executive member of the Young Liberals in
Montreal, and as a member of the speaker's committee of the Liberal
party of Canada.

Together, Coyne and de Grandpré plotted strategy, researched the
more than 270 briefs formally presented to the commission and conducted the cross-examination of witnesses. Although the schedule was
arduous — the commission heard testimony over a six-month period in
1956 during fourteen sessions of public hearings in twelve cities — de
Grandpré says it was "enjoyable work." In particular, he seemed to
delight in the rigorous form of cross-examination he and Coyne adopted for the hearings. Their strategy was "to take a view opposite the
view expressed by the witness. . . . We were never going to agree with
anything said by the witness, even if it meant asking, 'Why do you say
the sky is blue?' We were taking the contrarian view in order to bring
out the arguments of the witness." But more than that, the strategy also
served to insulate the commission from any charges or perception of
bias. Says de Grandpré: "It came as a shock at the very beginning
because there was, I think, a feeling that the CBC would be the darling
of the commission and that the private broadcasters would be the enemies. And when we started to cross-examine, people didn't know who

were the friends and who were the enemies. We were always mixing up our cross-examination. But I think it produced good results."[123]

De Grandpré and Coyne were invited by the commissioners to help write the final report.[124] They all met for marathon writing sessions in a room at Ottawa's elite Rideau Club. "I was a stickler for grammatical sentences, much to the delight of some of the other people who would tell us, 'here's a French Canadian telling us how to write English.'"[125] De Grandpré was delegated the task of writing the commission's suggested draft of a new Broadcasting Act, which was eventually published as an appendix to the report. Following their sweeping majority in the 1958 general election, the Diefenbaker government declared that creating a new broadcast regulator was a priority and introduced a bill similar to the Fowler commission's suggested draft on August 18, 1958.[126] Among its most important provisions was the divestiture of the CBC's regulatory functions and the creation of the Board of Broadcast Governors, the precursor of the Canadian Radio-television and Telecommunications Commission. Ironically, it was this federal regulatory agency that later locked horns with de Grandpré over the regulation of Bell Canada.

De Grandpré's associates at Tansey de Grandpré were able to handle his absence in part because the law firm took on another partner, Anthine Bergeron, in 1955. Although his colleagues coped well, the Fowler commission hearings were very rough on Hélène, who was pregnant at the time with their last child. "I was on the road again, and as soon as I left, my kids started to fall sick one after the other with contagious diseases, you know, mumps and chicken pox, and my wife was literally physically in the house, except to do some errands, from the end of September to the following Easter. . . . We had some help at home, so she could spend most of her time with the kids, but she went without . . . adult conversation for that long."[127] After the hearings ended, de Grandpré decided not to return to the exclusive practice of insurance law, the legal field that had played such an important part in his family's life.

The death of Quebec premier Maurice Duplessis in September 1959 was a watershed event in the history of the province. Scholars agree that Quebec's political modernization lagged behind the rest of Canada under the corrupt and autocratic regime of Duplessis's Union Nationale. The defeat of Duplessis's successor, Paul Sauvé, in July 1960 is cited as the beginning of the Quiet Revolution, a dramatic set of social, political and economic changes that transformed Quebec into a modern, secular state. Under the slogan, *Maîtres chez nous* (Masters in our own house), the government of Liberal premier Jean Lesage channeled French-Canadian nationalism into support for a sweeping program that removed the remaining power the Church held over Quebec society. Guided by two royal commissions, one on the constitution and one on education, the Lesage government developed programs aimed at convincing Quebeckers that their social and economic needs could be better met by the provincial government, rather than the Roman Catholic Church.[128]

"Reform was the order of the day," wrote sociologist Hubert Guindon in his 1978 analysis of the Quiet Revolution. "Education, financed by the state, was to expand rapidly, welfare was to be professionalized, health services were to be secularized, the Church was to retreat from its secular roles into matters of private rather than public concern. This coming out of the 'Dark Ages,' as the ancient dispensation used to be labeled by many, was widely acclaimed. Quebec had entered the modern era."[129]

On coming to power, the Lesage administration had enjoyed the support of Quebec's union leadership. The Catholic unions had ended their religious affiliations in 1960 and banded together to form the Confédération des syndicats nationaux (CSN). Anxious to maintain the support of labor, the Lesage government facilitated the unionization of government employees and the employees at the Catholic institutions that the government had assumed authority over. The CSN was the chief beneficiary of public sector unionization and became a strong proponent for social and economic change. By 1970, 60 per cent of Quebec's unionized workers were employed by the government. More than half of CSN's membership in that year were government employees, compared to 15 per cent of the membership of the international-trade-union-dominated Quebec Federation of Labour. A third union, the Corporation des enseignants du Québec (CEQ), or teacher's union, was made up solely of public sector employees.[130]

The coalition between the State and organized labor was destined to break down. Although anxious to unionize public employees for self-interested political reasons, Lesage refused to give the unions greater power. Their pleas for collective bargaining rights were initially rejected by the premier in the now famous phrase, "The Queen does not negotiate with her subjects."[131] However, he was soon obliged to change his stance to keep the peace with the growing public sector unions and, in 1964, the government adopted a new labor code that ensured the financial independence of the unions and gave public employees collective bargaining rights and the right to strike. The passage of these new legal rights altered the management prerogatives of school boards, hospitals and other local institutions.

It was against this backdrop of profound change that de Grandpré recognized a new opportunity and set out to further diversify his firm's legal practice. "There was a lot of activity in labor relations. I realized that there was no firm, in Montreal at least, that was prepared to negotiate for clients. They were prepared to advise clients on legal matters, but they were not prepared to get their hands dirty in negotiations. I felt that there was a very big vacuum there because very few firms can afford economists, advisors and lawyers to take care of labor relations. It takes a big firm because these contracts are only negotiated every few years. You need a big base."[132]

De Grandpré was, in fact, one of the "pioneering labor lawyers in Quebec," says O'Donnell.[133] Yet that aspect of his legal career has been overlooked in most profiles published about him, even though it placed him at the center of the maelstrom of change that swirled throughout Quebec society at that time. It also placed de Grandpré in opposition to many advocates of change, such as Trudeau, who had acted for workers whose civil rights were trampled by the Duplessis, then Lesage, governments. Unlike the Citélibristes, who were galvanized by the infamous asbestos strike of 1949 in Thetford, Quebec, de Grandpré worked for the system. His outlook led to retainers to act for employers, rather than employees. For example, he represented the Montreal Transportation Commission (MTC), which operates the city's buses and subways, when the Brotherhood of Railway Employees was replaced with the CSN. It was a "very difficult period" with "strike after strike," de Grandpré recalls. The Quebec government turned increasingly to its legislative powers to end the strikes by imposed mediation, which in turn spurred on the radicalization of the CSN. As time went on, a greater proportion of the negotiations that de Grandpré was involved in broke down into strikes.

His work for the MTC led to a retainer from the Montreal School Board to negotiate with the teachers' union. That work then led the Quebec government to retain de Grandpré as a trouble-shooter in their province-wide negotiations with the teachers. In addition, he represented several private sector companies in labor matters. As that part of his practice expanded, the firm took on more partners, including Claude Lavery, another of Quebec's first labor lawyers.

Throughout that period, the mounting frustration experienced by Quebec's union leaders in their dealings with the province led them to perceive the Quebec government as no different from private sector employers and as equally hostile to workers' interests. The perception that the provincial government itself contributed to the inferior position of francophones in Quebec was given credence by analyses published by the federal Royal Commission on Bilingualism and Biculturalism between 1966 and 1969.[134] Those studies revealed that unilingual anglophones in Quebec earned more than either bilingual anglophones or bilingual and unilingual francophones. That economic disparity lay at the heart of criticisms of Lesage's Quiet Revolution and led to increased efforts by organized labor for still more change. De Grandpré is critical of the unions, which he believes hampered the province's, and Canada's, competitive position by "demanding more compensation, more benefits and less work." He carried that attitude with him to Bell Canada, which, coupled with his past work as a government mediator, made him a tough negotiator who gave up few concessions to the unions. Needless to say, he was not seen as a sympathetic boss by labor leaders.

His criticism of the government's management of the changes wrought by the Quiet Revolution reveals both his elitist sentiments and a romanticized vision of Quebec that comes close to resembling the stereotypical views of the province sometimes found in English Canada — a view echoed in his earlier statements that the Church had hampered Quebec's commercial development:

> The evolution has happened too quickly, in my view, in the province of Quebec after Duplessis died and Lesage came to power. This explosion of democratization of the system has been too fast, not thought through well enough, and it has created some strains and stresses within the system with which we live today. Again, you know, even twenty-five years later, we are experiencing some of these problems — the tremendous growth of the unionization within the public sector, for instance, compared to the unionization of the private sector. . . . We can see the difficulties with the nurses. We can see the difficulties with the public transportation employees . . . the hospital support staff. We can see the num-

ber of strikes that have taken place in the education sector or
the police force.[135]

Educational reform was the second important element needed to
change Quebec's social structure because it is the vehicle by which
culture is transmitted. Prior to the Quiet Revolution, education had
been controlled by the ecclesiastical elite. The rationale for the
sweeping changes made to the *collège classique* system by the Lesage
government was succinctly summarized by John Porter, a Carleton
University sociologist, in his 1965 work, *The Vertical Mosaic*: "The
educational system which has developed in French Canada has not
conformed to the democratic industrial model. . . . It may be true, as
is so often claimed, that the Christian humanism which pervades
French-Canadian education is to be highly valued, but on the other
hand its classical orientation has prevented French Canadians from
making their full contribution to Canadian society."[136]

De Grandpré, on the other hand, believes that the educational
reforms went too far in rejecting the curriculum of the *collège classique:*

> I think that we've taken the wrong turn. I think that we've
> abandoned the ideal of educating generalists. We're trying to
> specialize the students too early in life and giving them choices
> of taking their options before they know exactly what the
> future holds for them and what's out there, [or] what their
> tastes will be. I've seen amongst our children and our friends'
> children situations where they've opted to go one way and
> then they decide that "that's not for me" and they had to back-
> track and waste a year . . . I think it's very destructive.[137]

Extending that view to the Quiet Revolution as a whole, he says that it
"went too fast, it went too far, and we're still paying the price for it." In
the same 1990 interview, he argues that those changes were to blame for
much of the conflict that took place in Quebec during the Sixties, and his
critique again echoes an elitist and outmoded view of Quebec history:

> For a society like Quebec to be changed from high population
> growth to the lowest birth rate in the industrialized world is a
> dramatic change, which is the result of probably. . . a liberty or
> freedom that was not acquired over a long period of time, but
> rather was given to them overnight.

And that's what has happened, you know. The churches
were full, now the churches are empty. The houses were full of
kids, now there are no children. The family was really a very
essential element of our society. Today, families in the
province of Quebec — 50 per cent of them, or something like
that — end up in divorce within a few years. These are not
only dramatic, but traumatic, changes for society.

I always make the analogy between this and the eighteen-
year-old who inherits a million dollars. You have to learn to
manage money. Newly acquired liberty has to be managed,
and you don't get this often at the same time.[138]

Although the slowing of Quebec's birth rate is a more recent phenomenon, the notion that Quebec was somehow unprepared to manage political or social change is a view that was discredited by the
Royal Commission on Bilingualism and Biculturalism. The commission also rejected the notion that the province was an archaic, rural
society and stated in a 1965 report that: "The fact of the matter is that
Quebec has never been the most rural province in Canada; it is also a
fact that urbanization and industrialization have been taking place
there at a steady, at times very rapid, pace; that the majority of the
population has been living in cities since 1921; and finally that Quebec
today is more urbanized than Canada as a whole."[139]

In the Sixties, in addition to his work as a labor lawyer, de
Grandpré was involved in a number of professional activities that
helped increase his own stature and the reputation of Tansey de
Grandpré. In 1961, in recognition of his work, the provincial government awarded the forty-year-old lawyer the designation of Queen's
Counsel. It also appointed him to the board of directors of l'École des
Hautes Études Commerciales (HEC), the leading French-language
business administration program in Quebec, which has trained many
of the leading members of the province's new entrepreneurial elite. De
Grandpré was also appointed a member of the council of the Quebec
bar in 1961.

Meanwhile, Louis-Philippe de Grandpré was receiving his own share
of special appointments. In 1963, he was involved with the federal
inquiry that examined the disruption of shipping on the Great Lakes and
St. Lawrence River system caused by rival labor unions. He served as
counsel to Judge Gabriel-Édouard Rinfret of the Quebec Superior
Court, who was appointed trustee of the Seafarers' International Union

of Canada in the wake of revelations that its iron-fisted president, Hal C. Banks, had waged a violent campaign of intimidation against non-union workers and shipping companies.[140] De Grandpré's older brother enjoyed a reputation as "one of the leading members of the bar in the entire country," O'Donnell says. That view is shared by Francis Fox, former federal cabinet minister in the Trudeau government, who says, "Philippe is one of the most impressive civil lawyers in Quebec of his generation."[141] Louis-Philippe was president of the Montreal bar in 1968 and 1969, president of the Quebec bar and president of the Canadian Bar Association in 1973 prior to his appointment to the Supreme Court of Canada. The mention of his name still instills awe in young lawyers, who irreverently refer to him as "Louis-Philippe de God."

Both Tansey and de Grandpré's brother held teaching positions at McGill's faculty of law in the Sixties, which allowed them to recruit the best young talent for their firm. One lawyer who worked at Tansey de Grandpré then, and who asked to remain anonymous, says Louis-Philippe assumed the role previously held by Brais: "He is a very special, unique man who was probably the mentor of more pre-eminent lawyers in his time than anyone else." Both brothers challenged their students to be their best and often gave them tough assignments for the weekend. "We never minded because they took the time to teach us things and to show us how to be the best. The partners at Tansey de Grandpré took enormous time in developing people through their apprenticeship, reflecting the de Grandprés' view that, while you can do anything in life if you specialize, you don't need to specialize if you know how to think."

The former student recalls that de Grandpré was as demanding as his older brother. "But he was not the tough guy of the two. Jean came across as warmer and more extroverted — not as tough and as arrogant as he seems to have become now." He was also struck by de Grandpré's incredible memory: "I once told him he was quoted in court and I told him what the lawyer cited. De Grandpré said, 'No, I couldn't have said that!,' and he got the exact text and, sure enough, there was one word different."[142]

Both brothers were regarded as very bright and as people whose integrity was beyond reproach, Fox says. Few people ever challenged de Grandpré's work or advice and anyone who did would be confronted with his abundant self-confidence and steel will. "It didn't happen often, but I remember a client questioning an opinion he had offered," O'Donnell says. "He turned the library upside down to establish that he had been right."[143]

Despite its diversification, the firm maintained its leading position in insurance law and, in 1965, de Grandpré became the first Canadian lawyer selected to serve on the executive of the International Association of Insurance Counsel, a leading American organization. The following year, he delivered the keynote address to welcome members to their annual meeting in Vancouver — the first time the conference had been held in Canada. His speech turned out to be a fitting farewell to his practice of insurance law. A short time later, he decided to leave his law firm to begin a new career as a corporate counsel and executive at Bell Canada. It was at the telephone company that he was able to put to use the traits nurtured by his ambitious parents — his decisiveness, his pugnacious tenacity, but most of all his ability to lead.

Chapter 2
Pa Marries Ma

The only bad decision is the one that's never taken.
— A. JEAN DE GRANDPRÉ, MARCH 12, 1986

A telephone call from one of Montreal's most influential lawyers in the late summer of 1964 led de Grandpré down a path toward a major mid-life career change. It came from John L. O'Brien, who asked whether de Grandpré would be prepared to tackle a long and demanding job for "a very important Canadian company? I asked him a lot of questions and he said, 'I can't give you the answers because it's confidential.' I said, 'Well, when you can give me the answers, call me back.'"[1]

In addition to practicing insurance law, O'Brien was the senior out-side lawyer retained by Bell Canada to prepare and fight their rate cases before the telephone company's federal regulator. O'Brien "had always been a godfather to us," says Louis-Philippe de Grandpré. "He was a brilliant lawyer and he had the Irish charm."[2] As their "godfather" in the legal world, O'Brien looked out for the de Grandpre´ brothers and would turn legal cases their way whenever he had a conflict of interest and could not act for a client.[3]

The relationship between Tansey de Grandpré and O'Brien Home Hall & Saunders deepened throughout the Sixties. "We had a very friendly relationship and so, in the 1970s, we merged the two firms," says J. Vincent O'Donnell, a former partner of Tansey de Grandpré and senior partner at Lavery O'Brien, where de Grandpré and O'Brien's son, Robert, practice law today.[4]

When O'Brien telephoned Tansey de Grandpré about the "mystery" client, seniority would normally have dictated that the case be offered to Louis-Philippe. But, says de Grandpré, O'Brien "knew that my brother could not do it on account of his health."

A few weeks after his first cryptic call in August, O'Brien phoned de Grandpré again and asked him, " 'Have you read the papers this morning?'

"I said 'Yes.'

"O'Brien said, 'Did you see that the Board of Transport Commissioners [which was the regulatory body at the time] is launching an investigation into the earning levels of Bell Canada? I am asking you to become involved in this because I have been in the hospital.' "[6]

Although the ostensible purpose of the Board's hearing was to review the financial affairs of the country's largest telephone company, the regulator also had a second, more sweeping purpose in mind — to assess the very basis by which Bell's allowed earnings were calculated.[7] As the Board stated, it had no intention of examining "the propriety of existing rates." Rather, Bell's regulator contemplated changing the way it had set subscriber rates since it began regulating telephone rates on July 13, 1906. It was that part of the case that O'Brien asked de Grandpré to "take charge of."[8]

It was a major decision for de Grandpré because he would have to devote most of his attention to the proceeding for the better part of a year. However, his partners were supportive — Bell Canada was an important client and recent staff additions meant their firm would be able to handle the increased workload — so after consulting his major clients, de Grandpré agreed to take the case. In representing Bell, he was following in the footsteps of his mentor, F. Philippe Brais. Bell not only retained Brais in 1949 to fight the company's first rate case since 1927, but also appointed the lawyer a vice-president of the corporation.[9] The 1950 hearing that Brais fought was the only rate case in which the regulator did not make some reduction to the company's revenue request.[10]

De Grandpré's first Bell hearing, on the other hand, was not a routine affair. With a mandate to address the company's financial future and to alter the status quo, the stakes for venerable Ma Bell were potentially enormous. It would be a major test not only of de Grandpré's abilities as a lawyer but of his intellectual depth as well. Although the success of Bell's case would be dependent on his arguments before the Board, the case would be grounded much more in the principles of corporate economics than in any principles of law.

De Grandpré began working on the case in the fall, helping O'Brien and Bell's own corporate legal staff prepare their arguments and evidence. Initially, he was spending about 10 per cent of his time on the case, but that quickly increased to more than 75 per cent of his time as the deadline for submitting Bell's supporting evidence on November 30, 1964, approached. By early spring, the Bell case had become a full-time job.

As a prelude to the May 1965 proceeding on Bell's financial future, the Board conducted a brief hearing on Bell's employee stock plan.[11] O'Brien gave that case to de Grandpré to prepare, and he presented the company's argument to the Board at a two-day hearing in Ottawa on March 23 and 24. The Board's hearing room resembled many other formal courtrooms, complete with witness box — a feature that has since been abolished by the current regulatory agency. Although the hearings were conducted in the adversarial fashion of a trial proceeding, Bell's regulator was not bound to base its judgment solely on the evidence before it. The experience at the stock plan hearing gave de Grandpré a valuable opportunity to get a feel for the Board's style and procedures and, more importantly, to become a known entity to the commissioners and their staff before the important rate review hearing began on May 4.

That inquiry was completed on June 29 after twenty-two days of hearings, which is lengthy even by today's standards. A good portion of it was spent rehashing issues that had been raised by other interested parties, known as intervenors, at every Bell Canada proceeding since 1927 — Bell's relationship with Northern Electric Company, the service contract with minority shareholder AT&T and adjustments to the company's basic financial ratios in order to obtain lower subscriber rates.[12] The Board's decision to review how it calculated the amount of money Bell could make was precipitated by the tremendous economic changes of the postwar "baby boom" era. From 1950 to 1958, the number of telephones installed in Bell's territory almost doubled, rising to 3.1 million subscribers, an increase one-and-one-half times as great as the growth achieved in the preceding twenty-two years. This growth led to an insatiable demand for money. Bell's annual capital spending tripled between 1949 and 1958, exceeding $1 billion and leading Bell to seek four rate increases in that nine-year period.

In the six-year period between the Board's notice of the review and the last rate increase in 1958, Bell experienced a further acceleration of growth. Even with the previous rate increases, it was having difficulty meeting its growing demand for capital, which was required by the urban expansion into the suburbs. Says Robert Scrivener, who was then Bell's executive vice-president of operations, "The biggest problem we had then was this expansion and the backlog of demand for service that had built up during the war. The problem was to get new facilities built, which was a question of money and manpower — both of which were in short supply. But then there was even greater change. I don't think people expected the country to take off like a rocket." He adds that this was "still the age of prewar technology."[13]

The most important function of Bell's regulatory agency is setting subscriber rates. However, nowhere does the law setting up the agency state how it is to establish appropriate rates. Rather, it declares in typically broad fashion that all telephone rates are to be "just and reasonable" and are to be free from discrimination. Regulatory agencies throughout North America have been given broad discretion in deciding what economic measures can be used to set subscriber rates. Many methods can be used, so long as they meet the broad philosophical principles enshrined in the law and allow the company to earn a fair return on profit for its investors. In practice, though, regulatory agencies have been very conservative and have tended to share the same methods.[14]

In setting rates, the regulator must balance two competing interests: that of the subscriber and that of the company. It begins the process by examining how much money the company brings in from revenue, then subtracts total expenses. Next it looks at the company's revenue requirement — the money the company cannot obtain from its existing rates or finances, but must borrow or raise on stock markets to finance its growth. This leads to the bottom line, or profit, which is the money that the utility makes. Regulatory agencies then apply a financial indicator that tells the company how much profit it is allowed to keep. This magic figure is used to determine the rates paid by subscribers by working backward through the company's income statement.

Since 1906, the Board of Transport Commissioners had used share profit as the test of "reasonable rates" and the measure of what Bell Canada could earn.[15] Share profit is calculated simply by dividing profit after taxes and expenses by the total number of shares outstanding. One problem with this method of rate regulation is that share profit is not a static figure. In fact, it varies greatly any time the company issues or buys back more common shares. This meant that any time Bell proposed to sell more stock, the Board had to approve the sale and readjust the permissible share profit figure.

Under early regulatory principles, courts determined that reasonable rates should include a premium to reward investors. Two important U.S. Supreme Court cases, both decided in 1922, defined the additional components of reasonable rates that regulators were expected to take into consideration in making their rate decisions. The principles of those cases were also adopted by Canadian courts and were applied by Bell's regulator to the setting of telephone rates. In the first case of *Southwestern Bell Telephone Company versus Public Service Commission of Missouri,* the U.S. court stated that a regulated utility was not only allowed to earn compensation for all reasonable costs of conducting its business, but was also allowed to earn an additional "allowance" to cover its risks, plus a premium to allow it "to attract capital."[16] It was that case that gave rise to the term "cost-plus," used to describe the method of rate regulation used by Bell's board prior to 1966. In the second case, known simply as the *Bluefield Waterworks* case, the U.S. court went further in developing what it felt should be included in the "plus side" of the concept. The court stated that a public utility is entitled to earn the same return on its property "which it employs for the convenience of the public" equal to the return being made by other businesses with "corresponding risks." Although the court also stated that a utility has no right to the same level of profit as made by a highly profitable or speculative enterprise, it declared that a utility's financial return should be sufficient to allow it to maintain its credit and to attract new capital.[17]

What Bell wanted to achieve at the 1965 hearing was a radical departure from the share profit method of calculating an allowed level of earnings. Bell's top leadership took a strong interest in the case because they saw an opportunity to attain a greater degree of financial freedom. Behind the scenes, Bell's grand strategy was plotted by Scrivener and his deputy, Alec Lester, who worked closely with O'Brien and de Grandpré. The thrust of the case that they wanted de Grandpré to argue involved much more than a change in a traditional way of measuring permitted earnings. The alternate method sought by Bell's management entailed discarding the cost-plus method of rate regulation in favor of the diametrically opposed profit incentive method.

The cost-plus method assumed that regulation would always strictly limit profit and that any increase in allowable profits could only flow from increased rates. "You had to guarantee a certain dividend, and the regulators were offering something like a 10 per cent premium over our dividend to shareholders, and that was what you were supposed to earn per share," de Grandpré explains. "It had nothing to do with the risk of the company — the financial or the business risks. So we were trying to make the company more like the other industrials because we were competing with the other industrials to raise money in the marketplace."[18]

Bell's preferred method was to set a permitted level of earnings using a measure called the rate of return on average common equity. Unlike the simple share profit calculation, which takes into account only the *number* of shares held in a company, the rate of return method includes the *value* of the shares. It is calculated by dividing net income (or profit after taxes and preferred dividends) by average common equity (or the book value of common and preferred equity). Switching to this method meant rates would be empirically tested in relation to the level of earnings on invested capital.[19]

As de Grandpré says, it was "a very, very significant departure from the old way of regulating the company." It meant that the basic financial ratios recognized by the investment community would be considered in setting Bell's allowed level of earnings. "This new regulatory environment was more akin to the marketplace,"[20] de Grandpré says. In his final argument to the Board, de Grandpré outlined the economic conditions that had led Bell to seek this approach. Foremost among its concerns was the increasing need to raise more equity capital to expand service. Based on economic forecasts that showed competition was increasing for scarce investment capital, it was essential to allow Bell to realize competitive financial returns. "Regulated industries are part and parcel of the total industrial complex of Canada," de Grandpré argued. "Bell is not in a niche by itself, cut off from all other business activities. . . . Bell competes for labor, material, management, consumers' dollars and capital funds. It is even in competition with other communications media offering similar services or alternate methods of communication."[21]

The lawyer also invoked shareholder interests to advance his argument: "It is so obvious that we are sometimes tempted to forget that the investor is the source of services offered by the company," he told the Board. "Without his capital the plant required would not exist. We cannot think in terms of providing the services demanded by the public without thinking in terms of the provider of funds."[22]

Bell's corporate leaders have not always held the "provider" of capital, the simple investor, in such high regard. Without detracting from de Grandpré's argument, it is worth noting that investor relations is the corporate equivalent to patriotism. Although many things are done, or are expected to be done, in the name of promoting shareholder interests, noninstitutional investors are in fact anonymous, lowly financial foot soldiers — and often treated the way nations treat their armies. "Shareholders are entitled to their say one day a year, please and thank you,"[23] says Scrivener.

As de Grandpré argued, an increasing number of shares were being held by more sophisticated institutional investors. Pension funds, investment companies, insurance companies and other institutions had collectively increased the market value of their common stock holdings in the preceding decade by 380 per cent to $3.6 billion from $750 million. By the end of 1964, institutional investors held 35 per cent of Bell's stock, compared to 28 per cent in 1958.[24] Viewed in this light, it is easy to see how some corporations have increasingly come to view the average shareholder as a bother.

The new method of rate regulation had as its basis the profit motive. "This formula will be an incentive to management to operate outside the cost-plus method. It will be a goad for management to operate more efficiently," de Grandpré argued, referring to the testimony of one of Bell's expert witnesses. "The progress of mankind is not made by reducing the return of capital but by reducing the cost of production," he said in comparing the two methods.[25]

Then came the veiled threat, a device de Grandpré would employ with greater force in subsequent years when he was not a lawyer arguing a case, but an executive on the stand. "If after having displayed skill, talent, efficiency, vision, wisdom and foresight to improve its service in the face of changing needs, this improved performance produces results that the company and its subscribers cannot reap, management will gradually sink into the routine cost-plus approach." But worse, and now the greater threat, "If the earnings level cannot match that of the industries with which it competes, its strength will dwindle and the private subscriber will probably be the first affected by such a decline."[26]

As a lawyer trying to make a case for a client, de Grandpré suppressed any urge to respond to statements by intervenors, whose lawyers also testified before the Board, with the biting sarcasm that became a hallmark of his testimony as a corporate officer in later years. Less than thirty minutes into his all-important final arguments to the Board on June 17, 1965, de Grandpré cited an order of the Quebec Public Service Board and read it into the record in French. He didn't flinch when Lovell Carroll, the lawyer for the Canadian Federation of Mayors and Municipalities, rudely quipped, "You have a very good accent."[27] His remark was met simply with silence from the commissioners, reflecting the attitude prevalent in Ottawa in the era before the Official Languages Act. Such an insensitive outburst would not likely be tolerated by any federal tribunal today.

Continuing on with his final argument linking Bell's financial success to the fortunes of the nation, de Grandpré articulated a view of the relationship between telecommunications and economic competitiveness that only recently has come into vogue. "Good communications are essential to . . . business if the latter wish to operate swiftly and efficiently. Without rapid transmission of information . . . Canadian business will not be in a position to match the competition of foreign exporters, and we know from history that exports are at the source of growth of the Canadian economy. Bell will not be in a position to play its role in the development of the country if its construction programme is curtailed by lack of funds."[28]

De Grandpré won his case. The Board sided with Bell's major arguments in favor of regulating rates on the basis of an allowed rate of return on common equity (or ROE). Instead of being allowed to earn share profit of $2.43, as set in the 1958 rate decision, Bell would be permitted to earn a rate of return on average common equity of between 6.2 and 6.6 per cent. Bell's management would now have the flexibility to retain more profit if it succeeded in curbing expenses. Under the old method, the allowed share profit acted as a fiscal straitjacket. There was no incentive for the company to become more efficient. If it was able to curtail its expenses, thereby increasing its share profit, it would exceed its fixed limit and would have to lower its telephone rates, benefiting subscribers but not shareholders. Under the percentage ROE method, the company would be able to retain any extra profit gains achieved through efficiency and, provided it remained within an allowed range, distribute those gains to shareholders, which in turn would make Bell's stock more attractive to investors.

But the change also meant that the investor's return would fluctuate with any increase or decrease in the number of shares of Bell's stock held by a shareholder. An increase in the book value equity of the company would produce a greater return for each share held by an investor. Therefore, the new method simultaneously fostered Bell's strategic objective to be a widely held corporation because it rewarded shareholders for buying more shares. This was critical in preventing hostile takeover bids for the phone company's shares. There is an important distinction that de Grandpré makes between increasing the number of shares held in the company as opposed to increasing the total number of dollars invested in it. "Under the old basis of regulation . . . if you issued one share at $100 you would be entitled to only $2.43 on your investment because there was only one share. If you issued four shares, then you would be entitled to four times $2.43 — so that the number of shares had a direct bearing on the total earnings of the company under the old type of regulation.

"Under the new type of regulation that was instituted by the Board . . . the number of shares became absolutely irrelevant. It was the total number of dollars invested in the business that became the yardstick, if you wish, to determine the level of earnings."[29]

Although the Board agreed to change the method of regulating rates, it was not prepared to give Bell the full freedom it wanted. De Grandpré argued that as long as the Board was setting a minimum floor and still establishing reasonable subscriber telephone rates that would allow the company to earn that return, then the company should have limitless discretion in seeking to achieve the highest return possible. The Board concluded that it would be more appropriate to set a "narrow range" with "precisely defined" boundaries.[30]

"I still think that there are better ways to do it today," says de Grandpré in commenting on the impact of the 1966 decision. "I think it was a reasonable step in the right direction at the time. It was a big step for the [Board] to take. And still today it is basically the basis of regulation and for setting the revenue requirements. My main criticism against that, though, is that there is very little reward for the productivity improvements under good management. In my view, it's a lot easier for Bell or for the regulated utility to have the freedom to set the rates that they want to set, provided the rate of return does not exceed a certain level. And the expenses would be subject to review by the commission. But I don't see how a commissioner who has no experience in the marketing of services can determine whether that rate should be at this level or that level. To me, that's a marketing decision. As long as the *total rates* are just and reasonable — as long as the total rates do not exceed a certain level — then the components of that global rate structure should be left to the discretion of the marketers."[31] No doubt the regulator feared then, as would interest groups and politicians today, the prospect of basic rate increases at the telephone company's discretion. But as de Grandpré sees it, the commission is not benefiting users because the utility can't reduce rates at its discretion either: "Why should we wait six months to reduce our rates? It's beyond me. And we have. We have, in some cases, waited a year and a half to reduce rates because this reduction had to be approved."[32]

At his first rate hearing, de Grandpré made use of his formidable knowledge of economics — an expertise that is virtually a prerequisite to arguing any rate proceeding on behalf of a regulated enterprise. That knowledge helps explain his later attraction to the business world. "There are few businessmen that have such a keen interest or detailed understanding of economic issues," says Bruce Scott, a professor at the Harvard Business School. "His grasp of economic issues and public policy is very comprehensive."[33]

One of the stories whispered throughout Bell Canada's empire, from the polished granite corridors of the corporate headquarters at 1050 Beaver Hall Hill in downtown Montreal to the isolated repair stations in the Northwest Territories, is that A. Jean de Grandpré had the temerity to agree to join the company in return for a promise that he would be given the presidency of the corporation. Like the proverbial glass that is half-full and half-empty, the legend is not quite true, yet not patently false. Although promises were made to woo him, the man who many career telephone company employees perceived as a brash outsider had to navigate the tortuous and mysterious road of corporate succession in competition with other candidates, as had insiders before him.

Even before de Grandpré appeared before Bell's regulator to argue the rate review case in the spring of 1965, the top officers of the utility had the lawyer in mind to head up Bell's legal department after P. Charlemagne ("Charlie") Venne, Bell's vice-president of law, retired in September 1966. De Grandpré had several qualities that appealed to them. He had extensive experience in several areas of the law, he was familiar with corporate issues and interested in management, he had top political connections, he was bright and tough — and, in addition to being a francophone, he was perceived as having a broad, cosmopolitan outlook. "They went after him," says H. Rocke Robertson, former principal of McGill University and retired director of Bell. "He didn't apply for the job."[34] The irony was that Bell Canada had been one of de Grandpré's pet peeves, says his son, Jean-François. "It was not love at first sight."[35] The younger de Grandpré recalls a running battle his father fought with Bell for two years before he joined the company over a malfunctioning switchboard leased to the family's law firm.

"As I was spending more and more time on Bell business . . . the president called me one day and he said, 'Aren't you getting tired of coming up the hill? — because I was on St. James Street and they were on Beaver Hall Hill."[36] Bell's president at the time was Marcel Vincent, a tall, studious and soft-spoken career employee who had joined the utility in 1927 as a commerce graduate. Although Bell had served Quebec since 1880, Vincent was the first francophone to head the corporation when he was appointed president in 1963. He asked de Grandpré if he would meet with Bell's head of personnel, Jim Hobbs, to discuss the prospect of joining the company as Venne's replacement. Vincent knew that Bell's proposal was a giant step for the lawyer to take and so he wisely afforded both Venne and de Grandpré the luxury of more than a year-and-a-half's worth of leeway before Bell had to find someone to actually fill the slot. Like an old-fashioned courtship, the wooing got off to a slow, gradual start as each party explored what the other wanted — and was prepared to offer — before more serious commitments were made.

De Grandpré met with Hobbs several times. At their first meeting, sometime in late 1964, he indicated to Hobbs that if the position "was an opportunity to become a member of a management team, I would be rather interested, but that . . . if I were to make a lateral move to become the head of a law department to stay there for the rest of my career, it was far less attractive." Although he says he knew that the company couldn't give him an ironclad commitment, he asked for an "assurance that I would be considered as a candidate for a management job outside the law department within the reasonable future."[37] Bell's management was not able to comply at that time, but the situation changed in less than a year. In the late summer of 1965, after the rate review hearing had ended, Vincent asked de Grandpré to meet with him in his office. Vincent began the discussion by asking him if he would still consider coming to Bell if there were an opportunity to join management.

"I said, 'Yes. I don't want any special treatment, but I would like to be considered as a possible candidate for a transfer or a promotion to some other part of the organization.'

"So he said, 'Well, I could not give you this assurance last year, but this year [we know] there are a number of retirements that will take place within the next four or five years. And if you're prepared to come in, I can assure you that you will be considered as a candidate for a job outside the law department.'

"That's why I came," de Grandpré says, adding that he turned down the offer to draw up a contract. "No, I said, if I don't like you, I want to have the liberty to walk out and if you don't like me, I want you to have the liberty to cancel the deal."[38]

As to the perception about de Grandpré's hiring, Robertson says, "Of course, the board couldn't promise him the presidency outright." However, he adds, the hiring was fully vetted by the management committee of Bell's board, "which has been very careful over the years to think of the future and of succession and to cultivate top executive managers. They — both the board and the chairman [Tom Eadie]— decided they needed better legal and regulatory talent. They also decided that Jean de Grandpré was just the man they wanted — that he had the proven abilities and he was a francophone. So they brought him in, and it was made pretty plain to him that he'd be a good candidate in the company for promotion."[39] While the presidency of Bell was not handed to de Grandpré on a platter, he was pretty much guaranteed a shot at the top job. "I never, never assumed that I was the only logical heir. I realized I was one of them . . . but never was there any indication given to me that this was so."[40]

Bell announced de Grandpré's appointment as corporate general counsel on October 25, 1965. He would start his new job on January 1 with Venne helping to smooth the transition by retaining the position of vice-president of law until his retirement in September 1966. Although his departure from Tansey de Grandpré was a major loss, his partners were psychologically prepared for his departure, says Louis-Philippe, because "we had learned to get by without him the two times he was away."[41]

De Grandpré's job at Bell was to manage the company's legal affairs and deal with regulatory and governmental affairs. As a manager, Robert Scrivener described Bell's newest executive as "a capable guy. He could take a problem and analyze it. He had knowledge; lots of knowledge. He learned the business. . . . He could understand balance sheets. . . . But he saw the problems as a lawyer sees a problem. Good lawyers couldn't run a business, but they can understand the problems. And they tend to look at legal solutions to problems."[42]

The second part of his job — protecting and promoting Bell's corporate interests in Ottawa — was a task that not only preoccupied him for the next seventeen years, but, some would later say, obsessed him. Eadie and Vincent needed a tough man for a tough job; a lawyer who not only knew the ropes in Ottawa, but someone who could get Bell's way with the regulator. The cozy, gentlemanly peace between Bell and the federal government which was a hallmark of the eighty-seven-year period between the company's incorporation and de Grandpré's arrival came to an abrupt end during his tenure. More frequent clashes were inevitable as Ottawa sought to extend its authority over the utility and its growing stable of subsidiaries, while at the same time an expansionist Bell sought to break free of what it perceived as the strictures of the federal government's constraining regulations. Management needed a fighter, someone who could be a son-of-a-bitch, if necessary, to keep Ottawa at bay and to prevent nettlesome intervenors from interfering with corporate goals. Venne was not a candidate for the job because he was not aggressive enough. "Charlie was a nice guy," says Scrivener, "but he wasn't a de Grandpré. He never was going to be. And Marcel retired him."[43]

Bell had never devoted much attention or resources to government relations, says de Grandpré, but now it was serving notice to Ottawa "that we were not going to accept all their dicta without putting up a good case for the company. Mr. Scrivener wanted me to be tough. . . ."[44] And he was. It was a job that he would shine at. Not that de Grandpré particularly relished being thrust into the role of public defender of Bell's monopoly, rather, he was conditioned to fight the good fight. "He has a tenacity," says Louise Brais Vaillancourt. She defends de Grandpré's toughness, arguing that "there has to be an instinct of going for the jugular"[45] when it comes to protecting the corporate interest.

Lured by the prospect of advancement, de Grandpré admits he did not know enough about Bell itself to fully appraise the "pros and cons" of accepting the position. After only a few months at the utility, he was on the verge of quitting. "I almost walked out. Oh boy, I was so discouraged," de Grandpré says, exhibiting a human frailty he rarely shows and that runs counter to his image as an abundantly self-assured giant who is undeterred by adversity. He knew few people within the organization, other than the key people who had worked with him on the rate review and, from January through to the summer, he faced the daunting and boring task of familiarizing himself with almost a century's worth of legal opinions relating to Bell's business. "You're given documents that thick," de Grandpré says, pointing to a hefty binder, "and you're reading and reading and the telephone never rings." After reading through the documents and opinions on one rule, he was handed another set of opinions to "know what has been said in the past. So you do that for days and months, and you say, 'What the heck am I doing here?'"[46]

As a regulated company, Bell faces a litany of legal obligations that other companies do not. And for every rule and regulation, there is a mountain of three-ring binders containing Bell's collected experience with that rule. Take, for example, the provision of the Railway Act that states Bell must submit any contract or agreement with any other company that operates a telephone system to its federal regulator for approval. Bell's lawyers have a binder of material containing all of the company's dealings with sub-section 320 (11) since the provision was adopted by Parliament in 1906. In addition to pages of arcane memos and opinions regarding that particular law through the years, there are pages of court decisions and, for a little extra reading, excerpts from legal reference texts and other laws that a Bell lawyer could use to make yet another opinion or to argue a case against applying the sub-section to something Bell wanted to do.[47]

Reading through just one of those binders, it is easy to see how anyone's eyes would glaze over. As dry and boring as their contents may be, however, those binders are the equivalent to gourmet fare for the legal chefs in the law department who, by pulling each ingredient off a shelf, can quickly concoct a skillfully crafted case with elements all but forgotten or consigned to oblivion outside Bell. The fact that Bell has always been a formidable adversary in any proceeding is in no small measure due to this meticulous collection of materials and the fact that the corporate counsel is expected to read through every page.

None of those binders, though, describes Bell's true history which, far from being boring, is vibrant and rich. More than just a prelude to his own era, it is the legacy that de Grandpré inherited and that his predecessor transformed.

Chapter 3
The Making of Ma Bell

Command large fields, but cultivate small ones.

— VIRGIL

Alexander Graham Bell wanted to make a telegraph wire speak. It was a logical preoccupation for the inventor and third-generation speech therapist who taught elocution and vocal physiology at Boston University. Tall and slender with dark, wavy hair, the bearded scientist was only twenty-seven years old when he discovered the means to send and receive sound transmissions electrically over wire in 1875. His subsequent invention of the telephone was a consequence of perfecting a device he called the harmonic multiple telegraph system, which allowed multiple telegraph messages to be sent simultaneously to a single receiver. It, in turn, led Bell to an experiment that reproduced a faint sound in a coiled wire by using an intermittent electric current. On March 1, 1875, he excitedly reported his findings to Joseph Henry, a physicist and head of the Smithsonian Institution in Washington. "You have the germ of a great invention," Henry replied when Bell told him of his idea to make a telephone to transmit speech electrically.[1]

Bell realized that, unlike the harmonic telegraph, a telephone had to rely on a continuous current of electricity as opposed to an intermittent current. He discovered that the key to producing sounds lay in the principle of variable resistence: a magnetized reed attached to a diaphragm that simulated the human ear could generate a current that varied in intensity, just as air varies in density when sound passes through it.[2]

The breakthrough came on June 2, 1875, at Bell's Boston lab at 109 Court Street. Bell and his assistant, Thomas A. Watson, were in different rooms preparing an elaborate experiment with the harmonic telegraph. Bell was tuning the reeds of three transmitters and three correspondingly tuned receivers connected by wire to similar devices tended by Watson in the other room. One of Watson's magnetic reeds was screwed down so tightly that it stuck to its electromagnet. When Watson plucked it to free it, Bell heard distinct twangs in his receiver instead of the usual whine sent out by the vibrating transmitter. Yet Bell's receiver had vibrated without any steady transmitter current. The transmitter reed plucked by Watson had instead transmitted an undulating, or alternating, current over the line. Both the receiving and transmitting reeds acted as a diaphragm, like a human ear drum, and reproduced sound with a miniscule level of induced current. When the current was increased, the sound became louder. Watson made two primitive telephone sets that night, which they then tested the next day to transmit and receive recognizable voices. They continued the experiments for most of the summer and, by September, Bell started to write the specifications for a patent.[3]

It was in this period that a Canadian businessman was offered, but passed up, one of the greatest economic opportunities ever offered an individual. If he had seized the opportunity to acquire world rights to Bell's patents, a corporate equivalent to Bell Canada could have had a virtual global monopoly on the telephone by owning the rights to the Bell patents in every country except the United States.

Although eager to patent his inventions, Bell was also anxious to sell the rights. He had already sold the U.S. rights to his future father-in-law, Gardiner Green Hubbard, a wealthy Boston lawyer. Hubbard told Bell that he should not rush off to patent his invention in Washington if he wanted to sell the foreign rights as well. The most lucrative foreign patent was a British patent, which included the territories of the British Empire and most of the Commonwealth. By a typical quirk of British law, an application could not be granted if a patent had already been granted in the United States, so Bell had to find a buyer for the foreign rights and have that person file for a British patent before he could apply for a U.S. application.[4]

In early October 1875, Bell traveled to his father's homestead outside Brantford, Ontario, in search of a buyer. He didn't have to look far. Below Alexander Melville Bell's backyard lay the Grand River estate of George Brown, a Liberal member of Parliament and influential owner and editor of the Toronto *Globe*. The detailed accounts of Bell's experiments in Brown's *Globe* helped to garner widespread interest in the inventor's work, including an account of his subsequent long-distance telephone work published the following year in *Scientific American*. Drawing up his courage, Bell asked Brown for an introduction to Sir Hugh Allan, a prominent financier and transportation magnate who was head of the Montreal Telegraph Company. Instead, Brown made his own offer to acquire the rights rather than arranging an introduction to a political rival. He agreed to make an initial payment upon receipt of Bell's specifications and not to do anything to jeopardize the acquisition of a British patent.[5]

Throughout the fall, Bell anxiously waited, in vain, for any answer or payment from Brown. Hubbard was growing increasingly impatient. He was eager to avoid a dispute with rival inventor Elisha Gray, whom Bell suspected of spying on his laboratory. Finally, at the end of November, he gave Bell an ultimatum. He demanded that he make a choice — his teaching and speech work or the patents and his fiancée.[6] The patient but assertive Bell convinced Hubbard to cool off. He met with Brown shortly after Christmas in Toronto where the two men reached a conclusive agreement.

Brown wrote his wife, Anne, on December 30 about the agreement with Bell. Although he excitedly described the multiple telegraph, he made no mention whatever of the telephone.[7] Bell delivered the specifications to Brown in New York the day before the publisher set sail for London on January 26, 1876. He asked Brown to search for any competing patents at the British Patent Office and, if none were found, to file an application and cable him immediately. Brown did not perceive the same sense of urgency. He and his associate in London, James McFarlane Gray, found no infringing patents after a search on February 16. Although Gray urged Brown to apply immediately, Brown wanted to meet with a British scientist first before making a decision on whether to acquire the rights to the invention.[8] Instead of cabling Bell with a reply, he wrote to his wife that evening to tell her of the outcome of his discussion with the scientist on the harmonic multiple telegraph:

> The result of the meeting appears to be that Bell has made a step in advance of everybody else — but that it is doubtful

whether it would be workable on the ocean cable and quite
certain that it would be of no commercial value on this side of
the Atlantic (whatever it might be in the U.S.). Until I see Tom
[Brown's brother-in-law, Thomas Nelson] I shall not finally
decide — but my present inclination is to go no further in the
business.[9]

Although he applied for a patent in Bell's name, as requested, Brown
seemed to have little understanding of Bell's work, for he was interested in the the less significant of the two inventions. Nor did he have the
courtesy to relay any of this to Bell until the end of the month.[10] Not
only was his dismissal of the telephone a lost opportunity of momentous economic proportions, but Brown's lassitude also jeopardized
Bell's chances to patent his invention in the United States. Hubbard
had again grown tired of waiting for an answer from Brown and,
unknown to the inventor, had gone ahead and submitted the U.S.
patent application in Washington on February 14 while the British
patent filed by Brown was pending. In one of the most astounding
coincidences in the history of invention, and one of the most rewarding
acts of serendipity, Hubbard's application guaranteed the primacy of
Bell's patents in the face of a second competing application filed by Gray
only two hours later.[11] Granted by the U.S. Patent Office on March 7,
1876, Bell's patent No. 174,465 was destined to become the single most
lucrative patent ever awarded. Along with a second issued in Bell's name
on January 30, 1877 — No. 186,787 for a box phone combining a transmitter and receiver, metallic diaphragm and permanent magnet — the
patents soon spawned a new form of business organization and, almost
a century later, ushered in a second Industrial Revolution. When Brown
turned down the foreign rights to those patents, he set the stage for
Canada to become a branch plant in telephony, dependent upon a U.S.
patent holder for access to the inventions.

In the summer of 1876, Bell perfected long-distance telephony,
which further stimulated commercial interest in his patents. He made
the first long-distance telephone call to his father's house over a thir-
teen-kilometer-long Dominion Telegraph Company line between
Brantford and Paris, Ontario, on August 10, 1876. Thomas Swinyard,
general manager of the telegraph company, initially rejected Bell's
request to conduct the experiment but he was persuaded to change his
mind by his young assistant, Lewis Brown McFarlane. Almost forty
years later, McFarlane was appointed president of Bell Canada.[12]

Although dismissed by some as a novelty and opposed by others as a competitive threat, the telephone was vigorously promoted to a curious world by a new breed of marketers. Throughout the first half of 1877, Hubbard set up a company to sell his son-in-law's invention and retained agents across the United States to sell telephones and start up service. As much as the young Bell craved the success and fame that he believed his invention would bring, he had absolutely no interest in developing a commercial enterprise or in organizing a telephone system. After Brown turned down the foreign patents, Hubbard became trustee of all past and future patents, and Bell carved the world into franchises that he distributed to his friends and relatives.

The Bell Telephone Company was created by Hubbard, Thomas Sanders, Bell and Watson as an unincorporated voluntary association on July 9, 1877. Although the five thousand shares were split equally, they had no value because the association had no stated capitalization.[13] The following day, Bell assigned 75 per cent of his Canadian patent rights to his father and the remaining 25 per cent to Charles Williams, a Boston-based equipment maker, in return for 1,000 telephones to be delivered to his father to help launch the new venture in Canada.[14]

Melville Bell appointed an old friend from Paris, Ontario, the Rev. Thomas Henderson, as his general agent. Despite a lack of capital, expertise, organization and connections — political or mercantile — the elocutionist and the preacher devoted their full energies to the fledgling business for two years. They quickly learned the first immutable law of telephone economics: that the value of a telephone — and the price charged for service — grows in direct proportion to the number of other telephones a subscriber can call.[15] For this reason, Henderson followed Hubbard's example and solicited other agents to sell the new invention under temporary licenses. One of the most enthusiastic was Hugh C. Baker, a Hamilton businessman who ran a telegraph dispatch business and promoted the Hamilton Street Railway. Baker helped organize one of the first public demonstrations of the telephone in Canada on August 29, 1877, between three homes in Hamilton. The following month, Bell and Baker installed a private telephone line in Ottawa linking Prime Minister Alexander Mackenzie's office with Rideau Hall, the official residence of the Governor-General.[16] The telephones were hooked up in November and the two transmitters and two receivers rented for $42.50 a year. The first telephone lease in Canada was signed on October 18 for a connection that hooked Baker's home in Hamilton to the homes of two of his colleagues for $45 a year.[17]

The early policy to exploit the patents by renting, as opposed to selling, the telephone equipment to subscribers proved to be crucial in launching the telephone business. Hubbard's shrewd legal mind is credited with the idea.[18] Like a bank deposit that accrues compound interest, telephone leases would, over time, grow into ever-expanding sums that could sustain greater growth. That policy was not overturned in Canada until de Grandpré's tenure more than a century later when the federal telecommunications regulator issued a ruling in November 1982 allowing subscribers to own their own telephones.

Henderson and Bell's father had recruited agencies to supply telephones in Ontario, Quebec, British Columbia, Manitoba, Nova Scotia and New Brunswick by the end of 1878. Interested businessmen also organized companies to offer Bell's service in Toronto, Hamilton, London and Windsor.[19] Bell, Sr. soon exhausted the supply of phones he had obtained from Williams, who could not even keep up with the demand for orders in the United States. At Bell, Sr.'s urging, a Brantford hardware store merchant and inventor, James Cowherd, traveled to Boston to learn from Williams how to make a telephone. In December 1878, Cowherd began making the first Bell telephone for Canadian subscribers.[20] Although he was the first licensed manufacturer of Bell phones in Canada, he was not alone. Thomas Ahearn copied Bell's design and placed some phones in service in Ottawa, as did Cyrille Duquet, a Quebec City jeweller who patented a modified version of Bell's transmitter in 1878.

The Bell companies in both Canada and the United States aggressively pursued infringers and tested their patents in some six hundred separate legal challenges in the United States alone in the eighteen years between 1879 and 1897. Bell's lawyers won every case that was not dropped by a defendant.[21]

Neither Hubbard nor Bell, Sr. possessed enough control over their new markets to come anywhere near forming a monopoly. Far from being orderly, the introduction of the technology was chaotic with a large number of entrants vying for a share of the market. These included entrenched telegraph monopolies offering competing telegraph service or telephone service from competing suppliers. For the telegraph company, telephones were a loss leader, leased at low prices or given away free in a bid to thwart the competition.

Intense competition, coupled with a severe shortage of capital, almost brought the early telephone entrepreneurs to their knees as their debt mounted. Hubbard brought in outside investors to avoid financial ruin in July 1878. The Bell Telephone Company was reorganized as a corporation and capitalized with $450,000 worth of stock.[22] Because Hubbard was preoccupied with retaining control of his company and battling competitors and patent infringers, he also hired a new general manager.

The hiring of Theodore J. Vail was a brilliant stroke on Hubbard's part. Portly, bespectacled and mustached, Vail was the spitting image of Theodore Roosevelt. Hubbard persuaded the thirty-two-year-old general superintendent to quit the government's Railway Mail Service in the spring of 1878. Having started out as a mail clerk for the Union Pacific Railroad, Vail's career took off after he organized a new system for fast mail delivery.[23] But his greatest achievement was the creation of the strategy and organization behind the telephone company that would become the world's largest corporation, American Telephone and Telegraph Company (AT&T). Although widely hailed as the architect of the Bell system, Vail's legacy extends beyond the telephone field to big business in general. His organizational methods are still in use today.

Vail joined the Bell Telephone Company when it was on the verge of bankruptcy. To avoid ruin, Hubbard merged his company with the New England Telephone Company in March 1879 to create the National Bell Telephone Company of Boston. The merger almost doubled the company's capitalization to $850,000. The younger Bell left the board and Hubbard, after surrendering his majority control of the company, was ousted by a group of monied shareholders led by Boston financier William H. Forbes, who took over as company president. Vail was kept on as general manager in charge of operations.

In that same year, the company elicited the support of Alexander Graham Bell to defend his patents in a case brought against National Bell by Western Union, the telegraph giant controlled by William H. Vanderbilt — the Vanderbilt scion who entered the history books with the famous quip, "The public be damned; I am working for my stockholders."[24] The case pitted two great patent lawyers — James Storrow for National Bell and George Gifford for Western Union — against each other. Aided by Bell's testimony in July 1879, National Bell defeated Western Union, despite its money, experts and power. Vanderbilt opted for a settlement with the phone company, rather than a protracted struggle in the marketplace that would have driven Forbes and Vail into an alliance with competing telegraph companies. In return for 20 per cent of the revenue for seventeen years, Western

Union agreed to give up the telephone business, including its telephone agent subsidiary, the Gold and Stock Company, and to assign all of its patents to National Bell.[25] Thomas Edison also found himself adversely impacted by the tussle because Western Union had acquired the rights to a competing voice transmission system that he had invented. In the wake of the Western Union lawsuit, The Bell Telephone Company legally blocked the issuance of a patent to Edison, with the result that he had to wait until 1892 for his invention to be patented — by which time AT&T had command of the market.

Meanwhile, Bell's father sought an alliance with Dominion Telegraph Company to counter Western Union's ally in Canada, Sir Hugh Allan's Montreal Telegraph Company, which had acquired the rights to Edison's system only a few months before the legal fight began. Bell was tired of the business and told his son that it was "very up-hill work in Canada."[26] He offered Thomas Swinyard, general manager of Dominion, exclusive rights to the telephone patents for a paltry $100,000. Swinyard tossed the offer in the garbage, forfeiting an opportunity that he would later regret. Forbes and Vail could not let the Canadian territory pass into enemy hands and so, late in the fall of 1879, National Bell purchased the rights.[27]

To build a Canadian telephone utility using Vail's blueprint, Forbes hired a crusty but energetic forty-five-year-old former sea captain, Confederate soldier and insurance company executive named Charles Fleetford Sise. The stocky and stern-looking Sise was an exceedingly tough taskmaster and an effective executive. He and his two sons would rule Bell Canada and its manufacturing subsidiary from 1880 until 1944.

The son of an importer and shipowner in Portsmouth, New Hampshire, Sise first took to the sea at the age of sixteen. At twenty, he was captain of his own ship and, for the next seven years, ran his father's worldwide shipping and cotton business headquartered in New Orleans and Mobile, Alabama. After the U.S. Civil War was declared on April 12, 1861, the young sea captain served as private secretary for his father-in-law's neighbor — Jefferson Davis, the president of the Confederacy.[28] In May 1863, he was sent to Liverpool, England, as a Confederate agent with orders to procure a steamship and run the North's naval blockade on his return, but he never got the chance. The Confederate leaders reassessed the risk in early 1864 and asked him to invest the money instead.[29] Sise made a substantial profit for the Confederate leaders while waiting out the end of the war in Liverpool as a shipping agent. He took to the sea two years later and circumnavigated the globe in his clipper from September 1867 to October 1868.[30]

Sise returned to the United States in late 1868 and settled in Boston, where he held positions in the insurance industry for the next twelve years. He was head of the U.S. subsidiary of the Montreal-based Royal Canadian Insurance Company when it suspended its U.S. operations in the late fall of 1879. A friend referred him to Forbes and Vail. Like the Royal Canadian Insurance Company, they overlooked his Confederate past and chose Sise because of his financial expertise and, equally important from the standpoint of the owners at 95 Milk Street in Boston, his business and political connections in Canada.

Unlike the Boston business establishment, which held Sise in disrepute because of his Confederate ties, Montreal's business leaders feted him. In his first year as president of Bell, Sise brought Jefferson Davis to Montreal as a guest of the prestigious St. James Club. During his address to Montreal's business elite on May 31, 1881, Davis referred to his former private secretary's services to the South. "No one would ever know how devoted the service had been,"[31] he said.

Forbes and Vail had already set the political machinery in Ottawa in motion to incorporate the Bell Telephone Company of Canada by the time Sise was hired in early 1880. The government of Sir John A. Macdonald recognized that the tasks of nation-building required a process to expedite the incorporation of companies involved in transportation, communications and finance. Bell Telephone retained Melville Bell's agent, Hugh Baker, to petition Parliament for a Special Act that would empower the company to make, sell and lease telephones and "to build, acquire or lease and maintain and operate lines for the transmission of messages by electricity or otherwise in Canada, and elsewhere, and with such other powers and privileges as may be proper or requisite for or in connection with the operations of the business of the company."[32] It was the power to construct facilities that required the sanction of the federal government. In the case of Bell, the company needed the power to erect lines and poles along or under any public highway, waterway, street and bridge. Bill 17, introduced in Parliament on February 23, 1880, also sought the power to amalgamate with any company operating a telegraph or telephone line.[33]

Sise arrived in Montreal on March 9, 1880, with special instructions from his new employers to guide the bill through the legislative machinery and to secure "a favorable charter." Once there, he was to stay and run the company. At least that's how Forbes portrayed the mission of National Bell Telephone's new special agent in his first letter to Sise.[34] But that description hardly does the task justice because there had never been such a company before. Although there were dozens of rivals in the telegraph business, along with an army of charletans, all eager to cash in on the invention of the telephone, no one else was expected to single-handedly eliminate the competition in his territory and build a monopoly from scratch. As Forbes described the task, Sise was "to harmonize the conflicting interests in the Telephone business" and to buy out both the Dominion and the Montreal telegraph companies' telephone subsidiaries in return for selling them stock in the new telephone monopoly.[35]

Sise's mission was part of Vail's strategy, which was to build more than just a telephone company: it was to build a *system.* Vail learned about networks when he managed the U.S. government's railway and telegraph mail system. As he saw it, networks led to efficiency. In the case of the telephone, he realized that network efficiency could lead to greater profits, particularly in a monopoly situation. "It is more the system established in connection with the telephone, than the telephone itself, that makes the value of the telephone,"[36] he stated in his deposition filed in the Western Union case. That 1899 statement became the gospel of the Bell system until the breakup of AT&T in 1984. And it is still the gospel at Bell Canada.

The occupation of territory was the key to building the system and gaining the desired monopoly. "What we wanted to do," said Vail, "was to get possession of the field in such a way that, patent or no patent, we could control it."[37] Contrary to popular belief, however, Bell Canada has never been granted an exclusive monopoly. When Western Union delivered the bulk of the rival telephone patents to National Bell, Forbes and Vail were well on the way to achieving their objective in the United States, but there was no such agreement in Canada.

Some Canadian legislators feared the creation of another monopoly in communications, given the existence of the vast Anglo-American telegraph combine. Although members of the House of Commons limited their concerns over Bell's Special Act to the company's power to cut down "shade trees" and to interfere with private property,[38] more substantial misgivings were raised in the Senate. Several senators voiced fears that the creation of a second monopoly in telephony could lead to an even larger monopoly through a merger with the telegraph combine. "There is nothing which this House ought to look after with greater jealousy than the establishment of large monopolies,"

Senator Haythorne stated on April 8, 1880. "There is a distinction between the legitimate enterprises which we all admire, and the mode of conducting them," he said, adding, "we should be abandoning our duties in this House if we let this Bill pass without some guarantee against the evils which may arise from the amalgamation of companies."[39]

Haythorne lost his bid to curtail Bell's corporate powers of amalgamation. The sentiment of the majority of legislators and the popular opinion of the day was expressed by Senator Dickey, who was also chairman of the Senate Committee on Railways, Telegraphs and Harbors: "So far from such a bill calling for obstruction in this House, it is one that should commend itself to our most favorable consideration, because it initiates a new species of legislation, in connection with one of the most interesting of modern scientific discoveries."[40] Dickey's view typified society's glorification of the inventor-entrepreneurs of the late nineteenth century and helped explain the speed with which Bell Canada's bill was passed: it took less than nine weeks for it to clear all three readings in both Houses of Parliament and to be debated in committee, before receiving royal assent on April 29, 1880.

One of the most important issues that arose from Bell's corporate charter was the question of how the new company should be organized. It was in resolving that issue that Sise not only left his imprint on the Act, but also established the legal context behind de Grandpré's corporate reorganization a century later.

Forbes and Vail wanted Sise to adopt the corporate structure of the National Bell Telephone Company of Boston, which was, in effect, a holding company. Their instructions were to create two corporations. The first, a Canadian holding company called "Company A" would acquire and hold patents. It would, in turn, lease rights to those patents to local operating companies, including an operating utility owned by it called "Company B," for which the Special Act was sought. In a memorandum accompanying Forbes' letter of instructions, Vail proposed that the Canadian holding company should be incorporated by a provincial charter obtained from the Ontario legislature.[41] Sise wrote back advising against a provincial charter. He argued that the holding company would have greater freedom if it were incorporated by a federal charter, but told Forbes that he planned to obtain more comprehensive advice the following day from one of Canada's leading corporate lawyers.[42]

Sise journeyed to Ottawa, where he met with John Joseph Caldwell Abbott, the Member of Parliament for Argenteuil and prime minister from 1891 to 1892. Abbott questioned the need to form "Company A" at all, and told Sise that "Company B" possessed the full powers contemplated for the holding company. While he ag.eed with Sise that any such holding company would need to be incorporated under federal law, he rejected the idea of a Special Act for "Company A" as redundant.[43]

Forbes and Vail still wanted a parent company, however, so that their capitalization in the operating company could be reduced. Sise's solution was to incorporate the Canadian holding company as a limited company by a federal letters patent. The holding company, known as the Canadian Telephone Company, was capitalized with $300,000 worth of stock, of which the founders subscribed to half. The operating company bought the remainder of the stock in consideration for the patent licenses.[44]

For the first two years of its existence, until it was folded into its operating subsidiary, the Bell Telephone Company of Canada, Canadian Telephone operated under a corporate structure that resembled the arrangement de Grandpré set up a century later. Had the incorporators sought a statutory charter for their holding company, as Forbes had requested, it is interesting to speculate whether such an arrangement would have been as easily altered a century later or whether "Company A" would have, over time, evolved into a modern-day conglomerate resembling Bell Canada Enterprises Inc.

Sise was also instructed to place $250,000 of stock in Bell Telephone. National Bell acquired 25 per cent through the holding company. The remainder was sold in equal thirds to the two major competing telegraph companies that Sise was to acquire in return for stock, to the founding incorporators of the phone company and to "other desirable parties in Canada" that Sise was to recruit for Bell Telephone's board of directors.[45]

For his outside directors, Sise turned to his friends at the Royal Canadian Insurance Company. He persuaded his former boss, Andrew Robertson, to become the first president of the new telephone company. A prominent dry goods importer, chairman of the Montreal Harbour Commission and former head of the Dominion Board of Trade, he was "in every respect one of the best men in the Dominion," Sise wrote Forbes, adding, "his influence with the government alone would make him desirable."[46] Sise assumed the positions of vice-president and managing director of Bell Telephone, as well as the presidency of the holding company. He recruited three other Royal Canadian directors to Bell's board: the Honourable J. Rosaire Thibaudeau, a

Liberal senator and director of La Banque Nationale; Duncan McIntyre, an importer and member of the Canadian Pacific Railway syndicate; and Thomas Davidson, a tinware manufacturer. Hugh Baker and Richard Alan Lucas, both of Hamilton, Ontario, were also appointed at Bell Telephone's first board meeting in Toronto on June 1, 1880. One other founding incorporator, Henry S. Strathy, resigned from the board after its first meeting because of other obligations.[47] Bell's first board, like those today, was composed of members of the corporate and financial establishments who represented both major political parties and both the English- and French-speaking communities.

Deciding upon a corporate structure and setting up a board were simple tasks for Sise compared to his assignment to acquire competing telephone businesses. While Dominion Telegraph capitulated on June 1, 1880, after driving a hard bargain,[48] Montreal Telegraph was a different matter. Since it was not a direct subsidiary of Western Union, the U.S. settlement was of no consequence in brokering a deal in Canada. Sise's boast to Forbes on March 31 that "if Sir Hugh Allan should be inclined to give us any trouble, I can arrange with the Montreal Exchange to come into our arrangements; and leave him out in the cold"[49] proved to be premature bravado. The Montreal Exchange, which was the telephone subsidiary affiliated with Montreal Telegraph, called Sise's bluff. He was forced to write Forbes on May 15 that "the Montreal Exchange claims that they are independent, and that the Montreal Co. cannot sell them out. The Montreal Tel. Co. say they will sue for possession if we deal with the Exchange direct." Sise dismissed the dispute as "a family quarrel."[50]

The American Bell Telephone of Boston (reorganized and renamed when a new charter was obtained on May 19, 1880), on the other hand, was becoming impatient. It complained that the Canadian acquisitions consumed too much precious cash instead of being struck for shares. Forbes confronted Sise over the cost of acquiring the rival telegraph companies in a terse exchange of letters between July 4 and 13, 1880. "You can hardly afford to leave there the money you now have paid out without making an effort by rendering that investment remunerative, and thereby creating a demand for the stock," wrote Sise. "The stock will be of value in Canada only as representing the Monopoly of the entire Dominion — and as you have gone so far in the matter, I think policy and prudence both will suggest the completion of your scheme in its entirety."[51] Sise added that the acquisition of Dominion Telegraph, coupled with the purchase of competing telephone companies in Toronto, Windsor, Hamilton and London, placed Bell Telephone close to achieving its objectives. He argued that holding out for a slightly better price for the Montreal Telegraph subsidiary would only sustain a money-losing competitive battle. But Forbes was adamant that he would not put more cash into the Canadian company and reminded Sise that American Bell expected to get their money back. Sise could pay Sir Hugh his $75,000 price, but only if he could raise the money through the sale of more Bell Telephone stock. Sise's reply was that shareholders would find Bell Telephone stock attractive only if the company amalgamated with the telegraph companies' subsidiaries.[52]

The debate pointed to a problem inherent in managing a multinational enterprise, particularly one in which the parent company wants to encourage the autonomy of the local business units.[53] Policy differences over tactics, or the means to achieve a common objective, are almost inevitable. As Forbes and Sise quickly learned, such differences can be divisive. Forbes may have complained about the price Sir Hugh asked for his company, but, ultimately, he either had to adhere to the doctrinaire policy of his head office or grant an exception in favor of his Canadian subsidiary. On October 12, he traveled to Montreal to participate directly in the final negotiations for the acquisition of Montreal Telegraph's telephone subsidiary. The $75,500 deal was signed on November 16, 1880. The assets went to American Bell Telephone in a cash and stock transaction similar to what Forbes had opposed earlier. He then transferred the Montreal Telegraph Company assets to his Canadian subsidiary in return for the same amount of Bell Telephone stock.[54] The deal also included the issue of an exclusive license to Bell Telephone and the sale of Western Union's telephone companies in the Maritime provinces in return for $10,000 worth of Bell Telephone stock. After Western Union sought but failed to secure two seats on Bell Telephone's board in early 1882, it sold its shares.

Sise achieved the bulk of his objectives to consolidate his grip on telephony throughout Canada and to thwart rival telegraph interests during his first year with the company. At the end of 1880, Bell Telephone operated thirteen exchanges and leased 2,100 telephones, which grossed revenue of $29,670.58. Although Bell's capital costs to construct lines and poles were lower than the costs incurred by hydroelectric and gas utilities — the company spent almost $20,500 on poles and lines in its first year — Sise consumed almost nine times that amount, or $180,000 in working capital, to acquire rival telegraph companies.[55]

The death of Bell Telephone's sole telephone maker, James Cowherd of Brantford, on February 27, 1881, caused a temporary shortage of phones. Sise turned to American Bell's primary supplier, the Gilliland Electric Company of Indianapolis, which had contracted to make telephones and switchboards for Bell licensees in 1879.[56] Knowing this added dependency on the U.S. parent was a potential liability, he hired an associate of Charles Williams in Boston to set up Bell Telephone's own workshop in Montreal, which later became the Northern Electric and Manufacturing Company.[57] First, however, he had to seek broader manufacturing powers from Parliament by requesting an amendment to Bell Telephone's Special Act. The federal government agreed, and in addition to authorizing the establishment of a manufacturing subsidiary to make telephones, Bell was granted the authority to make "any other electrical instruments and plant the company may deem advisable."[58]

The strong opposition Bell faced from municipal governments, which saw telephone poles as unsightly pollution and road hazards, led Sise to also seek a declaration from Parliament under the British North America Act that Bell Telephone was a work for the "general advantage of Canada." Again, his request was granted. While the declaration did not give any special powers to Bell Telephone, it warned all other levels of government that the company's affairs could be regulated only by Ottawa. When Ottawa's power to invoke the declaration was challenged in the courts, Sise ensured Bell Telephone's powers another way. His lawyers obtained legislative charters in each province that Bell served in the event that the court found that the federal government did not have the power to let the company put up its poles. Although Ottawa's declaratory power was upheld, the fight over the regulation of Bell Telephone's poles and wires continued for the next twenty-five years as various cities in Ontario and Quebec sought to curb Bell's rights by challenging Ottawa's constitutional powers over the phone company in the courts.

The need for a Canadian manufacturing arm was hardly inspirational on Sise's part. Rather, it was a necessity dictated by Canadian patent law, which would protect Bell's patents in Canada only if the telephones were made there. It also played a part in Vail's corporate strategy. For some time, Vail had coveted the Western Electric Company of Chicago, the manufacturing arm of Western Union, to extend American Bell's monopoly over the pricing and distribution of telephone equipment. "The decision [to buy the company] was an integral part of a larger set of strategies for advancing the business and

technological development of telephony on a national scale under Bell
Company control," George Smith wrote in his 1985 history of Bell and
Western Electric.[59] Such a deal made sense for several reasons. The
acquisition would further cripple the telegraph companies' combine by
removing their manufacturing arm. And, in the long term, it would
allow American Bell to meet growing demand, as well as extend its
control over a wider range of patents and technologies. On February 6,
1882, Vail successfully completed the takeover of Western Electric after
its parent company was acquired by Jay Gould, an unethical takeover
artist from Toronto[60] who was more than willing to sell the subsidiary
for the right price.

The takeover was part of Vail's ambitious dreams for American Bell.
He envisaged it as a management holding company with decentralized
operations at the local exchanges, "giving to each exchange all the benefit of one giant organization with general experience and abundant
capital."[61] The acquisition of Western Electric allowed Vail to consummate his vision of the Bell system — a triad consisting of three organizational cornerstones: regional operating companies that provided
service; a parent holding company that plotted strategy, expanded the
monopoly's territory and performed all centralized planning; and a
manufacturing and research arm that held a technological monopoly
over the telephone patents.

American Bell is a textbook case of how the business consolidations
and technological changes of the 1880s gave rise to the development of
the modern corporation. While the company had expanded horizontally
in the past by acquiring competing telephone companies, the Western
Electric deal transformed it into a vertically integrated organization that
controlled every facet of its telephone patents. The legacy of late-nine-
teenth-century American capitalists was the invention of vertically integrated companies like Bell. By dominating every sequence of the
processes of a company's business, these large, decentralized corporations could realize the level of efficiency they needed to expand into,
and perhaps control, a national market. In theory, such control sup-
planted the chaotic and unpredictable "Invisible Hand" of free market
forces described by Adam Smith in his 1776 treatise *The Wealth of
Nations* with the "Visible Hand" of management described by business
historian Alfred Chandler in his 1977 work, *The Visible Hand: The
Managerial Revolution in American Business.*[62]

Although scholars of the Bell system emphasize that Vail did not
immediately perceive the full benefits of the structure that American

Bell adopted after the Western Electric acquisition, his vision and managerial skills did lead the corporation to make certain choices, one of which was to seize control of equipment manufacture and distribution. The integration that ensued allowed American Bell to amass financial, administrative and technical resources on a scale never seen before or rivaled since. Those resources permitted Vail to realize his dream of a single nationwide telephone network.

The documentary record of both the U.S. and Canadian companies reveals that Forbes, Vail and Sise not only possessed a long-term plan, but were also driven by a sense of urgency about their mission. All decisions were made with an eye toward 1894, when Bell's patents would automatically expire. Their goal was to achieve "control of the field," as both Vail and Sise expressed it, before that date. They knew that Bell's hold on the patent monopoly was tenuous at best; they were reminded of the threat competition posed each of the six hundred times they had to defend the Bell patents. Success was dependent on the ability of their organization to occupy large urban markets, develop long-distance service in competition with the telegraph companies and ensure an uninterruptible supply of equipment.

The precariousness of their position, as well as the first significant test of their new organizational strategy, came in Canada in 1885 when Bell Telephone lost its major court battle with the Toronto Telephone Manufacturing Company over its patents. The rival telephone maker asked the Commissioner of Patents to negate Bell Telephone's Canadian patents on the technical grounds that it had not manufactured equipment in Canada, as required by the Patent Act. Although Sise had set up a manufacturing workshop in Montreal, Bell Telephone's subsidiary could not make enough equipment to meet demand and Sise had imported a significant amount of equipment manufactured in the United States since the death of Cowherd in 1881. Faced with that evidence, Joseph Taché, the patent commissioner, had little choice but to side with Toronto Telephone.[63] Sise believed that Taché was ordered by Sir John Carling, the minister of agriculture who also had authority for patents, to nullify Bell's patents as a prelude to the nationalization of the telephone industry.[64]

The pooling of a majority of competing patents through the Western Electric acquisition and the takeover of rival telephone systems staved off the ruin of American Bell's Canadian subsidiary in the face of what was otherwise a disastrous legal decision. Vail and the new president of American Bell, John Hudson, met with Sise in Montreal on February 1 and 2, 1885, to prepare a strategy to deal with anticipated attacks from Bell Telephone's rivals.[65] With its interests at stake and its survival in question, Bell Telephone proved to be a bold and ruthless competitor.

The utility even competed against itself in a bid to annihilate a rival and to maintain its monopoly hold on territory in Manitoba. That extraordinary episode revealed how calculating Sise could be in his battles against new entrants. The struggle began in the fall of 1885 when a group of Winnipeg businessmen formed a cooperative telephone system to provide cheaper telephone service. To undercut their efforts, Sise secretly formed a third company to compete against both his agent, F. G. Walsh, and the upstart Wallace system. Walsh was kept in the dark about the whole affair. Only G. H. R. Wainwright, Sise's go-between in Winnipeg, was told about the plan in a confidential letter Sise wrote November 10, 1885:

> "The People's Telephone" — George D. Edwards, General Agent — will shortly make its appearance in Winnipeg and will commence a Guerilla War upon all other companies. His advent will be announced by the "Evening News", and will doubtless stir up Walsh. Of course it is all important that we should not be known in the matter. . . . Should he want money I will, on hearing from you authorize you to draw on us and supply him. I don't wish him to go too far — that is I don't want him to spend any more money than is necessary. . . . [66]

Edwards was a former classmate of Walsh, and Sise told Wainwright that he hoped the two men would not "come to blows." Sise's tactics won the battle and the Wallace system sold out to "The People's Telephone Company" the following April. Sise explained the need for secrecy to Walsh later that month. "I did not wish to tell you this before, as I wished your opposition to be in earnest, and in good faith . . . had you known all the facts, you would have had some difficulty in maintaining your position as the opponent of both the 'Peoples' and the 'Wallace'. . . . "[67]

Sise's methods resembled the underhanded tactics of the U.S. robber barons. The Baronial Age had seen the rise of heavy industry and the growth of huge money trusts to satiate the appetites of growing conglomerates and monopoly utilities. By the late 1880s, however, only the backers of the great Canadian Pacific Railway syndicate had the economic power and resources to take on Bell Telephone in Canada. In

1885, as an encore to completing the country's transcontinental railway, the syndicate had acquired a shipping firm and begun construction of the first Canadian Pacific Hotel. Sise feared that the CPR's backers would soon set their sights on the telephone business.

He was right. His worst competitive nightmare came true in April 1888 when the syndicate created a rival telephone network within Bell's urban territory. That month, several principal members of the CPR syndicate — Sir Donald Smith, William Cornelius Van Horne, C. R. Hosmer and Richard Bladworth Angus — incorporated the Federal Telephone Company and received permission from the city of Montreal to string up their poles and wires in competition with Bell Telephone. Because their rates were cheaper, Sise was forced to lower his rates below Federal's. As it turned out, in an important lesson for today's long-distance competitors, the price reduction stimulated demand to such an extent that Bell Telephone could not expand its Montreal network fast enough to keep up with it.[68] In Toronto, the threat of competition from Federal led Bell Telephone to negotiate an exclusive contract with the city — a tactic adopted by the utility with municipalities throughout its territory as another weapon in its arsenal to thwart competitors.

Over the next three years, neither firm could unseat the other. Finally, Sise approached Federal with a bid. Although it was spurned by Smith and Van Horne, Bell Telephone's management thought the rejection might be a bluff and that the syndicate was merely holding out for a higher price. Nevertheless, the mighty Sise cowered at Van Horne's feet out of deference to the great power of the Canadian Pacific Railway:

> . . . we were all of the opinion that with the ability to turn the Federal Co. over to the C.P.R., they could cause us to lose a great deal of money, even if the C.P.R. made nothing.
>
> On the other hand, we were assured of the cordial support and cooperation of these people, who to all lookers on, appear to own Canada . . . . We must shortly go to Parliament for an increase of capital, or must issue new bonds to replace those now maturing . . . and it would be impossible for us either to sell stock or issue bonds if we are engaged in competition with the Canadian Pacific Railway.[69]

Van Horne countered with a take-it-or-leave-it proposition on June 19, 1891, that gave Sise and the board of directors little alternative but "to concede much more than we intended."[70] Bell agreed to acquire 52 per cent of Federal's stock in return for $188,500 worth of Bell shares and the assumption of $75,000 of Federal's liabilities.[71] Although the deal seemed costly, it was a small price to pay to keep Bell's monopoly intact. In fact, Bell Telephone saved money in the long term because Federal's network allowed it to accommodate new demand without having to build as many new facilities. The deal also put to rest any prospect of further rivalry between Bell and the CPR in Toronto or the Northwest, and cleared the way for the two companies to form an alliance, like the one Bell had cemented with the Grand Trunk Railway in 1885. Forging alliances with the railways to freeze out competitors was an important part of Sise's and Vail's strategy for Bell. Their plan was to trade free telephone service to the railways in return for exclusive rights to install phones in their depots and for access to railway rights-of-way to construct long-distance telephone lines.[72] The CPR granted exclusive rights to Bell to put telephones in its stations in 1904.

The elimination of competition in most utility markets led a handful of monopolists to "succeed beyond the wildest dreams of avarice," historians Armstrong and Nelles wrote in their history of Canadian utilities, *Monopoly's Moment.*[73] Success, however, took its toll on Vail's health. He left New York and AT&T in 1887 and spent the next six years traveling before settling down briefly in Vermont where he developed advanced agricultural methods at his model farm. Sise, on the other hand, remained invigorated by the challenges of running a utility. He wrote Vail's successor, J. E. Hudson in New York in early 1891 that Bell Telephone had built up a cash surplus of over $1 million and that, in preparing a confidential income statement, he had deducted $88,000 from the company's revenue "to avoid making too good a statement."[74]

Bell Telephone's success was realized after Sise conceded two parts of Canada to other players. The company simply could not afford to maintain a national monopoly. He relinquished British Columbia to a local company shortly after entering that market in 1887 because Bell Telephone could not raise all the money needed to finance the network that had to be built to serve the province. The following year, he sold Bell Telephone's Maritime affiliate, the Nova Scotia Telephone Company, and supported the creation of the New Brunswick Telephone Company by local businessmen. By giving up the less profitable parts of the country, Bell was free to commit its limited resources to expanding in the more lucrative markets in central Canada and the Prairie Provinces. That decision led to the development of a divided regulatory jurisdiction over telephone companies.

In 1885, the burgeoning need for capital led to another corporate reorganization in Boston. American Telephone and Telegraph Company (AT&T) was formed to manage American Bell's long-distance business. It became American Bell's corporate parent through a two-for-one stock swap and moved its head office to New York so that it could be closer to capital markets. The reorganization made sense from an operations standpoint as well because AT&T had already become the central point of contact with the Bell system's operating companies. The U.S. parent's interest in Bell Telephone peaked at 48.8 per cent in 1885.[75] However, further share issues steadily diluted AT&T's stockholding in its Canadian subsidiary.

The financial success of the telephone utilities was aided by the doctrine of economic laissez-faire and the lack of government regulation, which permitted utilities to set their rates at will. Bell Telephone set its own rates until 1892, when its Special Act was amended by Parliament. The Senate added a clause to Bell's proposed amendments that stated "existing rates shall not be increased without the consent of the Governor in Council [cabinet]." The intent of the clause, explained the committee chairman, was to protect the public from exorbitant rate increases.[76] In fact, it was a classic case of too little, too late.

The public had had enough of the fierce and sometimes unethical competition that characterized capitalism at the turn of the century, with its predatory pricing and secret takeovers. Following the 1904 re-election of Laurier's government, hostility over high rental charges and concern over Bell Telephone's railway and municipal contracts finally boiled over with demands for the nationalization of Bell Telephone. In the United States, the trust-busters and populists of the Progressive movement renewed their quest for a beefed-up antitrust act. AT&T's image at the turn of the century was that of a "ruthless, grinding, oppressive monopoly,"[77] and Bell Telephone's image in Canada was not much better. Monopolies were easy targets for civic and agrarian populists who organized protests on both sides of the Canada-U.S. border against unregulated development. They considered it to be the root cause of most of society's evils, including the terrible working and living conditions endured by many inhabitants of the rapidly growing cities.

The beginning of the Progressive era can be traced to September 14, 1901, when Theodore Roosevelt was sworn in as president of the United States. An equally significant date, however, was the incorporation of

United States Steel Corporation six months earlier on March 3. U.S.
Steel was the first corporation with stock valued at over $1 billion. It
was capitalized at $1.4 billion by J. Pierpont Morgan as a trust created
from the merger of his steel firms with those of Andrew Carnegie.[78] The
creation of that industrial behemoth, which controlled more than half of
the world's steel business, dramatically illustrated the concerns of the
Progressives, whose fundamental platform was the rejection of the laissez-faire determinism that governments had adopted since the end of
the Civil War. Their concerns were underscored by the statistics from
the first great wave of business mergers in the last years of the nineteenth century. The creation of national markets, aided by the telephone
and the railway, dramatically increased the number of mergers from 69
in 1897 to over 1,200 in 1899.[79] In the wake of that economic transformation, Progressives sought "to use the government as an agency of
human welfare," William Allen White, the Kansas journalist and leading
Progressive, wrote. "That was the real heart of the movement."[80]

In Canada, urban and rural populists demanded reforms to alleviate
housing shortages and provide sanitation and sewage services, public
health programs, clean water, public schools, parks and clean government. The aim of "civic populism" was simple: public ownership and
local control. Many local businessmen and chambers of commerce
added their voices to the calls for reform and recalcitrant local governments found defeat at the polls.

Because of their size and the fact that they were unregulated monopolies, utilities were perceived as the most threatening of corporations
and were frequently the targets of campaigns for public ownership. One
of the greatest victories of the civic populists in Ontario was the creation
of a publicly owned hydroelectric system, Ontario Hydro. Not surprisingly, Bell Telephone was a key target of populist organizers and would
later accept regulation as the lesser of two evils, a view shared by many
U.S. utilities, including AT&T. One of Canada's leading populists, the
mayor of Westmount, W. D. Lightall, was instrumental in organizing
the disparate local groups. With Toronto mayor Oliver Howland,
Lightall created the Union of Canadian Municipalities, which held its
first meeting on August 28, 1901. The meeting attracted fifty-two
municipalities, and a year later the number of members doubled. By
1907, the Union had been reconstituted as a federation of provincial
organizations.[81] From its inception until the late 1960s, it was Bell
Telephone's chief opponent at all of its regulatory proceedings and frequently called for nationalization of the telephone company.

One of the first tests for Bell on this issue came on January 6, 1902, when a plebiscite was held by the Ottawa city council on the question of public ownership of Bell. Although a majority of ratepayers — 4,220 to 3,058 — voted in favor of a municipal system, the issue divided council, which went on to debate the merits of a municipal telephone system for the next three years. Unable to reach a consensus, the city finally surrendered and negotiated a new contract on more favorable terms with Bell in 1907.[82]

However, in 1902, the twin cities of Port Arthur and Fort William (now Thunder Bay) did build a publicly owned system to compete with Bell. Two years later, Sise used his agreement with the CPR to prevent the towns from placing a telephone in the local railway station. That action led the towns to apply to the newly created Board of Railway Commissioners for a ruling striking down Bell's exclusive contract with the CPR and allowing them to set up a second phone at the train station. The Board ruled in favor of Bell. The chief commissioner, A. G. Blair, the former federal minister of railways, concluded the contract was not a restraint of trade and that the communities could put in a phone only if they paid Bell. It didn't help their case that Blair was also a former president of Bell Telephone's affiliate, the New Brunswick Telephone Company.[83]

That sort of cozy relationship between Bell Telephone and senior public officials did not help the monopoly's image. And it certainly gave the populists every reason to doubt the effectiveness of the federal cabinet's regulation of the telephone company. In 1902, in a move to placate the populists, Sir Wilfrid Laurier's government strengthened its scrutiny of Bell Telephone by making three major changes to its Special Act. The first introduced the principle of "universal service," which required Bell to provide service to any individual within its territory, "with all reasonable dispatch." The provision not only recognized the utility's social responsibility, but also enshrined the prevalent view of society that the telephone was no longer a luxury but an essential instrument of both social convenience and economic progress. The government also amended the rate regulation provision and required Bell to obtain approval for any reduction or increase to the rates it charged consumers for telephone service. A third change made the regulatory process more impartial. Although the cabinet still had the final authority to rule on a rate change, the government was given the power to appoint a Supreme Court judge to conduct an inquiry into any rate application and to report to the cabinet on whether a change should be granted.[84]

No other president of Bell Telephone, until de Grandpré, took such a keen personal interest in the company's lobbying activities as Sise. His role as Bell's chief parliamentary "agent" was augmented by the personal friendships that some of his officials had with cabinet officials. One of Bell's greatest assets at that time was the close friendship between Robert Mackay, the company's vice-president, and Laurier. Mackay didn't hesitate to use that friendship to advance his corporate interests. He once enlisted Sir Wilfrid's "good services" on behalf of Thomas Ahearn, owner of the Ottawa Electric Railway. Ahearn was "one of the best friends" Bell had, Mackay wrote to the prime minister, and stated that Ahearn's interests were being attacked by the Hull Electric Railway, which had applied to run its trains on the tracks of the Ottawa line. Mackay's arguments closely resembled the view taken by Bell against competition for most of this century: "I think that you will agree with me regarding the impropriety of legislation which places the capital of one company at the mercy of another and rival company, and that you will use your influence to have the bill thrown out."[85]

Sise's connections also reached Washington. A former business associate and head of the Cotton Exchange once asked Sise to lobby either President Chester Arthur or Postmaster-General Walter Gresham on some matter. In his reply, Sise voiced his disapproval of the U.S. consul general in Montreal:

> The present incumbent not long ago persisted in taking precedence of Lord Stanley, the Governor General, in speaking at a public dinner because, as he told Sir Donald Smith and myself, "The eyes of my country-men are upon me" . . . I made up my mind that the eyes of one of his country-men were upon a damned ass.[86]

Although Sise promulgated a restriction that barred Bell Telephone from making any formal political campaign contributions, he allowed employees to take special leaves of absence to work for political candidates. In defending one such case, he argued, "We are not supposed to know how he occupies himself during his vacation." Yet he added, "We cannot appear to take sides for or against either party,"[87] a position that Bell management has adhered to ever since.

Not long after the Ottawa city council held its plebiscite on municipal control of Bell Telephone in 1902, the Hamilton city council launched a committee to study the telephone question. After visiting twenty-two U.S. cities, the committee came back with a strongly worded judgment against competition, which stated that the costs of two separate telephone systems greatly outweighed any benefits. That view echoed Bell Telephone's advertising slogan, "Two Bells means two bills." The phone company argued that competition resulted in either inferior service or higher costs.[88] In a 1903 federal cabinet memorandum submitted by the minister of justice, Charles Fitzpatrick, he agreed, saying, "Effective competition among telephone companies is not only impossible but highly undesirable."[89] Although the memo voiced suspicions surrounding a government-sanctioned monopoly, Fitzpatrick's officials believed that "the nature of these enterprises as well as evidence collected as to their working points to the necessity of their being a monopoly." The memo concluded that Bell Telephone had a natural monopoly and, more importantly, rejected the view that such a monopoly was inevitably "injurious to the public interest," as the British government had concluded when it expropriated long-distance telephone lines in 1892.[90] Following a timeless Canadian political tradition, Fitzpatrick figured a royal commission could put the matter to rest. He recommended cabinet strike an inquiry to investigate the operation of telephone systems and seek evidence of how the system could be improved.[91]

However, Sir Wilfrid's postmaster-general, Sir William Mulock, held an opposing view. In a June 18, 1904, reply to a letter from the prime minister, who had passed on a letter from a leading populist, Frank Spence, he said, "An announcement this session that the government was considering the question of nationalising the telephone system to the extent at least of bringing it within the reach of rural districts would be extremely popular at least throughout the province of Ontario."[92] His advice was hardly dispassionate or free of self-interest, as he no doubt expected that a nationalized telephone company would be run by the postmaster-general, as in Great Britain. Mulock wrote the prime minister again in February 1905 and urged him to create a special committee of the House to study the telephone question. Other powerful voices also spoke out against Bell Telephone's monopoly, including the Windsor and Toronto boards of trade, which passed resolutions urging the nationalization of the phone company.[93]

Bell was not entirely taken by surprise when Sir William stood up in the House of Commons on March 17, 1905, and announced the creation of the Select Committee of Inquiry on the Telephone Question, "to consider and report what changes, if any, are advisable in respect of the methods at present in force for furnishing telephone service to the public."[94] Although the Speaker of the House of Commons, N. A. Belcourt, was paid a retainer by Bell Telephone to appraise the company of any matter of interest coming before Parliament, he had not received any advance notice of Mulock's announcement. However, Sise's logbook records that Sir William himself had notified company officials of the announcement on March 14.[95] Sise was vacationing in Genoa, Italy, at the time and was summoned back to Montreal by the president of AT&T, Frederick Fish, on March 28. The committee was keenly watched by AT&T because Bell Telephone was the first member of the Bell system to face the threat of nationalization.

The Select Committee held hearings from March 20 to July 15. Sise and Lewis B. McFarlane, a Bell vice-president, were the company's chief witnesses. Sise appeared five times and at one point characterized his interrogation as "venomous." In fact, much of the questioning was basic fact-finding. When the committee's lawyer asked him about his role in organizing the Bell Telephone Company and its companion holding company in 1880, he replied: "Well, I can hardly say they were organized. I had a very insignificant part in getting the thing started. I could hardly say that I organized it, although I do not know that anybody else did."[96]

Sise plotted an elaborate strategy to defuse the potential crisis. He hired a well-connected Toronto lawyer, Allen Aylesworth, as Bell Telephone's chief counsel at the hearings. He then summoned Herbert Webb, a British telephone expert, as chief witness to challenge the glowing reports about the state-owned telephone monopoly in the "mother country." Webb testified that it was a "blight" and received a private interview with the prime minister.[97] The third element of Sise's strategy was to lobby committee members. He told Fish in mid-May that six of the fifteen members could be considered "friendly" and that a majority of the committee was "within reach."

Aylesworth summed up Bell Telephone's position in his closing argument, which rejected the view that the utility had a natural monopoly. He stated that there was, in fact, no telephone monopoly and to support his claim pointed to the hundreds of municipal and independent telephone companies operating in Canada. His position on regulation, though, equated Bell Telephone's corporate interest with the national interest: "I am urging that the shareholders of the Bell Telephone Company, whose money has been put into it just as legitimately as any man's money has been put into his farm, want and are entitled to manage their property according to their own best interest. Those interests are identical with the best interests of the public."[98]

Sise won his victory over the civic populists. The committee made no recommendations and buried the "telephone question" in two thick volumes, which simply reprinted the 2,000 pages of testimony verbatim.

Politicians still make headlines when they call for Bell Canada's nationalization. In asking how the company has avoided being expropriated, two possible reasons leap to mind: first, the government has consistently lacked the political will or inclination to do so; and, second, the utility has in the past co-opted enough politicians to swing a vote or defuse the threat. Both those factors came into play in 1905.

A lack of political will on the part of the Laurier government is the strongest factor in explaining why the strong sense of moral outrage that characterized the urban populists failed to translate into action. Sir Wilfrid's reluctance to act reflected a lack of public consensus on the issue. Although many populists and business groups called for the nationalization of Bell Telephone, there were just as many local governments and other interest groups who viewed state enterprise as a worse evil, mired in inefficiency and corruption. Sir Wilfrid himself shared that view.

Bell's influence in the Laurier cabinet and ability to curry favor with the federal government was strengthened when Allen Aylesworth was named postmaster-general after Sir William Mulock resigned for health reasons on October 11, 1905. Sir William received an appointment to a judgeship.

The urban populists also lost a major court case against Bell Telephone in 1905. Four years earlier, the City of Toronto had launched a suit against Bell over the telephone company's right to erect poles on sidewalks without a permit. Bell claimed it had an unfettered right to place its poles wherever it chose because of its federal charter. Had the city won the case, the populists would have scored a major constitutional victory, which would have given the provinces and municipalities significant jurisdiction over the telephone company. However, the Judicial Committee of the Privy Council, which was Canada's court of final appeal until 1949, unequivocally dispelled any doubts about Ottawa's legislative authority over Bell Telephone.

The judgment hinged on the interpretation of the 1882 declaration in Bell Telephone's Special Act that stated the company was a work for the general advantage of all of Canada. In the opinion of Lord Justice Macnaghten, that meant the utility was a federal undertaking.[99]

Jurisdiction over its business could not be carved up by different levels of government with some parts regulated by one level and other parts by another. The court's decision was later used to uphold the federal government's exclusive powers over radio, aeronautics and cable television, but earlier in the century it was used by Ottawa to strengthen its regulatory powers over the phone company.

While AT&T and Bell Telephone mulled over their legislative strategy for the coming session of Parliament in the summer of 1905, Sise formulated a crucial refinement to Vail's guiding principle of the Bell system — that it should have a monopoly within its territory. Having already relinquished the B.C. and Atlantic markets to others for economic reasons, he concluded that it was impossible to hold on to every inch of territory in view of the large number of small, independent telephone companies that operated throughout Bell Telephone's territory. In a strategic *quid pro quo* aimed at preempting legislation forcing Bell to grant interconnection rights to all comers, Sise agreed to tolerate connection of Bell Telephone's network to noncompetitive, independent companies and accepted government regulation in return for the continuation of Bell's monopolies in urban and rural areas it chose to serve. Interconnection rights would be denied, however, to companies that were in competition with Bell in those territories. His strategy was embraced by AT&T and adopted in the United States the following year.[100] Under the new policy, rate regulation and competition policy were deliberately and inextricably linked. "The regulated natural monopoly approach to the telephone industry mandated a perfectly hermetically sealed network,"[101] wrote Hudson Janisch, a professor of administrative law at the University of Toronto faculty of law.

The *quid pro quo* was enshrined on July 13, 1906, in a one-paragraph amendment to Bell Telephone's Special Act. It simply declared that the Bell Telephone Company was subject to the Railway Act, which automatically placed the company under the regulatory authority of the newly created Board of Railway Commissioners.[102] The Railway Act preserved Bell Canada's monopoly by stating that the Board had authority to set the rules on interconnection to Bell Telephone's network. The Board was the first independent regulatory tribunal created in North America. Although regulation represented the end of an era of unfettered competition, Sise stated he would rather be governed by an independent, nonpartisan body than by the federal cabinet or Parliament. In fact, Sise termed the type of instrument chosen by Laurier as "congenial."[103] His remarks support the critical stance that many observers have since adopted toward regulation. The earlier model of regulation held that such agencies were capable of articulating and defending the public interest; the revised theory holds that those agencies are "captured" by the clients they regulate.[104]

The regulatory changes in Canada coincided with corporate changes in 1906 at AT&T's head office. While Sise worked on the Railway Act amendments, J. P. Morgan and Company wrestled AT&T's financial business away from Boston's Kidder Peabody.[105] After the financial panic of 1907, J. Pierpont Morgan led a corporate takeover of AT&T that was, in effect, a bankers' coup.[106] Fish was forced to resign as president and Vail agreed to come back from Argentina, where he had worked setting up railway and telephone utilities since 1903, to run AT&T.[107]

Morgan personified the autocratic power of the giant trusts and monopolies. In Edward Steichen's 1903 photograph, the stern-looking financier is shown with his hand clenched around the brightly polished arm of a chair, which resembles a dagger. Morgan's corporate power was unmatched: he and his partners held seventy-two directorships in 112 corporations and held the accounts of 78 of the largest U.S. corporations.[108] One wonders if there would have been a greater outcry from the populists at the 1905 Bell Telephone hearings had Pierpont Morgan acquired AT&T and an indirect stake in its Canadian subsidiary prior to that hearing.

Little of the wealth created by the telephone company's monopoly trickled down to the workers. In 1907, Bell Telephone's operators in Toronto, who had no union, went on strike. The strike was provoked by the utility's announcement that it intended to end an experimental five-hour work shift and revert back to an eight-hour schedule at correspondingly less pay per hour. The operators, who wanted better pay, working conditions and health improvements, had much to protest. A company manager wrote that "operators cannot earn enough to pay for their board and clothing."[109] They were paid according to length of service — a chief operator could hope to earn only $23 a week, while a student starting out was paid $7.

Working conditions were also atrocious in most locations. The work was physically taxing because of the three-kilogram headsets operators wore. Many early exchanges were located in the back corner of other business establishments and conditions were often appalling. One operator worked in an office heated by a coal stove, which she had to stoke herself, and recalls "the exchange was as hot in summer as it was cold in winter. I remember one manager who used to reduce the temperature in summer by sprinkling the floor with a watering can."[110] Although the flagship Montreal exchange at the company's head office was well heated and ventilated, Sise was not averse to firing an operator if he did not consider her dress or demeanor up to the rigid standards of the Victorian era. Women operators were held in low esteem by the company, which two decades earlier had debated whether to employ women at all.

The strike was short, unlike most of that period, which were protracted and bitter. A federal government mediator was appointed even before the operators walked out and, after the strike, Ottawa appointed a Royal Commission. The Commission changed little for the operators. The inquiry ended when William Lyon Mackenzie King persuaded his fellow commissioner to accept a compromise from Bell Telephone's Toronto manager to reduce the work schedule to seven hours spread over a nine-hour shift to allow for more relief breaks. Despite widespread public sympathy, the operators did not receive a raise and the International Brotherhood of Electrical Workers failed in its attempt to organize a union at the phone company.

Sise also ran into growing opposition from the agrarian populists in the Prairie provinces. It was a battle Bell Telephone and AT&T had tried to win since the fall of 1905 when Manitoba premier Rodmond Roblin announced his intention to form a publicly owned telephone system for the province. His plans to take over Bell Telephone were thwarted by his attorney-general, who told him that the company's Special Act precluded expropriation by a province.[111] Undeterred, Roblin held a plebiscite on forming a provincial telephone company as an alternative to expropriation, but the idea was rejected by a majority of communities. Although Sise refused to countenance the sale of its Manitoba business, Bell Telephone had experienced difficulties raising enough money to fully meet the burgeoning demand brought about by the opening of the West.[112] Winnipeg's population alone jumped from 42,000 in 1901 to almost 180,000 two decades later.[113] Bell Telephone's inability to tap that growth was caused by a shortage of capital. The utility had reached the statutory limit of its allowed share capital. Under its Special Act, Bell Telephone could not raise its limit without obtaining permission from Parliament. The crunch was relieved in June 1906 when its authorized capital was tripled to $30 million. Sise could have spent a substantial portion of that money in the West, but Bell could earn more money from each dollar spent in its more lucrative eastern markets.

Financial necessity finally forced Sise to reevaluate the company's position with respect to all three Prairie governments, which, unlike the federal government, were eager to acquire Bell Telephone's business. They had each included state ownership of utilities as a central plank in their populist programs. Yet, as Armstrong and Nelles state, "In Alberta as in Manitoba, the telephone had been thrust onto the public agenda by politicians, not by aggrieved consumers or angry farmers."[114] For many western politicians, the battle to own the provincial telephone system symbolized their fight for control of their dependent economies, which brought with it the attendant rhetoric associated with taking on an eastern monopoly. On February 14, 1907, the discussion of the Alberta government's plan to build a competing provincial telephone utility let loose an orators' frenzy in the legislature. Few topped John T. Moore, the MLA from Red Deer, who declared, "This is Emancipation Day in the very broad sense. Alberta kindles today a beacon of light of wise legislation that will illumine the pages of Canadian history for many years to come. . . . Yonder in Montreal there is a man named Sise, the Napolean of the Bell Telephone Company. . . and today the Bell Telephone Company meets its Waterloo."[115]

Sise did not actually face defeat for another eight months, and he met his Waterloo in Manitoba, not Alberta. On October 4, 1907, Vail recommended the sale of the Manitoba business for $4 million. The sale of the Alberta business for $650,000 followed on April Fool's day in 1908 and the Saskatchewan business for $367,500 in May 1909.[116] The politics of early telephony, in producing a regulatory commission at the federal level and public ownership in the Prairies, left a legacy of policy division over telecommunications which has lasted to this day.

The Canadian market was a crucial proving ground for Morgan and Vail. Aware that cutthroat competition would lead to political difficulties, Morgan wanted AT&T to grow by acquisition, so Vail launched a program to either buy an independent service or to concede the field to it. In one of his first messages to AT&T shareholders in the 1908 annual report, Vail stated AT&T's goal as "one system, one policy, universal service."[117] Telephone service was a natural monopoly and competition was incompatible with Vail's goal of universal service. Two years later, in AT&T's 1910 Annual Report, he said:

> It is believed that the telephone system should be universal, interdependent and intercommunicating, affording opportunity for any subscriber of any exchange to communicate with any

other subscriber of any other exchange. . . . It is believed that
some sort of a connection with the telephone system should be
within reach of all.

It is not believed that this can be accomplished by separately
controlled or distinct systems nor that there can be competition
in the accepted sense of competition.[118]

Vail also believed that the "combination" of the smaller companies
would lead to all telephone companies being closely associated under
one centralized organization. He even extended his design to overseas
telegraph communications when he engineered the acquisition of
Western Union in 1909.[119]

That acquisition and AT&T's "combination" with a small, independent telephone company in the Pacific Northwest led to an antitrust
case in 1910 that threw into question Vail's new strategy and led to the
divestiture of Western Union in 1913. Unlike Bell Telephone, AT&T
was not subject to one single regulatory agency in the United States.
Although new state regulatory agencies tried to make sense of the competing local telephone companies, the U.S. Congress created a void by
not indicating how the federal government was to take charge of interstate monopolies like AT&T. The Sherman Antitrust Act simply left it
to the Department of Justice and the newly created Interstate
Commerce Commission to work out between themselves whether a
monopoly was to be prosecuted or regulated. That gave rise to the now
famous out-of-court settlement known as a consent decree.

Under AT&T's first consent decree, the company agreed to stop
buying up the independent companies and to provide them with connection to AT&T's long-distance lines, as Sise had done in Canada
seven years earlier.[120] The deal brought peace and could be reconciled
with Vail's goal of a single system for the simple reason that interconnection assimilated the independents into the Bell system under
AT&T's rules and procedures. Like Sise, Vail also conceded to regulation, having previously stated that:

It is contended that if there is to be no competition there
should be public control. It is not believed that there is any
serious objection to that control . . . provided that capital is
entitled to its fair return . . . and enterprise its just reward.[121]

Vail refined that sentiment in a statement that became a regulatory battle cry echoed by every Bell system president since and that de
Grandpré, in particular, seemed to take to heart sixty years later: "Let
regulators regulate and managers manage."[122]

Regulation was seen as a substitute for the market and, in Vail's words,
was "a bulwark against future economic disturbance." Indeed, the chief
commissioner of the Board of Railway Commissioners believed that
since Bell Telephone had a natural monopoly, it ought to be protected
from competition. The head of Bell Telephone's new regulatory agency,
J. P. Mabee, echoed Vail's sentiment when he stated that "competition in
connection with telephones never appealed to me."[123] That belief led him
to side with Bell Telephone in its dispute with the Ingersoll Telephone
Company over connection of their systems for long-distance service.
Mabee saw his task as reconciling two competing and legitimate interests: that of the Ingersoll subscriber for service and that of the Bell
shareholder for profit. To that end, he ordered Bell to provide access to
its facilities, but ordered Ingersoll to pay a fifteen-cent premium over
and above the regular charge for each telephone call.[124]

The Ontario government fueled the rivalry between Bell Telephone
and the independents the same year when it passed legislation that permitted municipal telephone systems to extend service to rural areas not
served by the utility. Bell Telephone's board of directors feared that its
dominant position was threatened by the legislation and earmarked
funds to buy small rural telephone companies in 1915. It reverted to a
policy of nonexpansion in 1922.[125] By that time, interconnection with
smaller companies under the terms of the Ingersoll decision had mitigated the threat, allowing Bell Telephone to pick and choose which
rural areas it wanted to serve and liberating the utility from its obligation to serve the most marginal markets. Successive regulators grappled
with the high cost of expanding rural service and developed a variety of
economic subsidies in Bell's rate structure as an incentive to encourage
expansion. Over the years, Bell has either embraced or rejected the
acquisition of smaller companies depending on economic conditions.

Lewis Brown Macfarlane became the third president of Bell
Telephone after Sise retired in 1915. Shortly after Sise's death on April
9, 1918, the company faced a financial crisis due to the high inflation
after the First World War. Macfarlane applied for the utility's first general rate increase on September 11, 1918. Bell Telephone's four previous rate hearings had all been called by the Board. Although the Board
could not find evidence of a financial emergency and rejected the com-

pany's request for a 20 per cent increase in local rates, it granted half the increase on May 14, 1919.[126]

As opponents appealed the decision for being too generous, the company's board of directors received forecasts that the utility would have a deficit of almost $225,000 for the current year, even with the 10 per cent rate increase it had just received. Bell had also exhausted its share capital limit, requiring it to return to Parliament in early 1920 to more than double its limit to $75 million.[127] AT&T recommended that Bell Telephone take the bold step of submitting a second rate increase application. It was filed in July 1920, before the appeal of the previous request was disposed of, which precipitated a loud outcry at the second round of hearings starting that September.[128] In addition to a 22 per cent increase in long-distance rates and a whopping 31 per cent gain in local rates, Bell Telephone applied to introduce pay-as-you-use billing for business customers. The long-distance increase was allowed along with a 12 per cent local rate increase on April 1, 1921.[129] The pay-as-you-use proposal was rejected and was not dusted off until de Grandpré floated it again almost sixty years later.

But even that increase was not enough.

Bell Telephone's stock issue did not do well, according to the underwriters, because the regulator had not provided the company with sufficient income to guarantee what investors considered to be a rewarding rate of return. This caused the utility to come back with yet a third request for a rate increase less than six months after the second decision. That request was completely rejected, and the company was criticized in the Board's 1922 decision for pursuing a "temporary emergency" to the point of "pure fiction."[130] Bell did not attempt a set of annual pilgrimages to its federal regulator for consecutive rate increases for another fifty years — until de Grandpré's tenure.

The Roaring Twenties brought a new boss to AT&T. At a time when many businessmen were trying to get into the movies, Walter Gifford began his tenure in 1925 by pulling AT&T out of the film industry and selling its broadcasting operations to NBC. He also made the profound decision to sell Western Electric's international operations to International Telephone and Telegraph Corporation. The two companies reached a *quid pro quo*: ITT would not compete against Western Electric in the U.S. market and AT&T would stay out of ITT's international telecommunications equipment marketplace.[131]

The exception was Canada, where Western Electric controlled Bell Telephone's manufacturing arm, Northern Electric Company. It was created on February 19, 1914, from the amalgamation of two of Bell's manufacturing subsidiaries: Northern Electric and Manufacturing Company, which in 1895 took over the manufacturing operations of Bell's mechanical department, and the Imperial Wire & Cable Company, formed in 1899 to make wires, coils and cords previously made by Northern Electric. Western Electric acquired control of both companies in return for granting Bell Telephone access to its technology and patents. The amalgamation gave Western Electric 43.6 per cent of Northern Electric's equity while Bell Telephone held 56.4 per cent.[132] Both companies maintained their respective equity positions at that level until 1957.

Gifford's most important contribution was the formation of the Bell Laboratories, which he created by combining the research and development labs in AT&T's engineering department with Western Electric's labs.[133] Northern Electric paid a fixed percentage of its annual revenue to AT&T, in addition to its annual dividends, in return for access to all of Bell Labs' patents and research discoveries. A similar service agreement between Bell Telephone and AT&T gave the Canadian utility access to all of AT&T's operating, administrative and maintenance services, training and advice.

Those ownership links and agreements greatly influenced the evolution of Bell Telephone's network. Equipment made by Northern Electric was predominately of U.S. design. From its incorporation until the late Fifties, "Northern had been operating as a branch plant of Western in the full sense of the word, deriving its technology totally from the Western Electric Company in the United States," Donald Chisholm, later chairman and president of Bell-Northern Research Laboratories, stated in testimony to a federal inquiry in 1980.[134] In addition, Bell Telephone's network was built to U.S. specifications, making it a branch plant equivalent of the U.S. Bell system.

The Twenties was the era of great engineering breakthroughs. The advent of new long-distance technology allowed AT&T to make its service truly national and universal. In Canada, the fragmented authority over telephony created a leadership vacuum. Through its privileged access to the standards, methods and technology of the U.S. Bell system, Bell Telephone stepped in and assumed the pre-eminent role in the Canadian telephone industry and guided the planning of Canada's national long-distance network.

To orchestrate the construction of that network, Canada's public and private telephone monopolies put aside their historical differences and forged a unique consensus that led to a feasibility study commissioned from Bell Telephone. After the report was unanimously approved in 1930, the major telephone companies in Canada formed a cooperative consortium called the TransCanada Telephone System — now known as Telecom Canada — to build the first national long-distance network. The $5-million network began carrying messages across the country in 1932.[135] TCTS became the de facto hub of Canada's "Bell system," responsible for building, operating and planning national networks and services. In the quest to communicate through a single network, functional integration occurred with the adoption of common facilities, technical specifications and accounting procedures.[136]

The TCTS consortium is a microcosm of Canadian federalism. Every province is represented through its major telephone company and each member has a single vote. As in federal-provincial politics, unanimity must prevail for a collective decision to be taken. TCTS also redistributes income from "have" to "have not" telephone companies through a revenue sharing plan which divides up long-distance revenue. The plan subsidizes the construction of the national networks and ensures that the smallest telephone company has sufficient funds to provide the consortium's national services.

From the Thirties through to the Sixties, the telephone industry in North America was a highly integrated monopoly. According to the renowned economist John Kenneth Galbraith, probably no other form of organization would have ensured the great success the phone companies achieved in pursuit of their goal to provide universal service. The complete subordination of the market to comprehensive planning, Galbraith concluded, allowed that objective to be realized.[137] A new corporate culture also emerged in the telephone companies that placed an emphasis on engineering standardization and on scientific management principles.

The mission of Bell Telephone and its brethren was service. That core value was pursued with a messianic zeal by executives, managers and engineers alike, who perceived their role as contributing to the greater good to which all must yield and submit. The objective of attaining universal service helped the phone companies and their regulators achieve a high degree of congruence. That shared outlook, in turn, fostered the popular perception of the phone company as a sovereign corporate state. As Manley Irwin, a U.S. expert on telecommunications and a professor at the Whittemore School of Business and Economics at the University of New Hampshire, says: "Company policies became more than management decisions; they now possessed the force and power of the government. Violation of company practices transcended management decisions and provoked the police power of the state."[138]

By the early Thirties — more so in Canada than in the United States — a gentlemanly peace had developed between the telephone companies and their regulators. For the next thirty-five years, the regulation of Bell Telephone was modest and nonconfrontational. The equilibrium that characterized the period spanning the Great Depression through to the postwar era has been attributed to the shared objectives of the telephone company's leaders and their regulators. But it also had much to do with the personal styles of Bell's bosses and the moderate rates of economic and technological change. That consensus would dissipate in the Sixties with the arrival of new executives who were forced to deal with profound, accelerated rates of change.

Like many watershed events in a corporation's history, the full impact of the second consent decree between AT&T and the U.S. Department of Justice, signed on January 24, 1956, was not immediately felt by Bell Canada. Yet that event, which settled a seven-year-old antitrust investigation into the vertical integration of AT&T and its manufacturing arm, Western Electric,[139] proved to be one of the most important milestones in the corporate development of Bell and Northern Electric (now Northern Telecom). Although the decree left the U.S. Bell system intact, the resolution of theWestern Electric case "changed everything" for the two Canadian companies, says J. Derek Davies, executive vice-president responsible for corporate strategy at Northern Telecom.[140] The settlement marked the beginning of the long process that led to the eventual Canadian ownership of Bell Canada and Northern Electric. The first step was the development of a strategy to make the two companies independent of the U.S. Bell system. This, Davies says, led Northern Electric to develop its own technology and triggered the Canadian manufacturer's eventual entry into the international marketplace.

The consent decree settled a 1949 complaint by the Justice Department, which accused AT&T and Western Electric of conspiring to monopolize the telephone equipment market, of restraining trade by excluding other manufacturers from selling to the Bell system and of earning monopoly profits. The U.S. government directed its suit against Western Electric and asked the court to order the divestiture of Western Electric from the

Bell system and to break the manufacturer into three competing companies. The government also wanted to open the Bell system to competitive suppliers by ordering the court to direct the phone companies to procure equipment through competitive bidding.[141]

The complaint was based on concerns held over from the time of Franklin Roosevelt's New Deal, when AT&T's new regulator — the U.S. Federal Communications Commission (FCC) — had launched an inquiry into Western Electric's prices. Although the FCC had issued a report to the Congress in 1939, the inquiry was shelved because of the War.[142] The differences between the Justice Department and AT&T boiled down to opposing views of the public interest. The antitrust lawyers argued that vertical integration was inimical to the public interest because it thwarted the attainment of the goal of universal service and contributed to higher telephone costs, reduced efficiency and impaired technological development. AT&T countered that the benefits of economies of both scale and scope had allowed it to double service in the first decade after the Second World War launch a program to link every subscriber with unassisted direct dialing and embark on an accelerated research program characterized by an unprecedented level of innovation.

In the end, it was the nuclear bomb rather than lawyers' arguments that prevented the breakup of AT&T and Western Electric.

Western Electric's critical role in the postwar U.S. nuclear weapons program, coupled with the support of the U.S. Department of Defense and the powerful U.S. Atomic Energy Commission (AEC), helped garner AT&T a favorable settlement. Western Electric's Bell Laboratories was asked by U.S. president Harry Truman on July 12, 1949, to take over the management of the strategic Sandia Laboratories in Albuquerque, New Mexico, under contract to the AEC. Sandia was, and still is, responsible for nuclear weapons design research and nuclear warhead manufacture and stockpiling.[143] That role made Western Electric essential to U.S. national security during the height of the Cold War, a fact that was not lost on the conservative administration of President Dwight D. Eisenhower when the consent decree was settled. Breaking up Western Electric was not only inimical to U.S. security interests, but also ran counter to the government's desire to strengthen big business.[144] In reaching a settlement with Western Electric, the government took the view that AT&T was a national asset and that economic harm to the company would economically and strategically impair the nation.

In terminating the suit against AT&T, the government stated that the corporation was allowed to retain its manufacturing arm.[145] In return, Western Electric could only make equipment required for telephony, and members of the Bell system were confined to their business as "common carriers," precluding their entry into the emerging computer data-processing industry. The third condition of the settlement compelled Western to grant nonexclusive licenses on all of its present and future patents to any U.S. applicant. That condition was similar to one in a consent decree reached the same year with computer giant International Business Machines Corporation (IBM), which was also required to grant nonexclusive licenses on its patents.[146]

It was that last condition that dramatically altered Northern Electric's special relationship with Western Electric. As Northern officials explained, both AT&T and Western Electric were concerned that if they continued to furnish Bell and Northern with automatic access to all technical information on the Bell system and from the Bell Labs, the terms of the consent decree would require the same access to be extended to any competing U.S. equipment maker. Western Electric therefore made a corporate decision to terminate its special relationship with Northern, beginning with a reduction in its Northern holdings to 10 per cent and the sale of the shares to Bell Canada in 1957.[147]

That decision was of no small concern to Bell Canada, given Northern Electric's sizable contribution to Bell Canada's revenue and profit. Although Northern Electric did not report its own financial figures until the early 1960s, it contributed almost two-thirds of Bell's total revenue in the first half of the 1950s, Davies says.[148] U.S. ownership had also insulated Bell Canada and Northern Electric from federal government inquiries into their arrangement. Bell Canada had signed an exclusive supply arrangement with Northern Electric in 1939, which was similar to the agreement AT&T's operating companies had with Western Electric. That supply contract made Northern Electric the utility's preferred supplier, in return for most-favored-customer status, and made Northern the central procurement arm of the telephone company.[149] Through Bell Canada's role in the TCTS consortium, Northern Electric was also the de facto central supplier for the TCTS long-distance network. Without access to Western Electric's technology and the Bell Labs complex, Bell Canada recognized that Northern Electric would have difficulty sustaining its business.

At that time "Northern was not too popular" and faced a host of problems it would have to solve before it could become independent, says Robert Scrivener, who was then a Bell Canada manager and later served as president and then chairman of both Bell Canada and Northern Electric.[150] The company's image was certainly not helped by the 1953 report of a federal government combines investigation, which detailed Northern's leadership role in a major illegal price-fixing scheme organized by ten manufacturers in the wire cable industry. In 1912, Northern Electric had helped create the cartel, which set prices for 93 per cent of the wire and cable produced in Canada during a forty-one-year period.[151]

Northern Electric had also been complacent about research and development because it was wedded to "Western's hind teat," as Scrivener says. Little wonder: Bell Labs has turned out, on average, more than a patent a day since its formation in 1925. Supporting its reputation as one of the world's preeminent and most prolific research institutions is a list of important inventions that includes the laser, the electronic transistor, the semiconductor, stereo sound and fiber optics.[152] Bell Canada's executives did not set out to replace the expertise of the U.S. Bell Labs, and the flow of information did not end overnight, but they anticipated that the amount of information they obtained from Western Electric under their technical assistance agreements would be curtailed, creating a void for Northern Electric. They also perceived an opportunity to develop an in-house R&D capability in order to design systems that could better meet Canadian needs and that would have been given low priority by AT&T. Thomas Eadie, then Bell Canada's president, launched a set of planning studies in 1957, which recommended the creation of Northern Electric's first indigenous research and development division.[153]

The company named Brewer Hunt as its first R&D director and opened a lab with a staff of four at its Belleville, Ontario, plant in 1958. The first separate lab building was built in Ottawa in 1960. A second Ottawa lab, along with an advanced devices center to conduct research in the emerging science of solid-state transistors, was constructed in 1964 — a decision that proved to be auspicious.[154] The effect of those changes alone led Douglas Parkhill, the former assistant deputy minister of research at the federal Department of Communications, to characterize the 1956 consent decree as the single most crucial development for Canada's high-technology industry in the 1970s and 1980s.[155]

Northern Electric was the only domestic company that had access to Bell Labs' transistor technologies. (The transistor was invented in 1947 to replace the vacuum tube.) But a second source emerged in the late

1950s at the height of the "Missile Race." The Department of National Defence was eager to attract the support of domestic industry in its bid to build the first Canadian-designed guided missile. To support the Velvet Glove project, the Defence Research Board obtained cabinet approval to transfer technology and expertise from their electronics and telecommunications labs to civilian defense contractors and to set up a training course on transistor techniques.[156] That policy influenced the decision of many high-technology companies in the Fifties to locate near DRB's Shirleys Bay labs in Ottawa's west end, including Northern Electric, which located its R&D labs a kilometer away at Moodie Drive and Highway 7. The DRB and Northern labs became the nucleus of Canada's "Silicon Valley North" in what is now the satellite community of Kanata, Ontario.

Like the military scientists who recognized that transistors radically altered modern weaponry, the engineers at Northern Electric knew that solid-state electronics would revolutionize the telephone business. DRB's offer to share its expertise meant that Northern had an alternate source of technical advances and know-how if Western Electric halted the company's access to Bell Labs' new transistor technology. In fact, no sooner did Northern open its first labs than management's view of the new conditions in the wake of the consent decree were confirmed. In 1959, Western Electric negotiated a more restricted technical information agreement with Northern, which was no longer of general application but was instead strictly related to specific equipment. The term was limited to five years and Northern Electric was charged a higher price for access to less.[157]

The first problem posed by these new circumstances was choosing priorities and finding a focus for Northern Electric's new R&D labs, says Walter Benger, former executive vice-president of marketing and technology.[158] Both Northern Electric and Bell Canada had to coordinate their R&D planning, as well as choose carefully what technologies or products they wanted to buy from Western Electric under what was one of the last agreements signed between the two manufacturers. However, merely attaining technological self-sufficency would not be enough. To make products from its own designs rather than reproducing products from Western Electric's blueprints. Northern Electric required a broader manufacturing base. At that time, however, the company was "a financial mess," Scrivener says. It was losing money and kept asking Bell for more, "as if it was an ice cream cone."[159]

Shortly after Scrivener was made vice-president of finance at Bell in 1961, he was handed responsibility for helping Eadie, then Bell Canada's chairman, determine Northern Electric's future. A second-generation Bell career officer, Scrivener joined the utility as a telephone operator during the Depression after he received a degree in history from the University of Toronto. When he ended up in the finance job, he jokingly says that he "didn't know a bond from a debenture." Nevertheless, Scrivener's training gave him a big-picture outlook, which led him to believe that corporate problems were rarely financial problems: they were marketing problems or personnel or strategic problems. "If the other things were done properly, your financial problems were zip. Even I could see that."[160] A corollary to that outlook was his belief that problems rarely get solved by throwing money at them. And that perspective colored his views in dealing with Northern Electric's problems when Eadie asked for a report evaluating the company and weighing Bell's options.

"I had a report . . . which in effect said, you've got two options: either fix it or sell it. Then I explored the two options: What did 'fix it' mean; what did 'sell it' mean; and what would be the implications of either?" While Scrivener opted for fixing the company, his report contained a tough message in "simple, Shakespearean prose," as he described it, in which he rightly concluded that "Northern Electric could not exist downstream solely as a supplier of Bell Canada. Northern Electric had to get out and get its balls beat about a bit in the marketplace."[161]

Scrivener called his report "an ego buster." It rubbed some people at Northern Electric the wrong way, including the manufacturer's president, General R. Holley Keefler. Scrivener pushed Keefler to get the company's costs under control and to modernize its facilities in order to be competitive. The view from Bell Canada's corporate headquarters on Beaver Hall Hill was that if Northern Electric couldn't be competitive, "then we sure as hell shouldn't be buying from them," Scrivener says. "Northern Electric would either have to stand on its own feet or the fucking thing wasn't worth a damn."[162]

At Bell Canada, Eadie stuck Scrivener's report in a drawer and made a quiet executive decision that the phone company would keep Northern Electric and fix it. He acquired Western Electric's remaining 10 per cent shareholding in Northern Electric in 1962 and "we found ourselves owning 100 per cent of a clunker," Scrivener says. "I don't know whether we were backed into it and didn't know better, or whether we were smart and made a crucial management decision. But I don't remember any epic day when we toasted our glasses to fix it."[163]

In the wake of the changes wrought by the consent decree, AT&T began to question the efficacy of maintaining its equity position in Bell Canada. Since the turn of the century, Bell Canada's successive share issues had diluted AT&T's shareholding. Rather than maintaining its position, AT&T decided to let its stake decline further, which ultimately culminated in the complete buy-back of Bell's shares and its independence from the U.S. Bell system. Such an event is both rare in the annals of Canadian business and an achievement of which Bell's management can be justifiably proud. With the exception of the patriation of Alcan Aluminium of Montreal from the Aluminum Company of America (Alcoa), it is hard to think of any other major example.

In the mid-1960s, though, Bell Canada and Northern Electric were not yet ready to exploit the opportunities presented by the prospect of independence. The two companies not only differed over research priorities, but also possessed divergent conceptions of what their relationship to the U.S. Bell system should be. Those differences were spelled out in a 1965 report on relations between Bell Canada and its manufacturing arm authored by W. C. MacPherson of Bell Canada and Vernon Marquez of Northern Electric. Dubbed the "M&M Report," the document was part of "a soul-searching exercise" and was circulated to senior officers in both companies for comment. In describing the differing views over the Bell-AT&T relationship, the authors stated:

> Bell thinks they should keep the maximum possible Bell systems design to maximize the value of the Bell-AT&T Service Contract and ensure compatibility with the continental network. Northern is of the opinion that Bell Canada will have to move away from the Bell system designs to a greater degree than Bell is prepared to contemplate.[164]

As Northern officials later stated, the difference in orientation between the two companies was due to the fact that "Bell is service oriented while Northern is profit oriented."[165]

Bell Canada believed that AT&T product designs, such as the black rotary dial telephone or the new touch-tone push-button phones, could be adapted to the Canadian market. They saw this as the least-costly method of product development. Although Northern Electric would incur costs related to modification, such a strategy would buy the company more time until it could develop its own products. As Scrivener stated in a May 30, 1966, memo:

> Canada must use Bell system experience in design and manufacture to the maximum possible extent so as to minimize its own costs of development. Canadian development costs would essentially be adoption to fit smaller production runs and special conditions plus slight differences in outward appearances of (telephones) for P.R. (public relations) purposes.[166]

But Northern Electric officials complained that Bell Canada's standards were too rigid and too expensive. They feared that the extra costs incurred in designing products to Bell Canada's specifications would make Northern Electric uncompetitive and jeopardize its goal to be financially self-sufficient. "Bell requires a level of quality or refinement greater than other customers are willing to pay for, which makes items built to Bell's specifications difficult to market elsewhere,"[167] MacPherson stated in the notes of a September 1965 meeting with a counterpart at Northern Electric.

A study by Harvard University researchers commissioned by Northern Electric and based on interviews with eighty middle-level managers also stated that there was a general perception that "Bell frequently wants a Cadillac when a Ford would do as well."[168] Northern Electric's managers believed that Bell Canada's higher standards often precluded subcontracting component manufacturing, which further added to the equipment maker's costs. Marquez told Keefler that he suspected that "the Bell lab philosophy of designing their own components specially rather than building product design around commercial components has been carried into our own R&D labs."[169]

However, the Harvard report also pointed to the need for improved joint decision-making bodies. As Scrivener saw it, the challenge facing Bell Canada's corporate leadership was to get both organizations "in sync."[170] As the Harvard study stated, Bell had "the most influence over the initiation of R&D projects" — a situation that annoyed Northern's scientists and engineers, who felt that they were the people best qualified to determine what the research priorities should be.[171] Yet, as both owner and customer, it is hard to argue that Bell Canada should not have had a significant degree of input.

Both sides were forced to learn the intricacies of managing the innovation process. The integrated corporate structure of the two companies meant that internal mechanisms had to be found to reconcile the competing views held by the manufacturer, the developer (the labs) and the customer. Each product represented a compromise.[172] The task of hammering out those agreements fell to Alex Lester, Bell Canada's vice-president of engineering, Scrivener says. Yet executives in both organizations hand Scrivener the credit for guiding the two companies in a common direction.

No sooner did Bell Canada launch its strategy to make its manufacturing arm technologically self-sufficient than it faced opposition from government officials who perceived the vertical integration of Bell Canada's manufacturing and research arms as a threat to domestic competitors. In 1966, the federal government secretly prepared to embark on an inquiry that would ask the same questions of Bell Canada and Northern Electric that the U.S. government had asked of AT&T and Western Electric. Ironically, it was the U.S. consent decree that necessitated the closer ties between Northern Electric and its utility parent.

As Northern Electric's officers saw it, the government should have encouraged other companies to emulate their effort to maximize the strengths of the firm. To back their claim in a subsequent brief to the federal government, Northern's lawyers jokingly cited U.S. president Abraham Lincoln's reported response to a complaint that his successful commander, General Ulysses S. Grant, drank too much: "Find out what brand of whiskey he drinks and tell our other generals."[173]

The fight between Ottawa and Bell Canada over vertical integration was but one battle in a protracted war over many fronts that Jean de Grandpré found himself waging within his first year at the utility. That particular battle, though, would preoccupy him for the next sixteen years.

Chapter 4
Running Ma Bell

This is a dull business.

— A. JEAN DE GRANDPRÉ, 1990[1]

De Grandpré's boredom with his new job as Bell Canada's general counsel soon passed after the legal and regulatory front "exploded," as he says, in the four-month period between the summer and fall of 1966. Although occasionally given to hyperbole, this time he was not exaggerating. Between June and September of that year, he simultaneously prepared and argued a Supreme Court of Canada case that challenged the province of Quebec's minimum wage law, planned Bell's legal defense against a federal combines inquiry in the wake of a police raid on Bell's corporate headquarters, engineered and defended a set of controversial acquisitions of several independent telephone companies and drafted the most comprehensive set of amendments to Bell's statutory charter ever proposed by the company.

By coincidence, the first court case de Grandpré fought for Bell as general counsel was a landmark case in labor law, one of the legal fields he specialized in. The *Commission du Salaire Minimum v. Bell Telephone Co.* case was also the last court case de Grandpré argued before the Supreme Court of Canada.[2] Bell challenged the Quebec government's minimum wage law on the constitutional grounds that it did not apply to Bell Canada because the utility was solely within Ottawa's jurisdiction. At the time, there was no federal minimum wage law. The Supreme Court upheld Bell's argument that it was immune from the provincial law on the ground that compliance would affect "a vital part of the management and operation of the undertaking."[3] The court based its judgment on the 1905 *City of Toronto v. Bell* case, which defined the nature of a a federal undertaking under the Constitution Act of 1867. The court declared that, like any other part of such an undertaking, rates of pay and hours of work could not be subject to the jurisdiction of another level of government. Although the provincial law remained valid, the *Quebec Minimum Wage* case determined that provincial labor laws could not be applied to federally regulated industries.

Many legal scholars and constitutional law experts have criticized the judgment for "coming down on the wrong side of the issue" and for not siding with concurrent jurisdiction, states Peter Hogg, professor of law at Osgoode Hall Law School in Toronto, in his text, *Constitutional Law of Canada*. He asked, "Is there any reason why residents of Quebec should be denied the protection of a minimum wage law because they happen to be employed in a federally regulated industry? Is there any reason why that industry should not pay wages regulated by provincial law, just as it pays for its accommodation, equipment and supplies under provincial law?"[4]

In the years that followed the *Quebec Mimimum Wage* judgment, the case would be remembered by the Quebec government as an attack by Bell on its political prerogatives and seen as a regressive step by the labor unions that represented Bell's unionized employees.

Later that fall, Bell found itself on the other side of the law. The federal anti-combines watchdog, D. H. W. Henry, the Director of Investigation Research, launched a Combines Investigation Act inquiry into allegations that Bell was "foreclosing important markets to competitors."[5] Armed with search warrants, a team of investigators from the Royal Canadian Mounted Police raided the company's executive offices on November 29, 1966. "After that, I never stopped," de Grandpré says. As vice-president of law — a position he had assumed after Charlie Venne retired on September 1 — de Grandpré had authority over a department that was a key target of the Mounties. "I told [Bell's] officers, well, we're going to be involved in this for maybe ten or twelve years," de Grandpré says. "They started to laugh because they thought I was cockeyed. But it lasted sixteen years."[6]

The senior officers of Bell also had reason to chuckle at the government's expense. As Robert Scrivener says, "There was nothing on my desk." In fact, there was no overly sensitive material in or on any officer's desk. "We out-smarted them," boasts Scrivener, still wiry for his seventy-five years as he draws on a cigarette and flashes a cunning grin. From his perspective as executive vice-president of Bell at the time, the raid was more "a pain in the ass" than a serious problem.[7] That does not mean Bell did not help the investigators, who remained on the premises collecting and copying documents until December 9. The company "cooperated to the fullest possible extent" and made "any requested document available to them," according to an official account of the raid in a memo to de Grandpré from his legal staff.[8] The investigators could have anything they wanted — they just had to know what, and where, it was.

The DIR's target was Bell's relationship with Northern Electric and the goal of the investigation was to force the breakup of the two companies. Henry had decided in September that "the integration between Bell and Northern had and was likely to spread monopoly into the unregulated area of the telecommunications industry." He initiated his inquiry on the grounds that "he had reason to believe" that Section 33 of the Combines Act was either being or was about to be violated. In particular, he believed the two companies were in violation of the provision that makes it illegal to form or operate a monopoly "to the detriment or against the interest of the public."[9]

But that was not the full story behind Henry's raid.

The investigation was precipitated by a flurry of acquisitions of independent telephone companies, which Bell had launched that June to expand its system. The objective of Bell's management and board, Scrivener says, was to ascertain "how much of the telephone business in Canada could we, or should we, control?"[10] For the first time since Sise's tenure, Bell was going for broke in a bid to dominate the telephone business in Canada.

Scrivener was Bell's key player in its bid to expand across the country. He had already gained experience fighting the CPR and its chairman, Norris Roy (Buck) Crump, for the Fort Francis telephone company in the early Sixties. Crump, maintaining a tradition begun by William Van Horne, had a private railway car, named the *Laurentian*.[11] Scrivener recalls riding in it with Crump from Parry Sound in the lake country north of Toronto, where both men had cottages. "I told him the best thing he could do for the Canadian Pacific Railway would be to sell their telegraph business."[12] Crump laughed off his traveling companion's advice and, in 1967, amalgamated the CPR's telegraph arm with the Canadian National Railway's telegraph business in a joint venture named CNCP Telecommunications.

Scrivener did, however, win the battle for the Fort Francis telephone exchange, which gave Bell a connection in the Kenora area north of North Bay. He then set his sights on Northern Telephone of New Liskeard, Ontario, a large independent telephone company serving 81,000 subscribers between Dryden and Kenora. The company was originally incorporated as Temiskaming Telephone Company in 1904 by Angus McKelvie and Tom McCamus, two former prospectors.[13] Bell made a $19-million takeover offer for the company on June 9, 1966. Although Northern was only about one hundredth the size of Bell, its service area extended across the massive expanse of northwestern Ontario from the Manitoba-Ontario border to the shores of Hudson Bay and into northern Quebec.

Scrivener battled Crump and the CPR a second time over Northern, and he recalls that the sparring between the head offices spilled into the bush. Two line crews, one from Bell and the other from Northern Telephone, met in the woods and ended up taking "pot shots" at each other. "I don't think we actually carried side arms, but they sure as hell had rifles in their trucks," Scrivener says, adding, "we fought a lot of guys, but we tried not to make enemies of them."[14] In the case of Crump, it was undoubtedly better to be his friend than his enemy. The cigar-chomping railroader had a passion for gun collecting and talking tough. When Hal Banks, the violent head of the International Seafarers' Union, gave him a veiled threat, Crump upstaged him by declaring, "Look, Hal, I'm an old gun man and in my fraternity there's an old saying, 'God created man but Colonel Colt made them equal,' and any goddamn time you want to start anything I'm all set."[15]

In early June 1966, Scrivener also went after Community Telephone by seeking equipment financing of $750,000 for Community from Northern Electric.[16] A month later, Bell officials began their acquisition of the family owned Telephone Arpin. But Bell lost out on the opportunity to buy one of Quebec's largest independent telephone companies, Québec-Téléphone of Rimouski, which, like Northern Telephone, was one of the jewels among the independents.

Québec-Téléphone's territory was not only wholly contiguous with Bell's territory, but also covered 40 per cent of the province of Quebec, which made it a choice candidate for acquisition by Scrivener. The company was formed in 1927 after the Honorable Jules-A. Brillant bought the Compagnie de Téléphone Nationale from its founder, Dr. J.-Ferdinand Demers. Through numerous mergers and acquisitions, Brillant extended Québec-Téléphone's coverage from Trois-Rivières west of Quebec City to Matane on the Gaspé Peninsula.[17] By 1966, the company served 109,000 subscribers and its territory covered both sides of the St. Lawrence River estuary in eastern Quebec, running south and east of Quebec City into most of the Eastern Townships and including the entire Gaspé Peninsula on the south shore and the bulk of the north shore of the Gulf of St. Lawrence up to Labrador and along the Quebec-Newfoundland border.

The need to finance the increasing capital investments required to meet postwar expansion and to buy new technology led Brillant's son, Jacques, to seek a buyer for the family's interest in the company in the early spring of 1966. Although Scrivener and Vincent had preliminary discussions with him about buying Québec-Téléphone,[18] to both men's surprise, the company announced on March 17 that it had signed a deal to sell control to the Anglo-Canadian Telephone Company, a Montreal-based holding company owned by the second-largest telephone holding company in the United States, General Telephone and Electronics Corporation (GTE) of Stamford, Connecticut. Through Anglo-Canadian, GTE also owned a 50.3 per cent majority stake in Canada's second-largest telephone company, British Columbia Telephone Company (B.C. Tel) of Burnaby, British Columbia. Scrivener later approached GTE's president, Theodore Brophy, to acquire their interest in B.C. Tel, but "he wasn't interested in selling a goddamn thing except at an exorbitant price."

Bell and Québec-Téléphone soon became more spirited rivals as Scrivener battled with their new president, Basile Bénéteau, for several small independents that neighbored both their territories:

> One time, I remember a little telephone company coming up for sale on the south shore of the St. Lawrence adjoining Québec-Tél's territory. So we put in a big, stupid bid. Goddamn thing was a piece of junk. And we paid for it like it was a gold mine. But, anyway, we did it. Jesus Christ, Bénéteau was on the phone as soon as he got word of it, then Brophy called me from New York.[19]

Scrivener and Bénéteau often met in the Laurentians north of Montreal to go snowmobiling and to drink and swear with each other over their running feud. A decade later, Scrivener hired Bénéteau away from B.C. Tel to work at Northern Telecom.

Bell used GTE's acquisition of Québec-Téléphone to defend their next, and most controversial, acquisitions in the summer of 1966. On

August 18, only two days after it had completed its deal with Northern Telephone,[20] Bell made an offer to buy control of two major telephone companies that served Nova Scotia, New Brunswick and Prince Edward Island. Having already acquired 99 per cent of the shares of the Avalon Telephone Company (now the Newfoundland Telephone Company) of St. John's, on July 24, 1962, Bell wanted control of both the Maritime Telegraph and Telephone Company (MT&T) of Halifax and the New Brunswick Telephone Company (N.B. Tel) of Saint John. MT&T, in turn, owned 56 per cent of the Island Telephone Company, which provides service in Prince Edward Island. If Bell's offers were approved, the company would increase its stake in each company to 51 per cent from the 35 per cent it already held in N.B. Tel and the 6 per cent it held in MT&T. Scrivener estimated the deal would cost about $60 million and require 1.2 million Bell common shares. Bell would have to obtain approval from its federal regulator, however, to issue 1.19 million new shares to consummate the transaction — a task that fell to de Grandpré. Bell proposed to swap three of its shares for five MT&T shares and five for eleven of N.B. Tel's shares.

As Scrivener says, "We tried to take them out of play for GTE."[21] In a public statement, Marcel Vincent, Bell's president, stated the deals would preclude further foreign takeovers, and he invoked the national interest as a justification for Bell's action: "The directors of Bell Canada feel strongly that, in the best interests of the progressive development of this essential Canadian industry, we must do all we can to insure that as many as possible of the key elements of the industry remain permanently under Canadian ownership and control."[22] With its offers to shareholders set to expire on September 8, Bell did not seem to expect any major resistance.

But Robert Stanfield, then premier of Nova Scotia, was livid at Bell's attempt to acquire control of MT&T. Contrary to Stanfield's popular image as a meek politician known for his underwear business and, later, as federal Opposition leader, for fumbling a football, Stanfield proved to be far more dangerous to Bell than a rifle-toting rival in the bush. Armed with the Nova Scotia legislature, the premier was determined to use the force of the law to stop Bell in its tracks.

Stanfield also had a second and even more potent weapon in what became a bitter and intense campaign against Bell's scheme. The DIR received complaints about the utility's purchasing practices and relationship with Northern Electric immediately following the announcement of its bid for MT&T. Henry used that complaint as the pretext for launching his investigation and obtaining the search warrants to raid Bell's offices later that Fall.[23] Stanfield directly linked both events when he threw down the gauntlet over Bell's bid in his statement released on August 22, 1966:

> It is important that the Maritime Telegraph and Telephone Co. Ltd. be free to acquire telephone equipment that is best suited to our needs and will give the best service. The Maritime Telegraph and Telephone Co. Ltd. should be free to buy telephone equipment from any supplier. Clearly, therefore, the Maritime Telegraph and Telephone Co. Ltd. should not be controlled by a company which manufactures telephone equipment, either directly or through a subsidiary. . . . If the Bell company were to acquire control of the Maritime company, the Maritime company would not long be free to exercise any independence of decision in acquiring telephone equipment.[24]

Stanfield's retort to Vincent's invocation of the national interest was that MT&T was "a creation of the Legislature . . . established to serve the people of Nova Scotia."[25]

"They looked at it as Balkanization and as an invasion of Atlantic Canada," de Grandpré says of the Nova Scotia government twenty-four years later. "You have no idea how vicious it was. . . . You would believe there was a war between Ottawa and Halifax,"[26] he says, referring to the front-page headlines of the Halifax *Chronicle-Herald*.

As corporate counsel, de Grandpré did much of the backroom legal work for all of the acquisitions that summer. But the souring of the MT&T deal led Scrivener and Vincent to turn to him as a trouble-shooter in their abortive eleventh-hour attempt to get the deal pushed through. Out of anger and believing they had nothing to negotiate with the provincial government, Bell's officials spurned Stanfield's request for a meeting to work out a compromise, leading the premier to recall the provincial legislature into session on Friday, September 9. He made good on his threat to introduce special legislation severely curtailing the voting rights of MT&T shareholders; the amendment to MT&T's Act of Incorporation received final reading and royal assent the next day. Although that provision allowed Bell to purchase any number of MT&T shares, it prevented the company from increasing its voting control beyond one thousand shares — even though it now owned 51 per cent of the company.[27] Outside the legislature, Stanfield declared, "During my ten years as premier, I have never encountered men who pursue their own interests so ruthlessly while expressing patriotic sentiments."[28]

De Grandpré, on the other hand, termed the amendment "an infamous law," which was "the equivalent of expropriation without compensation."[29] Stanfield's tight schedule left Bell little recourse, however, and relegated the officials that had finally been sent to Halifax to virtual spectators in the legislative gallery. Bell's team included: Scrivener; James Hobbs, an executive vice-president and director of both MT&T and N.B. Tel, who was Bell's chief witness before the Committee of the Legislature; de Grandpré; and Halifax lawyer Gordon Cooper.[30]

After an unsuccessful effort to block Stanfield's bill by lobbying MPPs and appearing before the committee, de Grandpré flew to Ottawa, where he appeared on Monday before the Board of Transport Commissioners to obtain permission to issue the 1.19 million shares Bell needed to complete the deal. He raised the specter of a legal challenge to Stanfield's legislation in his testimony to the Board, stating, "There is considerable doubt in many people's minds that this legislation is constitutional or valid."[31] The Board gave its approval two days later[32] on September 14, a forty-fifth birthday de Grandpré never forgot. Bell obtained 52.4 per cent of MT&T's shares, but despite the lawyer's opening remarks to the Board, Bell never challenged Stanfield's legislation in court. "We learned to live with it," de Grandpré said in 1990.

Aside from Bell, the RCMP and the Registrar-General's Office (which was responsible for the DIR's office), few people knew of the November raid on the phone company's head office, much less of the link between the Nova Scotia acquisition and the combines investigation. The public, in whose interest Henry was purportedly acting, was kept entirely in the dark, as is the custom with combines cases. The existence of the inquiry was not disclosed until December 1967, and was revealed only because Henry was asked to testify before a parliamentary committee that was considering a sweeping set of amendments to Bell's Special Act. Stanfield could not have exacted harsher revenge on Bell. The complaints to the DIR over the MT&T acquisition not only provided ammunition for the combines inquiry, but also jeopardized Bell's objectives in Parliament.

The amendments were introduced in a private member's bill sponsored by Russ Honey and tabled before Parliament on October 14, 1966. Getting the proposed amendments Bell sought through the legislative machinery would occupy de Grandpré for the next year-and-a-half. The bill died after second reading when the session ended, but was reintroduced as Bill C-104 at the next session and referred to committee in May 1967. The Commons Standing Committee on Transport and Communications took the unusual step of announcing on June 13 that it would hold public hearings on the bill to investigate Bell and that it would be calling combines investigation officials to testify.[33] The company was criticized for applying for a bill while it was the subject of a combines investigation by the DIR. De Grandpré addressed those concerns in his opening remarks to the Commons committee on February 1, 1968, by stating that Bell did not reveal the existence of the inquiry because it had a legal duty to protect the confidentiality provisions of the Combine Investigations Act.[34]

The revelation that officials from the DIR's office would testify served notice to Bell's officers that their plans could be jeopardized if they did not go to greater lengths than usual to seek support for their legislation. There was also more at stake because of the nature and magnitude of the changes sought by Bell. Among the amendments were clauses to expand the company's powers to permit it to offer services by any means of modern communications, including computer communications; clauses to authorize it to invest in any company either engaged in businesses similar to Bell's or involved in telecommunications research and development — a provision which, notwithstanding the DIR's investigation, would ensure Bell had the powers to direct Northern Electric's nascent research efforts; and further provisions to increase the company's capitalization and to give it greater flexibility in raising additional funds.[35]

Another change sought by Bell, which was close to de Grandpré's heart, was a direct consequence of the rate case decision that he had argued the preceding year. Bell wanted Parliament to remove the stipulation in its Act that had required the company since 1929 to seek regulatory approval for any increase to its share capital, as well as approval of the terms and conditions of any share issue.[36] Under the new method of rate regulation adopted by the Board of Transport Commissioners in May 1966, the number of shares no longer had a direct impact on the calculation of Bell's subscriber rates. Bell viewed the existing clause in its Special Act, which had been upheld in 1957 after an intense controversy over its new share issues, as "a redundant regulation," because, as de Grandpré stated in his cross-examination before the Commons Committee, "The price of the issue became an irrelevant factor so far as the subscriber was concerned."[37]

Vincent and Scrivener agreed to de Grandpré's request that he should be personally involved in lobbying parliamentarians and senior mandarins. Although he already had the responsibility for overseeing the drafting of the bill, he says the subsequent extent of his involvement came as a surprise to senior officers. He and his wife, Hélène, took an apartment in Ottawa, and de Grandpré spent the better part of fifteen to sixteen months "trying to convince people in Ottawa that these changes to our statute were justified and were for the advantage of the corporation and also for the advantage of the country." In briefing MPs, cabinet ministers and caucus committees, de Grandpré was the first Bell executive since Sise to be personally involved as a parliamentary agent on a draft bill. "It was something that had never been done aggressively, I suppose, the way I was planning to do it. And they thought that maybe I was putting too much emphasis on the lobbying aspects."[38]

Henry's testimony on December 7, 1967, dispelled their reservations. In addition to publicly disclosing the existence of the formal combines inquiry into the manufacture, distribution, purchase, supply and sale of communications equipment, the DIR said Bell should not be granted any new powers. In his brief to the committee, Henry urged Parliament to defer any decision on extending Bell's powers to invest in or acquire other companies until his inquiry was complete, which he estimated would take two years. He also suggested that Parliament should broaden the powers of Bell's federal regulator in recognition of the utility's movement into new spheres of unregulated businesses. Henry succinctly described the conflict that would be at the heart of the company's relations with the federal government for the next sixteen years when he stated that the crucial issue for Parliament to decide was: ". . . the extent to which the Bell Telephone organization, including its subsidiaries, should be subjected to the authority of a regulatory agency and the extent to which activities of the organization should continue to be regulated by the forces of the market."[39]

It was not an easy time for de Grandpré, particularly as he recognized that Bell faced opposition not only from organized interest groups, but also from segments of the federal government. Louise Brais Vaillancourt, a close friend of de Grandpré's and now a BCE director, remembers seeing him at a party in Outremont at about that time:

Jean had just come back from Ottawa and he was staring out
of a window. His physical appearance showed how tired he
was, but his shoulders and jaw told the whole story of his
mind, that there was a problem but that he was going to get to
the bottom of it. His whole appearance said "he can do it" and
showed the stubbornness.[40]

The appearance of Bell's officials as the last witnesses to testify before
the House committee on February 1, 1968, opened with an allegation
that the company had harassed a witness. The charge had been made
by Michael Holt, the manager of a Toronto-based company, DCF
Systems, who had been highly critical of Bell in his testimony before
the committee on November 30.[41] Holt openly admitted to having a
non-Bell phone attached to his residential phone line in violation of a
Bell rule that required customers to either lease their phones from the
utility or pay a charge to have any foreign attachment tested by the
company before it could be hooked up. Immediately after Holt gave
his testimony, Bell representatives called him and then visited his
home after monitoring his line. (According to Scrivener, Bell had the
ability at this time to run remote electrical tests on subscriber lines to
determine how many phones a subscriber had connected "over and
above what our records show[ed] that there should be on the line.")

De Grandpré vigorously defended Bell's actions and told the committee that the company had a duty to make sure Holt complied with
the tariff:

> If you happen to be Mr. Holt's neighbor and you have an
> Ericafon set on which you have paid $25 for testing, $5 for
> connecting and $1.25 a month for service, you may say,
> "Why am I being treated differently from the fellow who is
> violating the tariff?" I think we have no alternative but to act
> against a subscriber who, in fact, is violating the tariff. That is
> the position we have taken, because if we do not do this it is
> an invitation to violations and it is discrimination against those
> who are prepared to abide by the law.[42]

While de Grandpré fielded questions about Holt that Thursday morning,
Pierre Elliott Trudeau, who was then the federal justice minister and
attorney-general, called in person on Hélène de Grandpré at the couple's
Ottawa apartment to ask if he could meet with her husband. After she

told Trudeau that he was on Parliament Hill testifying before a committee, Trudeau sent a message to the committee room saying he needed to talk to de Grandpré. He enclosed a pass to allow his former classmate into the members' gallery in the House of Commons during Question Period, which Trudeau was leading that day as acting government House leader. Afterwards, the two men met in the private lounge located on the other side of the velvet curtains behind the government members' desks.

Although de Grandpré says he was neither a close friend nor an intimate advisor of Trudeau, on this day the cabinet minister asked for his advice on a crucial personal matter. Trudeau had been in a dilemma since December, when Prime Minister Lester B. Pearson announced he was retiring. His friends in the Cabinet, including Jean Marchand, Marc Lalonde, Edgar Benson and Walter Gordon, were coaxing him to run for the Liberal leadership. Trudeau still had not made up his mind after a two-week holiday in Tahiti in January. On his return, Pearson called him into his corner office on the fourth floor of the Centre Block to tell him he was the only possible French-Canadian candidate for the job. But Trudeau remained skeptical and continued to ask numerous friends and insiders for advice. Now it was de Grandpré's turn.

"Should I run?" he asked him.

"No, don't run," de Grandpré told Trudeau, thinking that his former classmate was not cut out for politics. "I told him he was very shy, that he'd always been very private . . . and that he didn't like backroom politics. He'd been a writer for so long and in politics for so short a time, so I said, don't do it."

Trudeau replied that he "kind of agreed" with de Grandpré's assessment, but that he felt he had little option, given the absence of other French-Canadian candidates.

"So I said, well — although it is with great reluctance — step in, if you have to."[43]

After he chaired the First Constitutional Conference of federal-provincial ministers the following week from February 5 to 7,[44] Trudeau made up his mind on Valentine's Day that he would seek the leadership. He explained in a televised interview that he did not want the job very badly and cited the Greek philosopher Plato, who wrote that "men who want very badly to head the country shouldn't be trusted."[45] Trudeau was elected leader at the Liberal Party Convention in Ottawa on April 6, 1968, and sworn in as prime minister on April 20. In contrast to the "Trudeaumania" that swept the country at this time, de Grandpré remained ambivalent about Trudeau's political abilities. Later his ambivalence turned to hostility, reflecting the sentiments of the mainstream corporate world that Trudeau consistently acted against business interests.

After his tête-à-tête with Trudeau, de Grandpré and his fellow officers returned to the committee room to field the final questions on Bell's bill and to commence clause-by-clause debate. Bell had to agree to some important concessions to gain support for the amendments it most coveted — the right to expand its acquisition powers; to invest in R&D; to update the definitions of its business; to carry broadcast signals; and to issue new stock without regulatory approval. However, the road to compromise had been worked out by de Grandpré well in advance of that Thursday appearance through a series of meetings he had had in January with the committee chairman, Joseph Macaluso, and informal meetings with the steering committee responsible for drafting the bill. Once again, as on the Fowler Commission ten years earlier, de Grandpré was able to hone his considerable skills in legislative drafting. This time, though, he shepherded the provisions through the legislative machinery himself. His draft amendments took into account the views expressed by all interested parties at the previous committee hearings and were circulated by Macaluso to the DIR, to the federal Department of Transport and to Bell's regulator, the Canadian Transport Commission, for modification and approval.[46]

Bell had to relinquish the portion of clause 8 that sought to give it the powers to acquire companies that were engaged in the same business as Bell. That provision was perceived as extending the utility's powers too far and was traded off when Henry stated he was opposed to it but that he did not object to the portion that granted Bell the powers to invest in telecommunications research and development companies.[47] Although the provision seemed to go against the grain of Henry's inquiry by sanctifying the vertical integration of Bell's telecommunications business, the DIR viewed it as simply enshrining the status quo. What he was opposed to was horizontal integration through acquisition, and he had won his bid to curb Bell's rights in that area.

Asked why Bell sought clarification of its rights to invest in R&D companies, de Grandpré said clause 8 was needed to deal with the "attacks" on Northern Electric and suggestions that Bell's investment in Northern "was unlawful because it was an investment in a company that was unauthorized under our original act."[48] Vincent cited a second, more compelling reason when he declared to the committee that, "it might be preferable if the R and D work were done by a separate organization, not directly under Northern nor directly under Bell, but a separate corporation which would, perhaps, be jointly owned by the two."[49] De Grandpré later revealed to the committee that Bell was weighing the possible advantages of transforming Northern's R&D department into a separate corporate entity. "This is a new avenue that we have never explored before, and we wanted to remove any possible doubt that we have such a power so that we would not be under another kind of a cloud."[50] It would be three years until Bell completed its study, but the passage of clause 8 ensured that Bell and Northern Electric had the legal power to create a separate R&D subsidiary.

A further restriction was added to the clause, which stipulated that Bell could buy any company engaged in R&D related to its business *provided* that the R&D company did not make products for Bell or for sale to other customers. The purpose of that proviso was to preclude Bell from buying into manufacturing companies that competed with Northern. An interesting exchange revealing a loophole in the proposed clause transpired when Macaluso asked de Grandpré to explain how Bell would deal with a company it wanted to invest in that happened to have combined telecommunications R&D and manufacturing operations:

> Mr. de Grandpré: Were we ever confronted by this situation, Mr. Chairman, the way to do it would be to tell a research and development plus manufacturing company such as RCA, or those with whom we wanted to become partners, that the charter of the company would not permit our investing, but that they or we would have to set up a research and development company out of their research and development department so that we could become partners with them in the R&D section.
>
> Mr. Lewis: It is very simple. I suggested that to the Chairman while you were out.
>
> Mr. de Grandpré: Great minds meet.
>
> The Chairman: I want to say, though, that I agreed with you.
>
> Mr. Rock: That is what I like about lawyers. What they cannot do legally they will do illegally and indirectly!
>
> The Chairman: You are smartening up![51]

Bell was also questioned on the impact that its relationship with Northern Electric had on its phone subscribers — an issue that had been scrutinized by its federal regulator. In response to a question from David Lewis, then deputy leader of the New Democratic Party, de Grandpré went beyond the stock answer and alluded to a countervailing impact that the holding company concept — which he would later embrace — would have on the subscriber. He explained that as long as Bell owned Northern Electric, the manufacturer's profit would flow back to the phone company and benefit telephone subscribers by offsetting the need for rate increases. If Bell operated as a holding company that owned both the telephone company and the manufacturing arm, however, "the profits of both the operating company and of the manufacturing company would flow to the holding company and there would not be that flow of profits between the manufacturing company and the operating company."[52] That is indeed what happened fifteen years later after de Grandpré completed the corporate reorganization that created Bell Canada Enterprises Incorporated.

Bell's bid to update the language of its Act by replacing the word "telephone" with "telecommunications," and to ensure that it had the powers to use any technology to send and receive messages, forced the company to appease numerous interests. Radio and television broadcasters, cable television companies and the data processing industry, all feared Bell might use such a provision as a pretext for an incursion into their fields. The issue was identical to the problem Bell faced in 1948 when it sought an amendment to its Special Act that would grant it the power to use new radio technologies. High frequency microwave radio communication was then a new technology at the heart of transcontinental long-distance services, and Bell wanted to ensure that it had the legal right to use those techniques. Section 5 of the 1948 bill declared that Bell "has and always has had the power to operate and furnish wireless telephone and radio-telephone systems and to provide services and facilities for the transmission of intelligence, sound, television, pictures, writing or signals."[53] The verbal assurance of Frederick Johnson, then Bell's president, that the company only wanted to relay television signals for broadcasters and that it had "no intention of entering the field of television" was enough for the legislators.[54]

Bell had three compelling reasons to substitute "telecommunications" for "telephone" in its Act. First, it needed to protect its growing business leasing coaxial cables and telephone poles to cable television companies. In 1968, as Scrivener testified, Bell was replacing its obsolete cable, which he described as "a one-way hydraulic type of system where you pump a signal in from one end and it squirts out the holes down the cable," with more powerful systems able to carry two-way interactive services that could simultaneously transmit telephone messages and video signals.[55] Second, the company was participating in the federal government's space program to launch a domestic communications satellite — in effect, a giant microwave tower in the sky. Finally, the growing demand for computers to communicate with each other and with data-processing centers had led to visions of new "electronic highways," and Bell was determined to solidify its position as the pre-eminent "common carrier." "If you are in the jet age, you should have jet definitions,"[56] de Grandpré stated.

The new provision, entitled "power to operate communications system," gave Bell the right, subject to both the Radio Act and the Broadcasting Act, to "transmit, emit or receive and to provide services and facilities for the transmission, emission or reception of signs, signals, writing, images or sounds or intelligence of any nature by wire, radio, visual or other electromagnetic systems. . . ." It also declared that Bell could use "any improvement or invention or any other means of communicating."[57] "Under this clause we have the power to be a telecommunications company,"[58] de Grandpré told a Senate committee the following month.

De Grandpré went to great lengths to assure the committee, government officials and interest groups that Bell was not going to become either a radio or television broadcaster, a cable television operator or a publisher "moulding the thinking of the country."[59] But Bell had to give more than verbal assurances to appease its critics in 1968. The provisions that clarified its powers had to be qualified by two other clauses, one to strengthen the 1948 restriction forbidding Bell from entering the broadcasting business, and a second forbidding it from influencing the content of a message.

The second restriction made a crucial distinction between transmitting a message and altering it. By stipulating that Bell "shall neither control the contents nor influence the meaning or purpose of the message emitted, transmitted or received,"[60] Parliament not only barred the company from the publishing business, but from the data-processing business as well. This took care of the concerns expressed by the emerging data-processing industry, which wanted to prevent Bell from gaining a monopoly in their field.[61]

The utility also faced intense and intractable opposition from a number of parliamentarians and several interest groups over its bid to remove the provision that required its regulator to approve any new issue of its stock. "If it makes people uneasy, although we still feel that it is unnecessary regulation, we are quite prepared to drop this clause . . . from our bill and leave the Canadian Transport Commission the regulatory body over the issue of shares," de Grandpré told the committee.[62] The restriction was kept in the Act and, sixteen years later, the court battle over Bell's corporate reorganization and the creation of BCE was fought over the interpretation of a single word in it — "disposition."[63]

Several other changes put Bell on a par with other large corporations governed by the Canada Corporations Act. These included: the power to issue preferred shares, which are nonvoting debt-equity issues that generally pay shareholders a higher return than common shares; the ability to issue capital stock for cash or by installments; and, to the board of directors, the power to delegate certain duties to an executive committee. The company was also authorized to use the short form of its name, Bell Canada, as its legal name. At the conclusion of the committee hearing on February 1, Macaluso was effusive in congratulating the officers of Bell, perhaps because all fourteen clauses were reworked by de Grandpré and approved by 6:00 p.m. in time for dinner. The appearance of de Grandpré, Scrivener and Vincent before the Senate Committee on March 6 was all but perfunctory, and the bill was sent to the full Senate without amendment. Although de Grandpré was forced to make some significant tradeoffs in the House to ensure its passage, the new Special Act prepared Bell for the emerging Information Age.

Scrivener and de Grandpré were both promoted shortly after Roland Michener, the governor-general, gave royal assent to the legislation that amended Bell's Special Act on March 7, 1968. The retirement of Thomas Eadie as chairman of Bell on August 1 cleared the way for the board of directors to appoint Vincent as the new chairman and to elevate Scrivener, a thirty-year man, to the presidency. The board also approved the promotion of de Grandpré to executive vice-president of administration. Every leader of a well-managed corporation must prepare for his own succession and, for the next decade, Scrivener presided over the grooming of de Grandpré as the two rose together through the executive ranks.

The presidency of Bell was considered the top job in corporate Canada at that time. Few business leaders had more power than the person in the wood-paneled corner office on the nineteenth floor of Bell's headquarters. No company owned more assets than Bell or paid more taxes, and only one company in 1968 — International Nickel — earned more profit. "If there is an Establishment, I guess I'm in it. I belong to the right clubs," Scrivener said in a rare interview published in 1969.[64] For $100,000 a year, plus a car and driver, Scrivener often put in a six-teen-hour day at what he described as "a killing, demanding job."

Scrivener assumed the task of setting Bell's strategic directions, a task that he not only relished, but at which he was particularly adept. He loved to set goals, saying, "Everyone needs them," and not caring if that sounded corny or unexciting to others. For him, business was a creative pursuit which derived its excitement and heart from the goals that were set. Trim and erect, the six-foot, mustached executive radiated candor and sincerity. Behind the ever-present cloud of cigarette smoke and the horn-rimmed glasses were eyes that saw big dreams. Under his guidance, the paths were blazed that de Grandpré would later follow and that led both Bell and Northern Electric into the future.

"There are many contractors who implement change, but few archi-tects of change," says James Kyles, a career Bell executive and former head of the company's international contracting arm. "De Grandpré was a contractor, but Scrivener was *the* architect of change."[65] That assessment is shared by many of Scrivener's colleagues, including Walter Light, the mild-mannered engineer whom Scrivener would later appoint as president of Northern Electric. "Scrivener moved to what we called 'BIG' goals, which stood for 'Best in the Industry Goals,'" Light says. "Under that program, we moved from the bottom of the Bell system to the top."[66]

Paradoxically, although Scrivener weaned Bell and Northern from their links with AT&T, there is no other Canadian telephone company executive who was so revered by the Bell system bosses at 195 Broadway — the address of AT&T's corporate headquarters in New York. Preceding the 1984 breakup of AT&T, Scrivener was considered something of a legend by the top managers within the companies of the U.S. Bell system because of his vision and his blunt, outspoken style.

The genesis of Scrivener's exalted status goes back to a classroom at Dartmouth College in the summer of 1964 — four years before his appointment as president. Located in Hanover, New Hampshire, near the idyllic Green Mountains, Dartmouth is one of the oldest Ivy League liberal arts colleges in the United States. It is perhaps best known for the Amos Tuck School of Business Administration. It was there for some fifteen years, between the late Fifties and mid-Seventies, that AT&T took over a classroom for its own two-month-long summer program, nicknamed "Bet the Job." Its students were the elite managers of the Bell system: those executives considered to be potential candidates for the presidency of a Bell operating company. Each company, including Bell Canada, could send one student.

The Dartmouth program was designed to weed out the less-promising of the chosen and to determine who could best handle pressure. "It was two months of celibacy and unbelievable quantities of booze and stress," Kyles recalls. "And there were heart attacks and deaths," noting that two members of his class had had heart attacks and one had died. An open bar operated twenty-four hours a day. "You could spot the guys who wouldn't make it because they were the ones who hit the bar all night." Before Kyles left for Dartmouth in the summer of 1971, Scrivener telephoned him with some advice, "You can play golf or you can become a lush!"[67]

"It was a hell of a good time," Scrivener says. "As a matter of fact, that's where I started to take up golf. I was a tennis player, but I started to take up golf seriously there. I was going over to the Dartmouth Country Club on the weekend, and then we'd have four lobsters and about three dozen oysters, and then we'd go downtown loaded for hay."[68]

The course work was exceptionally broad, unpredictable and grueling. Whatever the seminar or subject, AT&T would bring in the best person in the field to give guest lectures from 8:00 a.m. till noon. There was a two-hour break for lunch, exercise in the afternoon and dinners at the Dartmouth Inn. Essays were due the following morning on subjects as diverse as philosophy, history, psychology, economics or international affairs. "Occasionally, one of the instructors would come around the dormitory at about 9:00 p.m. to say the subject of the paper had been changed. It was a bit of a trial by fire," Kyles recalls. And, all the while, for the full two months, there were three representatives from the New York headquarters in the back of the class constantly evaluating the performance of each student and writing notes "every time you picked your nose."

A good portion of the course, however, was taken up with presentations and discussion of issues relevant to the Bell system. One afternoon, the theme of Scrivener's class turned to the future directions of the Bell system and how the operating companies should respond to the changes created by emerging competition and technological change. One of the guest speakers was Frederick Kappel, the chairman and chief executive officer of AT&T. The story goes that Scrivener voiced his disagreement with the views expressed by the head of what was then the biggest corporation in the world and, in fact, challenged most of the orthodox wisdom held as gospel at 195 Broadway. Foreseeing the impact of the data communications explosion and how the business of the telephone company would be forever altered by the computer, he argued that simply rethinking AT&T's dogma as decreed by Vail at the turn of the century would not suffice; profound organizational changes were also required. Otherwise, Scrivener told the class, the changes being debated that day "would tear down the Church" and lead to the breakup or destruction of the Bell system.

As Scrivener saw it, AT&T would have to focus not only on marketing, but also loosen its central control of the operating companies to allow them to confront new competitors. Both notions, from the standpoint of Vail's principles, were akin to heresy. "They never forgot Bob Scrivener for that class," Kyles says, and recalls members of his Dartmouth class asking him, "How's Scriv?," more than seven years after Scrivener went through the program. Much of the strategy that he subsequently formulated for Bell and Northern had its origins in the ideas he explored at Dartmouth.

By the time Scrivener became president of Bell, Haakon I. Romnes had replaced Kappel as chairman and CEO of AT&T. Romnes was generally regarded as technically brilliant, having orchestrated AT&T's development of the national long-distance direct-dial scheme, but managerially inept. His successor later described Romnes as a deep thinker, but not a vigorous or charismatic leader. He was simply unable to galvanize AT&T's gigantic bureaucracy to respond to the changes of the late Sixties.[69] His weaknesses were, in fact, Scrivener's strengths. Having survived "betting his job," Scrivener soon turned to gambling his company.

He wagered the future of both Bell and Northern on the new technologies of electronics and computerized switching. Scrivener's greatest contribution to both companies, according to his peers, was in perceiving the implications of those new technologies for the telephone company. "The whole push from analog to digital was managed by Scrivener. It was Bob who really saw that we had to go down the digital path,"[70] Light says.

When Scrivener was appointed president of Bell, Northern was presided over by Trinidadian-born Vernon Marquez. The son of a prosperous cocoa plantation owner, he immigrated in 1929 to

Montreal, where he found his first job soldering wires in switchboards at Northern. He stayed and became the first president of the company who was not an engineer. Like his boss at Bell, Marquez was a generalist with a flair for marketing. He also shared Scrivener's vision that Northern's future lay in taking risks. As head of Northern, Marquez took to the hustings, preaching the twin virtues of risk and R&D. "Knowledge without the courage to make decisions is at best an academic benefit," he told the annual convention of the Canadian Industrial Management Association on June 14, 1968. "We have become so obsessed with the maxim that we should look before we leap that we spend an inordinate proportion of our time looking, and scarcely any time at all leaping. We are chronic sufferers from that most common of Canadian management diseases, paralysis by analysis." But his real message to the audience was that Canadian managers could be every bit "as ingenious as their American counterparts, if only they would permit themselves to be innovative and if only the corporate atmosphere in which they live would encourage instead of inhibit innovation."[71]

Together, Scrivener and Marquez set out to shake off that paralysis and to prove that what they preached was not just hollow rhetoric. Both men recognized that the key to Northern's success lay in intensive research and in tapping international markets. Only by entering foreign markets could Northern find the manufacturing economies of scale needed to compete and to finance continued R&D. Both men also recognized that Northern needed something new to sell in order to break out of the chicken-and-egg dilemma, and that it would be found by pursuing the new Holy Grail of telephony — digital, or fully computerized, communications.

"The real catalyst of change was Bob Scrivener; he set the course for Bell," Light says. "He had the conceptual skills to sense where the Bell system was going and saw the need for new organizational change."[72] Scrivener's description of his vision was deceptively simple: "The subject was what kind of a switching network do you visualize and what should you be doing today to get there?"[73]

He dragged Northern Electric kicking and screaming into a new realm of switching technology, and spearheaded the acquisition of critical technology from Western Electric. After much prodding from Bell Canada, Northern Electric obtained a license from Western Electric in 1963 to make the No. 1-ESS switch — a large device to route subscriber telephone calls. It was the last Western Electric product licensed to Northern Electric. Under that final agreement, which was signed the following year and gave effect to the 1956 consent decree, Northern Electric could pick one last proprietary product that it could buy the rights to make. Scrivener could not have chosen a better one. The No. 1-ESS was not just another switch: it was the world's first electronic switch to use a stored program, or software, to control the switching machine's cumbersome mechanical relays. The No. 1-ESS project was the most costly R&D effort ever undertaken by the Bell Labs to that point, and access to it gave Northern Electric's new research division a head start in the emerging technology of computer-controlled switching.[74]

"With the Western/Northern flow of technical information drying up, this looked like a last chance to get help on this important development," Alec Lester, then retired, stated in a 1980 report prepared for a federal inquiry into the relationship between Bell and Northern.[75] But under the final five-year renewal to its Technical Information Agreement with Western Electric, Northern Electric paid more to receive less information than it had received in the past,[76] Lester added. In less than ten years, Northern Electric's relationship with Western Electric had changed "from a preferred one to the opposite extreme, in which we pay more and get less than we could arrange with almost any other manufacturer,"[77] Marquez wrote in a June 21, 1965, memo to Keefler.

Northern almost squandered its opportunity to exploit the No. 1-ESS switch because the company did not want to give the project very high priority. Company officials later said they were cajoled by Bell Canada into working on the project and admitted that "Bell was right because of the knowledge Northern Electric acquired from developing the system which enabled it to design its own system."[78]

The computerization of telephone company switches is one of the most significant achievements in telecommunications since the invention of the telephone itself, because the switch, which routes calls between subscribers, is the heart of the telephone company's network. With the advent of microelectronics, engineers gave that machine a new and improved heart. With software, they dreamed of giving it a brain. Scrivener's embrace of those technologies not only changed Northern Electric's manufacturing business, but also radically transformed Bell's network.

Until the advent of Western Electric's No. 1-ESS switch in 1965, telephone switches were awkward, monstrous mechanical devices, which consumed huge amounts of power, occupied enormous amounts of valuable real estate, required constant maintenance and were prone to failure. Prior to mechanical switching, telephone calls were connected manually by operators who made the connections by plugging flexible cords into jacks on a switchboard. The first automatic telephone switch was invented in 1889 by Almon B. Strowger, a Kansas City undertaker. The Strowger step-by-step switch allowed customers to use dial telephones to complete local calls without operator assistance. Lifting the receiver activated the first selector in an available rotary switch unit. Each unit had ten rows of ten sets of contacts. As each digit of a telephone number was dialed, a corresponding number of pulses was sent to the switch. Dialing a six, for example, sent six pulses, causing the selector on the switch to move up a shaft to the sixth row of contacts and then to rotate along the row until an available relay was detected.[79]

Although Strowger installed his first switch in 1892 in La Porte, Indiana, it was several years before the Bell system began introducing the technology and displacing local operators. Bell Canada waited until 1924 to install its first step-by-step switch in Toronto's Grover exchange.[80] A decade later, the U.S. Bell system began developing a second generation of automatic switch known as the crossbar switch.

Although first invented in Sweden in 1910, the first working machine was not developed until 1930, when the Swedish manufacturer, Televerket, developed a small crossbar switch for rural exchanges.[81] The crossbar switch takes its name from the rows of horizontal and vertical bars that contain the relays used to complete a call when a number is dialed. This matrix of two hundred contacts, which occur where the horizontal wires intersect the vertical ones, forms a rudimentary form of wire logic. Crossbar switching also incorporated a new advance called common control switching. Rather than the switching being performed directly by the digits that are dialed, the digits are stored as they are received and the number itself identifies the line of the called customer. Once the entire number is received, the common control equipment then makes the connection by identifying any available path through the matrix of relays. Unlike the step-by-step switch, the relays that are connected do not correspond to the digits of the dialed number. One of the most significant features of the crossbar switch was the freeing up of the common control equipment for other calls as soon as a connection was made.[82] That process, known as time division, allowed expensive circuitry to be shared, giving crossbar switches numerous advantages over step-by-step switches by making them faster, more reliable, smaller, more efficient and more economical.

Despite those advantages, crossbar switching did not go into widespread use until after the Second World War. The U.S. Bell Labs' first crossbar switch, introduced in 1938, was a behemoth intended only for large metropolitan areas. They did not develop a smaller, cheaper version until 1948. That switch, unceremoniously dubbed the No. 5 crossbar, became the workhorse of the postwar Bell system, and was first installed by Bell Canada in Chatham, Ontario, in 1956.[83]

The No. 5 switches had hardly been depreciated when Scrivener acquired the No. 1-ESS switch. Yet securing access to that technology in 1964 was only the beginning of the journey toward the full computerization of switching. The No. 1-ESS switch was still only a hybrid. The relays were replaced with programmed instructions contained in the software inside a central processor, which kept track of telephone numbers and routing patterns. It still used an electro-mechanical switching matrix, though, to connect the circuits between two callers.[84]

Northern Electric sold its first No. 1-ESS to Bell for installation at the Expo '67 world's fair in Montreal. Eleven others were installed between 1967 and 1973.[85] However, manufacturing and selling the No. 1-ESS under license was just one part of Scrivener's plan. He also sought to develop a smaller and more economical version, which would be better suited to the Canadian market and, potentially, to export markets. To this end, Northern's research division set up its own development group to explore stored program control technology two years before the company obtained the license to the No. 1-ESS, says Walter Benger, who retired as executive vice-president of Northern Telecom in 1989 and headed the switching development team at Northern from 1962 to 1968.[86] That effort led to the development of the SP-1 family of electronic switches, heralded as one of the most important achievements of the Bell group — and one of the costliest.

It is difficult to appreciate twenty-five years later the degree of risk that Scrivener assumed in propelling Northern ahead on the SP-1 project, Michiel Leenders, a professor of management studies at the University of Western Ontario, stated in a Bell-sponsored case study on the project. The project was a sizable economic and technological gamble, which cost $30 million before a single sale materialized, at a time when Northern's own resources were meager.[87] The whole project ultimately cost more than $90 million, but in the end it was worth it.

The SP-1 became a big hit with independent telephone companies in the United States. Total sales over the ten-year life of the product from 1971 to 1979 — before it was eclipsed by the next generation of switches — was more than $900 million, or ten times its development costs.[88]

Greater than any financial return from the SP-1 project, however, was the expertise that Northern's engineers gained for their work on a smaller, more sophisticated product, which went even further in incorporating the new technologies of silicon and software. Dubbed the SG-1, the device was essentially a scaled-down telephone switch adapted for business use. In industry jargon, such a switch dedicated to the use of a single customer and located on the business' premises is called a PBX, or private branch exchange. The SG-1 was one of the first electronic PBXs made. It moved Northern much closer toward the development, ahead of any of its competitors, of a fully digital, or computerized, switch.

The telecommunications industry had always regarded digital switching as being technically possible but too expensive. By 1968, however, progress in applying advances in semiconductor electronics to the development of solid-state memory devices, coupled with advances in software engineering and new techniques to transmit computerized speech signals, brought the goal of digital communications within reach. The idea for the SG-1 was conceived at the same time that the first SP-1 was being tested at Bell's Britannia switching center in Ottawa's west end. Northern determined that it needed a larger-sized electronic PBX that could connect between one hundred and six hundred telephones. Based on input from Benger's team, Northern also opted to use the SG-1 as an engineering test bed for the development of solid-state circuit technology and designs for digital switching.[89]

Digital switching not only uses a computerized switch to route subscriber calls, but also computerizes, or digitizes, the signals that it is processing — whether they are voice, data or image. Electronic switches, like the No. 1-ESS, can process only conventional analog voice signals, which are wave-like. A digital signal, on the other hand, consists of streams of pulses that are either "on" or "off." Those pulses are represented by the binary digits, 0 or 1, which are the binary language of computers. Stringing pulses together into eight-bit sequences, called bytes, allows any letter or symbol to be transmitted in a continuous bitstream.

Digital telephony could potentially pay for itself many times over by generating new sources of revenue and through cost savings. The advent of large computers in the mid-Sixties, such as Sperry's UNI-VAC and IBM's System 360, had led to a burgeoning demand for more efficient means for computers to talk to each other. Because digital switching systems speak a computer's language, they were seen as holding out the prospect of super-efficient, high-speed data communication networks. Better yet, all signals at any frequency could be converted into the simple, universal machine language of binary "ones" and "zeros." Because the telephone network was engineered to transmit low-frequency analog voice signals, it was unable to transmit high-frequency communications, such as television signals, on regular phone lines. Digital systems, on the other hand, would operate without the constraints of the frequency spectrum. In addition to opening up new markets, digital voice networks could lead to greater efficiencies and economic savings through the transmission of a greater number of signals at faster speeds. Converting the voice to bit-form, along with the network control signals accompanying them, could also make whole new services possible, as well as more efficient methods of tracking and billing calls.

The idea of digital communications had its roots in a top-secret military research project, which began at the outset of the Second World War. Although researchers at Bell Labs transmitted digitally encoded information as early as 1940,[90] there was still a major hurdle to conquer before it could be applied to the civilian telephone network. The widespread use of digital communications required the development of complex devices to convert the analog signal of human speech to digital bits and then back again. A key component of that task, which had also been developed in 1940, was a technique called Pulse Code Modulation (PCM) to transmit a digital signal.

Both PCM and digital signal transmission were developed for "Project X," one of the most closely guarded projects of the war. Its aim was to design a secret system to encode and broadcast speech signals, which would be immune from interception. All patents and details of this pioneering digital speech system were classified military secrets until 1975.[91] However, the usefulness of "System X" was confined to over-the-air radio communications. A telephone network would need PCM coders in the switch to convert every subscriber's analog telephone line into a digital one. Such a device was called a Codec, the contraction of Coder-Decoder, and would have to be economical to make digital communications practicable — a criterion that had eluded telecommunications engineers for almost thirty years.

The invention of semiconductor transistors, and their subsequent miniaturization into microelectronic circuits, heralded the development of a cheap Codec.

Semiconductors are really miniature switches of purified sand, or silicon. Because they have no moving parts or wire filaments, they are not prone to failure. They are called "semiconductors" because they can either be conductors or insulators, depending upon whether they are made to move or stop the flow of electrical current. The secret of semiconductor manufacturing is to mix impurities in with the pure silicon by a process called "doping." A semiconductor will either conduct or insulate according to what element is mixed in with the silicon, hence the term "solid-state" transistor.

Researchers at Northern Electric's Advanced Devices Centre in Ottawa joined forces with scientists from DRB and the National Research Centre in 1962 in an intensive program to develop new semiconductors, giving Northern a domestic edge in semiconductor technology. Brewer Hunt, head of Northern Electric's research division, gave top priority to the work and by 1966, the Centre made 260 different types of semiconductors and had 550 employees working in its factories.[92] The SP-1 switch was the first product to use some of those early transistors, but it was the SG-1 design team that in 1968 perceived the importance of advanced chip technology and circuit design.

Those early transistor devices were only a crude beginning. The first big breakthrough was the invention of the solid-state memory device in California's Silicon Valley in 1967. The invention of the computer memory chip by Robert Noyce and Gordon Moore meant computer makers could replace the huge maze of wire circuits needed to store and move data with tiny chips, which would make more powerful computers smaller and easier to make and run. Noyce and Moore set up a new company named Intel Corporation. In 1968, Intel turned out its first Random Access Memory (RAM) chip, which could hold 1,024 bits of data or 1K, using the scientific shorthand for one thousand. Lloyd Webster, then director of switching development at Northern's research unit, recalled that the RAM chip was the key to realizing the engineer's dream of digital telephony because "it made all other ingredients practical."[93]

The quest for digital telephony and the development of stored program control switching led the scientists at Northern's Advanced Devices Centre on the same pursuit as the computer makers. Through a $12-million license agreement signed with Intel in 1969, Northern took a crucial time-saving shortcut by obtaining access to Intel's MOS manufacturing processes.[94] That step not only enabled Northern to design and manufacture its own proprietary memory chips but, more importantly, it also meant that it was now technically possible and economically feasible for the company to make a digital Codec chip.

In view of these rapid technological advances, the systems engineering group in Northern's research division decided to examine the status of digital switching and consider when market entry might be viable. They reported their findings in January 1969 and concluded that impending advances in Large Scale Integrated (LSI) circuit technology would shift the competitive balance in favor of digital switching within five to ten years.[95] They called for the formation of an engineering team within Northern to establish a development program and recommended that: "Northern Electric envisage very seriously the development of an integrated switching system to be available for domestic sales within seven to ten years. . . . In parallel, Bell Canada should study the needs and the availability of integrated switching [and] how it can be introduced in its long-term plans. . ."[96]

The SP-1 switch was still two years away from market, but was expected to meet Bell's needs for at least fifteen years, according to corporate estimates. Although the systems engineering report was viewed with skepticism by some managers, it was supported by Scrivener and Marquez. Senior executives struck a joint Bell and Northern team to prepare a study on the development of the telecommunications network in the 1975 to 1985 time frame. The study would become a crucial planning document.[97]

Northern Electric also became an instrument of statecraft in microelectronics at this time. The federal government saw microelectronics as a powerful new fuel and Northern as the engine to drive a new industrial and high-technology policy. Eager to reduce Canada's burgeoning trade deficit in electronics, which was running at a whopping $300 million a year in 1968, Ottawa was willing to spend money to help finance what it termed a "world-scale microelectronics facility"[98] that would export Canadian-made integrated circuits. To that end, the Department of Industry, Trade and Commerce (ITC) solicited proposals from potential partners for a government-backed joint venture. Only Northern responded. ITC accepted the proposal and, as in several other joint ventures with high-technology firms — such as with Telesat Canada in satellite communications, Canadair in aerospace and Consolidated Computer Incorporated in computing hardware —

annointed Northern as the "chosen instrument" of its policy, making the company the prime beneficiary of its largesse and attention.

In what was destined to become an abortive scheme — some say fiasco — Northern and ITC became allies in setting up Microsystems International Limited (MIL) on April 1, 1969. The Advanced Devices Centre was spun off from Northern as a separate subsidiary and, not long after, taken public with a share issue on the Toronto Stock Exchange. The objective was to create "an internationally competitive" production operation that would break even within four or five years.[99] Scrivener hired Austin Olaf Wolff, a chemist and vice-president of research from the smelting and fertilizer giant, Cominco Limited, to be MIL's first president. Northern's initial investment was $20 million, and the company promised to invest a further $24.5 million over the next five years. ITC threw in a substantial financing package of almost $40-million worth of conditional grants, interest-free loans and loan guarantees to stimulate Northern to go beyond its own needs and to become a merchant computer chip maker — a manufacturer of integrated circuits for export markets. In addition, the government awarded MIL a $12-million interest-free loan to equip its facilities.

The changes at Northern Electric in this period illustrate how the role of innovation in a company changes when a firm is released from subsidiary status, as occurred following the 1956 antitrust consent decree. Between 1960 and 1970, Northern's design capability increased in origin from 10 per cent to 99 per cent Canadian; its in-house design rose from 5 to 80 per cent; and its use of Canadian-made components rose from 62 to 85 per cent. It employed more Canadian workers, increased its exports and established its own foreign subsidiaries to find and protect offshore markets,[100] the authors of a 1978 study on technology for the Science Council of Canada wrote. Those trends would soon accelerate as Scrivener moved Northern further down the road toward technological sovereignty.

There is no set curriculum to teach an executive how to run a telephone company. Most Bell presidents earned their prerequisites by following a life-long career path that took them through a variety of positions in both the operations and business sides of the utility. De Grandpré's boss, for example, had worked at Bell for thirty-one years before being promoted to president. Although Scrivener described his first job as an operator in Alexander Graham Bell's hometown of Brantford, Ontario, as teaching him how to say "number please" and "thank you," it led to his first management position, running the traffic department in Montreal. From his first day on the job, he was steeped in the operations of the company — the nitty-gritty of what Bell's business is and how it, and its thousands of employees, work.

In contrast, de Grandpré's entry-level position was only two levels below the top job, and, because it was related to his own life's work in law, it gave him little insight into how a telephone company really works, much less the peculiarities of its business. To help his new executive fill in the blanks, Scrivener enrolled him in a high-level crash course, which began when de Grandpré was made executive vice-president of administration effective August 1, 1968. As Scrivener explained, "We gave Jean some experience, and we talked to him a lot about the telephone business after work because he hadn't been in it, and we gave him some jobs in the region or something like that. Just to get him to know a little bit about what happens out there so he wasn't somebody who'd never been outside . . . Beaver Hall Hill."[101]

De Grandpré's direct move into operations did not occur for two more years. As head of administration, he was still confined to headquarters and had little direct say or influence over the company's actual day-to-day operations; that job was in the hands of Arnold Groleau, whom Scrivener had appointed as his executive vice-president of operations. "That's where the real nut-cracking takes place," Scrivener said, "because the whole company, aside from a few head office departments like accounting and finance — almost everybody — reports to you."[102] The mandate of de Grandpré's department was to support operations. Nevertheless, the appointment was significant because it made it clear that de Grandpré was one of Scrivener's lieutenants.

The functions of Bell's administration department were similar to those of most large corporations. In addition to personnel, accounting, law and public affairs, however, they included managing mangement itself, controlling the organization and disseminating the company's institutional values. It was the heart of Bell's large bureaucracy and the keeper of corporate policy, which is contained in the utility's twenty-plus-volume manual of procedure, compiled of general circulars on every conceivable detail of Bell's business. In one volume, there is even a circular on the general circular.

At first glance, there seemed to be little, if anything new, that de Grandpré could contribute to Bell in his new position since he lacked operational experience in the telephone company and the administrative functions were divorced from his legal experience. He soon discovered, however, that he had the power to shape Bell's organizational framework. Subject to the wishes of his boss and the board, he could, if he so chose, shorten or lengthen, and trim or fatten, the company's corporate structures as long as Bell's formal hierarchy remained unchanged. As it turned out, this was in accord with de Grandpré's own style and proclivities. He was himself an "organizational man" when the notion of the "organizational man" was widely popularized. Conformity had been a core institutional value at Bell Canada since its formation, strengthened by the rigid Bell system principles of one system, universal service and quality. Employees even joked about the extent to which conformity permeated the company: stay too long and you would develop a Bell-shaped head.

The best way to view Bell's corporate structure inside is from the outside — by standing on the southeast corner of Beaver Hall Hill and La Gauchetière, just kitty-corner from the dark granite monolith that is Bell headquarters. Although neither Montreal's tallest nor most modern building, as it was in 1929 when it was built, 1050 Beaver Hall remains a monument to Bell's pyramidal hierarchy. Dominated by a large, long base, the pyramid-shaped building stretches up twenty floors. It narrows sharply at the top, where an isolated, two-story, tomb-like structure stands, containing the inner sanctums of the president and the chairman of the board and the corporate boardrooms. It is a long elevator ride to the top — and many years of dedicated work before there is any chance to take it.

Sociologists have said that it is the shape of the hierarchy, not merely the number of ranks, that conditions the behavior and nature of an organization.[103] Bell's corporate organization chart is shaped like its building, which has important consequences on the decision-making process and the flow of strategic intelligence vital to the business. Its structure is characteristic of other large institutions and, because its organization is so enormous, it provides what Harold Wilensky, an organizational sociologist, has described as "a long promotion ladder for a few."[104] Those few in Bell who reach the assistant vice-president level or higher, called "tier A" managers, numbered ninety-six in 1989, which makes the organizational chart of those levels alone a wall-sized poster. The vast multitude of middle-level managers at the directors level or below are destined to be immobile. That part of the organization can be characterized as defensive, having little motive for either acquiring or passing on strategic information, and as being highly resistant to change. Senior executives who want to effect change or who require access to vital information must either build new structures or work outside existing channels.

De Grandpré overcame those organizational barriers to decision-making, and enhanced the power of his new position, by retaining direct control over Bell's regulatory and policy affairs. Shortly after his appointment, he created Bell's first department devoted exclusively to regulatory affairs. He consolidated his position by having its vice-president, Orland Tropea, report directly to him. The role of the new department was further expanded by adding the existing rates department to it. Tropea became de Grandpré's right-hand man in regulatory affairs for the next sixteen years, until his retirement from BCE in 1984. By continuing that reporting relationship, de Grandpré maintained personal control over all regulatory and policy activities of the corporation during his entire tenure at Bell and BCE.

A second organizational tactic adopted by de Grandpré was decentralization. He had formed the opinion that Bell "was too much of a monolithic block" when he first joined the company's law department two years earlier. He believed that having "one big monolithic" law department impeded the utility's ability to deal with "more modern, more fast-moving change."[105] His solution was to reorganize and decentralize it, creating, in effect, three legal divisions: one for corporate-wide affairs and one in each of the two operational regions in Ontario and Quebec. He also decentralized his newly formed regulatory matters department by locating it in Ottawa, away from the all-powerful operations branch at Montreal's headquarters.

Despite that decentralization, power resided at the center. A strong sense of hierarchy pervaded the regulatory matters department, which was accentuated by a strict adherence to rigid protocol in Bell's dealings with government officials. That protocol extended throughout the company and dictated that a corporate officer could only deal with officials who were on a comparable level. For example, only the chairman and CEO dealt with the prime minister; only the president dealt with cabinet ministers; executive vice-presidents from operations dealt with provincial ministers and vice-presidents with deputy ministers, and so on. Access was viewed as a corporate perk and, within the organization, was perceived as a further measure of an officer's status or rank. Occasionally, Bell officers adopted an arrogant view of their privileged access and sought to impose their sense of hierarchy on outsiders of supposedly lesser status. A former assistant to a newly appointed Ontario cabinet minister recalls a visit from Bell's executive vice-president in charge of Ontario operations:

> He seemed surprised that my minister did not ask me to leave the room. He told me that his discussion with the minister was private and he said that, in the past, all such meetings had been conducted alone, so could I please leave the room? I told him that this was a new government and we did things differently.[106]

The creation of a regulatory department was also a corporate response to a major change in the regulation of Bell made in late 1967. In the last months of the Pearson government, a new regulatory agency, the Canadian Transport Commission (CTC), came into existence. It was created by the National Transportation Act (NTA), a controversial piece of legislation that had taken the government almost four years to get through Parliament. One of Ottawa's most powerful politicians, John Whitney Pickersgill, had been handed the responsibility for the act's passage when he was appointed minister of transport in 1964. Despite criticisms that it would be unworkable, Pickersgill was determined to create a single large agency that would have jurisdiction over all modes of transport and federally regulated telephone companies.[107] In creating this super-regulator, Pickersgill ensured that the NTA granted it sweeping new powers to make policy decisions. In fact, there was virtually no policy-making role assigned to the minister of transport under the NTA..[108]

The Act also gave a controversial appeal power to the federal cabinet. In effect, it granted the government a statutory veto over the CTC, which allows any decision of the commission to be changed or rescinded by cabinet at any time. That provision remains in the NTA today and has been the subject of considerable angst in the telecommunications field over the years.[109] It gave the government a trump card over Bell and other clients of the agency, removing, in Bell's case, the significant degree of predictability that had characterized its past relationship with the federal regulator. Critics have considered the cabinet appeal provision abhorrent because cabinet secrecy rules apply to any such appeal, allowing the government to overrule a quasi-legal judgment without having to give any written reasons and without giving any parties the right to be heard.

Perhaps the greatest controversy arising from the NTA and the creation of the new commission, however, was the appointment of the

CTC's first president. The task of recommending the name of a suitable candidate to the Pearson cabinet fell to Pickersgill as the minister in charge of the new agency. "Sailor Jack" had spent forty years "in the service of himself and his country," Peter Newman wrote in a 1976 essay.[110] Knowing that the end of the Pearson era was at hand, Pickersgill engaged in an act of unrivaled patronage and put his own name forward to his cabinet colleagues. He was handed the $40,000-a-year job on September 19, 1967. As a confidante and political advisor to three Liberal prime ministers, he had been the archrival of former prime minister John Diefenbaker, who once remarked, "Parliament without Pick would be like hell without the devil." As head of the CTC, Pickersgill quickly became de Grandpré's nemesis.

De Grandpré still bristles at the mention of Pickersgill's name: "The worst president of the CTC we ever had was Jack Pickersgill. Guy Favreau told me . . . all he cared about was tripping the Opposition. He is a politician in the narrowest sense."[111]

That hostility can be traced to Bell's first encounter with the CTC in the spring of 1969, which followed the utility's request for a general rate increase on December 5, 1968. No doubt de Grandpré would not have such disdain for Pickersgill if the CTC boss had granted Bell what it wanted. The increase was the first the utility had requested since 1958. It was also the first rate application decision to employ the new method of rate regulation set by the CTC's predecessor following de Grandpré's first Bell hearing three years earlier. Bell had asked for an additional $83.6 million in revenue, which would have increased subscriber telephone rates by an average of 9.9 per cent.[112] Bell argued that it needed a higher rate of return to encourage investors to buy more stock, which, the company stated, it had to issue to pay for its growing construction and research programs.

Although no one disputed that the company's intensified R&D effort required access to ever-larger amounts of capital, Bell landed in hot water over the appearance on Bell's behalf of the chairman of the Science Council of Canada, O. M. Solandt, at the hearing on June 16. This public relations ploy was vintage Scrivener. The former head of the Defence Research Board was eminently qualified to judge Bell's R&D program and to testify on the social and economic benefits of increased R&D spending. But the stunt backfired when Opposition politicians raised the matter in the House and challenged the propriety of a public official appearing before a regulatory agency at an applicant's request. Although the government defended Solandt's appearance, Bell risked tarnishing the scientist's reputation and the CTC's independence in order to make its case. Opposition leader Robert Stanfield, never a Bell supporter, made his point in the House of Commons, and Bell has never again asked a government official to appear on its behalf at a hearing.[113]

After forty-four days of testimony, which generated six thousand pages of transcripts, the CTC turned down all the proposed increases to basic residential and business rates. However, the commission also ruled in the September 29, 1969, decision that Bell could raise an extra $27.5 million by increasing the rates charged for several other services, including long-distance rates, fees for service charges and requests for unlisted numbers.[114]

The CTC's ruling was noteworthy for setting two new precedents.

First, Bell was granted the discretion to change its annual capital spending budget if its earnings were too low to attract a sufficient level of new capital from investors. The CTC stated that it was not convinced that Bell needed to spend all of the $2.5 billion that it claimed it had to over the next five years. It believed the utility could postpone some of those expenditures without drastic results.[115] De Grandpré interprets that part of the decision as Pickersgill saying, "You know, you could make a deal," as though Bell could trade off some elements of the construction program to compensate for its inability to raise its prices. "We struggled for years with the effects of that decision," de Grandpré added.[116]

Bell was not used to an assertive regulator flexing its muscle. The 1969 ruling seemed to mark the passing of the old, gentlemanly style of regulation that had been in place since the beginning of the century. Scrivener described the contrast between the old and the new when he recalled an attempt by an unnamed member of the CTC's predecessor in the mid-Sixties to cut a back-room deal with Bell: "It was a little collegial. I won't mention any names, but there was a member of the Board of Transport Commissioners who came to me and said that he could see Bell's side of the argument very well," provided Scrivener agreed to hire him.[117]

The second precedent in the CTC decision underscored the divergence between the old and the new styles of regulation. Bell was ordered for the first time to prepare certain studies that the Commission stated it would follow up at a future date on its own initiative. One was a study of possible methods to separate the costs and revenue of Bell's regulated utility business from its unregulated ventures. The issue of cost separation, as it is now called, had been raised at the hearing by lawyers for the Ontario government, who argued that revenue from Bell's monopoly subsidized its other business operations. Analysts observed that requesting a study on this issue showed a greater willingness of the CTC than its predecessor to act on its own initiative.[118] In hindsight, the collision course that Bell and its regulator navigated for the next decade over the utility's nonregulated ventures can be traced to the concerns raised at the 1969 hearing.

However, on September 29, Bell's management was not as concerned with that element of the CTC's decision as it was with news of the rejection of the general rate increase. After a company car arrived from Ottawa with copies of the decision, a crisis atmosphere pervaded the meeting of top management in the new boardroom on the twentieth floor. The reaction of most officers around the table that evening matched Groleau's, who had declared earlier, "It's the worst judgment we've ever had." Because Bell's rate decisions are always released late in the day after the stock markets have closed at 4:00 p.m., the timing gave the executives a night to sleep off their indignation and self-pity.

As head of administration, de Grandpré was responsible for drafting the company's response to the CTC decision. Early the following morning, he met with Scrivener to review the final text. The statement, which was attributed to Scrivener, called the ruling a "disappointing decision." It concluded with both a threat and a promise:

> In the final analysis, the adequacy of the award will be judged by the investing public and by the company's customers. Should the company be unable to raise from investors on reasonable terms the capital required, its ability to continue to furnish the amount and quality of service expected by the public will be jeopardized.
>
> With regard to the future, it is to be noted that the level of earnings has not been stated in rigid terms, and that the commission has left the door open to further action for relief.[119]

At 8:30 a.m., Scrivener and de Grandpré were joined by Groleau and Light, who was then vice-president of operations in charge of the construction program, to discuss the implications of the decision. The four men moved across the hall from Scrivener's office to the ornate old boardroom with its traditional baize-covered table, high plaster ceiling and plush red carpet, where Bell's directors still prefer to hold their meetings. Bell's directors have never taken to the newer high-tech boardroom, which is a windowless room that occupies almost an entire half of the top floor next to the executive dining room. Entered through smoked-glass doors off its own private reception lobby, the meeting room is both intimidating and futuristic. It is dominated by a gleaming, jet-black, oval-shaped table that sits underneath a white, identically shaped, suspended ceiling. In its center stands a console to control the room's sophisticated audiovisual equipment and multiple rear-screen projectors, which make the meeting room ideally suited for the company's high-level management meetings. To use Scrivener's phrase, the room is where the real "nut-cracking" takes place.

More relaxed in the older boardroom that morning than in the newer room the night before, de Grandpré expressed his opinion that if Bell could not live with the decision, it could simply go back to the Commission. The prospect of future relief from the CTC was the cloud with a silver lining that Scrivener also grabbed that Tuesday morning, leading him to be far less pessimistic than the previous evening.[120] He argued that maybe it was a good idea, after all, that the company was not hemmed in by a set rate of return. But, in the meantime, Bell had to plot its strategy, and Scrivener had to decide what measures were required to cope with the revenue shortfall.

De Grandpré briefed his colleagues on the preliminary findings of his rate experts, who had stayed up all night putting the figures from the CTC decision through a computer model. Their analysis indicated that the rate changes allowed by the Commission might not produce the amount of revenue estimated in the decision. The group then turned its attention to considering what parts of the construction program could be trimmed back. Should production of the new Contempra telephone be halted? Or should Bell cancel some of its orders for Northern's new electronic switch, which was slated to go into production by the end of the year?

Light marshaled several compelling reasons for not slashing the switch purchases. Bell would face a multi-million-dollar cancellation charge, for one. A delay would also result in a subsequent financial penalty due to rising copper prices.[121] And finally, given Bell's equity stake in Northern, Light provided a broader rationale for maintaining the switch expenditures, which took into account the strategic implications of Bell's program on Northern's operations. Cancellation of the orders could, in a worst-case scenario, jeopardize Northern's entire switch R&D effort and lead to massive layoffs — a major setback for Scrivener's plans for the equipment maker. Or, at best, Northern would simply produce the switches for other customers, which would freeze Bell out of Northern's production plans for months or years. The meeting continued for almost four hours as Light pored over every item of the $425-million capital spending budget and suggested nonessential items to cut.

At the end of the evening, Groleau and Scrivener had not reached a conclusion on how to present the CTC decision to the rest of the officers. Groleau argued that simply stating that the judgment required more expense reductions would not "buy us a thing." Scrivener mulled over the thoughts of his advisers until Friday, the day he chaired his monthly president's meeting. The meeting would provide a forum for Bell's officer corps to thrash out the details of where, and what, to cut.

The general management meeting, held in the new boardroom, resembled a government cabinet meeting and brought together two dozen of Bell's most senior officers, as well as Zbigniew Krupski, the chairman of the TransCanada Telephone System, and Vernon Marquez, the chairman of Northern Electric — both of whom had significant stakes in the outcome of the CTC decision. De Grandpré and Tropea opened the meeting with a review of the Commission's decision and outlined several errors in the CTC's assumptions that Bell was seeking corrections on. Scrivener then briefed the officers on the earlier discussion with Light and Groleau and delivered his main message — managers would have to try to do more with less. Although there would not be any drastic cuts, he stressed there was no guarantee that Bell would be able to finance all of its construction program. He wanted support staff reduced by 10 per cent through attrition and no new projects started unless another project was stopped.

"Okay, so your mouth was set for filet mignon and it came up hamburger — but you still ate," Scrivener told the group.[122]

The CTC decision served as a strong signal that the unbridled growth and prosperity of the Fifties and mid-Sixties had gone the way of the Edsel and mini-skirt. Along with the upheavals of the Vietnam War came rampant inflation, accentuated by the shock to the Western economies caused by the devaluation of the American dollar after it was removed from the gold standard. Scrivener told the meeting that Bell had been caught by "bad political timing" and Ottawa's efforts to control inflation. Although the utility could cut costs under the guise of fighting inflation, he said basic service could not be jeopardized: "It would be fatal to get into the frame of mind that we're going to show the public how lousy the service is going to be because we didn't get all we asked for. That would just be cutting our throat for the future."[123] De Grandpré agreed. As he saw it, the company could not gain public and political support for its plans to invest in new satellite and computer communications ventures "if we can't provide plain service in Chibougamau."[124]

Scrivener then laid out Bell's basic regulatory strategy for the next decade. "I think we have an invitation to go to the Commission every time we're doing something that doesn't pay for itself," he told the meeting. But, he added, "Before we go back to this commission we have to have some evidence to indicate that we have responded, the implications are undesirable and a different course of action is required."[125] Scrivener and de Grandpré took that strategy to heart and, under their joint leadership, Bell went back to its regulator for rate increases eight more times in the next ten years — a frequency unmatched in the utility's history.[126]

As the meeting broke up, Scrivener thought about the financial implications of the CTC decision for Bell's 250,000 shareholders and sarcastically mused that it would be a great annual meeting that coming spring.[127] At the previous annual meeting six months earlier, the company's management had been threatened with a proxy fight by angry shareholders unless they improved Bell's diminished returns and increased the share dividend payments.[128] To make matters worse, Bell lost the double-A rating on its bonds in the wake of the CTC decision, which raised the cost of its debt.

To accommodate the "poor bastard" who held Bell stock, as Scrivener characterized the company's shareholders at the end of the year,[129] Bell decided to fund its external financing for 1970 through a convertible preferred-share issue, which also included a more speculative share in the new Microsystems subsidiary. (Convertible preferred shares, like bonds, are debt instruments. They yield a fixed-rate dividend, but can be converted into voting common stock at a future date at the owner's discretion.) That stock issue was a departure for the company, which had raised its external financing solely through bond issues since 1966 and through fourteen common-share issues since 1946. Groleau also announced on December 1 that Bell planned a $55-million reduction in its capital spending program for 1970, to $405 million.[130] However, most of that cut existed only on paper because Bell had announced an increase to $460 million from the previous forecast of $425 million in the company's third-quarter financial results released in October 1969.

But there would be no cutting back at de Grandpré's new regulatory affairs department. In fact, Scrivener declared, "We're going to give priority to functions directly related to the regulatory and political climate."[131] The formation of the new department paralleled the creation of the federal government's Department of Communications (DOC) in March 1969. The DOC reflected Trudeau's consuming interest in cybernetics and systems theory, Allan Gotlieb, the department's first deputy minister, says.[132] Trudeau's fascination with organizational theory also led to the creation of a new post in the powerful Privy Council Office of assistant secretary to cabinet for the machinery of government.

A more significant change, however, had been launched by Trudeau within two weeks of his appointment as prime minister, on April 30, 1968, when he announced an overhaul of both the cabinet machinery and the Prime Minister's Office. Although many observers have stated that those changes reflected the new prime minister's rational side and Jesuit training, they actually stemmed from a basic insecurity. Just as his former classmate at Bell needed a new regulatory department to consolidate power, Trudeau needed new organizations in Ottawa to help him master an enormous bureaucracy that he did not fully understand. In a move that affected every branch of the government and every business that dealt regularly with it, including Bell, Trudeau revamped the entire cabinet committee system. He declared that it was necessary to centralize the functions of those committees and to delegate decision-making powers to them. This meant that for the first time, cabinet committees would be empowered to do more than just recommend courses of action to the full cabinet.[133] These changes not only made the cabinet process more orderly, but also gave greater power to individual cabinet ministers and made the process more predictable for outsiders.

The first minister of the DOC was Eric Kierans. In his first parliamentary speech as minister, he alluded to the noted Canadian communications philosopher Marshall McLuhan when he declared that "Our responsibility will be with the medium, not with the message."[134] Ironically, McLuhan's belief that modern communication technologies would tear down the boundaries of sovereign nation states and erode cultural barriers ran counter to the federal government's plans.[135] Through the DOC, the government sought to harness and control those technologies as tools of social change and nation-building. Given the nature of Canadian federalism and the country's vast geographic distances, it was natural that telecommunications and broadcasting were perceived as having important roles to play in the country's development. Yet McLuhan taught that the very nature of those technologies made it futile for any state to attempt to control them.

Contrary to that wisdom, the Trudeau government sought to use communications as a central tool to strengthen national sovereignty. In the summer of 1969, Canadian territorial sovereignty was breached by the provocative incursion into Arctic waters of the U.S. supertanker *Manhattan.* The oil tanker's voyage to find an economic sea route through the Northwest Passage led to a flurry of responses, including a conference in Yellowknife on development of the North and the unilateral extension of Canada's territorial sea limit to twelve from three miles. The *Manhattan's* journey dramatically underscored Trudeau's quest to assert Canadian independence in international affairs. Gotlieb described three roles that communications played in the federal government's policies: extending and defending national sovereignty, both over the Arctic and as a bulwark against cultural domination by the United States; promoting economic and regional development; and fostering national unity by promoting Canada's dual cultures and Trudeau's vision of federalism in French and English Canada.[136] All those elements had, in fact, been linked together in the federal government's statement of policy on federalism following the 1968 constitutional conference co-chaired by Trudeau when he was justice minister in Pearson's government:

> The Government of Canada believes it must be able to speak for Canada, internationally, and that it must be able to act for Canada in strengthening the bonds of nationhood. . . . Internally it seems to us to imply an active federal role in the cultural and technological developments which so characterize the 20th century . . . indeed cultural and technological developments across the country are as essential to nationhood today as tariffs and railways were one hundred years ago.[137]

Gotlieb had greatly impressed Trudeau when, as head of the legal division at External Affairs, he served with the team of civil servants that advised Trudeau on constitutional matters in preparation for the 1968 conference. Described as a brilliant Renaissance man,[138] Gotlieb joined External in 1957 after completing his legal training at Harvard

University's law school and at Oxford University. Author and journalist Richard Gwyn, who was Kierans' executive assistant and then director-general of the socioeconomic planning branch of DOC until 1973, recounts in *The Northern Magus* that when Trudeau asked Gotlieb for his ideas on foreign policy, the lengthy reply became the catalyst for Trudeau's 1968 foreign policy statement espousing the principle of enlightened self-interest.[139]

Kierans got a taste of Gotlieb's razor-sharp critical faculties at their first meeting, which was a dinner arranged by Montreal lawyer Carl Goldenberg, who knew Kierans was searching for a deputy minister. Gotlieb thought he had blown it after he angrily told Kierans that the minister's criticisms of Marcel Cadieux's federalist views were off-base. Cadieux was one of the few senior francophone officials in Ottawa at the time — and one of Gotlieb's superiors at External.[140] Impressed by his dinner companion's candor, Kierans succeeded in lobbying Trudeau for Gotlieb's appointment as his deputy minister, catapulting him over the heads of three levels of superiors in the civil service.

The communications portfolio dovetailed with Kierans' strong nationalist sentiments, says one former official. The task of building the department's bureaucratic empire fell to the professorial Gotlieb.[141] Together, the two made Bell the most prominent target for their policy schemes. "We believed in one big national system and we believed in the chosen instrument concept, which required fairness to that entity," Gotlieb says. "Bell was our chosen instrument." In other words, DOC would seek to use Bell as its principal tool to influence communications policy. "It was a brave new world and we felt that Bell and the government needed to work together as forces of change," Gotlieb says. "The tradeoff was they would have a monopoly, so we had to regulate them; it was the price they had to pay for being the chosen instrument." Although Gotlieb maintains that the DOC was "not hostile to Bell, we were aggressive about regulation,"[142] Scrivener and Kierans developed a running, gentlemanly feud. In response to a suggestion by Kierans, who was also the postmaster-general, that the post office should take over the telephone company, Scrivener told an interviewer in 1969, "I got a better idea. We'll take over the post office instead."[143]

That attitude certainly helped condition de Grandpré's response to the regulators and aroused his instincts to hit back at the bureaucrats in Ottawa: "People say about me that 'he's always fighting,' that 'he's difficult,' or 'he's scrappy.' You're damn right. I had to fight. I may not like it, but what are you going to do? . . . If you'd been a milquetoast individual, you'd never have gotten anywhere. If you're forced into a fight, you have to fight."[144]

DOC's first fights, however, were with other parts of the government. "There was a great dichotomy and fragmentation at that time because of two different philosophies about communications technologies," Gotlieb explains. On one side were the "software and anti-technology" bureaucrats, led by Pierre Juneau, a close friend of Trudeau. He had been appointed the first chairman of the Canadian Radio-Television Commission (CRTC), which was created in April 1968 to regulate the broadcast and cable television industries.[145] "That body was in love with the cable television industry, but was anti-Bell," Gotlieb says. On the other side of the debate were the scientists and former Department of Transport officials at DOC, which Gotlieb describes as "a department of wires, lines and balls in the air. We weren't supposed to be in the software part of the business, but the hardware and technology were exploding. So, politically, it was very difficult." Bell was stuck in the middle of these two warring factions, and "de Grandpré and Scrivener had to try to cope with those frictions."[146] Gotlieb says his own consuming passion over the next several years was to end the feud by integrating the technological and the cultural nationalists into one department and a single regulatory agency.

First, though, he had to deal with the failure of the government to set any concrete goals for the DOC. Other than the priority of launching the first domestic satellite communications system, the department lacked a mandate. To rectify the problem, Gotlieb hired the head of the Ford Mitre Corporation's military satellite communications project, Douglas Parkhill, who took a leave of absence from the Boston-based think tank to become assistant deputy minister of planning. To help the DOC define its mission, Parkhill convinced Kierans and Gotlieb to launch a massive study project, called the Telecommission, which was announced on September 18, 1969. "It is still the most comprehensive study of computers and communications conducted anywhere in the world at anytime," he says.[147] In addition to holding six seminars and two national conferences on communications topics, the Telecommission project published forty-six studies on virtually every aspect of communications technology and regulation. These were authored by industry members, consultants and academics who were all brought together in the cooperative venture.

A promising young university administrator from the University of Montreal, J. C. de Montigny Marchand, was hired as director of research and co-secretary of the Telecommission project. The suave lawyer is de Grandpré's first cousin on his mother's side, the son of Françoise Magnan. Like Gotlieb, Marchand's star rose quickly during the Trudeau administration. As deputy secretary to the cabinet for operations from 1975 to 1979, Marchand became one of Trudeau's four top advisors and, along with Michael Pitfield, Jim Coutts and Dick O'Hagan, met daily with the prime minister.[148] De Grandpré's older brother indicated there was little contact between the cousins, and he lamented the fact that Marchand never once called him or invited him for dinner during the three years he lived in Ottawa while serving on the Supreme Court bench.[149] Although that estrangement may have resulted from Marchand's senior position and his desire not to put himself in any perceived conflict of interest by meeting privately with a high court judge, their nineteen-year age difference more than likely contributed to Marchand's distance.

Not to be outdone by the bureaucrats, Scrivener instructed de Grandpré's department to commission Bell's own authoritative study on the telecommunications industry and policy in Canada. Bell retained H. Edward English, an economist and director of the school of international affairs at Carleton University in Ottawa, to put together a team of professional researchers to perform the work. Dubbed the Telecommunications Research Group, members included Carl Beigie, head of the conservative think tank C. D. Howe Research Institute, and William Lederman, former dean of law at Queen's University and an expert on constitutional law. Their study, published in 1973, was "complementary" to the Telecommission studies, according to English.[150]

The Telecommission project culminated in the publication of the *Instant World* report in 1971.[151] The report had the immediate aim of providing advice to the government for a forthcoming White paper on communications policy, which had been announced in the Speech from the Throne on October 8, 1970. However, Parkhill says *Instant World* also "blazed the intellectual trails on those subjects for the entire OECD (the Paris-based Organization for Economic Co-operation and Development)."[152] Its discussion of the emerging Information Society was certainly prophetic and almost ten years ahead of works such as Nora and Minc's influential 1979 report to French president Valéry Giscard d'Estaing.[153]

One statement in the report represents a lost opportunity for Bell. The authors expressed concern that the growing dominance of the telecommunications and computer industries by large U.S. firms was a threat to Canada's economic independence:

> Canada is only one relatively unimportant segment of the global market for computers and computer-services, which is almost totally captive to United States interests. Against them Canada has only one high card to play — the existence of a telecommunications industry that is largely Canadian-owned and has, in one instance, a corporate tie with a Canadian manufacturing undertaking sufficiently large and diversified to benefit from economies of scale in production, research, and development.[154]

DOC, in effect, declared its support for the principle of vertical integration that joined Bell and Northern Electric together. Surprisingly, Bell's officers failed to capitalize on the fact that DOC's position ran counter to the policy being pursued by the combines officers at Consumer and Corporate Affairs, whose investigation into Bell's corporate links with its manufacturing arm was then underway. De Grandpré, in fact, still dismisses the DOC's role in telecommunications policy and regulation over its twenty-year existence. He not only says it failed to do anything for Bell, but describes it "as another attempt to exercise power over the telecommunications industry. I'm still wondering what they have done since their creation in 1968? We, and I, had constant contact and discussions with the ministers and the deputy ministers of DOC for twenty years. I repeat, I don't know what good it has done for the country."[155]

In the wake of *Instant World,* Parkhill says the DOC attempted to pursue its dream of making Canada a world leader in the integration of distinct and competing information networks into one global information utility, was still the quest of the telecommunications industry in 1991. Kierans, in a speech to the Canadian Broadcasting League, declared:

> If our policies are to fulfill the promise of the technology, the promise of a better and more equitable distribution of information power, public policies must look beyond artificial boundaries, vested interests, and specialized knowledge. . . we must have the courage to accept new facts, and to bring these facts together in a coherent and flexible policy.[156]

Despite the minister's lofty rhetoric, however, the vision of *Instant World* was doomed to failure, its ideals skewered on the stake of bureaucratic politics.

In the back rooms, DOC was mired in bureaucratic infighting and failed to produce the requisite political consensus to achieve its goals. Coincident with the release of the Telecommission's final report, DOC took a Yellow paper to the federal cabinet asking for funding for a national institute that would work toward the development of a new computer communications network. The department received tentative funding for the project from the Treasury Board Secretariat — the cabinet committee that dispenses largesse to government departments — but then DOC got cold feet, Parkhill says. Despite all the analysis conducted for Telecommission, DOC "needed more studies on the business and economic side."[157] An industry task force on computer communications, headed by one of Parkhill's planners, Herbert Von Bayer, was created and produced the two-volume *Branching Out* report in 1972.[158]

The studies did not stop there. They were used as the basis for a Green paper on a proposed computer communications policy released in April 1973 by DOC's second minister, Gérard Pelletier.[159] The timing was bad, Parkhill says: "With Kierans gone, the department was adrift and the recommendations that lots of money be spent on developing a national public data-communications network with DOC in a role to coordinate planning was torpedoed by the first oil crisis." Bureaucratic rivalry also consigned the department's goals to oblivion. "Within the cabinet, DOC was greeted with hostility by ITC [the Department of Industry, Trade and Commerce] and TBS [Treasury Board Secretariat], who both viewed DOC as an upstart." However, Trudeau's cabinet ruled by consensus. The failure of his ministers to agree on which department should rule the electronic highway led cabinet to slough the dispute off to a new interdepartmental committee of officials. Parkhill was named chairman and he still bristles with anger as he describes the committee as "an exercise in frustration."[160]

The DOC's early work on computer communications led it to arrive at a similar conclusion as the CTC concerning the need to develop rules to regulate the telephone companies' growing activities in that unregulated area. Although DOC supported Bell as a chosen instrument, it also advocated new regulations to deal with the plethora of new technologies and services. The most important of these dealt with competing private networks and the requirement that the telephone companies establish "separate subsidiaries," which the CTC had discussed at Bell's

1969 rate-increase hearing. Kierans shared the CTC's concern, but for quite a different reason, as he explained in Parliament during November 1969. Private networks, which were used primarily by data-processing service bureaus to connect business users to their large, centralized mainframe computers, were the fastest-growing telecommunications service in the late Sixties. They were also the most competitive segment, with CNCP Telecommunications deriving about 75 per cent of its revenue from private networks at that time. But as Kierans stated:

> It is inherently unsatisfactory for a body such as the CTC to be charged with regulating only one portion of a particular company's business. In terms of the rates charged to users, the relationship between the regulated and unregulated operations of a company can be decisive.[161]

To resolve that regulatory anomaly, the government introduced an amendment to the Railway Act requiring the CTC to regulate private networks under the same rules that govern monopoly networks.

As both Opposition critics and de Grandpré pointed out, it would have made just as much sense to deregulate the competitive segment of Bell's business. To realize its vision of one giant computer communications network, the federal cabinet had approved CNCP's acquisition of a 51 per cent stake in Computer Sciences (Canada) Limited earlier that year,[162] despite the protests of members of the data-processing industry who feared conquest by the telecommunications carriers. David Orlikow, the NDP member for Winnipeg North, lambasted Kierans for the apparent contradiction in government policy that resulted from the official sanction of that deal:

> One section of the government apparently has some doubt about the good sense of permitting Bell, a communications carrier, to own Northern Electric while another section of the government, in this case the Cabinet, has permitted CN-CP Telecommunications to purchase a computer company, which I submit is in precisely the same position in relation to it as is Northern Electric to Bell Canada.[163]

Kierans responded that regulation of the carriers' computer business was better than no regulation at all because of fears that, if unregulated, telephone companies would engage in anticompetitive behavior, such as predatory pricing and discrimination in favor of certain customers.[164] The government had already sanctioned Bell's entrance into that business through the amendments to Bell's Special Act the preceding year. Kierans stated in Parliament that a key task of the Telecommission studies would be to deal with the ramifications of the merger of the two industries.

The entrance of the carriers into the data-processing field also raised the specter among competitors, and public interest groups, that subscribers of basic telephone service would be used to unfairly subsidize the telephone companies' competing businesses. The *Instant World* report echoed the CTC's concerns that more stringent methods were needed to scrutinize Bell's cost accounting to prevent cross-subsidization. The DOC also perceived a need for the telephone companies to compete through separate corporate subsidiaries because, as the report's authors stated, "the most stringent regulatory procedures may be circumvented unless the regulatory authority has some control over corporate structure."[165] Despite that concern, it took the DOC more than five years to draft a coherent policy on Bell's entrance into the computer business that could be agreed to by other government departments.

To de Grandpré, the whole exercise was an expensive, and specious, waste of time: "I've seen so many papers from that department — Green papers, Red papers, White papers and Orange papers — but what really burned me up with all of those papers was that they would say, 'Bell is a good company,' it provides 'good service,' it has 'excellent engineering,' and it 'spends lots of money on R&D.' They would be glowing for the first two or three pages, but then there would be the buts. Even though they would say the thing was not broken, they would always say we should fix it. My God, just think what we could have done with the money we spent to refute those things."[166]

He was even more intensely opposed to the notion of competing through separate subsidiaries and any government scheme that permitted regulated competition. He consistently argued that forcing Bell to compete through separate, but regulated, ventures was unfair to the company and its shareholders.[167]

The debate between Bell and the DOC over the next six years from 1969 to 1975 on those points was tame, compared to the dustups that occurred between de Grandpré and Bell's regulator in subsequent years at various proceedings. Stripped down, the issue of separate subsidiaries was little more than haggling over corporate structure. Like the veteran poker player that he is, de Grandpré eventually attempted to end the debate by reorganizing Bell Canada's corporate structure himself. In looking back, however, it is worth asking whether separate subsidiaries protect subscribers and competitors, or whether they even matter? The U.S. government has wrestled with that question much longer, and in greater depth, than Canadian authorities have. In a comprehensive analysis of competition in the telecommunications industry, the U.S. General Accounting Office (GAO), which is the investigative arm of the U.S. Congress, concluded in a 1981 report that separate subsidiaries have little efficacy:

> . . . separate subsidiaries are not a panacea, a cure-all or a self-sufficient solution to the problem of monopoly power and its abuse. . . .
>
> Imposing a separate subsidiary requirement on a dominant firm does little or nothing to alter the incentives of the overall firm or make the incentives of the separate subsidiary significantly different from those of the corporate parent. . . .
>
> A separate subsidiary merely serves the function of drawing a line of demarcation, a boundary between the parent and its affiliate.[168]

Separate subsidiaries, the GAO stated, are merely accounting devices which cannot change the behavior of a monopoly firm or eliminate any incentive the firm might have to exercise its power to the disadvantage of either a consumer or a competitor. Federal policy makers in Ottawa may have thought differently about separate subsidiaries had they remembered Bell's experience at the turn of the century with the People's Telephone Company in Winnipeg when Sise incorporated a separate subsidiary to compete against itself in order to annihilate a rival.

Unlike Bell's first boss, who forged the company's monopoly by extinguishing competitors, Scrivener and de Grandpré ran a utility that had not had to deal with a serious competitive challenge for sixty years. That would soon change. In addition to learning how to run a phone company, de Grandpré would educate the utility on how to compete.

Competition from companies seeking to string private lines between computers was not the only competitive challenge Bell faced in 1969. Technology made possible another new means of communicating that could, if permitted, put Bell out of business or, at best, relegate it to providing money-losing local service. The new development was satellite communications. The elimination of that threat began with policies developed by Scrivener. But the fight between Bell and the federal government over satellites was a long, drawn-out affair with two major rounds fought, and won by Bell, during de Grandpré's tenure. How Bell mitigated the threat posed by communications satellites is a uniquely Canadian story and one that hearkens back to Bell's response to competition in the previous century.

Although Canada was the first country in the world to launch a commercial venture to use satellites for domestic communications, Trudeau's embrace of that scheme in the interests of national sovereignty presented DOC with yet another contradiction in government policy. Kierans was obligated to promote the satellite program at the same time that he embraced the country's largest telephone company as a chosen instrument. The conflict between those two policy objectives plagued Ottawa's satellite policy for almost two decades.

Space was considered the preserve of the nuclear superpowers in the Fifties and Sixties, yet Canada was also a major space power.[169] Canadian defense scientists at DRB's Shirleys Bay telecommunications labs (DRTE) near Ottawa were among the first researchers outside of the Soviet Union to monitor the transmissions and thereby determine the orbit of *Sputnik I* — the Soviet satellite that ushered in the Space Age — on October 4, 1957.[170] Sputnik threw the U.S. national security establishment into a crisis.[171] In the wake of the launch, the Space Science Board of the U.S. National Academy of Sciences called for research proposals in July 1958 for an experimental satellite program. DRB submitted a proposal to design and build a satellite that would carry a special probe, called an ionosonde, into the upper atmosphere. The DRTE scientists wanted to study the physics of the ionosphere, a conducting layer of charged ions that envelopes the Earth at an altitude of between 60 and 3,500 kilometers and makes the transmission of radio waves possible. With their satellite ionosonde, DRTE hoped to learn more about the effects of solar winds and sunspots, as well as develop expertise in satellite communications.

DRB's satellite proposal, spearheaded by John H. Chapman, was rooted in an existing atmospheric research program that DRTE scientists had begun during the Second World War. Their work was considered important because many military communications systems operate in the high-frequency range, a region of the electromagnetic spectrum that is greatly affected by disturbances in the ionosphere. By the late 1950s, however, DRTE scientists had virtually exhausted the potential of gaining more data from high-altitude balloons and rocket-fired sounders. Their satellite ionosonde proposal also attracted widespread interest in the United States, particularly from military planners, because the advent of offensive intercontinental ballistic missiles and defensive early warning radar systems necessitated a greater understanding of the vagaries of the atmosphere.[172]

The DRB proposal was approved by NASA in April 1959 as a joint Canada-U.S. effort. DRTE engineers designed the satellite, named *Alouette I,* and NASA provided free rocket vehicles and launch services. The satellite, launched in 1962, was an unbridled success, producing some of the most important data on the ionosphere since radio research began almost fifty years earlier. Within three months of its launch, NASA and the federal Department of National Defence (DND) announced an ambitious expansion of the project. A series of satellites capable of recording measurements through at least one entire eleven-year cycle of sunspot activity would be launched.[173] The subsequent *Alouette* satellites were renamed *"ISIS"* — the acronym for the new project, which stood for International Satellites for Ionospheric Studies.

"Alouette and *ISIS* didn't do much in themselves to further the commercial use of space," David Golden, chairman of Telesat Canada, said in a 1987 interview. "But a number of government people, particularly John Chapman, wanted a Canadian manufacturing presence in the space industry through a Canadian-owned company."[174] That domestic space manufacturing industry was launched when Chapman decided in August 1963 that DRB should spin off its *Alouette* satellite technology. As its chosen instrument, DRTE picked de Havilland Aircraft of Canada in Toronto, which acquired the Crown patent for the novel retractable antenna developed for the satellite.[175] De Havilland was then selected as subcontractor to RCA Victor of Montreal, which had obtained the contract to build the second satellite, *ISIS 1.* After de Havilland spun off its special products and applied research division in a 1968 management buy out, the subcontract for the *ISIS* satellites was taken over by Spar Aerospace of Toronto.[176] Spar achieved fame in the late 1970s for its manufacture of the Canadarm remote manipulator for NASA's space shuttle fleet.

The success of the Alouette-ISIS program led Chapman to push for a broader commercial space program for Canada. In August 1963, the same month that he announced DND's technology spin-off program, the federal Department of Transport and NASA signed an agreement for Canada's participation in a project to launch an experimental communications satellite.[177] Unlike *Alouette*, a communications satellite has to be lobbed into orbit precisely 35,580 kilometers above the Earth's equator in a special region known as the geostationary arc. Any object placed there has the same orbital period as the Earth, completing a revolution every twenty-four hours, and gives the illusion that it is suspended above the planet.[178] This allows a satellite to act as a radio relay with its antenna fixed on any antenna within range on the Earth's surface. The *Syncom 3* satellite, placed in geostationary orbit in August 1964, proved the feasibility of operating a satellite-based communications service and paved the way for the international Intelsat system, of which Canada was an initial member. Northern won a subcontract in the mid-Sixties from Hughes Aircraft Company of El Segundo, California, to make a repeater system for the Intelsat satellite.[179]

Chapman outgrew the defence department, says Edward J. Bobyn, the former chief of research and development at DND, and pushed the government to find a proper home for the nascent commercial space program. The science secretariat of the Privy Council Office put him in charge of a special task force to study the government's space programs in May 1966. After holding public hearings across the country that fall, Chapman submitted a secret report to PCO and a public version, known as the Chapman report, was released in February 1967.[180] He concluded that the primary objective of the government's program should be telecommunications and remote sensing. The report marked a turning point in Ottawa's space activities, because it urged the government to shift its focus away from basic scientific research to commercial applications. Chapman also called for Canadian communications satellites to be built and launched by 1970 or 1971.[181]

But some businesses did not want to wait that long.

As soon as the Chapman task force began its hearings in October 1966, Power Corporation of Canada, the large Montreal-based conglomerate which was then headed by Peter N. Thomson, combined forces with Niagara Television and applied to the federal broadcast regulatory agency, the Board of Broadcast Governors, for permission to use satellites for a third national television network. They proposed that Cansat, which would be a publicly traded company controlled by Power Corp., would build, launch and use the satellites.[182] The telephone companies were taken aback by the surprise application. Although Bell intervened at a special one-day hearing held by the Board in March 1967 to consider Power Corp.'s application, the regulator rejected the Cansat proposal because of a lack of policy direction from the federal government.[183] The Cansat proposal and the Chapman task force prompted the government to create a high-level interdepartmental committee on space headed by Industry Minister Bud Drury.

Although the Cansat bid had failed, the telephone companies countered in kind later that spring. Their competing satellite bid was not their first foray into the field, however. As early as December 1962, Bell had lobbied the federal government, asking that the telephone carriers be allowed to build and own Canada's main Intelsat ground station at Mill Village, Nova Scotia.[184] Instead, that right was granted to the Canadian Overseas Telecommunications Corporation, the Crown corporation that was Canada's sole overseas communications carrier, now known as Teleglobe Canada Incorporated. Bell and Northern Electric had also tested for the Department of Transport an experimental ground station to receive satellite signals in the Arctic.[185]

The new satellite bid was submitted jointly to Transport Minister Don Jamieson by TCTS, the consortium of monopoly telephone companies headed by Bell executive vice-president Z. H. Krupski, and rival CNCP Telecommunications.[186] TCTS and CNCP had put aside their competitive rivalry when faced with the greater, mutual threat posed by satellites. Satellite communication was more than just a rival technology: it threatened to undermine the economic underpinnings of the telephone companies by eliminating the costs of distance. The cost of transmitting a message by satellite is distance-insensitive, meaning that it costs the same to send a message across the street by satellite as it does to call Inuvik from Halifax.

The carriers hoped to neutralize that threat by creating their own wholly owned satellite company, which would be a "consortium's consortium." Under the terms of the proposal, their satellite carrier would have no direct dealings with customers and would act solely as a wholesaler to the carriers. In vintage Scrivener style, the proposal invoked the public interest and precluded any equity participation from either the federal government or the public on the supposed altruistic ground that such a venture was fraught with risk.[187] The proposal formed the underpinning of Bell's corporate policy toward satellites.

The pressure on the government to develop a satellite policy intensified later in the summer of 1967 when RCA, eager to win the contract to build the commercial satellites, went public with an aggressive lobbying campaign. Northern Electric and Hughes — RCA's major U.S. rival in the satellite business — also turned up the heat when they submitted a joint study to DOT that June. Although hardly impartial analysts, Northern and Hughes were asked to advise DOT on the conditions under which a domestic commercial satellite system could be financially viable.[188]

The newly created Science Council of Canada jumped into the fray in July 1967 with a report urging the federal cabinet to create a central space agency, which would manage all space-related activities.[189] Instead, Chapman was moved to DOC, where he was put in charge of space programs and research, and where he was "destined to become alienated and discouraged because of the government's failure to create a space agency," Bobyn says.[190] That failure plagued the satellite communications industry for more than two decades as government departments feuded over competing space programs and budgets. Twenty years to the month after the Science Council published its first report, the Mulroney government decided to create such an agency. Whether it should be located in Montreal or Ottawa became a controversial patronage issue. Patronage, and Montreal, won out.

As if Ottawa did not face enough pressure over satellites in 1967, Quebec attempted to negotiate a separate agreement to receive satellite signals from France. In response to this welter of conflicting interests, the government issued a White paper that was intended to be the definitive statement on satellite policy. The Department of Industry's White paper, released on March 28, 1968, concluded that a domestic satellite communications system was of "vital importance to the growth, prosperity and unity of Canada and should be established as a matter of priority."[191] In addition to the lure of high-tech spin-offs, the report's authors described how satellites would realize the Canadian dream of establishing communications and promoting development in the North. The ability of a satellite to beam a signal of equal strength over the entire Canadian land mass meant telephone and broadcast services in both official languages could be extended to remote communities. Satellites thus appeared to be the perfect tool to promote national unity and Canadian culture.

The White paper attempted to reconcile the apparent conflict between a new satellite operator and the telephone companies by providing assurances that the new space system would complement, rather than duplicate or replace, existing networks.[192] It recommended that the satellite operator take a unique "mixed" corporate form and envisaged a company jointly owned by government and the private-sector telecommunications carriers. Share ownership by the public would be reserved until an undetermined time in the future. The proposed equity structure squeezed out rival players, such as Power Corp. and the Quebec government. By declaring that the different systems had to be free to "compete effectively in those areas where competition is appropriate,"[193] the White paper also flatly rejected the telephone companies' bid to control a satellite company.

Trudeau moved quickly on the report. The following month, the federal Department of Industry invited two consortia to submit design proposals to make the satellites. RCA headed one group and the other was made up of Hughes Aircraft, Northern Electric and Canadair of Montreal.[194] Hughes and RCA were the only two companies in the world that had developed the critical antenna stabilization platforms for civilian satellites. Called the "bus," this platform keeps the satellite's antenna stationary. Despite the government's rhetoric that it wanted to foster a Canadian space industry, the failure to finance the development of a Canadian-made satellite bus meant that the government and its Canadian contractor would be dependent upon U.S. aerospace companies for a major component of any satellite they developed. In the years that followed, that dependency hampered the attempts of domestic contractors, such as Spar, to win export sales.

Scrivener played a key role in the debate that ensued over the design of the satellite and what kind of system the government should procure, Gotlieb says. "Scrivener said, 'Let's put up a Volkswagen, not a Cadillac.'"[195] He wanted a satellite with a limited number of small radio channels, called transponders, so that the satellite could be made in Canada. Both RCA and Hughes submitted similar proposals to the government on November 25, 1968. Each proposed to build two "Volkswagen" satellites, each with six channels to handle television or telephone signals. In the end, RCA won the contract to help define the program's specifications.

Meanwhile, Kierans had hired Robert M. MacIntosh, then deputy chief general manager of the Bank of Nova Scotia, as a consultant to prepare a study on how the satellite venture should be financed and incorporated. Although his report was kept secret, it was later revealed that MacIntosh recommended a tripartite corporate structure in which the government, the common carriers and the general public would hold 30, 30 and 40 per cent of the satellite venture's shares respectively.[196] Kierans declared in Parliament that share ownership of the three parties "should be approximately equal" and stated that public participation would "ensure accountability."[197] However, that is not what he proposed when the legislation to create Telesat Canada, Bill C-184, was introduced in early 1969. The new company would be unique. Rather than being a Crown corporation or an agent of Her Majesty, it would be a private commercial venture. Its equity would be owned jointly by the federal government and the telecommunications carriers.[198] Although the Telesat Canada Act contained a provision for share ownership by the public, no block would be reserved for public subscription. Instead, the government opted to set aside a single share to be held by the president of Telesat in the name of the public.

The bill was assailed by the senior lawyer from the Department of Justice, who testified before the parliamentary committee considering the Telesat bill that the law incorporating Telesat was drafted in such a way that "there is no firm commitment to a tripartite share ownership or to a dual share ownership."[199] He also testified that the bill gave too much discretion to cabinet because it could dramatically alter Telesat's share structure without legislative amendments.[200] The government clearly reneged on its promise of public ownership to appease the phone companies. They had responded to the previous rejection of their "consortium's consortium" proposal with another brief in July 1968 that stated the satellite company should be owned jointly by the government and the carriers.[201] One TCTS member, Alberta's Telephone Minister R. Reierson, who was head of the provincially owned Alberta Government Telephones, admitted in his testimony to the parliamentary committee that the satellite company would be perceived as a far greater competitive threat to the phone companies if public shareholders or other outside investors were brought in by the government.[202]

Dual ownership, however, granted the telecommunications carriers a virtual veto over satellite policy on Telesat's board. Alex Lester, Bell's executive vice-president of engineering, confirmed the widely held perception that Telesat would be hampered at the hands of the phone companies when he testified that: "I do not really think that the long-term future of telephone facilities is via satellite. I think this lies more in some of these other directions of wave-guides and lasers."[203]

Gotlieb says the cash-strapped government had little choice but to bring in the phone companies. As he saw it, Telesat was a financially unattractive proposition for most investors because of the high proportion of debt relative to Telesat's equity. "With the carriers, we could put a position of weakness into a position of strength. The carriers were the key to the success of Telesat."[204] And Scrivener was "instrumental in making it work out," Gotlieb says. "He was . . . front and center in putting the telephone companies' positions across." De Grandpré appeared in Ottawa often with Scrivener during the formation of Telesat and was his "solid and sensible backup."

Having previously been rebuffed in their bid to control Telesat, the phone companies delivered an ultimatum to the government at the hearings. On May 6, 1969, TCTS threatened to pull out of Telesat unless the government accepted a proposed amendment that would have restricted the use of the satellites to the phone companies.[205] Kierans refused to write such a provision into the Telesat Act, arguing that TCTS wanted to put the control of Telesat "right into the boardroom" of Bell.[206] However, despite his public statements, the minister capitulated and, following an *in camera* meeting, gave assurances that Ottawa would embrace a policy that Telesat was to act solely as a "carrier's carrier."[207] TCTS agreed to withdraw its amendment, and Kierans inserted a clause in Telesat's service agreements stipulating that, except for the CBC, the carriers would be Telesat's sole customers. A letter from Kierans was read into the record of the House of Commons Standing Committee on Broadcasting, Films and Assistance to the Arts, in which he restated the principle that Telesat was to "operate as a complement, not as a competitor, to the common carriers." He added that:

> Canada has not enough capital available for it to bear the luxury of duplicating investments in the same area. Further, I will say plainly that it would be inequitable to invite such companies to compete against themselves.[208]

Kierans' declaration was the *quid pro quo* for the telephone companies' support and meant that Telesat was effectively co-opted by the phone companies. Scrivener had won the first round and curtailed Telesat's influence even before the Telesat Act received royal assent on June 27, 1969.[209]

Telesat opened for business on September 2, 1969, the day after the government named David Golden, a former lawyer, bureaucrat and lobbyist, as its first president and CEO. Golden came to Telesat from the Air Industries Association of Canada where, as president since 1964, he had been the aerospace industry's chief lobbyist.[210] In 1959, as deputy minister of defence production, Golden had implemented Diefenbaker's directive to cancel and destroy the Avro Arrow jet. Telesat was seen by the aerospace industry as a chance for Canada to heal some of the wounds inflicted by that controversial decision.

Golden obtained financing in the first year from a $10-million advance from the government, which he supplemented in the next four years with an additional $25.5 million borrowed from a line of credit guaranteed by Ottawa. Telesat raised an additional $60 million in that period through the sale of six million and one common shares to the thirteen participating carriers and the government. These were split equally between the two groups, minus the single share held by Golden in the name of the public.[211]

A major portion of those funds was earmarked for the purchase of Telesat's first satellites. The award of that contract was one of the most controversial decisions made by Kierans. RCA Canada's president, John Houlding, was a good friend of Michael Pitfield, deputy secretary of cabinet and one of Trudeau's closest advisors.[212] Behind the scenes, much of the senior mandarinate supported RCA because of a belief that its satellite would have higher Canadian content, Gotlieb says. "The Department of Industry, the Prime Minister's Office, Privy Council and Chapman were all behind RCA,"[213] as was Telesat's board of directors. To say that RCA had the inside track is an understatement.

RCA submitted its first proposal to Telesat's board in April 1970 and promised delivery of two six-channel satellites at a cost of $63.5 million. But the company lost several key allies when an unsolicited proposal from Hughes arrived the same day that promised the same thing for $28 million.[214] Kierans, Gotlieb and Golden switched their support to Hughes and were backed by Finance Minister Edgar Benson. "Kierans was very tough in pushing for Hughes and felt RCA wasn't playing fair,"[215] Gotlieb says.

The Hughes bid presented Kierans with a dilemma because, despite its lower price, it had significantly lower Canadian content. Hughes could afford to be paid less because it in effect offered an assembly line satellite that had the specifications already worked out. RCA, however, planned to build a new design to original specifications with a greater percentage of more expensive, Canadian-built components. The two companies revised their proposals in June. RCA knocked its price down to $42.1 million for satellites with 55 per cent Canadian content. Hughes offered an off-the-shelf U.S.-made system for $28 million or satellites with 12 per cent Canadian content for $29.7 million.[216]

Kierans faced significant opposition from cabinet colleagues over the satellite bid, particularly from economic nationalists and Quebec ministers who, like Bryce Mackasey, favored RCA's bid if only because the satellites would be made in Montreal. Kierans convinced

Trudeau to support the Hughes bid after he reportedly threatened to resign from the cabinet if RCA won the contract.[217] In the end, price won out, and the cabinet authorized Golden to award a $30-million contract to Hughes in September 1970 for three twelve-channel satellites. The government boosted the Canadian content to 20 per cent, and Northern Electric and Spar were awarded electronic and communications subcontracts.[218] Julie-Frances Czapla, a Montreal supermarket clerk, won the national contest to name the satellites. A panel of judges, including Marshall McLuhan and songwriter-poet Leonard Cohen, selected her entry of *Anik,* the Inuit word for "brother."[219]

For Bell, the decision meant a sizable chunk of work for Northern's new Aerospace Communications Lab in Ottawa. However, Telesat rejected Scrivener's idea of an economy-sized satellite. "Golden's push for a full-service communications satellite swept Scrivener's model for a smaller satellite that could be built in Canada away,"[220] Gotlieb says. Although Northern had told the government in two separate sets of studies that it only needed satellites with three to six channels each,[221] Telesat's specifications claimed greater redundancy was needed to provide backup radio channels. It doubled the number of channels for each satellite, which greatly escalated the cost of both building and launching the satellites.[222]

Scrivener still bristles when he talks about Telesat opting for a Cadillac instead of a Volkswagen, and critics maintain that the decision contributed to Telesat's subsequent financial woes.[223] Hamstrung by the restrictions on its business and with so many of its channels orbiting the Earth empty, Telesat was forced to rely even more closely on the phone companies during de Grandpré's tenure. The carriers had become Telesat's largest customer, and their use of satellites was deemed "essential to the long-term viability of Telesat," DOC officials stated in a secret cabinet memo.[224] Unprepared to ante up more money for Telesat and fearing the consequences if Telesat lost the phone companies' financial backing and their business, Ottawa had a vested interest in policies that served the phone companies. Ultimately, Telesat was stripped of any remaining autonomy.

Bell Canada's regulatory affairs were greatly complicated by the crisis over Quebec sovereignty and national unity that erupted in the late Sixties. For the next decade, from the imposition of the War Measures Act during the October Crisis in 1970 to the defeat of the separatists' motion on sovereignty-association in the 1980 Quebec referendum, de Grandpré found Bell caught in the middle of the struggle.

Despite the intensity of his repeated clashes with federal politicians and mandarins, de Grandpré believes in cooperative federalism in which a strong federal state shares powers with the provinces. As a federalist, he consistently decried the intrusion of either level of government into the jurisdiction of the other. Such conflict was, of course, bad for business. He rejected calls for Ottawa to rewrite the Constitution to satisfy Quebec and, instead, urged the federal government to "re-examine the path it has taken lately into fields that were intended to be provincial property." In the same 1977 speech, de Grandpré remarked, "If each province ensures that its own air is clean, can there really be any dirty federal air?"[225]

Quebec nationalists viewed Bell, which had been dominated by anglophones since its formation, as an illustration of the problems facing Quebec society. Until Vincent's appointment as president in 1963, few francophones had been promoted to Bell's top executive ranks, an oversight Vincent set to rectify with the promotion of officers like Groleau and de Grandpré. And resentment still lingered among many francophones who remembered not being able to get a French-speaking operator in Montreal between the Thirties and Fifties, before automated telephone switches were adopted.

When confronted with the arguments of Quebec nationalists, de Grandpré showed more than just impatience. He exhibited a characteristic defiance, as if on a personal mission to convince his political foes by example that economic success was possible for Quebec-based enterprises within the federal system. He dared francophone entrepreneurs and enterprises to match, or rival, his own and Bell's successes as global players. "It is pointless to shout from the housetops against the invasion of foreign capital, or to decry economic colonization by largely English-speaking capitalists during our history," de Grandpré told a lunch-time audience in a speech to the Montreal Chamber of Commerce in February 1977. "I think that it is no exaggeration to say that our businessmen have for a long time been imbued with a sort of fear, mixed with distrust, at the thought of their businesses extending beyond their own areas. It should be the opposite: we should be aiming to extend ourselves abroad."[226]

Under Premier J. Jacques Bertrand, the Quebec government established its own Department of Communications in 1969. It quickly became one of the most active communications departments in Canada as Quebec City began an intergovernmental battle to assume jurisdiction over Bell Canada. Gaining control over the communications industry has been perceived by every provincial government since Bertrand's as integral to Quebec's quest to achieve cultural sovereignty.

De Grandpré resisted Quebec's bid for control. "For years there has been a tug of war between Ottawa and Quebec City as to who should control and regulate Bell," he says. "I had discussions with all the premiers of Quebec during that period — Mr. Johnson, Mr. Bertrand, Mr. Bourassa and Mr. Lévesque — indicating to them that there were serious dangers in doing this as far as Quebec was concerned because the province of Quebec would probably have to charge more for telephone rates than the province of Ontario. And it could place the province of Quebec in a more difficult position from the competitive standpoint."[227]

Québec-Téléphone attempted to make use of the federal-provincial dispute in its own battle with Bell Canada over long-distance rates. The independent provincially regulated company was angered by the adverse effect on its affairs caused by federal regulatory decisions that governed Bell Canada's telephone rates. The CTC decision that approved Bell Canada's announced long-distance rate reductions on May 10, 1968, would produce a $580,000 loss at Québec-Téléphone.[228] Although that decision compelled Québec-Téléphone to change its rates for long-distance calls made between the two companies' territories, the small utility refused to adjust its rates on the grounds that its provincial regulator, the Quebec Public Service Board, had not approved of the changes. Because that board had no authority over Bell matters, however, Québec-Téléphone opted to force the jurisdictional issue in court by applying to the Quebec Superior Court on July 4, 1968, for an injunction prohibiting Bell from altering the rates it charged Québec-Téléphone for connection of their networks.

Bell filed a motion that challenged the jurisdiction of the court to interfere with its regulation. The case wound its way through the appeal process for three years and attracted the attention of the Quebec government, which intervened as an interested party in 1971. Although Bell lost its bid to quash the injunction request, the lower court eventually ruled that it could not interfere with the utility's federal regulator. Despite that legal victory, Bell's vociferous arguments against provincial regulation and authority had moved the company further down the path toward confrontation with Quebec over federal-provincial jurisdiction over telecommunications. De Grandpré and Scrivener knew that Quebec's objectives, if met, could lead to a profound change in the regulation of their company. Provincial regulation would drastically change Bell's operations and require a division of the company along political lines.

Rather than wait for such a change to be forced on Bell, de Grandpré made it himself. After being appointed executive vice-president of operations in August 1970, he split the powerful operations department into two halves, an eastern and a western region. He took the title of executive vice-president, eastern region, and assumed responsibility for the company's operations in Quebec on September 1. James Thackray, a career Bell official and engineer, was appointed executive vice-president, western region. Coordination of operations was placed in the hands of a new executive committee.

The restructuring came at a time when the separatist movement in Quebec was growing. Observers remarked that Bell's reorganization prepared it for the prospect of Quebec's separation. De Grandpré agreed with that assessment in a 1990 interview, saying, "There was no doubt that the decentralization also became a plus, and the company would be well equipped to go one way or the other."[229]

As rising inflation further eroded Bell's profit in 1970, the utility applied for permission to raise monthly residence and business rates by more than 6 per cent. The CTC granted half of the request in its December 1, 1970, judgment.

In 1971 Bell applied for yet another general rate increase that would have added $78.1 million to its 1972 revenue. After twenty-two days of hearings, the CTC awarded a $47-million increase in a decision released on May 19, 1972. Although it granted the requested long-distance increase, the CTC turned down the request for a local rate increase.[230]

The 1971 hearings were noteworthy for the vigorous participation of the Quebec government. Following the release of a Green paper on communications by Quebec's minister of communications, Jean-Paul L'Allier, in 1971, the province embarked on an aggressive program of intervention at Bell Canada's rate proceedings. The Green paper outlined the province's interests in communications, including its desire to eliminate what L'Allier described as a flood of American television programming and to provide universal, low-cost telephone service to unserved areas. "The telephone company was very slow in providing service in Quebec outside major cities," Raynold Langlois, a partner at Langlois, Trudeau and Tourigny and former lawyer for the Quebec government, says. "We believed Bell's quality of service was not satisfactory."[231]

Broadcasting, telecommunications and the preservation of French language and culture all came together at Bell Canada's poles. Quebec wanted to launch its own French-language television services to be delivered by Quebec cable companies, but that required access to Bell's poles because federal rules gave the utility a monopoly on pole ownership. "Quebec felt Bell Canada was using its position as pole owner to block access to cable distributors and new entrants," Langlois says. As one of the lawyers responsible for arguing Quebec's case at the CTC hearings, Langlois found that the province was pitted against an adversary that had "greater strengths than we could have." To overcome that weakness, he says, "There was a great deal of cooperation, or a conspiracy, between Ontario and Quebec. We developed our questions and a computer database together to confront our common enemy."[232] That cooperation marked the dawn of effective, organized resistance to Bell Canada, a resistance that became more vocal in years to come.

De Grandpré was appointed to Northern Electric's board on March 5, 1970. He interpreted the appointment as a leg up on Bell's succession ladder. "It started to give me some indications that I was certainly one of the considered candidates."[233] The appointment also put him in a better position to oversee the manufacturer's regulatory affairs. In the CTC's 1970 judgment, the commission stated that it was concerned that Northern Electric appeared to make more money from Bell than from other customers. It stated it was not satisfied with Northern Electric's reporting procedures and served notice it would examine the subsidiary much more closely at future hearings.

Northern Electric's regulatory affairs assumed critical importance as Scrivener moved both companies ahead on the digital technology effort. The joint Bell-Northern Electric task force on switching had issued its report on the evolution of the telephone network, in which it endorsed a move to digital communications by both companies that fall. It recommended that Northern Electric develop a family of digital switches and make the first one available by 1978. And it suggested that all new transmission systems for Bell's network be digital.[234] Any doubters within the organization were hammered by a second document issued by Northern's systems engineering group in October 1970. Its covering letter described the document as most unusual because it did not define a product or a market. Rather, it concluded that "due to rapid technological advances in memories, logics, etc., our SP-1 system will be obsolete for new local and toll [long-distance] installations by 1980." The SP-1 was still one year from market and already five years had been shaved off its forecasted life expectancy. Together, the two reports had the effect of lighting a fire under the feet of Northern's executives and the research division.[235]

The twin strategies of Scrivener's plan — to assault the U.S. market and to bolster Bell and Northern's R&D program — were inextricably woven together, because Northern's entry into the U.S. market would lead to the immediate termination of all research agreements between Northern and the Bell Labs. Bell also faced three other compelling reasons to rationalize and strengthen its R&D programs: first, both the SP-1 and the SG-1 had substantial ongoing development requirements; second, Bell had an increased stake in applying the new technologies to its own network; and third, the work in new electronic chip and digital technologies cut across all existing operational or product boundaries, requiring improved coordination.

Those combined imperatives led senior management at Bell and Northern to spin off Northern's research division into a separate corporate entity in January 1971, named Bell-Northern Research Laboratories (BNR). A tri-corporate board was created to run it. The manufacturer initially owned 49 per cent of the equity (it now holds 70 per cent) and Bell held 51 per cent (now 30 per cent) — of what has become the largest private research complex in Canada and one of the largest in the world.

The creation of BNR was also part of Northern's response to the recommendations of a task force into the company's operations by a leading management consulting company, McKinsey and Company of New York.[236] Scrivener and Marquez took the advice of the consultants, who told them to junk Northern's existing structure, which had the company organized along functional lines. That meant Northern had pursued a shot-gun approach to marketing, selling everything everywhere, J. Derek Davies, now executive vice-president for strategy at Northern Telecom, says.[237] Instead, the consultants told management to reorganize the company by product line and to create internal profit centers that would operate as small business units. Responsibility for working out the new structure, as well as input on the decision to create BNR, fell to Arnold Groleau, Bell's executive vice-president of operations; Alec Lester, Bell's executive vice-president of R&D; and Brewer Hunt, Lester's counterpart at Northern.[238] Groleau was later appointed the first chairman of BNR's thirteen-member board.

The product-line management scheme, which is still in place, meant that each product group would have to pay for its own R&D out of the business unit's budget. Setting up a separate R&D arm, instead of having the money channeled to an in-house division, meant that each unit could be treated as a research customer of BNR and finance development work essential to its own mission. The new organizational methods also meant there would no longer be just a single Northern R&D budget with a single set of priorities hammered out after much internal bickering. Projects of universal scope or strategic importance, such as the digital effort, could be financed at the corporate level by both Northern and Bell.[239]

To run the new lab complex, Scrivener repatriated a senior Canadian-trained scientist from the U.S. Bell Labs. Donald Chisholm was actually hired to be BNR's first president almost a year-and-a-half before the new research company was created. Scrivener lured him to Ottawa in June 1969 to be vice-president of Northern's R&D division with the promise of making him president of BNR once the plan was fully worked out. Chisholm had headline-grabbing scientific credentials; one newspaper story called him the "model of a modern master scientist."[240] He was both an electrical engineer and a physicist, who never used the title "doctor." Brilliant and irreverent, the tall, bearded scientist eschewed convention — he preferred sandals to shoes, for example. When Scrivener found him, Chisholm was managing director of the Bell Labs' Washington-based Bellcomm Incorporated subsidiary, where he oversaw AT&T's efforts for the U.S. manned space program's Project Apollo.[241] By the time *Apollo 11* astronaut Neil Armstrong took his first step on the lunar surface on July 20, 1969, one month after Chisholm joined Northern Electric, Project Apollo had not only lost its initial challenge, but had also become too bloated, bureaucratic and immersed in politics to keep top engineers like Chisholm. Although he called that program "exciting," he said it was like any engineering job that, once done, simply had to be repeated.

John de Butts, who was then vice-chairman of AT&T, played an instrumental role in Chisholm's hiring. Because de Butts had responsibility for the Bell Labs, Scrivener visited his golfing companion's New York office in 1969 to ask him for a list of recommended candidates for BNR's first president. De Butts called the president of the Bell Labs, James B. Fisk, to set up a meeting with Scrivener. After mulling the request over for a few minutes, Fisk, who was also one of Scrivener's golf pals, scrawled two names onto a piece of paper and handed it across his desk to Scrivener. "Take your pick," he said, adding, "Either of these two scientists could succeed me as president." Chisholm wanted to move back to Canada, so he got the BNR job, and the other candidate, William O. Baker — a leading polymer chemist — stayed in New Jersey and became head of the Bell Labs in 1973. "That's what having friends is about,"[242] Scrivener says, which belies the brilliance of his executive search methods. In picking the brains of the Bell system bosses, Scrivener got Northern's potential rival to give up one of its best scientists, a leader who later presided over a research program that would unseat the technological supremacy of Western Electric and the Bell Labs.

Chisholm's hiring was a coup for BNR and, although not recognized at the time, was in fact a strong signal that the tide of Canada's great "brain drain" was finally being reversed. Although Canada's scientific and engineering communities never fully recovered from the mass exodus of talent precipitated by the Diefenbaker government's cancellation of the Avro Arrow project,[243] the curtailment of the U.S. space program, coupled with the Vietnam War, gave many of those expatriates a reason to come home, while BNR gave them something to come back to. When Chisholm returned, he carried a burning ambition with him. One former BNR engineer recalls the new president's first address to the scientists and technicians who worked at the Bramalea labs at Northern's sprawling plant west of Toronto in November 1970. "He told us he wanted a BNR scientist to be in line for a Nobel Prize within two years. He wanted to build the equivalent of the U.S. Bell Labs overnight."[244]

BNR's budget was $36 million in its first year, a substantial gain from the $26 million spent by Northern's R&D division in 1969. Northern's share of BNR's 1971 budget was $27.6 million, or 5 per cent of Northern's sales.[245] Chisholm led a staff of 1,800 scientists, engineers and technicians — an increase of almost one thousand employees from the 1967 staff level. Although most of its new staff were technicians employed to make transistors and integrated circuits, BNR also developed a growing appetite for computer programmers, electrical engineers and solid state physicists. When the company was launched, 675 members of its staff held degrees, of whom 155 had master's degrees and 55, doctorates.[246] Most of BNR's employees were based in the growing main complex nestled beside a wooded greenbelt in Ottawa's west end; however, additional research centers were set up at five other Northern plants in Montreal and Lachine, Quebec, and in Belleville, Bramalea and London, Ontario.[247]

A key feature of the new corporate R&D enterprise was Northern and Bell's agreement to invest in all research activities in concert. "One will not proceed down a path if the other is not proceeding down the same path with comparable work," Chisholm told an inquiry into the Bell and Northern relationship in 1980. If BNR proposed to explore the development of a new product idea, both of its owners had to be interested and agree to commit funds. Although Bell and Northern each have very different functions and interests, the process meant that neither partner could dictate to the other that something had to be manufactured or developed. "It is the synergistic relationship that gets each what it wants," Chisholm stated.[248]

Almost six months after Chisholm opened BNR's doors, a second element of Scrivener's plans fell into place with the hiring of Marquez's successor, a tough Minnesotan lawyer named John C. Lobb, on May 26, 1971. When Marquez's headhunters began wooing him, Lobb was the head of a Wall Street investment banking firm, Planter Management Company. He had only joined that firm in 1969 after leaving Crucible Steel Company of Pittsburgh, which he had turned around in less than two years. Before that, he had been an executive vice-president at International Telephone and Telegraph Corporation. Lobb earned his reputation as an aggressive "fix-it man" at ITT where, for a while, he was heir apparent to the king of conglomerates, Harold Geneen, ITT's chairman, president and CEO.[249]

Northern Electric had revenue of $480 million when Lobb was hired. He immediately launched an internal management shake-up and slashed unprofitable products from Northern's portfolio. After his first year, sales almost doubled to between $700 million and $800 million and profit nearly trebled, to $12 million from $4.5 million.

Lobb applied many lessons that he had learned at ITT in reorganizing Northern's corporate structure. In orchestrating his house cleaning, he also exhibited a management style that showed Northern's executives why he had been a candidate for Geneen's job. Although Marquez denied that Lobb was recruited as a hatchet man,[250] Northern's leadership "needed someone to tear the organization apart. That's what we got Lobb for," says Light, who later became one of Lobb's successors.[251] The new executive was mercurial and not always right, recalls Benger: "He could be very erratic and a lot of people were frightened by him."[252]

Not one to worry about people's sensibilities, Lobb fired subordinates the Geneen way: on the spot. Jim Kyles, the former head of Bell

Canada International, remembers when Lobb fired an official at a general management meeting — "It wasn't a question of taking you aside." And woe to the executive whose briefing to the boss was judged inadequate. "One day he turned to a guy and he said, 'You take those charts and you roll them up and shove them right up your ass and go collect your pay cheque.'"[253]

Lobb, however, had Scrivener's full backing, and Scrivener had powerful allies in the boardroom, Light says.[254] Their plans for Northern, including Lobb's shake-up, were championed by two of Bell's most influential outside directors: Herbert Lank and J. Angus Ogilvy. Lank, the founding president of Du Pont Canada, held interlocking positions on Bell's and Northern's boards and was a leading doyen of the Canadian Establishment. Ogilvy was the senior partner at Montreal's largest corporate law firm, Ogilvy, Cope, Porteous, Hansard, Marler, Montgomery & Renault — now called Ogilvy Renault. Scrivener defends Lobb's actions by pointing to his success in turning Northern around, adding that the task was "a thankless job" that required somebody who was "rough and tough."[255] Except for occasional lapses when Scrivener got "a little upset" at Lobb, he says the two developed a "beautiful" working relationship and became good friends.

After he assumed the presidency of Northern, Lobb kept his home in Ligonier, Pennsylvania, in "Arnie Palmer country," as Scrivener described it. Ligonier is a mecca for golfers in the U.S. business elite. Both Lobb and Scrivener were members of the Rolling Rock Club there, one of the most exclusive business clubs in the United States. Near Rolling Rock is the Laurel Valley Golf Course, whose eighteen-hole course was directly below Lobb's home. "It's amazing how many times our meetings had to be scheduled for late Friday evening with continuation on Monday morning,"[256] laughs Scrivener.

The upheaval launched by Lobb did more than rid Northern of unproductive people and unprofitable products. Its aim was to make the manufacturer one of the lowest cost producers of telecommunications equipment in North America. This was the only way the company could take on larger, entrenched competitors in the United States, Light says.[257] Selling products at the lowest price is still the only way to compete, and is all the more crucial to Northern's strategy today, as it strives to become one of Canada's only global manufacturers.[258]

"What Lobb did to Northern in 1970, AT&T did to Western Electric in 1984-85," says Edmund Fitzgerald, who retired as chairman and CEO of Northern in April 1990. In hindsight, he thinks one reason for Northern's success was that the company went through a shake-up more than ten years ahead of its competitors. By the time Western realized that it required the same organizational catharsis, it was more than a decade behind and had already lost a major portion of its market to its Canadian rival.[259]

Lobb launched Northern's first sales push into the United States. The details of Northern's U.S. strategy were sketched out by Marquez and Lobb at a press conference in Montreal on June 28, 1971, a month after he was hired. Lobb predicted that within five to seven years, Northern's sales in the United States would equal its domestic revenue.[260] Benger recalls: "Our big break was getting Lobb because he broke down the inertia within Northern and was keen to go after the U.S. market. He said, enough of the studies, here's what we're going to do."[261]

Lobb got busy right away in setting up Northern's U.S. organization. The first step was the incorporation of a U.S. subsidiary in September 1971. Northern Telecom Incorporated[262] based its head office and sales organization in Waltham, Massachusetts, near Boston. In early 1972, the technical information agreements between Western and Northern came to an end and NTI took possession of its first U.S.-based manufacturing plant in Port Huron, Michigan, where it made telephone sets.

In taking on AT&T in its own backyard, Scrivener laid down one basic rule. Bell and Northern were "not going to get into any pissing contest with AT&T."[263] To avoid such a showdown, he paved the way for Northern's entrée into the United States by advising John de Butts, AT&T's new chairman and CEO, of Northern's plans and marketing strategies. Scrivener assured him that Northern would only go after the business communication market and the independent telephone company business, which was not a direct threat to Western's monopoly over the AT&T Bell Operating Companies. Those terms did not offend the titan of telephony, which meant Northern would not be crushed in its initial foray.

In reaching their consensus, Scrivener was aided by his friendship with de Butts, a forceful and, at times, confrontational leader. AT&T insiders characterized an exchange with de Butts as walking into his office with your ideas and leaving with his.[264] But not Scrivener, who particularly recalls a discussion he had with de Butts over a few drinks at the Ocean Reef Club in Key Largo, Florida, in May 1972. The two men were talking about the evolution of the Bell system and, gradually, their conversation became more heated. "We got to arguing — nose-to-nose — and shouting at each other. We hadn't realized we were shouting and that the whole room had come to a standstill. All of a sudden, we looked at one another and I said, 'John, let's charge them for admission.' God, we laughed."[265] After Scrivener's retirement, the two men kept up their friendship through their joint membership on the powerful board of U.S. Steel Corporation until de Butts's death in 1986.

Scrivener had become aware of AT&T and Western Electric's "mammoth inflexibility" many years before he launched Northern Electric into the United States. Early in his career, he had read the famous Hawthorne studies done by Western Electric after the Great Depression. They are considered the most important analysis of manufacturing operations, worker motivations and ergonomic design ever written. Although Western was "smarter than hell," Scrivener says the studies made it clear that the company's cumbersome bureaucratic structures and shop-floor organization precluded quick decisions, particularly if new ideas were involved.[266] In fact, there was a general perception within the Bell system, prior to competition, which held that Bell Labs and AT&T introduced new products into the marketplace only when they were good and ready.[267] Such was the luxury of monopoly.

Although Scrivener modeled BNR's corporate structure after the U.S. Bell Labs, there were substantial differences between the two organizations in terms of R&D financing, product pricing and, indeed, in the very nature and degree of risk assumed by each. As Professor Michiel Leenders stated in his business case study, Western Electric and Bell Labs developed products for a single customer, AT&T, which represented 80 per cent of the U.S. telecommunications equipment market. It was not only a large market, but, for Western, was predictable, noncompetitive and virtually risk free. Western could price its products and recover its R&D expenses with almost guaranteed certainty. Northern, on the other hand, in seeking to capture only a portion of the 20 per cent of the U.S. market that was estimated to be held by suppliers to non-Bell companies, faced opposite conditions. The market was intensely competitive, much smaller, uncertain and fraught with risks. Northern's costs were governed by unpredictable market forces, which meant there could be no guarantee that a product would recover any of its R&D investment.[268]

The spirit of competition began to permeate BNR's Ottawa labs. During a visit to the complex by Soviet premier Aleksei Kosygin on October 18, 1971, the Russian leader, himself a mechanical engineer, pointed to a small electronic circuit and asked Chisholm what it was for. The scientist replied it was for a high-speed digital transmission system currently under development. Kosygin broke into a broad smile when Chisholm added that it was "our prize — we hope to beat the Americans on that one."[269]

What Kosygin saw was a preliminary version of the much-coveted PCM chip — a micro-miniature Codec device to convert analog signals to digital pulses contained in a single integrated circuit. BNR would spend the next couple of years working out the chip's software, which would allow the Codec to be used in the next generation of advanced digital switches. Each telephone line hooked up to such a switch would be connected to a printed circuit board, called a line card, containing a single PCM Codec chip that would convert both incoming and outgoing analog signals to digital pulses.[270]

Because the PCM chip functioned as a microprocessor, its software relied on the next quantum leap in semiconductor technology — the development of a programmable memory chip by Intel in the summer of 1971. Dubbed the Intel 4004, it was the first commercially available microprocessor chip. Called a four-bit microprocessor because it was able to process four steps at once, the chip allowed software programmers to put their own programs on its circuitry. Its inventor, Marcian "Ted" Hoff, had succeeded in putting all the logic circuitry of a computer's central processor unit (CPU) on a chip.[271] For $200 (U.S.) apiece, the chip allowed intelligence to be built into almost any device, which resulted in a proliferation of "smart" devices.

Although highly successful at incorporating the new integrated circuit advances into Northern Electric's own products, Microsystems International Limited (MIL) ran into trouble just as it attempted to gain a footing in international markets. Market growth was flat in 1969 and a steady decline occurred throughout 1970 and 1971. On the heels of Intel's memory chip breakthrough, overproduction of memory chips drove prices and revenue down and led to an industry shake-out. The bust-cycle shook investor confidence in MIL and was badly timed. It occurred as the chip maker launched its initial public share offering, which raised $18.8 million (Canadian) from investors in a Bell Canada rights offering. Bell raised a total of $93.4 million by issuing two million units that contained one $47 Bell preferred share and one $10 common share of MIL. Northern held an additional 2.6 million shares of MIL, or 56.5 per cent of the company after the public share issue.[272]

By the time of MIL's second annual shareholders' meeting on April 29, 1971, the company's stock was trading at $5.25 a share on the Toronto Stock Exchange — almost half of the listing price a year earlier.

The company was losing a dollar for each dollar of revenue that it brought in. Wolff reported MIL had a $1.8-million loss for the first quarter ending March 31, 1971, which brought its total losses for the first twenty-four months of operation to almost $13.8 million, an amount equal to cumulative sales for the same two-year period of $13.8 million.[273]

On top of those financial losses due to the price war and high start-up costs, MIL reported an extraordinary loss of $1,586,000 in its 1970 year-end results. That amount, which represented about 10 per cent of the cash that the company had set aside for future spending, was sunk when the Penn Central Transportation Company filed for bankruptcy on June 20, 1970. MIL had invested the amount in short-term commercial paper notes issued by Penn Central, which had a prime credit rating when they were issued. Although MIL sued the trustees and Penn Central for recovery, the company was forced to write off the investment. Yet *The Globe and Mail* reported that none of the shareholders present at the annual meeting had any questions for Wolff about the loss.[274]

Some members of the financial community and the business press, however, did ask questions and publicly expressed doubts about the company's ability to weather the storm surrounding the international semiconductor business. Those critics questioned whether it was a business that Canada should be supporting with its tax money. But Marquez, who was also chairman of MIL, stubbornly supported the venture as essential for Canadian high technology. He argued in a 1971 interview that "there are always risks and there are always people to say, particularly with hindsight, that you are wrong. But this is the only way to get out from following behind everybody else."[275]

Just how far behind Northern was in exploring digital switching technology, in comparison to its rivals, was outlined in a BNR review of international activities released to select officials in all three companies in June 1971. The review concluded that digital systems would be designed and installed in most advanced industrialized countries by the middle to late Seventies. It also included the dates that various countries had begun initial development work in the area. Canada was tied for seventh, and last, place with Australia for having started their work only a year earlier in 1970. The United States had begun in 1959, followed by Sweden in 1962, France in 1963, Japan in 1964, Great Britain in 1965 and Switzerland in 1968.[276]

Northern was eager to catch up. By late 1971, the digital switching project had taken on a life of its own and had its own formal organization within BNR. Although various tri-corporate groups that had sprung up for the SP-1 and SG-1 products were also involved, the digital switching project was driven by the central engineering departments in all three corporations. Northern's participation was managed by Wally Benger, vice-president of technology, and Ewart Bridges, vice-president of corporate planning. The latter was a Lobb protégé, who had come to Northern from ITT three years before Scrivener hired Lobb. "Like Lobb, Bridges struck fear into the heart of any engineer," says one former BNR engineer.

BNR completed a market forecast that summer for digital switching based on estimates of Bell's needs between 1978 and 1990. In addition, it began conceptual design work for a digital switch, circulated a draft outline of the requirements for such a switch and sought early approval to begin development work. A laboratory test bed was built by BNR in December 1971, where digital switching technologies were demonstrated and cost estimates obtained. The goal was to provide sufficient data to justify a decision in 1972 on the degree of digital techniques to be used in the next generation of switches.[277]

Northern Electric announced its arrival in the United States to the independent telephone companies with its SP-1 switch in early 1972. With that switch, Northern Electric competed head-on with some of the industry's biggest players, including GTE Automatic Electric, which introduced the No. 1 EAX the same year; NEC of Japan, which brought out the D-10 switch in 1973; and ITT, which unveiled its Metaconta L, also in 1972.[278] Northern's switch was considered more advanced than those of its competitors, and GTE's procurement arm, which had been under pressure to look for alternate suppliers as a secondary source for its independent telephone companies, chose the SP-1 for an evaluation trial conducted in 1972 with General Telephone of Everett, Washington, and Québec-Téléphone. Although GTE Service Corporation never recommended the SP-1 as an alternate switch, both Québec-Téléphone and B.C. Tel bought some SP-1 switches.[279]

Northern Electric developed another feature for the switch later in 1972 as part of a bid for a contract to sell the SP-1 to the Anchorage Telephone Utility (ATU), a subsidiary of RCA Alaska Communications. ATU had adopted a cordless operator system to automatically route calls requiring operator assistance from a switch to an operator's console. They told Northern Electric they would not buy the SP-1 unless it could be used with their system. Although the SP-1 required cord connections, a BNR team proposed a solution that not only made the SP-1 cordless, but that also eliminated the need for the separate switchboard to route the calls to the operator's console. A single SP-1 could thus perform both long-distance call routing and operator call distribution. ATU accepted the proposal, leading Northern to develop a new computerized system, which it called TOPS (Traffic Operator Position System).[280]

In its bid to expand into the U.S., Northern planned to take advantage of a 1968 U.S. regulatory decision, known as the *Carterfone* case. That decision had opened the business telecommunications market to competition by declaring that telephone company subscribers in the United States could own their own telephones. *Carterfone* prompted Northern to develop the SG-1 for the U.S. business market, and BNR engineers began technical trials of the electronic PBX at Northern's plant in Lucerne, Quebec, in May 1972. Northern's timing was auspicious because the PBX switch market was developing into one of the fastest-growing segments in the telecommunications equipment business.

Carterfone was not the only challenge to AT&T's monopoly. The Bell system was beleaguered with new competitive pressures on other fronts when de Butts convened the semiannual presidents' conference in Key Largo, Florida, on May 8, 1972 — his first as chairman of the board. The year before, the Federal Communications Commission had issued the *Specialized Common Carrier* decision, which had ordered AT&T to make local and intercity links available to independent carriers who wanted to provide specialty services to business phone users. That decision followed the *MCI* decision in 1969 granting an application made by Microwave Communications Incorporated of Washington to operate a common carrier microwave system between Chicago and St. Louis. It opened the way for direct competition in AT&T's lucrative market for private business lines — long-distance lines dedicated solely to the use of a single business customer — and extended the terms of a 1956 ruling that allowed companies to build microwave systems for private use or to serve business customers anywhere that AT&T did not provide business service. To add to the sense of gloom that permeated the meeting rooms in Key Largo, AT&T's earnings had stopped growing and its reputation as a company devoted to quality and impeccable service had been deeply scarred by a service crisis in New York City the previous year.[281]

The Key Largo conference was recognized as a turning point for the Bell system and became known for de Butts's proclamation that "we are going to decide."[282] By tradition, no decisions were supposed to be made at the conference, which brought together all the heads of the Bell operating companies with top AT&T management to consider joint issues. De Butts also tossed aside the convention that decreed that the conference would hear formal presentations only and threw the meetings open to discussion from the floor. In his own speech, de Butts rejected the view of his predecessor that competition was welcome and argued that regulated monopoly was far preferable. He admonished Bell managers to restore their earnings and to bolster their flagging commitment to providing quality service, which was a core value of the Bell system, as the best means to compete. Although he unwillingly embraced the need to make the Bell system market-oriented, he clung to the virtues of the public utility, which, because of its reliance on regulated competition, made it difficult for Bell system managers to compete effectively.[283]

The Key Largo conference gave Scrivener a strong feeling for the competitive forces that subsequently were brought to bear on the Canadian telephone system. Unlike de Butts, however, he had the luxury of not having to contend with those forces in Bell Canada's own market — yet. That task would soon fall to de Grandpré. In the meantime, Scrivener could unleash Northern Electric to compete in the United States and apply the lessons it learned there to Bell Canada's own business later.

Northern Electric's new market strategy meant that some regulatory decisions that might otherwise have been unopposed by Bell in previous years might now have far greater implications for the future of both companies. With so much at stake as they planned their joint evolution to digital communications, de Grandpré's task to promote and protect Bell's regulatory and policy interests assumed greater importance. This was underscored by his appointment to Bell's executive committee shortly after the Key Largo conference in July 1972.

The executive committee is the inner council responsible for forming Bell's corporate strategy. De Grandpré says the appointment gave him another indication "that there was something in the cards"[284] in his bid for the presidency. In addition to regular meetings of the executive committee, he was invited to sit in on Scrivener's private meetings with Vincent in the chairman's palatial office. "We would talk about all sorts of things involving the company from the strategic and tactical standpoint of what we should be doing, and they were taking me into their confidence at those meetings,"[285] de Grandpré says.

But Scrivener has a slightly different view of those meetings and says they were intended to keep the Bell chairman up to date on corporate affairs. The only time available was at the end of the day, so at 5:00 p.m. they would adjourn to the chairman's office and "sit around and smoke Marcel's cigars and have a drink,"[286] Scrivener says. As with de Grandpré's appointment to the operations job, his participation in those meetings was aimed at giving the president-to-be more knowledge about the telephone company.

Northern Electric's assault on the U.S. market was launched at an auspicious time. Sales of its first two electronic products, the SP-1 and SG-1, gave the manufacturer the additional revenue that it needed to embark on the more intensive digital research projects the company had set in motion in the months following the Key Largo conference. After four years of exploratory analysis, BNR finally issued a systems engineering prospectus in early 1973, which recommended that the companies proceed with a major research and development project to make a family of digital switches. The prospectus outlined the technical design of the system, which Northern Electric called DMS (Digital Multiplexed Systems), and provided cost information. It also stated that Bell Canada would provide information on network operations and planning.[287] At the same time, BNR pushed for approval of a project to develop a fully digital PBX, which it called the Business Communications System — later known by the code name, SL-1 — late in 1972. The SL-1 served as the development platform for the DMS switch.

The SL-1 was made possible by BNR's Codec chip research and "meant it was possible to move the all-digital technology . . . into other products with confidence," Leenders stated in his study. He concluded it was highly unlikely that any of those digital products would have been developed if Northern Electric had had to depend solely on Canadian sales. It would have taken ten years just to break even, by which time the products would have been obsolete, far surpassed by those of the competition.[288]

Few knew that, at the same time as Northern was marketing the SP-1, it had secretly decided to supplant it. Research into digital switching, even in the early conceptual stages, signaled a completely new direction in technology. "We made a conscious decision to slow down orders on the SP-1 and to displace it, even though that caused a hiccup in revenues," Benger said. "We were betting the company, but it was a good decision."[289] Northern Electric's unrivaled rise to assume the leading role in Canada's high-technology industry is the ultimate tribute to Scrivener's gamble to achieve technological self-sufficiency.

De Grandpré realized his quest "to be someone" on November 22, 1972, the day Bell Canada's board of directors met at the Bell–Northern Research Laboratories complex in Ottawa and approved his appointment as company president.[290] It was effective January 1, 1973 — seven years to the day he joined Canada's largest telephone company. The appointment, along with Scrivener's promotion to chairman and Vincent's retirement after forty-five years at Bell Canada, was made public on December 3.

Although de Grandpré was the first non-career Bell executive to assume the presidency, his quick rise to the top came as little surprise to those who knew him. "There are people who will rise to the top of any organization because of their intelligence, because of their drive and because of their hard-working characteristics,"[291] Francis Fox says. The new president, on the other hand, believed that his legal background had much to do with the appointment. "It's not by accident that the chairman of Canadian National, the chairman and president of Canadian Pacific, the chairman of Air Canada and the chairman of Royal Trust are all lawyers," he told Don Murray of the Montreal *Gazette* in one of his first interviews as Bell president.[292]

De Grandpré inherited Scrivener's traditional wood-paneled office on the nineteenth floor. The office had been unchanged since the Beaver Hall building was built and still had its original marble fireplace. The first change de Grandpré made as president was to redecorate it. Before his own furniture even arrived, he had a bank of fluorescent lights installed because he found the office too dark. Although he kept the large globe on the wooden stand that had been acquired by the company's first president, Charles Sise, Sr., he replaced most of the other furnishings with ones reflecting his own elegant, yet spartan tastes. An avid art collector, he hung bright, contemporary paintings by Quebec artists in the spaces between the windows, which were framed by heavy velvet drapes. Near one window in the corner of the office, he placed a large, elliptical marble table, without drawers, that replaced the solid wood desk used by his predecessors. The setting gave the impression of hurried efficiency; his table top was cleared as one visitor withdrew, only to be cluttered again following the arrival of the next one.

An early riser, de Grandpré was picked up by his chauffeur at his Outremont home at about 7:30 a.m. and was usually in his office by 7:45 to read the day's papers and to make phone calls. Robert Bandeen, the former president of Canadian National Railways, was also a morning person who liked to be at his desk early because he "treasured that quiet time." He recalls being surprised by a phone call one morning just moments after he had arrived at his office, only to hear de Grandpré's deep voice at the other end of the line.[293]

Tall and imposing, the balding and bespectacled president had a formal style and ran a rigid office. Unlike his early days as a lawyer, he did not keep an open-door policy; meetings were mostly by appointment only. And he never took off his suit jacket. A retired Bell Canada vice-president recalls being in de Grandpré's office with a reporter and a photographer from the *Toronto Star* during an interview session. The photographer asked his subject if he wanted to take his jacket off or roll up his sleeves. De Grandpré huffed, "I *never* take my jacket off."[294]

Elevation to the presidency brought with it numerous board appointments and responsibilities as senior spokesman of the Canadian corporate establishment. What excited de Grandpré even more than "being someone" was being at the heart of the action and "making decisions." His days were an endless stream of meetings with executive officers, almost a dozen corporate boards, government officials and politicians and countless other committees. One of Bell Canada's corporate jets was kept constantly aloft by his dizzying weekly schedule of travel around the country and abroad. An eager member of the business jet set, de Grandpré was efficiently whisked about in a French-made Dassault-Breguet Falcon jet. It was not uncommon for his pilot to be asked in a single day to fly his boss to a morning meeting in Toronto, an afternoon appointment in Ottawa and a dinner in New York, where he would spend the night before a meeting the next morning.

Although private jets were a popular corporate status symbol in the early Seventies, de Grandpré's Falcons became controversial. They belonged to a fleet of private jets owned by Tele-Direct, Bell Canada's publishing arm. Although originally incorporated to print telephone books, Tele-Direct also operated numerous leasing companies. Set up to take advantage of income tax provisions that allow leasing costs to be deducted from corporate taxes, these companies owned a variety of assets, such as the Falcon jets, which were leased back to Bell Canada. Tele-Direct was placed at arm's length from its regulated parent company in 1971, when it was sold to Capital Telephone, a small telephone company acquired by Bell Canada in 1969. Consumer groups and Opposition politicians used the arrangement as ammunition in their attempts to discredit Bell Canada's requests for rate increases.[295] Fully equipped, each plane had a range of 3,943 kilometers and carried a $2.25-million price tag. Northern Electric also leased back its own fleet of top-of-the-line, $5-million Grumman Gulfstream IIs,[296] one of which regularly shuttled Lobb between his home in Pennsylvania and the company's offices in Montreal and Boston.

Increasingly, de Grandpré traveled to help sell Bell Canada's, and then Northern Electric's, corporate financings on international money markets. To ease the disruption of frequent travel, he kept his own apartments in the financial capitals of Toronto and New York, which allowed him to be "at home" most of the time he was away from Montreal.

Insiders, certainly at the board level, viewed de Grandpré as "the right peer at the right time," to use Louise Brais Vaillancourt's phrase.[297] He assumed command at a time when the very nature of leadership was undergoing a profound transformation. According to Dr. H. Rocke Robertson, who served as a Bell Canada director for twenty-one years before his retirement from the board in 1985, that change characterized a crucial distinction between the Scrivener and the de Grandpré years: "In years gone by, it was easy to be a leader: you were told to do something and you did it. But it's not so simple now. Now, most people have input to decisions. A leader can't get his way by just saying it. You have to persuade people and go through committees and management. Leadership now comes by swaying a committee or a group to a line of thought. In older days, people were taught to obey. Now, that tradition of respect for authority and obedience has been set aside. But it becomes more difficult to lead people when leaders are subject to scrutiny."[298]

Leadership, in effect, required more political skills. However, rather than involving the art of compromise, Robertson says the proliferation of committees required corporate leaders to become adroit at the art of gamesmanship in order to attain their objectives.[299]

De Grandpré once joked that his management style was both "hands off, and hands on."[300] He perceived the Bell Canada of the pre-Scrivener era as less flexible because it was entirely service-driven and engineer-dominated. The adoption of new technology, coupled with the gradual introduction of competition, made the company more market-driven. "That meant the company had to be more flexible in order to respond."[301] As president, one way de Grandpré sought to attain that objective was to decentralize decision-making and to make every manager "a stakeholder" in corporate decisions. "I intervened only if the results were not what I expected or were undesirable." But, at the same time, he emphasized, "people should look to a leader to lead, not to follow the polls."[302]

Although de Grandpré knew what visions and opportunities he would pursue as head of the company, he did not bring any "grandiose schemes or road maps to the future" to his new job. Rather, he described his outlook as pragmatic: "I deal with things as they develop. I can't sincerely tell you I knew what I wanted to do." But in the same interview, he declared, "This is a dull business. There's no glamour in it. If you want to provide good service, you have to tell people they have to pay for it and that's not popular with the politicians."[303]

Revealed in that assessment was a sentiment that some career telephone company executives and managers below the officer level, particularly those who were threatened by him, interpreted as an arrogant outsider's disdain for the telephone business. His apparent lack of interest in the technical side of Bell Canada's business reinforced this view of the man. He did not fully tour Bell's own facilities until 1970, when his appointment as head of the eastern region demanded it.[304] Yet de Grandpré himself has admitted that it was not that he disliked this side of the business, but that he did not understand it.

The sniping wasn't widespread. In fact, few managers viewed his lack of interest in the technical or operational details of the utility's business as a negative. Many managers, particularly engineers, gave de Grandpré their loyalty because he deferred to them on those issues. Walter Light recalls that Scrivener handed him the job of helping to raise the lawyer's knowledge of the engineering and network side of Bell Canada's business when de Grandpré was promoted head of the eastern region. "I spent a lot of time with Jean on operations and I took him with me to AT&T's operations' meetings. He was a very fast learner and he had no trouble. I'd say he came to know the telephone company very well and the direction it had to go."[305] As president, de Grandpré delegated most of the operational aspects of running the telephone company to career officers so that he could preside over the financial, legal and regulatory strategies.[306]

Bell's hierarchy certainly provided little opportunity or time for him to deal directly with executives below the officer level. One former telephone company president says the Bell Canada boss showed his contempt for employees below that level by referring to them as "hirelings," a contraction of "hired workers" and "underlings." De Grandpré, however, vigorously defended his record with employees, arguing that he "tried to treat Bell's employees and its shareholders well." He then gave himself a plug for Bell's equal opportunity employment programs to promote more female employees. "I said, to do that, we may have to discriminate against men. But I am of the view that there is no one sex with an advantage over the other. In order to give women an opportunity to advance, we have to give recognition to women." He added that he "didn't think this was a sexist remark" when he was assailed for his views shortly after his appointment.[307] In reality, what landed him in trouble was his statement in a 1973 interview that women should not expect access to executive posts within Bell overnight and that it would take at least twenty years before progress would be made. He also claimed that women were partly responsible for the problem because so few had wanted to work in corporations or had engineering or business administration degrees.[308]

According to Robertson, who chaired the social and environmental affairs committee of Bell Canada's board, the utility moved toward a rudimentary pay-equity program in 1969 after the company's employment practices were reviewed by a federal government team. "Their report said Bell's upper management rank was devoid of women."[309] After de Grandpré's first year as president in 1973, only one position — an assistant vice-presidency — out of the eighty-two available in the top level-five management rank was held by a woman. Only 2,231 female managers occupied positions out of the more than 7,050 available in the bottom level-one supervisory tier. Ten years later, Bell Canada's data showed there was still much room for improvement: five women held level-five positions out of the seventy-two available, or just under 7 per cent and, in the same period, the percentage of female managers at the level-one tier rose slightly to 36 per cent, or almost 3,500 of the 9,695 positions filled.[310]

As a boss, de Grandpré was exacting and took a keen interest in any matter that he felt impinged on corporate interests. One former Bell Canada vice-president recalls being summoned into de Grandpré's office after the publication of remarks he had made to a reporter regarding an acquisition of a competing equipment maker by an offshore company. The associated newspaper story prompted an exchange in Parliament's daily Question Period, which resulted in a promise by the federal industry minister to look further into Bell Canada's position. De Grandpré launched into a blistering attack on his officer, not because of what the officer had said, but because he had been quoted.[311]

Once his attention was warranted, de Grandpré's quest for perfection and penchant for quick decisions became obvious to all involved. "If I have one quality, it is decisiveness," de Grandpré told an interviewer in 1982.[312] His favorite adage was that "the worst decision is the one that's never taken." "He made decisions a lot faster than many people I'm accustomed to. He doesn't like to fool around with a problem," says James Thackray, former executive vice-president of the western region who later served as president of Bell Canada under de Grandpré. "You felt very free to do your thing but always with the fear that you had to do your best. If it didn't work out or you struck out, you'd better have some pretty damn good reasons."[313] De Grandpré, like many tough executives, makes no apologies for how he performed his job which was, as he frequently remarked, "to make stress for others."[314]

De Grandpré took over a company which had just ended 1972 with revenue of $1.13 billion and profit after taxes of $149 million. He figured that Bell Canada's proposed 3 per cent rate increase was reasonable, particularly given his argument that the largest government-run company in the province, Hydro-Québec, earned only slightly less profit on about half the revenue that Bell made. Although the fight for increased profit and regulatory laissez faire was business as usual, another element was added to Bell Canada's regulatory struggle that year. The first serious challenge to its monopolies came with a fight between Bell Canada and several cable television companies.

News of de Grandpré's appointment competed with headlines about a vicious struggle between the utility and Transvision Magog, a small cable television company from Magog, Quebec, a town east of Montreal. The dispute had actually been in the works since 1958, when cable companies first began to push for the right to own their own equipment. The federal rules prohibiting anyone else from owning telephone poles forced cable companies to use Bell Canada's poles. In turn, the utility's rules prevented cable companies from attaching their own devices to the poles. Once a cable company in Bell Canada's territory signed what was called a "partial system agreement," the cable operator was required to consent to a host of restrictions that limited the use of the wires strung up on Bell's poles. For instance, the cable company could not distribute nonprogram signals, such as alarm services, or use any device capable of switching a signal to or from a subscriber. Although Bell did not provide those services, the telephone company argued that they were part of its monopoly.[315]

These arrangements gave Bell Canada tremendous power over the cable television companies. Toronto consultant Eamon Hoey, a former Bell Canada employee who worked on Scrivener's "cable desk" in Ottawa, described how the company carved up the service jurisdiction of the rival cable operators in Toronto: "We called all the Toronto cable operators to a meeting. My boss . . . put up a map of the city and told them to work it out. They couldn't, so we drew the boundaries."[316] By 1970, Bell Canada had been forced to amend its agreement with the cable operators. Although the new agreement eased the restrictions on what the cable companies could do with Bell's wires, and at what frequencies, the utility continued to be pressured to allow the cable companies to attach their own devices to Bell Canada's poles. In 1972, the cable companies forced the issue and applied to the CTC for a ruling that would compel Bell Canada to provide "pole attachment."[317]

Bell Canada's position on the pole issue was rooted in Vail's maxim to take "possession of the field" in order to occupy and defend one's territory, and in an underlying principle that Bell Canada would invoke in the many other competitive battles in the years ahead — "The public interest is best served when one carrier provides, for shared use, a single, integrated network capable of transmitting all telecommunications."[318] Echoing that philosophy, de Grandpré consistently defended Bell Canada's monopoly by arguing that an integrated network resulted in economic efficiency. It also produced peace in the marketplace. As Vail boldly declared in 1910, "Competition — effective, aggressive competition — means strife, industrial warfare; it means contention; it oftentimes means taking advantage of or resorting to any means that the conscience of the contestants or the degree of the enforcement of the laws will permit."[319] As Scrivener's protégé, de Grandpré not only shared that sentiment, but learned a few tactics from a master at defending Vail's monopoly faith. Scrivener demonstrated one of these tactics when Bell declared war on Transvision Magog less than a week before Christmas in 1972.

After the cable company refused to renew its contract to rent Bell's poles because of the cable companies' application to the CTC, the two sides entered talks aimed at reaching an eleventh-hour settlement. When those discussions broke down on the evening of December 19, 1972, Scrivener placed a call to Hoey on the cable desk in Ottawa. "I had a lineman on the pole in Magog hooked up to a special telephone line connected to Ottawa. When we were informed the negotiations had broken down, I had to give him the order to physically cut their cable."[320]

Nevertheless, Hoey maintains Scrivener was not a ruthless monopolist. In the early 1970s, when John Bassett [president and CEO of CFTO-TV Limited] and the Eaton family sold their interests in Rogers Communications, Ted Rogers was in a financial squeeze. Bell Canada didn't press him to pay his bills, Hoey said. It would have killed him. But, you see, Scrivener was a gentleman. In any event, it's almost humorous to hear Rogers now portraying Bell to be such a villain."[321]

Scrivener's order cut off television service to 28 of Transvision Magog's 210 subscribers. But of greater significance was the intervention of the cable television industry's national lobby association, which promptly applied to the CTC for an injunction against Bell Canada. Two days later, the CTC ordered the telephone company to restore service until both sides in the dispute submitted their arguments to the commission on the pole attachment issue.

The Canadian Cable Television Association's request for an injunction was filed by Gordon Henderson, the head of the Ottawa law firm of Gowling & Henderson. Although he also happened to be president of Ottawa Cablevision, Henderson was neither an industry hack nor a legal gadfly who simply dabbled in business on the side. Among his many accomplishments, which make his entry one of the longest in *Canadian Who's Who*, Henderson is considered Canada's pre-eminent patent and trademark lawyer. His life-long interest in intellectual property law also led him to specialize in competition law, which is the Canadian equivalent of U.S. anti-trust law. It was Henderson's expertise in both fields that led him to be retained by the DIR to represent the federal combines branch in the government's ongoing investigation into the Bell–Northern relationship.

That made Henderson about as close to being de Grandpré's nemesis as the devil is to God. Coincidentally, de Grandpré had met and worked with Henderson when the two were young lawyers in the late Forties. Brais Campbell, the Montreal law firm that employed de Grandpré, and the firm of Gowling Henderson sometimes helped each other out in insurance and liability matters. "We worked together on the purchase of a used airplane for B.O.A.C. [now British Airways]," Henderson recalled in a 1990 interview. "A year or so after the sale was completed, Jean told me the plane had crashed in the Mediterranean Sea. Then he said, 'Are you worried?' "I said, 'No, why?' He reminded me that we had certified the warranty for the aircraft."[322] Henderson added that de Grandpré told him the news at a Schenley Awards dinner. The two men served together as trustees of the Canadian Football League's annual awards gala, which meant "we were supposed to be drinking buddies, only I don't drink,"[323] Henderson says.

The two men were far from being legal buddies, either, later in their careers. Henderson was involved in nearly every competitive battle against Bell Canada during the 1970s, including the cable fight, the bid to break up Northern Electric and the battle over Bell Canada's bid for control over Telesat Canada. In addition, along with several partners at his law firm, he backed the two entrepreneurs who in 1973 launched a new high-technology company, Mitel Corporation of Kanata, Ontario. The company's spectacular growth made it the darling of Silicon Valley North during the high-tech boom of the Seventies as it gave Northern Electric and other international telecommunications equipment makers a run for their money. Despite their business rivalry, Henderson says he has the utmost respect for de Grandpré, whom he characterized as an "effective opponent" and a "builder and a leader in this country." He also reached the conclusion that Bell Canada reflected its president's character: "the company was efficient, far-seeing, but hard-nosed."[324]

During de Grandpré's first month on the job as president, the federal Department of Consumer and Corporate Affairs launched a formal inquiry into the Bell Canada–Northern Electric arrangement. Robert Bertrand, who took over as DIR from D. H. W. Henry in 1970, concluded that the evidence collected from the 1966 raid on Bell's headquarters during the Combines Investigation Act inquiry did not reveal that the companies had violated any provision of the Act. But he said his department's investigation had disclosed monopolistic practices that warranted an inquiry by the Restrictive Trade Practices Commission (RTPC). Bertrand filed a notice with the RTPC that moved the matter to that agency on January 23, 1973.[325] "Although the Bell–Northern inquiry began as an enforcement inquiry, he converted it to a research inquiry,"[326] Lawson Hunter, then a combines official and later Bertrand's successor as DIR, says. A copy of the DIR's notice to the RTPC was sent to both Bell Canada and Northern Electric, and the DIR then spent the next three years preparing his evidence for the commission.

Two months later, former federal finance minister, Edgar Benson, released the CTC's decision on Bell Canada's November rate request. Much to de Grandpré's relief, Benson had been appointed CTC president in August 1972 after Pickersgill resigned from public life. His pleasure at the appointment was no doubt compounded by the CTC decision to grant virtually all of the increases sought by Bell Canada.[327]

However, the euphoria was short-lived. A storm of controversy soon arose in Parliament over the CTC's judgment as Opposition politicians assailed the government's policies to curb inflation and attacked Bell Canada's relations with its regulator as too cozy.[328] Because Trudeau's Liberals had a minority government at the time, the Speaker of the House granted an extraordinary request by NDP leader David Lewis to adjourn the House for an emergency debate on the rate increase.[329] For three hours on the evening of April 2, Parliament debated Bell Canada's rate regulation. The Opposition's continued outcry over the next few days prompted federal Communications Minister Gérard Pelletier to announce a cabinet decision to defer the effective date of the CTC's decision, pending a review and clarification of the decision by the commission.[330]

The Opposition also drilled Benson on the increase when he was summoned to testify at the Public Accounts committee during its annual review of the CTC's budget. Although badly battered during his tenure as finance minister, Benson was an unflappable character. In a move that he described as "purely political,"[331] he showed up at the hearing room, pipe in mouth, pushing a wheelbarrow full of documents and transcripts.

Benson equated regulating Bell Canada with being the government's tax man. Just as he found it hard to educate the public on taxes, he says, "Many people don't realize that corporations consist of shareholders, and they need to be able to make enough money for them or there won't be enough shareholders to finance the business." He added that because Bell Canada is a utility, he believed it had to have a sufficiently high rate of return to be able to raise the large amounts of capital needed to finance the expansion and modernization of its network.[332] Pelletier lifted the suspension on June 27, 1973, and announced the cabinet had modified the CTC's decision and reduced the rate increase to $22.5 million.

Unmoved by public opinion, de Grandpré filed yet another rate application in August that asked the CTC for an additional $51.8 million in revenue for 1974. A wider range of interests was represented at that hearing than at any other previous Bell Canada rate hearing. Among the groups appearing for the first time were: the cable lobby and Transvision Magog; the National Anti-Poverty Organization (NAPO); the Consumers' Association of Canada (CAC); Action Bell Canada; the

Civil Liberties Association of Canada; the Centre for Public Interest
Law; and the Inuit Tapirisat of Canada. Individual citizens also
entered the fray. Carlyle Gilmour, a farmer from Covey Hill, Quebec,
launched a one-man vendetta against Bell Canada after he refused to
let the phone company build a microwave tower on one of his farms.
Gilmour was infuriated when the tower went up on neighboring land
across the road and has claimed ever since that the microwave radia-
tion from the tower neutered his roosters. He has showed up at every
Bell Canada proceeding since 1969 and, as an investor in Bell Canada's
corporate parent, continues to question management at least once a
year at the company's annual shareholders' meeting.

"The arrival of consumer groups represented a vote of no confidence in the tribunal's ability to 'protect' the public,"[333] T. Gregory
Kane, an Ottawa lawyer and former associate general counsel of the
Canadian Radio-television and Telecommunications Commission,
wrote in his 1980 study on consumers and regulators. Kane, who
had also served as a lawyer for the Consumers' Association from
1975 to 1978, when it began its regulated industries program, added
that it was common for consumer groups to be treated with indif-
ference or hostility at many regulatory boards. Benson still dismisses the Consumers' Association's involvement in the regulatory
process, saying, "They wasted a lot of money always worrying
about regulated industries."[334] De Grandpré was even more vocal. At
a public hearing of the Royal Commission on Corporate
Concentration in April 1976, he complained that Bell Canada's
lawyers had been outnumbered by lawyers acting for various public
interest groups by ten to one at the company's last hearing. He criticized the public financing of public interest groups, which allowed
them to retain outside lawyers, saying, "This does not add to the
protection of the consumers but breeds a new type of professional
intervenor living at the expense of the consumers."[335]

After forty-seven days of hearings, the CTC granted 90 per cent
of the increase in its August 15 decision. The commission then outlined a formula to implement de Grandpré's proposal for automatic
rate increases pegged to the rate of inflation and, despite the federal
cabinet's opposition to the idea, called for public comments on the
proposition. De Grandpré told a Conference Board of Canada
meeting that such a scheme would strengthen investor confidence in
Bell Canada and was needed because inflation rendered rate increases inadequate even before they took effect.[336] The CTC also rejected

an application by the government of Quebec to make Northern Electric subject to the CTC's jurisdiction.[337]

Northern Electric entered what it termed an expansionist phase in 1973, which coincided with the tri-corporate decision with Bell Canada and BNR to proceed with the DMS digital switches. De Grandpré helped place the company's first public share issue of 2.6 million shares in December. This had the effect of diluting Bell Canada's stake in its manufacturing unit to 90 per cent. The objective of that stock issue, de Grandpré later remarked, was to make Northern Electric a "household name."[338] It also gave Northern Electric more money to expand its old Montreal factory, to build new factories in Nova Scotia and Galway, Ireland, and to acquire Northeast Electronics of Concord, New Hampshire.

However, mounting financial losses at MIL, Northern Electric's microelectronics chip-making unit, eroded Northern Electric's financial position. Although MIL had come up with one of the world's first workable microprocessor chip designs in 1973, it was plagued by severe technical problems on its chip production line. Even worse, the downtime on the line caused the company to miss the six-month market boomlet that began that July.[339] MIL's president, Olaf Wolff, took a fall and officially "resigned" on November 19, 1973, while two hundred of MIL's 1,500 employees were fired in the first set of layoffs at the money-losing subsidiary.[340]

There was even worse news shortly after the company announced a second reduction of three hundred employees in September 1974. On October 7, minority shareholders were offered a choice of approving the sale of $9 million of MIL's fixed assets or seeing the company cease operations. The company planned to sell its Ottawa building to BNR and then lease it back. Minority shareholders approved the proposal at a special shareholders' meeting the following month. Again, none of the more than 150 shareholders present asked any questions of Northern Electric's chairman, John Lobb.[341]

With its debt and losses still rising, while its sales and credit were plummeting, MIL announced it was closing its doors on March 4, 1975, putting almost five hundred employees out of work.[342] The company's losses since 1969 totaled more than $45 million. Asked for an explanation of why Northern Electric was abandoning the merchant chip business, Scrivener told shareholders at Bell Canada's annual meeting the following month that:

> We found that the business could not be made viable in
> Canada. . . .
>
> So we decided enough was enough, and Northern Electric,
> in a responsible fashion, said, look, your company is bankrupt,
> and offered to exchange its shares for Microsystems shares.
>
> There is nothing improper, nothing sinister in this. We
> didn't try to lose the money just for kicks.[343]

The federal government was also criticized by shareholders for losing
the $29 million in grants it gave to MIL. Scrivener responded at the
same meeting that the company had paid back the $6.7 million owing
on the federal government loan and that the loss to the taxpayers was
not as great as the shareholders' losses.

The MIL venture serves as an important case study of why the federal government's innovation policy of the period was destined to fail.
One opponent of direct government support for private high-tech
ventures, Queen's University professor Kristian Palda, wrote that the
distinction between the picking of winners and the rescue of lame
ducks becomes "disquietingly blurred" under such schemes.[344] Rather
than encouraging R&D, such allocative measures often prop up troubled ventures. According to British economist John Jewkes:

> High-technology means exceptionally high-risk technology —
> projects to which companies, in close contact with realities,
> will not give their support because the chances of profit seem
> too small, problematical or remote, but where the government,
> for one reason or another, feels that it knows better. Private
> enterprise will not jeopardize the requisite shareholders' capi-
> tal but governments feel justified in risking the taxpayers'
> money.[345]

In the wake of the MIL collapse, Canada was again left entirely dependent on foreign semiconductor makers.

A Senate Special Committee report on science policy called the
period between 1972 and 1975 "the wasted years."[346] It noted that the
promised commercial spin-offs from federal science projects had failed
to materialize to the extent that the government had promised, that the
"make or buy" policy that favored a chosen instrument was a failure,
that the reduction in federal science expenditures was ill-timed and that
the departmental reorganization that created the Department of

Industry, Trade and Commerce had been a disaster. Other critics of the federal government's innovation policy have added that the problems of Canada's manufacturing sector were deeply structural and could not be cured by throwing chunks of money at single projects or by encouraging companies to do things that the marketplace had not already led them to do. Not the least of the problems found in the Canadian economy in the Seventies was a steep decline in most types of production. This indication that Canada was being de-industrialized meant that the country's dependency on foreign-owned enterprises was increasing, according to a 1978 Science Council of Canada study.[347]

MIL did produce eight direct spin-offs before being closed down. The most famous was Mitel Corporation, launched in 1973 by two British expatriate electrical engineers, Michael Cowpland, a former manager of circuit design at MIL, and Terence Matthews, MIL's former marketing manager. In 1986, MIL's high-tech progeny, including Mitel's twenty-three spin-offs, numbered thirty-six companies, which had exotic names like Dy-4, Mosaid, Omziq, Synapse and Teal.[348]

Cowpland lays much of the blame for MIL's failure on management, rather than government. As well as mismanaging some investments, he says they suffered "a loss of nerve" and threw in the towel after the first semiconductor recession: "It was a classic case of not being like the Japanese and spending during a recession." He is convinced that Northern could have parlayed its considerable in-house talent at microelectronics into a successful export business if it had had the will, and he refers to Mitel's still-profitable custom integrated-circuit business as a case in point.[349]

Extricating Northern Electric and Bell from the MIL mess was left to Scrivener, while de Grandpré struggled with the utility's sixth rate application in as many years. The CTC's own proposal for automatic increases fell victim to the storm that surrounded Bell's two-part application, which was filed in May 1975. The first part asked for urgent, temporary relief and an additional $28 million, effective August 1. The second part asked for the addition of another $10 million to revenue by the end of the year.

As Bell Canada's only witness, de Grandpré spent a week in the CTC hearing's witness box for the first phase alone. "Everyone was taking shots,"[350] he recalls. It proved to be the utility's most contentious rate hearing yet and led to a major fight with the government of Quebec. De Grandpré claimed that at no time since 1969 had the company earned an adequate rate of return on its equity.

He said the company could no longer continue to keep deferring modernization programs while maintaining previous borrowing levels with inflation running at between 10 and 12 per cent a year.[351] Bell Canada's profit after taxes was $224.4 million in 1974, compared to $205.4 million a year earlier. The previous rate increase generated only $13 million in additional revenue — far less than expected — and a new federal manufacturing surtax cost the utility almost $8 million in higher taxes.

On July 7, the opening day of a five-day hearing on the request for emergency relief, the lawyers representing the government of Quebec walked out of the proceedings, charging that the company had used improper procedure in asking the CTC for temporary relief. The lawyers acting for Quebec communications minister Jean-Paul L'Allier were also angered that the CTC had allowed little time for intervenors to prepare their case.[352] After they walked out, de Grandpré threatened that the utility would have to lay off as many as six thousand employees and cut its capital spending program by half if its request was not approved.[353] Although the CTC awarded Bell an extra $14 million on July 28, or half of its initial request, that was not good enough for de Grandpré. "In making this decision the commissioners are substituting their judgment for that of management,"[354] he said. Bell Canada issued a news release the following day stating that the rate decision would force the company to reduce its capital spending. Quebec responded the same day with a declaration from the outgoing L'Allier, who had lost his seat in the July 16 provincial election, that Quebec would resume its participation at the hearings and vigorously oppose Bell Canada's request for a further increase.[355]

De Grandpré made good on his promise in early August. The telephone company announced a $33-million reduction to its construction program and announced a hiring and purchasing freeze.[356] The utility even dragged Ontario premier Bill Davis into the fray when, as part of the construction program cuts, it halted work in a new subdivision in his Brampton riding. Gregory Kane, a lawyer for the Consumers' Association at the time, remembers "seeing pictures of empty reels that Bell dumped on the unfinished yards in the suburb."[357] Although the utility's spending program still totaled $803 million, it claimed that it had to halt work in the subdivision because of a shortage of funds. While that ploy gained the support of the Brampton city council, which passed a motion supporting the rate increase and urged the utility to accept an interest-free loan from a developer to finish telephone installation in the development,[358] others angrily accused the telephone company of blackmail. Those accusations prompted another news release from Bell Canada on August 21 and a statement by Gordon Inns, a Bell vice-president, who denied that the company was trying to force a favorable decision by witholding service.[359]

That denial sounded even more hollow after a spokesman for the air traffic control center at Montreal's Dorval International Airport claimed the utility had endangered the lives of the pilots of five aircraft by refusing to transmit landing instructions to them as they neared the LG-2 hydro-electric project at James Bay. The local controller relied on Dorval's air traffic controllers and relayed instructions to guide pilots attempting to land at the site. After the Dorval controllers discovered that their dedicated telephone line connecting them to James Bay was not working, they attempted unsuccessfully to have a Bell operator cut in on a busy long-distance line.[360] The same day, federal communications minister Pierre Juneau stated he would investigate the utility's cuts in service. This followed a similar action by the Ontario government that threatened Bell with possible prosecution under the Business Practices Act.[361]

De Grandpré termed the reaction of various politicians to the rate application as "emotional, and at times virulent."[362] Privately, he argued that L'Allier's public pronouncement was "a very aggressive position for the minister to take on a matter that was pending before the courts. It was almost unethical."[363] In the face of the name-calling and threats, Bell Canada's chief lawyer at the hearings feigned surprise and indignation at the reaction of the Quebec government: "Bell Canada is frankly disappointed and dismayed by the fact that the government of the province of Quebec, in the present case, has maintained the position of opposition to the company."[364] Scrivener admitted to reporters in September that the utility had curtailed its capital spending to attract public attention to its case, a rare admission that spoke volumes of Scrivener's public relations methods.[365]

De Grandpré had to address all of these issues when he climbed into the witness box for the second round of CTC hearings later that fall. They began on October 27 and ended twenty-six days later. He was supported by an army of officials and researchers from the regulatory matters branch, who worked out of an entire floor of the Chateau Laurier Hotel in Ottawa during the hearing. Robert Latham, who worked on that hearing and is now president of Bell Cellular, recalls discussing the day's proceedings with some colleagues in the lounge that Bell maintained for employees on the team. He wanted to know whether the information that two team members had stayed up all night preparing for de Grandpré had been useful. The remark was overheard by a woman who, they assumed, was a Bell employee. But a few moments later, when de Grandpré came into the room, "she lit into him, telling him exactly what she thought about that." The stranger was his wife, Hélène. "That's the kind of person she is — able to tell him what she thinks," Latham says, adding, "He took it in stride."[366]

De Grandpré began his testimony with a blunt and pessimistic analysis of the company's financial affairs in which he defended the utility's decision to reduce service that summer. He further inflamed opponents by telling the three-member panel that, even if the CTC approved the current request, Bell Canada would have to file yet another rate increase application the following year.

The former lawyer was a tough witness who treated cross-examination as a sport. "I found it very stimulating. You're fighting for your life."[367] Kane, who was representing the Consumers' Association at the hearing, says de Grandpré was a particularly formidable witness "because he would never back down or concede anything. And he would always anticipate a question and head off an entire line of questioning."[368] He recalls one memorable exchange between de Grandpré and Marc Cantin, one of the Quebec government's lawyers. In response to a particularly barbed question, Bell's president shot back with, "It's time you were instructed to say publicly the same things your minister tells me privately."[369] De Grandpré says he does not recall that statement, but he does remember Cantin's first long-winded question, which droned on for almost two pages of transcripts. "He asked me, 'Why don't you answer?' I said, 'Because you haven't asked a question.'"[370]

Later in the interview, de Grandpré added, "You must remember, you should never lose your temper, never do that, or you're at a disadvantage."[371] Yet many participants recall that he had a hard time heeding his own advice. Andrew Roman, the lawyer for the National Anti-Poverty Organization, found the Bell Canada president combative and hostile on the stand. "His mentality was that of a monarchist."[372]

De Grandpré did not hide his contempt for the consumer groups, which he viewed as Bell-bashers. Roman recalls being approached by de Grandpré at a coffee break during the hearing. "De Grandpré was a smoker then, too, and we were standing outside and he said, 'Look,

Roman, you're a good lawyer. Why don't you go after a real consumer problem. I just bought a new car and it's a real lemon. Why don't you go after that?'"[373]

According to Roman, many lawyers believed de Grandpré's hostility was directed at them personally. Others, like Raynold Langlois, senior counsel for Quebec, disliked his political interventions: "They fought everything tooth and nail; you couldn't even spit on a telephone wire without their consent."[374] Although de Grandpré denied the lawyer's charge that he went over L'Allier's head with direct appeals to Bourassa, that is how Quebec government officials interpreted his meetings with the Quebec premier, Langlois says. "Politically, we also believed Bell was exerting a lot of pressure at the federal level, as well." Like Roman, Langlois believed de Grandpré was driven by the bottom line. "Every action he took was for corporate self-interest."[375]

But as de Grandpré testified, "If you want to have the service, you have to pay for it."[376] The CTC agreed and, on December 22, 1975, granted Bell Canada all the increases it had asked for in full, which was calculated to give the company an additional $110.3 million in revenue in 1976.[377] NDP critics promptly called for the nationalization of Bell and demanded that cabinet roll back the increase. Although the new federal minister of communications, Jeanne Sauvé, a childhood friend of de Grandpré, said she would study the matter, she stated that the CTC "obviously decided that the increase is justified."[378]

Amidst the controversy of the MIL collapse and the rate hearing, other important and uncontentious decisions were made. AT&T sold its remaining 1.8 per cent equity interest in Bell Canada on January 27, 1975. The sale fully patriated Bell's stock after ninety-five years of foreign ownership. The $33.7-million sale of AT&T's 749,992 shares to Canadian institutional investors was arranged through a syndicate of New York brokerages, which was to have also managed a secondary offering of the Bell Canada shares. However, that issue was withdrawn because of the drop in Bell Canada's share price.[379]

De Grandpré and Scrivener listed Northern Electric's stock on the New York Stock Exchange later that year and sold the first U.S. issue of Northern Electric shares. De Grandpré believed that Northern Electric could expand its sales in the United States if he made both "Bell and Northern household names in international financial markets."[380] The additional shares reduced Bell Canada's holding in the manufacturer to 62 per cent.

More important than the share sale for the engineers in Bell Canada and Northern Electric, however, was the severance of the utility's formal ties with the U.S. Bell system when the AT&T–Bell Canada service agreement was terminated on June 30. That decision realized Scrivener's goal to cut the final cord that had made the Canadian companies dependent on U.S. technology. James Thackray was sent to AT&T's headquarters in New York to sign the termination agreement. He recalls that it was a logical step at the time because "we were beginning to diverge from AT&T on basic policies. We were recognizing that competition was coming down the road, and John DeButts was very strong that monopoly was the only answer. He'd travel around the country making very stirring speeches, and he was not happy that up here we were making somewhat different noises, although by today's standards they were relatively modest."[381]

The decision to terminate the AT&T agreement was one of the biggest decisions ever made by Bell Canada's board of directors,[382] Louise Brais Vaillancourt says. Former director Dr. H. Rocke Robertson links the move to "a vitally important and fundamental decision by Bell Canada to go digital in the Seventies. Bell Canada decided this was the way of the future and it was a very brave decision because no one in the world had done it before."[383] Embarking on that path, de Grandpré says, drove Bell Canada further away from AT&T's views of the telephone network. A final tri-corporate review of the DMS switching project in early 1975 concluded that the effort required a five-year R&D budget of about $100 million. But BNR's scientists also told the Bell Canada executives that it would cost twice that amount to delay the project by five years.[384]

Coincident with its push into digital telecommunications, Northern Electric updated its corporate identity and adopted the moniker of its U.S. subsidiary, changing its name to Northern Telecom on March 1, 1976. Shortly thereafter, the company's executive troika — John Lobb, chairman of Northern Telecom, Donald Chisholm, president of BNR, and Walter Light, who had been promoted to the presidency of Northern in 1974 after Marquez retired as chairman — opted to unveil the SL-1 electronic private branch exchange (PBX) in a special way. The small telephone switch to route calls between a large number of office extension telephones was unveiled at Digital World, the company's vision of telecommunications in the future, which was held at Disney World near Orlando, Florida, in May 1976.[385] Lobb wanted the company to make its pitch to corporate chief executives, so the event's marketers invited the CEOs of all the leading U.S. telephone companies to the launch,[386] a marketing tactic the company would replicate in later years.

Northern Telecom picked Disney World as its venue because the amusement park was a customer and had extensive computer communications networks that the company's sales representatives could highlight in their demonstrations. Lloyd Webster, the company's vice-president of switching, later rhetorically asked whether Digital World was just an advertising slogan?[387] Wally Benger admits, "It was lots of marketing hype, but it established our position in the market." Northern Telecom stole a lead of two to three years over its competitors with its public embrace of digital technology.[388] However, the SL-1 was only the rudimentary beginning of digital telecommunications. The company announced a timetable for the introduction of other products aimed at the telephone company market and opened new facilities at BNR in Ottawa for digital switching development at the same time.

Digital World coincided with a major executive change at the top of Bell Canada and Northern Telecom. Scrivener says he was bored with the telephone company, but enthralled with the changes at Northern Telecom. On March 24, 1976, the board agreed to a set of changes that Scrivener had requested. He wanted to relinquish the job of chairman of Bell to de Grandpré and move to Northern Telecom as its chairman, where he could preside over its assault of the U.S. market and the development of the digital switch. Lobb would stay in the United States as chairman and CEO of Northern Telecom's U.S. subsidiary and chairman of Northern Telecom International. He would report to Scrivener, as would Light, who would retain the title of president of Northern Telecom and head the company's Canadian operations. James Thackray would be promoted to president of Bell Canada. Scrivener says that when he and the board talked about these proposed changes with de Grandpré, his former subordinate asked him just one question, "You did mean it when you said you are working for me?"[389]

The Trudeau government also made a major change that affected Bell Canada in April 1976 when it delivered on a three-year-old promise hinted at in its 1973 Green Paper on telecommunications to move the regulation of Bell Canada away from the CTC to the newly constituted federal broadcast regulator, the Canadian Radio-television and Telecommunications Commission (CRTC). "The move wasn't made to obtain better regulation, but was purely a political decision to put regulation of communications under one body," Edgar Benson, former head of the CTC, says. "The move led to chaos for Bell because the CRTC was used to cultural policy and organizations and not economics and utility rate regulation."[390]

Allan Gotlieb, the architect of that reorganization, says, "I thought it was unhealthy to have two regulators, both in symbiotic relationships with their regulated industries. Juneau had the power and Benson did not. So the logic was to move telecommunications from the CTC to the CRTC."[391] But de Grandpré believed "it was a mistake to put the two, telephones and broadcasting, together because the talent and needs are different. They are two different arenas."[392]

The switch to a new regulatory agency involved some major procedural changes. One that irked de Grandpré was a statutory provision requiring collegial decisions. In other words, every commissioner appointed to the CRTC's executive committee is involved in making a decision, even if he or she did not sit on the public hearing panel that considered a particular issue. That requirement still exists and, as de Grandpré says, flies in the face of the legal maxim that "he who hears must decide."

Another major change was the CRTC's new rules of procedure for telecommunications hearings. "We wanted to open the hearing process to public intervenors and interest groups," says Charles Dalfen, the CRTC's first vice-chairman in charge of telecommunications and Bell Canada's head regulator. However, the CRTC had to reconcile its goal of openness with the wish of most parties to keep telecommunications proceedings formal. Dalfen attempted to strike a balance with a set of rules, still in place, that made the process clear and accessible while still maintaining the formal and mechanistic steps of legal procedure. His rules blended the two by allowing intervenors to decide how much they wanted to participate. Hearings would, in effect, be two-stage proceedings beginning with written comments that allowed anyone to air their views without having to participate in the formal part of the hearing. The formal hearing, on the other hand, still required written evidence and oral cross-examination of witnesses. Dalfen also instituted more informal regional hearings across the country to supplement the views heard at the formal hearings in Ottawa.

At the same time, the federal Department of Consumer and Corporate Affairs, which financed the Consumers' Association's regulated industries program, launched a more ambitious program to finance other public interest groups that wanted to participate in federal regulatory hearings. Andrew Roman, NAPO's lawyer, recalls being called to a meeting by Michael McCabe, then assistant deputy minister responsible for the program and now president of the Canadian

Association of Broadcasters. He told Roman that a couple of Bell officials had visited him to tell him why the utility believed NAPO's funding should be cut off. "So McCabe called me in to increase it."[393]

The proliferation of public interest groups and new procedures was a source of grief to de Grandpré. He says he was "not impressed by Dalfen's tenure" and added that the new procedures made the rate hearings "extremely long and complicated."[394] Already smarting from the added scrutiny, de Grandpré complained bitterly about the added costs and inconvenience of the numerous interventions. Asked how much Bell spent on regulation, Orland Tropea, Bell Canada's executive vice-president, told a parliamentary committee in 1977 that de Grandpré had put the figure at about $15 million a year.[395] The increasing number of intervenors was also a source of consternation to CRTC chairman Harry Boyle, as he told Gregory Kane, the Consumers' Association lawyer, at a 1976 hearing: "It seems to me that public interest comes up in regard to a regulatory agency in the same way that poor people come up in every political campaign. Everybody's got a concern for them. But it's very hard to define . . . ."[396]

The emergence and growing importance of computer communications networks led both the DOC and the new regulator to consider competition with greater interest. Granted, it was not like the United States, "where the White House spewed out a stack of papers urging more deregulation," Gotlieb says. "We needed a regulated monopoly."[397] But there was a marked change in attitude in Ottawa from the CTC and pre-Trudeau years, when the regulator was a willing guardian of the telephone companies' monopolies.

The Von Boyer task force at DOC had produced a White paper in 1975 that adopted the arm's length subsidiary approach as the preferred check on concentration of monopoly power. The paper recommended the telephone companies divorce their regulated telecommunications business from their public data-processing business to prevent cross-subsidization.[398] But Bell Canada had already taken action to insulate its computer services group, established in 1970 to serve large business customers through a group called Computer Communications and Network Services, from the federal regulator. Bell's computer group soon moved into the development of commercial computer networks and, in 1973, had developed Dataroute, one of the world's first commercial public computer networks. That same year it was renamed Computer Communications Group (CCG) and moved to the TransCanada Telephone System (TCTS), where it concentrated on building national data networks and servicing large national accounts, such as the chartered banks. Because TCTS was not a legal entity, the move kept CCG away from regulatory interference. The unit became a world leader after it developed Datapac, one of the world's first computer networks for public use, in 1977.[399] It used a Bell-developed technology, termed packet switching, which allows computer users to send and receive data over a regular telephone line, which is cheaper than leasing dedicated private lines. Datapac has since been adopted as an international standard.

One of the first rulings of the CRTC on competition that affected Bell Canada coincidentally involved the broadcast section of its jurisdiction. It followed a ruling by the CTC in late 1975 that granted the request by Transvision Magog and ordered the phone company to allow the cable company to put its own wires on Bell's poles. After Bell Canada filed its proposed "shared access" policy and rates for access with the CRTC in June 1976, the commission held a full public hearing on the filings. Access to poles proved to be the first chink in Bell Canada's armor.

The next assault was a broadside and came from Robert Bertrand, the assistant deputy minister of competition policy and the DIR, when he issued his statement of evidence, the *Green Book,* to the Restrictive Trade Practices Commission (RTPC) on December 20, 1976. "It blasted the Bell–Northern relationship,"[400] Lawson Hunter says, and called for the divestiture of Bell Canada's 62 per cent stake in Northern Telecom. The DIR stated that just as Northern Telecom "had been an economic colony of AT&T for too long," it had "also been an economic colony of Bell for too long."[401] Bertrand's report recommended that the RTPC begin public hearings into the Bell–Northern arrangement and concluded:

> It may be that the vertical integration between Bell and Northern Electric was at some time in the past a defensible arrangement but clearly this is no longer the case and the existing vertical integration between Bell and Northern would appear to be contrary to the public interest and indeed ultimately against the interest of both Bell Canada and Northern.[402]

Although the report was kept under wraps until well into the New Year, rumors of its contents had circulated for much of January, driving Northern Telecom's share price down. Bell received its copy of the

Green Book, along with a statement from the RTPC, on February 2, 1977. Although the government had not yet released the report to the general public, both companies disclosed the findings and issued statements to the media the next day. Northern Telecom called the report "outrageous," and de Grandpré told *The Globe and Mail* it was a "stupid conclusion," which he planned to "fight every inch of the way."[403] He added that the DOC had concluded in an earlier report that the industry had been "well served" by vertical integration.[404] What de Grandpré did not know was how the RTPC intended to proceed.

That was the subject of a meeting convened early the next morning, on February 3, by Tony Abbott, minister of consumer and corporate affairs, at the department's offices in the Place du Portage complex in Hull, Quebec. Also in attendance were Peter Connolly, Abbott's executive assistant; Sylvia Ostry, the deputy minister; Robert Bertrand and several of his officials, including Lawson Hunter; and Dennis DeMelto, an Industry, Trade and Commerce official. The group had hardly begun their discussion when de Grandpré burst into the room. "He was just livid," Hunter says, "and insisted on meeting with the minister. He said the report was 'crap' and he demanded that the report be withdrawn."[405]

De Grandpré says he doesn't remember crashing the meeting. But both Hunter and Connolly recall that Bell's chairman stormed angrily out of the room, then abruptly turned around as he reached the door and told Abbott, "I'm going to Europe and when I come back in ten days, I expect that report to be withdrawn and thrown in the garbage — where it belongs."[406]

Unfortunately for both Bell Canada and Northern Telecom, Bertrand's report was not tossed into the garbage. The RTPC waited until de Grandpré returned from his trip and, on February 16, the commission announced it would launch a public hearing to determine whether the corporate relationship was in the public interest. The three-member panel to conduct the inquiry was chaired by Luc-André Couture, vice-chairman of RTPC, who was joined by Robert S. MacLellan, a lawyer and formerly both a Tory member of parliament and the chairman of the RTPC from 1963 to 1970, and Frank Roseman, an economics professor.[407]

Hunter says the government officials were annoyed by de Grandpré's self-righteous attitude. "De Grandpré and Bell had some very valid concerns about the restrictive trade practices inquiry; about it being unwieldy and harassing them for years and years." But de Grandpré alienated the bureaucrats by acting like "General Bull

Moose," as Hunter called him, adding, "He had the same attitude about Bell, that what was good for Bell was good for Canada."[408]

Hunter was not formally introduced to de Grandpré until 1982, by which time he was the DIR and, as boss of the combines unit, had been in charge of making the government's case at the RTPC inquiry. He first met the Bell chairman at a party in Ottawa held by the Business Council on National Issues (BCNI), the powerful national lobby group for big business co-founded by Scrivener in 1976. The combines chief initially rejected the offer of a BCNI lawyer, Peter Vivian, to be introduced to de Grandpré. "He'll probably want to punch me in the nose," Hunter told his fellow guest. He guessed right. De Grandpré came over and, when the diminutive Hunter was introduced, he looked down at him and said, "I ought to hit you over the head."[409]

De Grandpré also assailed Sylvia Ostry when he ran into her at a cocktail party held by Consumer and Corporate Affairs, shortly after he made his futile plea that the DIR's *Green Book* be rejected. Alan Walter, then a Bell Canada vice-president who worked on the RTPC team, says, "Jean nailed her and asked, 'How can you accept this report and act on it?' After she replied, he said, 'Madame, you are either a fool or a liar. Or both.'"[410]

Couture's panel began its hearings in Ottawa on June 15, 1977. The DIR was represented by Henderson and his associate, Gordon Kaiser. After the procedural pre-hearing in Ottawa, the inquiry commenced a road show that traveled across Canada in September to solicit submissions from any interested parties in Halifax, Fredericton, Toronto, Winnipeg, Regina, Edmonton and Vancouver. "It was a strange phenomenon," Henderson says. "Early on, there was no interest, which surprised us because we felt it was of considerable importance to Canada and to the industry."[411] Henderson says both he and Bertrand came to the conclusion that many companies were afraid to publicly testify against Bell because they were also suppliers to the utility and feared retribution.

Bertrand went further and claimed to have obtained evidence that Bell Canada and Northern Telecom had pressured witnesses for three companies by telling them that their testimony to the commission could affect future business with their firms. After the allegations were reported in an industry trade newsletter in December, de Grandpré publicly denied Bertrand's charges as "long on innuendo and short on facts."[412] But the allegations of witness tampering and a charge that there were attempts to bribe the acting chairman of the RTPC were serious enough to be investigated by the RCMP. Although lawyers for Bell and Northern admitted that some witnesses had been interviewed before their appearances, the RCMP cleared the companies of all charges the following November after it said it could find no proof to support the allegations.[413]

During the fall of 1977, Henderson says the commission responded to the lackluster turnout by going "across the country several times. It took time, but once a few major suppliers appeared, others testified, too."[414] The hearings continued for the next four years and concluded on May 8, 1981, after more than two hundred witnesses were heard at 228 days of hearings, which produced over 35,000 pages of transcripts and 2,000 exhibits.[415] "We were not on a vendetta and tried to administer the law dispassionately,"[416] Hunter says.

His words provided no comfort to de Grandpré, who perceived just the opposite. "We were not just fighting our competitors or interest groups, but our own government. I don't understand their stupidity. They want winners, but they want to destroy one when they finally get one. It's frustrating. It doesn't speak highly of their intelligence,"[417] de Grandpré says, adding that the uncertainty created by the inquiry penalized Bell Canada and Northern Telecom. The hearings were ill-timed, coming just when Northern Telecom was making its major marketing push into the United States and trying to instill confidence in staid telephone company customers. One of the strengths the company was trying to sell was that the vertical integration between the manufacturer and its telephone company was intact, unlike in the United States where yet another consent decree proceeding was underway.

In accordance with the adage, "don't get mad, get even," de Grandpré took Bell's fight against the federal government's combines unit to a parliamentary committee which was convened that summer to review the government's proposed new competition policy. De Grandpré didn't miss the opportunity to assail the proposed Combines Investigation Act and to attack the competition watchdogs. He dispatched Tropea to the Senate Standing Committee on Banking, Trade and Commerce to deliver Bell's brief one week after the RTPC's pre-hearing. Bell opposed various provisions in the government's proposed new competition policy on the grounds that it constituted dual regulation of the utility. "If a regulatory body is to have the right to approve of mergers, acquisitions, market entry, incorporations, etc., it is submitted that the CRTC is the appropriate body to exercise any such supervisory powers,"[418] the company stated in its brief — a position it later repudiated when it came time for the corporate reorganization.

It was the summer of Bell Canada's discontent.

The DIR could not have asked for sweeter revenge when, on December 23, 1977, the CRTC delivered a precedent-setting pro-competition ruling, appropriately called the *Challenge* case, which dented one of Bell's most jealously guarded monopolies. The decision followed contentious hearings earlier that spring into a complaint from Challenge Communications, a company that leased mobile telephones in competition with Bell. Challenge argued that Bell's restriction forbidding customers from using their own equipment was illegal. "We wanted to break up Bell's monopoly on interconnection,"[419] Henderson says. Bell Canada's vice-president of law, Ernie Saunders, countered at the hearings, "Having gone to the trouble to take the steps necessary to make this service available to the public, Bell Canada quite understandably desires to reap the benefits of this new offering."[420]

The CRTC declared Bell's restrictions were illegal. The *Challenge* case was the first ruling to declare that the telephone company could not confer an undue advantage upon itself. Gordon Kaiser, Henderson's colleague, termed the decision "a competitor's 'Bill of Rights.'"[421] In terms of its impact, the *Challenge* case was comparable to the first Microwave Communications Incorporated (now called MCI) ruling in the United States.

De Grandpré next ran into opposition when he tried to amend Bell Canada's Special Act to give the utility greater financial autonomy. He proposed four major amendments that would raise the utility's share capital to $5 billion from $1.75 billion; issue and split its capital stock without the prior approval of the CRTC; give the company powers to invest in other companies with similar objectives and create separate, arm's-length subsidiaries; and allow the company to alter its powers and share capital in future by letters patent rather than by legislative amendment. However, the proposed bill, which was tabled in December 1976, was stalled by a fourteen-month-long procedural fight led by the NDP. The party's concerns over Bell Canada's repeated rate increases and diversification into other unregulated businesses led them to successfully delay a second reading of the bill, and it died when the House was prorogued on October 17, 1977.[422] A second version was re-introduced and, after substantial amendments, the NDP agreed to let the bill pass in March 1978. Those changes meant Bell Canada had to continue to ask Parliament for permission to change its corporate structure, that new shares could not be sold without the CRTC's permission and that it would not have the power to invest in other companies and set up separate subsidiaries.[423]

The utility was more successful in getting its way over Telesat Canada. A bitter fight erupted with pro-competition forces over the agreement of the satellite company to join the monopoly long-distance telephone companies' consortium, TCTS. Telesat signed the deal, named the "Connecting Agreement," on January 17, 1977. The agreement enshrined the restrictions on Telesat's business that the phone companies had previously sought from Parliament; namely, that Telesat could not sell its services directly to customers and could operate only as a "carrier's carrier." Through the Connecting Agreement, the phone companies in effect paid Telesat off. They agreed to subsidize the satellite operator by guaranteeing it a rate of return on its equity equal to the average rate of return of the two largest telephone companies — even if it lost money.[424] TCTS also agreed to lease a minimum number of satellite channels from Telesat to ensure its viability.

The proposed deal had been approved by the federal cabinet in early November 1976. Communications Minister Jeanne Sauvé stated the government's reasons for supporting it in a confidential letter to CRTC chairman Harry Boyle on December 14, 1976. She claimed that Telesat would be faced with "serious financial difficulties" and that the federal government would have to "provide substantial financial support" to Telesat if the agreement was blocked.[425] In fact, the Department of Finance had told Telesat's board early that May that it would not provide any further loans to the company.[426]

It did not take long for Telesat's non-TCTS shareholder, CNCP Telecommunications, to find out about the proposal. As expected, the revelation of the deal sparked a rift on Telesat's board. Canadian National and Canadian Pacific, the corporate parents of TCTS's major rival, sent their lawyers to meet with Telesat's CEO David Golden to voice their objections.[427] Their counterproposal, aimed at avoiding the creation of what they called an "inordinately large monopoly" by TCTS, was rejected. Eldon Thompson, president of TCTS, invoked the public interest and stated that the TCTS proposal was "somewhat altruistic" with "little direct benefit to TCTS but with long-term benefits to the public of Canada." He admitted, however, that the arrangement was possible because the two organizations "are not, and have not been, in competition. Telesat has chosen to operate essentially as a carrier's carrier and hence its development is in cooperation with that of the other carriers."[428]

Telesat's board approved the association with TCTS on November 24, and the company was accepted as the tenth member of TCTC by the consortium's management board on December 1. Telesat's management presented the agreement to its shareholders for ratification at a meeting on December 9.[429] Not surprisingly, Canadian Pacific's representative voted against it and, more significantly, read a scathing attack from CP's board of directors:

> The proposed association by Telesat Canada with the TransCanada Telephone System will amount to abdication by the shareholders, in particular by the Government of Canada, of their authority to direct the management of the corporation. The agreements we have before us for approval would delegate the management of Telesat Canada to a TCTS board of management dominated by Bell Canada. . . .
>
> Telesat Canada, formally independent in the national interest, will now become submerged in the TCTS, thereby extending the already virtual monopoly position Bell Canada together with the other members of the TCTS now have in providing telecommunication services to the Canadian public.
>
> When Canadian Pacific Ltd. invested in Telesat Canada it did not do so to promote a monopoly in telecommunications.[430]

CP's defection and the anticompetitive aspect of the Connecting Agreement made for a stormy public hearing, chaired by Charles Dalfen at the CRTC from April 25 to June 2, 1977. The CRTC refused to approve the agreement, declaring that it "would not be in the public interest." The commission stopped short of declaring the agreement to be illegal, but stated in its decision, released on August 24, that it was not consistent with the intent of the Telesat Canada Act.[431]

"I didn't think that agreement was in the interests of Telesat, that it be eviscerated by the telephone companies," Dalfen says. "Bell Canada wanted that connecting agreement to control Telesat and its facilities. Telesat wasn't allowed to compete. I wanted Telesat to be a retailer and we viewed that agreement as anticompetitive."[432] The CRTC ruling was significant because it was one of the first in which the commission stated that "competition policy was an appropriate public interest concern to take into account in exercising its jurisdiction under the Railway Act."[433]

But the matter did not end there. Telesat petitioned the CRTC decision to the federal cabinet on September 15, followed by a second petition from Bell Canada on September 23 and a third from B.C. Tel on September 29. Those petitions were vigorously opposed by Greg Kane on behalf of the Consumers' Association. He filed a leaked memo to cabinet from the Minister of Communications, which contained Sauvé's recommendations on the petition. Kane attached his reply to these recommendations and also wrote to Trudeau to oppose the procedure used by the cabinet in appeals of regulatory decisions. Although the regulatory decision being appealed was the subject of public participation by public interest groups, cabinet treated an appeal as a private matter between the cabinet and the petitioner, making it subject to cabinet secrecy.[434]

The cabinet overturned the CRTC decision and reinstated the Connecting Agreement on November 3, 1977. The cabinet simply concluded in its order-in-council that "the public interest will be better served if the Telesat Canada proposed agreement is approved."[435] In a statement reeking of condescension toward the regulator, Sauvé explained that the government had decided to reverse the CRTC ruling because "the range of factors affecting the policy issues is far wider than that which the CRTC could reasonably be expected to consider."[436] The government did not want Telesat's largest customer to bolt, leaving the bills for the next generation of satellites.

Ian Sinclair, the tough, barrel-chested chairman of CP, was not amused. He was de Grandpré's closest rival in the corporate world. Echoing de Grandpré's maxim about stress, Sinclair once declared, "I don't get heart attacks . . . I give them."[437] Not one to be deterred from getting his own way, Sinclair hired Gordon Henderson, the same lawyer who acted for the DIR — the federal competition watchdog — to launch a court action against Telesat and the Connecting Agreement in the Supreme Court of Ontario[438] one week after the order-in-council was passed. That challenge dragged on for five years, until the Supreme Court of Canada dismissed CP's motion for an appeal on May 31, 1982. "We won half the case,"[439] Henderson says, noting that the ruling agreed that the Telesat Act barred the company from entering into any partnerships. However, the court also declared that TCTS was not a partnership.[440]

On the financial front, Bell Canada obtained its first stock listing in the United States in August 1976 after the utility sold its first share issue to U.S. investors. But Bell's share price soon ran into stormy weather, buffeted by the election of the Parti Québécois headed by Premier René Lévesque on November 15, 1976. De Grandpré took to the hustings to fight the language and tax policies of Lévesque's separatist government and was the first Canadian leader, ahead of even Trudeau and Lévesque, to speak to U.S. financiers on Wall Street about the PQ victory. He told the institutional investors and brokers on December 10, 1976, that he did not expect the election to adversely affect the company and that Lévesque had assured him that the province had no plans to nationalize or expropriate Bell.[441] A week earlier, de Grandpré had told a group of Canadian financial analysts in Toronto that Lévesque "neither has the power nor the financial resources to do anything to Bell."[442]

On January 25, 1977, Lévesque made his pilgrimage to Wall Street, where he delivered a conciliatory speech to the Economic Club of New York. It echoed the theme of a paper he had written for the highly respected U.S. journal, *Foreign Affairs,* just prior to the election victory, in which he stated that Quebec wanted political sovereignty but an economic "partnership" with the rest of Canada. "One thing sure, is that Quebec will not end up, either soon or in any foreseeable future, as the anarchic caricature of a revolutionary banana republic which adverse propaganda has been having great sinister fun depicting in advance,"[443] he wrote.

Less than three weeks later, de Grandpré delivered the rebuttal of Quebec's big-business establishment to the PQ platform in a keynote speech to the Montreal Chamber of Commerce. The Bell Canada chairman accused Lévesque of a lack of realism in language policy and warned that businesses would move out of Quebec if the PQ failed to recognize English as the international language of commerce and refused to allow corporate head offices to work in English.[444] Despite his efforts, Bill 1, the PQ's controversial language law, was tabled in the National Assembly on April 27, 1977.[445] The government introduced a revised bill, the now famous Bill 101, on July 12, after a month of hearings before a committee of the National Assembly.

De Grandpré has been portrayed as the point man for the Quebec business establishment in its fight against the PQ's language law, which he likes to call "Bill 401" because of the exodus it prompted from Quebec.[446] But he wasn't alone. Other business leaders, particularly Paul Desmarais, the chairman and CEO of Power Corporation, and Alex Hamilton, president of Domtar, launched similar attacks and assumed key roles in leading the federalist fight against Lévesque.

De Grandpré presented Bell Canada's brief on Bill 1 before the National Assembly committee on June 21, 1977. As a representative of big business, he became a lightning rod for Quebec nationalists. In particular, they saw the utility's past record as part of the cause of the language problem. De Grandpré defended Bell's French-language record following the passage of the Bill 22 language law by Robert Bourassa's government and denied that Bell Canada had transferred large numbers of employees out of the Montreal headquarters to the national capital region to get around the new law.[447]

A PQ spokesman lashed out at de Grandpré and scathingly asked him, "How can anyone of your noble blood feel the aspirations of ordinary people?" De Grandpré responded that his family's name was Duteau. He explained that his ancestors were peasant farmers and that his family had lived in Quebec for more than three hundred years.[448] Camille Laurin, Quebec's Minister of State for Cultural Development and the architect of Bill 101, reportedly rejected de Grandpré's explanation and called him a vassal of the English.

"That remark really hurt him," Monic Houde, now an assistant vice-president at Bell Canada, says.[449] Journalist and author Gerald Clark recounts in his 1982 book on Montreal that after de Grandpré returned to his Quebec City hotel suite, he wrote a stinging rebuke to Laurin in which he reminded the minister that his mother's great-great uncle, François Marie Thomas Chevalier de Lorimier, was one of the dozen martyred *patriotes* hanged in 1839 for his role in the Papineau rebellion. "Where, sir, was your great-grandfather at the time?" de Grandpré countered. He never sent the letter to Laurin.[450]

Few people, including Clark, knew that the argument between de Grandpré and Laurin at the National Assembly was not their first clash over the heritage of their respective families. De Grandpré had, in fact, won a previous fight with the former psychiatrist. Robert Bandeen, then president and CEO of Canadian National Railways, recalls a 1976 meeting of Montreal's top ten businessmen convened on behalf of the minister of culture by his younger brother, Pierre, who was then dean of the École des Hautes Études Commerciales. The meeting was requested shortly after the election of the PQ, Bandeen says, so that Laurin could explain his government's platform and assuage any concerns that members of the business elite had about the PQ's policies: "Both brothers were sitting at one end of the table and then Camille Laurin launched into a big speech on the PQ and Quebec culture. Next thing you know, Laurin and de Grandpré got in a big argument and were shouting at each other over their ancestors and whose family had been in Quebec longer. Laurin ultimately backed down."[451] De Grandpré says that he does not recall this.

Once passed, the language law, which is administered by a separate
bureaucracy named Office de la langue française (OLF), created numerous changes and administrative headaches for businesses like Bell
Canada. De Grandpré told a Quebec Chamber of Commerce convention in Quebec City on November 12, 1977, that the province's language law had had "a more negative effect than people in certain areas
would have us believe."[452] This was underscored when Alistair
Campbell, the chairman of Sun Life Assurance Company of Canada,
stunned the nation by announcing on January 6, 1978, that the venerable
company had moved its head office from Dominion Square in Montreal
to Toronto. No move before or after had the impact of Sun Life's departure, because the company was the symbolic anchor of Montreal's financial establishment. By contrast, Northern Telecom's move in the first
weeks of 1980 from a high-rise tower on Montreal's Dorchester
Boulevard (now Boulevard René Lévesque) to a nondescript six-story
building in the sprawling Toronto suburb of Mississauga caused hardly
a ripple. De Grandpré pulled the company's head office out of Quebec
in a secret move acknowledged only in the small print of the company's
annual report released several months later.

According to de Grandpré, Northern Telecom's move was for "economic reasons," because less than 25 per cent of its domestic sales
were in Quebec, compared to Bell Canada, which made more than 35
per cent of its revenue in the province. "It did not create the kind of
reaction that would have resulted if we [Bell] had moved out of
Quebec," de Grandpré says, adding that he fought off pressure "from
all quarters" to relocate the parent company.[453] Nevertheless, coming
on the eve of the Quebec referendum on sovereignty association,
Northern Telecom's move was seen by some as a political statement.
De Grandpré had helped organize a corporate fund-raising drive with
Pierre Coté, chairman of Celanese Canada; Claude Castonguay, chairman and CEO of Laurentian Life Assurance Company; and Domtar's
Hamilton, which raised $2.7 million to finance the "Non" forces. "But
then I wasn't even here, I was in Japan, so I couldn't even vote in the
referendum!"[454]

Big business was equally opposed to the changes to Quebec's tax
law made by the PQ's Bill 45 in 1978. De Grandpré called the bill,
which increased the taxes on management salaries, a "punitive tax law"
and told shareholders at Bell Canada's ninety-ninth annual meeting on
April 17, 1979, that he had unsuccessfully lobbied Quebec finance
minister Jacques Parizeau to have the measure rescinded.[455] Later, Bell

Canada responded by transferring numerous executive jobs to the nation's capital so that the managers could pay Ontario tax rates.

A year later, de Grandpré stunned shareholders and reporters at the utility's 1980 annual meeting in Toronto with a proposal that called for provincial regulation of Bell Canada. De Grandpré explored the prospect for the next two years and told Quebec communications minister Clement Richard that Bell Canada would rather be regulated by the provinces than by the federal government, despite claims by financial analysts that such a change would result in subscribers paying higher phone bills.[456] The proposal dovetailed with de Grandpré's reorganization of Bell's operations department into two provincial halves more than a decade earlier and reflected both his mounting anger at Bell's federal regulators and his belief that the utility might do better at the hands of the provinces. His overture led the Parti Québécois to adopt a policy resolution at their convention in November 1981 calling for provincial regulation of the phone company. After the convention, the Quebec Department of Communications commissioned a study from Tamec Consultants of Verdun, Quebec, on the impact of a transfer of regulatory authority over the largest private-sector employer in Quebec to the province. The study was later shelved, however, and the idea of provincial regulation of Bell died when the PQ's mandate expired in 1985.

Meanwhile, the unrelenting march toward greater competition in the telecommunications industry had continued. Sinclair finally exacted his revenge. On June 14, 1976, Canadian Pacific applied to the CRTC for an order that would force Bell Canada to allow CNCP to connect its network with the telephone company's system. Dubbed the interconnection proceeding, it was the biggest battle yet fought by de Grandpré. Although CNCP had previously been allowed to compete in the private line market, it had to use its own networks and was not allowed to connect its customers to the monopoly networks. If Sinclair's request was granted, Bell Canada would have to give up its century-old monopoly on long distance, the phone company's most lucrative business.

The utility viewed CNCP's application as "a case of expropriation brought against Bell Canada by two of the most powerful corporations in the country, who are also direct competitors,"[457] explained Ernie Saunders, vice-president of law at the opening pre-hearing conference on January 10, 1978. At the CRTC hearings, which began in early April 1978, the Bell Canada chairman was even more direct than

Saunders in his attack on CP: "I told the CRTC that they want[ed] to put their trains down my tracks."[458]

De Grandpré defended Bell Canada's position on the stand at the CRTC hearings for a full day. His cross-examination began with a hypothetical question posed by CNCP's lawyer, to which de Grandpré replied that he believed it would be in the country's interest if CNCP did not exist and that its demise would save Bell, and its subscribers, money.[459] When cross-examined by Gordon Kaiser, counsel for the DIR, he asserted that CNCP's application was not an application for competition at all:

> If the application is granted then you don't have competition because they are competing on my tracks, as I said before. If they want to compete let them set up their own system and we will compete. That is not what they are asking. What they are asking is to compete in a very, very curious way. First of all, they don't want to have other competitors and, secondly, they don't want to compete by setting up their own switches. . . .[460]

Under further questioning, de Grandpré rejected the contention that two interconnected telecommunications networks were in the public interest: "That is why I say it is a monopolistic situation and it should remain monopolistic, but if you want to introduce competition, let's introduce true competition."[461] By that, he meant competition without regulation, under which he expected that Bell "could give CNCP a real ride for their money."[462]

His remarks did not impress Bell Canada's regulator, says Dalfen, who chaired the five-member panel that heard the case. "He wasn't pleased to be there and he had no love of regulation, which he showed in both the tone and content of his statements. He couldn't disguise his chagrin." Dalfen thought de Grandpré's testimony was tantamount to declaring "that Bell has a God-given right to a monopoly."[463] He, on the other hand, believed "a monopoly is an unnatural state of economics that is conferred upon the company by a democratic state, which uses that power very rarely and reluctantly. The government has an obligation and responsibility to ensure that the company is sensitive to the public they serve. The only criticism of Bell that I could make is that both Bell and de Grandpré would forget that they had a monopoly and that the state had conferred upon them certain powers and obligations."[464]

At one point during the day, de Grandpré contended that Bell Canada pursued the public, but not its own private, interest in opposing CNCP's application. One by one, Kaiser walked the chairman through the list of corporate supporters of CNCP's application and, each time he named one — the Canadian National Railway, the Bank of Nova Scotia, the Canadian Manufacturers' Association, IBM Canada or the Royal Bank of Canada — he asked whether de Grandpré thought that those organizations had placed their own private interests ahead of the public interest by supporting CNCP's bid. And, each time he was asked, de Grandpré answered "yes."

Then Kaiser delivered his punch line: "Why is it that all of these companies differ from Bell Canada in their perception of the public interest?"

To which his witness retorted, "Because we have an obligation to serve all across our territory."[465]

At the end of the afternoon, when Henderson and Kaiser returned to their office in downtown Ottawa, they found that de Grandpré had withdrawn all of Bell's legal business from the firm and had tried briefly, and in vain, to evict the law firm from the two floors that Gowling & Henderson leased in Place Bell Canada at 160 Elgin Street.

"That incident has been blown out of all proportion and our relationship with Bell is very good,"[466] Henderson said in a 1990 interview. However, Allan R. O'Brien, one of his partners, interjected with a laugh, "We didn't get kicked out, but we still haven't had any of Bell's legal work since then."[467]

Behind the scenes, the CRTC was concerned with the possibility that Bell Canada had intimidated some of CNCP's supporters and had pressured at least one major corporation into withdrawing its favorable intervention supporting the bid. On the second day of the pre-hearing on January 11, 1978, Dalfen had stunned everyone present when he read into the record a letter the CRTC had received from W. Earle McLaughlin, the chairman and CEO of the Royal Bank, on January 3. The bank's powerful boss unexpectedly asked permission to withdraw the Royal Bank's written evidence from the record of the proceeding. He also stated that the bank would not participate further in the hearing and asked to withdraw its October 31, 1977, intervention in support of CNCP.[468] The request was all the more surprising, given that McLaughlin held a seat on Canadian Pacific's board of directors.

Henderson and Kaiser, acting on behalf of Bell Canada's archrival, the DIR, took an unprecedented step and asked Dalfen for subpoenas that would compel the Royal Bank's witnesses to appear at the hearing. Dalfen granted the request — the first and only time that the CRTC has issued subpoenas to a witness — and the three managers, one of whom had been a Northern Electric and BNR employee, testified on March 15, 1978, as private citizens.[469] The DIR hoped to get public evidence that the Royal Bank had bowed to pressure, fearing reprisal from de Grandpré.

Clifford Downing, the Royal Bank's deputy general manager of the systems division and one of the authors of the bank's withdrawn submission, gave a brief summary of the document and testified that it "recommends competition, a maximum degree of competition, responsible competition in the communications field."[470] Although senior management disavowed the evidence to avoid jeopardizing the Royal Bank's relations with Bell Canada and to avoid taking any partisan position, Downing stated that the views in the document, which were now to be considered his division's only, "are entirely consistent with the views that have always been expressed by the Royal Bank." He also testified that the decision by top management to withdraw the bank's intervention came as "a complete and utter surprise" to his division and that he was not told of the reasons why the evidence was being withdrawn.[471]

Would-be competitors and their proponents, including the banks, had good reason to fear de Grandpré's wrath. Bell Canada threatened to cut off telephone service to the Bank of Montreal in 1975 when the bank attempted to install a special device to its private-line network. The bank had wanted to use the device, called divert-a-call, for its MasterCard credit-card authorization system. After Bell issued its threat, the manufacturer, a small company named Harding Communications Limited of Thornhill, Ontario, applied for an injunction in Quebec Superior Court to restrain Bell from interfering with its business.[472] The utility fought the court battle on the grounds that only its regulator, not the courts, could grant such an injunction. Bell lost its appeal in the Supreme Court of Canada in early 1979, and subscribers won a victory on the road to obtaining the right to own and attach their own equipment to Bell Canada's network.[473]

Among the expert witnesses that testified for Bell Canada at the interconnection hearing was Eugene V. Rostow, Sterling professor of law and public affairs at Yale University and former under secretary of state for political affairs during the administration of U.S. president Lyndon B. Johnson. He provided evidence on the advantages of monopoly networks and argued that because of economies of scale, "monopoly is the most efficient method for organizing the supply of all the services which the network can offer."[474] The CRTC rejected that defense of Bell Canada's monopoly. The utility lost the case, and the CRTC added in its 284-page ruling issued on May 17, 1979, that both Rostow and Bell Canada had failed to provide adequate empirical evidence to support their claim.[475] But it was not a total loss. Although CNCP was allowed to connect its private networks to Bell Canada's public networks, the CRTC preserved the public long-distance monopoly. The decision maintained the restrictions that prevented CNCP from using its private networks to provide service to the general public.

"I haven't changed my mind about it," de Grandpré said in a 1990 interview. "Competition and regulation cannot live side by side in any industry. True competition means competitors have to have the flexibility to serve a market or to withdraw from it. For example, if you're selling soap, or you try to introduce a product and you're losing your shirt, you'll get out of it. But in telecommunications, we have markets that we all know are unprofitable. I know if I only handle long-distance calls from Quebec City to Windsor, I'll make a lot of money. To me, if they allow only one company to serve the profitable areas and the other has to serve the unprofitable areas, there's nothing worse. It's not competition."[476]

While de Grandpré lost the battle, he still won the war, as Dalfen sees it. Despite the setbacks, he still managed to do two things that none of Bell Canada's counterparts in either Europe or the United States were able to accomplish by the mid-1980s. He kept Bell Canada and Northern Telecom intact and he retained Bell Canada's lucrative long-distance monopoly.[477]

Chapter 5

'Baksheesh! Baksheesh!'
Taking Bell Abroad

The combined efforts of the operating, research and development and manufacturing organizations have advanced this country to the forefront of world telecommunications technology. . . It is in large measure the close interworking of these vertically integrated members of the Bell Canada family which has led to the company's success. . . .
— A. JEAN DE GRANDPRÉ, 1982 [1]

It's a long way from Bell Canada's head office in Montreal to the palace of the house of Sa'ud in Riyadh, Saudi Arabia. In 1978, the acceptance of the bid by Bell Canada's consortium for the oil-rich kingdom's huge telecommunications modernization contract, which at $3.5 billion (U.S.) was the largest contract of its kind in history, made the Canadian company the envy of the world's phone companies and proclaimed its arrival as a global player in the telecommunications industry. Yet the precedent-setting treatment of the revenue from that contract by Bell Canada's federal regulator accelerated de Grandpré's collision course with the CRTC.

Bell Canada's first five-year Saudi contract, valued at $1.1 billion, fulfilled de Grandpré's ambition for the utility. All leaders have a guiding vision and his, inherited from Scrivener, was to make the telephone company a world leader. During a 1986 interview, he characterized himself as "a very strong internationalist." In the same discussion, he seemed to equate his job as Bell chairman to an ambassadorial post on behalf of Canadian business. "I'm trying to wave the flag and show a presence for Canada on my many trips abroad. That's good for Canada, but probably not good for my health."[2] He maintained that his roots as an internationalist stemmed from his upbringing : "In our family, we have always had an open mind about people. No single group has a monopoly on wisdom. I try to pick the brains of everyone."[3]

One of de Grandpré and Scrivener's greatest achievements is the high regard in which both Bell Canada and Northern Telecom have come to be held in the international marketplace. They were both ahead of their time in advocating the need of Canadian companies to become global competitors, a conviction that Scrivener and Marquez attempted to put into action when de Grandpré was first hired by Bell Canada. Only recently has global competitiveness become the objective of most businesses and governments, although still too few Western leaders have any inkling of how to achieve or foster it.

Bell Canada started its international consulting business as an adjunct to Northern Electric's nascent export business. Its entrance into the global marketplace coincided with the launch of Northern Electric's first overseas joint venture in Turkey in 1967. General Holley Keefler, who had just added the chairmanship of Northern Electric to his existing position as president, appointed Andrew Kovats, a Hungarian-born engineer, as the first director of the international unit formed to sell equipment abroad. Kovats's boss and co-architect of Northern Electric's international business division was Cy Peachey, the subsidiary's executive vice-president.[4]

The Turkish government had begun planning its tender for new automatic switches in 1963. It also wanted to develop a telecommunications equipment-making business in Turkey. As a result, the international tender was issued to supply the switches and establish a joint manufacturing venture.[5]

The Turkish Post, Telegraph and Telephone (PTT) announced that Northern Electric had won the $24.5-million order in late November 1966.[6] Under the five-year contract, Northern Electric supplied Canadian-made parts for crossbar switches and telephone sets that would be assembled in Turkey at a new plant to be built as a joint venture between the company and the PTT. With financing from the federal government's Export Credits Insurance Act, Northern Electric set up Netas (Northern Electric Telekomünikasyon A.S.) in February 1967. Northern Electric owned 51 per cent of Netas (pronounced

Netash), and the PTT the other 49 per cent. Kenneth Eadie, an engineer and nephew of Bell Canada chairman Thomas Eadie, had conducted the negotiations with the PTT and was appointed the first general manager of the new enterprise. He moved to Turkey and set up the Netas manufacturing plant in Umraniye, a suburb of Istanbul.

It was in that suburb that Northern Electric got its first taste of the intrigue and problems that often accompany an international contract. No sooner had the plant opened in 1968 than Northern Electric found itself embroiled in a dispute with a Turkish princess. The princess, who was related to the last caliph of the Ottoman Empire, laid claim to the same land that the new plant was built on. Unable to determine whether her claim was legitimate, but unable to relocate, Northern Electric negotiated a cash settlement with her.[7] Eadie's daughter recalls a dinner after the settlement was reached at the princess' palace, which was entered through a large foyer over which hung a crystal chandelier acquired by one of her ancestors from Napoleon Bonaparte.

As Eadie recalls, Northern Electric's move abroad was led by Keefler, but resisted initially by Bell Canada, "who was dragged into this screaming."[8] The establishment of the Netas venture made it necessary to train PTT employees in how to install and operate the new equipment. This led Northern Electric to ask its parent company to set up a companion international consulting arm, which Bell Canada established in July 1967. It acted as a broker, leasing experts from Bell Canada under contract to foreign governments. The new unit, headed by Russell Cline, assistant vice-president of consulting services, specialized in operations and maintenance contracts, dubbed O&M contracts in industry jargon, to help offshore state-owned phone companies run and fix their networks.

The Netas venture became a model for Northern Electric as it pursued contracts in other countries. The manufacturer soon set up a Greek subsidiary, Northern Electric Hellas S.A., and purchased a site near Athens to make equipment under the $9-million contract it received from the Greek government in 1969. Northern Electric also won contracts in Spain, the Philippines and Nigeria. It unsuccessfully pursued contracts in Iran and Peru. Bell Canada followed suit, and by the second year of its consulting service operations, Bell had forty-five employees working as advisers in a dozen countries.

But the utility soon discovered obstacles to expanding its consulting business abroad. "There is a strange resistance to paying for knowledge," Cline told *The Globe and Mail* in a 1969 interview. "Since most of the countries have a scarcity of capital dollars, they want to put them all in hardware."[9] Competition was also intense from more established rivals, such as L. M. Ericsson Telefonaktiebolaget of Sweden, which had operated abroad since 1912, and ITT Corporation, which had moved offshore in 1925.

Northern Electric learned numerous lessons about Middle East business practices working in Turkey, including, "Nobody works for nobody for nothing,"[10] as PTT chairman Emin Baser put it. That was also the case with an aborted contract bid in Peru in 1967. Northern Electric retained Nicholas Onassis, a Montreal business agent, to represent the company and help it in the bidding process. Onassis in turn represented Albatros Naviera S.A., a Panamanian company. After Northern Electric allegedly refused to follow their agent's advice to meet the conditions of the Peruvian government, Albatros Naviera sued Northern. It claimed the Canadian company had scuttled its bid by offering a higher rate of interest on its financing charges, allegedly costing Onassis $4.7 million in commissions. A subsidiary of ITT won the contract.[11]

Another lesson was learned in Iran — despite extensive contacts and preparation, a contract can be lost because of who you don't know, as much as won because of who you do know. That happened when Northern Electric and Netas wooed Mohammed Reza Pahlavi, the Shah of Iran, for a country-wide modernization contract in the early Seventies. Northern Electric had extensive contacts in Iran through its various Middle East agents and received a qualification to bid. Company officials believed they had the inside track after they were asked to perform a special favor for the Shah — to equip his palace with a state-of-the-art electronic telephone system. Northern Electric did the work, only to find itself double-crossed when the contract went to Siemens A.G. of West Germany. It turned out, Eadie says, that the Shah's sister was related, through marriage, to Peter von Siemens, the head of the West German telecommunications and computer equipment maker.[12] The loss, however, later turned out to be a blessing after Iran's King of Kings was deposed by the Islamic fundamentalist revolution that brought the Ayatollah Ruhollah Khomeini to power in 1979. "Something that you have to remember when you go out there is that you can be cut off overnight,"[13] says Jim Kyles, who took over from Cline as head of Bell Canada's international consulting services in 1974.

De Grandpré learned that harsh fact of international commerce after several months of intense effort in France. The French state-owned PTT, now named France Telecom, announced a grandiose network expansion program in 1975. The announcement led Thomson–CSF, a French telecommunications equipment maker, to seek a partner willing to sell it a license to make a suitable switching system. To Northern Electric, it seemed like Turkey on a grand scale. De Grandpré says he virtually took up residence in Paris to help René Fortier, a Bell Canada executive vice-president, negotiate the deal. A license agreement was eventually signed with Thomson–CSF after it had evaluated Northern Electric's SP-1 switch to ensure that it met the PTT's technical standards.[14] Technical information began to be transferred to the French company by the end of the year.[15]

However, the whole deal came apart in February 1976 after the French government decided that it wanted to finance the development of comparable technology by a wholly French-owned company. To that end, the government nationalized L. M. Ericsson's and ITT's subsidiaries in France and subsequently sold them to Thomson–CSF. Northern Electric was paid $1 million for its work.[16] Despite the loss, the exercise proved that Northern Electric's new switch was internationally competitive; if the French government had not intervened, the SP-1 could have been accepted as the standard for the French PTT.[17]

Shortly after Northern Electric lost its bid in France, Bell Canada won a prestigious contract from the North Atlantic Treaty Organization (NATO) to audit all the telecommunications networks of the four-teen-nation Western military alliance. The $1-million contract led to two other contracts in 1977 and 1978 to plan and then help start up the new networks for NATO's Integrated Communications System Organization.[18] Dubbed NICSO, the Brussels-based agency within NATO was created in 1971 to supervise the management of all the alliance's existing communications networks and to plan NATO's new satellite communications systems.[19]

Bell Canada's consulting unit also began to pursue extensive contracts throughout the Middle East in the early Seventies. One of their first jobs in the region was a small military contract in Iraq. Bell Canada's consultants began doing work there in 1975, four years before Saddam Hussein assumed command of the revolutionary Ba'ath Party. Kyles remembers the first job involved designing and making a special telephone set that could be put into a metal cabinet for field operations. Bell's consulting unit turned to Northern Electric to make the phones. "An Iraqi from the embassy was sent to my office on a Friday afternoon and he put a brown paper bag on my desk and he said, 'Okay, now you have been paid.'

"I opened the bag and there was $80,000 in uncirculated, hundred-dollar U.S. bills sitting on my desk. I didn't know what the hell to do with it. It still had the treasury bands on it."[20] Worse, the banks were closed and Bell Canada's consulting service's Ottawa office had no safe. Kyles took the money home with him for the weekend and joked that he could have filled the liquor cabinet extremely well.

"We also had a very good contract in Lebanon until they blew it up one day and we decided to come home,"[21] Kyles says, referring to the shell attack that demolished Bell Canada's office in Beirut. The company had won a major contract to help manage the installation of a new network in Lebanon in 1977, but cancelled the contract and pulled out of the country after the Lebanese civil war erupted again in 1980.

Although Bell Canada's consulting service lost out to its Swedish competitor in its bid in the early 1980s for a contract in Kuwait, more business from Iraq for both Bell Canada and Northern Telecom soon followed and, by the early 1980s, the two companies had done more than $40-million-worth of business there. Kyles recalls a visit to Baghdad in the summer of 1979 when a terrorist attack aimed at Hussein, who had only come to power that July, leveled the Baghdad airport. "The bomb came in on a plane in the luggage and it came in just after we left the airport."[22]

Bell Canada's consulting unit won civilian and military contracts to train Iraqi technicians and engineers. It also helped Northern Telecom win major contracts to sell small-size business switches, which were installed in various government departments,[23] and large digital microwave radio networks, which were the backbone of Iraq's telecommunications network — until they became a major target of Allied bombing missions in Operation Desert Storm in 1991. Bell Canada's consulting service set up a permanent office in Baghdad, which it maintained until a year before the invasion of Kuwait in August 1990. "We had a beautiful villa there," Kyles recalls. "It was a nice setup until the war started."[24]

The Saudi contract, in terms of both technical scope and financial magnitude, was in its own league. The proliferation of oil dollars that poured into Saudi Arabia in the wake of the energy crisis made the kingdom an economic superpower. When the Organization of Petroleum Exporting Countries (OPEC) raised the price of crude oil 70 per cent[25] on October 17, 1973, it both spurred the development of the country and unleashed the pent-up desire of Saudis for modern consumer goods. "Every Saudi wants to own a car and a phone," the *Arab News*, the kingdom's only English-language paper, stated. "If the roads are dug up or the lines don't work, there is fury."[26]

Prior to the energy crisis, Saudi Arabia was in most respects a backward country which, despite its oil wealth, lacked the basic infrastructure to diversify its economy and to provide its people with basic amenities, such as good education, public health facilities and roads. The citizenry lived under a conservative theocracy. The House of Sa'ud, the country's ruling monarchy, made Islam the law when the kingdom was forged in 1902 by Abdul Aziz ibn Sa'ud. It was Abdul Aziz who granted the first oil exploration concession a year after Saudi Arabia was established in 1932 and who set the country's moderate course.

Saudi Arabia was once described as the only family owned business to ever gain membership in the United Nations.[27] But if the kingdom is run like a business, the House of Sa'ud is the board of directors. The royal family exercises control through the powerful, unelected Council of Ministers, comprised of the king, ruling princes and key cabinet ministers appointed by the king. It was the rulers of the House of Sa'ud who decided that oil wealth should be used to trickle benefits down to the masses, thereby cementing their grip on the rule of the kingdom.

The Saudi government first raised the possibility that it might upgrade its telephone system in 1966. A year later, the international contract service at Bell Canada explored the possibility of obtaining a management consulting contract and hired a representative in Saudi Arabia. And in 1968, after a delegation from the Saudi state-owned telephone company visited Montreal during Expo '67, Bell Canada submitted a tender to operate and maintain a new telephone system. However, the kingdom declined all tenders, and Bell Canada did not submit a revised quotation the following year when the Saudis issued a second invitation. The contract was awarded to a European company.[28] When the kingdom cancelled that contract for nonperformance five years later, Bill Damon and Don McLean, Bell Canada's consulting services' employees responsible for the Middle East, urged the company to seek an invitation to bid on the Saudi contract. Bell Canada was asked to bid in 1974, and Kyles made a few trips to Saudi Arabia the following year to meet with officials to discuss their plans.

Meanwhile, as the period covered by the Saudi Planning Ministry's first five-year plan from 1970 to 1975 drew to a close, the ministry technocrats were planning a sequel designed to propel the country head-first into the twentieth century. The architect of Saudi Arabia's spectacular growth was Sheikh Hisham Nazar, who, as Minister of Planning, ranked as one of the two top ministers in the kingdom. His power matched that of fellow commoner Sheikh Ahmed Zaki Yamani, the Minister of Petroleum and Minerals, who financed the country's growth.

The statistics for the second five-year plan from 1975 to 1980 were mind-numbing. The kingdom planned to plow more than $140 billion (U.S.) into major infrastructure projects, including highways, airports, telecommunications, ports and desalination plants.[29] Compared with the $12 billion spent on the first five-year plan, the amount for the new plan seemed staggering, yet it represented only about one-and-a-half year's worth of income from Saudi oil exports, at 1980 prices and production levels, spread over five years.

Sheikh Hisham's first plan had budgeted for an expansion of the country's telephone network to 172,000 telephone lines in 1975 from the 134,000 estimated lines in service in 1970. With a rapidly expanding population of almost nine million people, the country's phone system was woefully inadequate. The second five-year plan hoped to make a dent in the demand for phone service by expanding the network by an additional 476,000 lines. But more important than increasing the number of telephone lines was the goal to extend the network to virtually every city and corner of the kingdom within five years. On top of that, Saudi Arabia also planned to buy the most modern, yet virtually untested, electronic equipment to build the world's first completely digital telephone network. The project received top priority for financing from King Khalid ibn Abdul Aziz.

Bell Canada, like most of the world's major telecommunications companies, learned of the Saudi plans to go ahead with the more comprehensive venture in late 1975 when the recently appointed Saudi PTT Minister, Dr. Alawi Darwish Kayal, retained the Boston-based consulting firm, Arthur D. Little Associates, to prepare an outline and detailed specifications for the five-year telephone expansion program. The new cabinet-level ministry for the PTT was created in October 1975. Foreign advisory groups played a crucial role in helping it and other ministries set objectives and, ultimately, evaluate foreign contract bids.[30] "They were sort of the Saudi's conscience," Kyles says of Arthur D. Little.[31]

The spadework for Bell Canada's participation in the Saudi contract was completed by Robert Scrivener during his visit to the kingdom in April 1976. He met with Dr. Kayal and made a verbal offer to establish a consortium to bid on the giant contract to equip, expand and operate the new telecommunications network. Bell and ITT planned a joint venture, under which ITT would sell and install the equipment and Bell Canada would do the management and training work. While in Jiddah, Scrivener met with Sheikh Ahmed Alireza, chief executive of the powerful Haji Abdullah Alireza & Company (HAACO), a diversified trading and contracting company and ITT's business agent in Saudi Arabia.[32]

Bell Canada and ITT had begun to court each other in late March, shortly before Scrivener's visit to the Middle East, de Grandpré says. Both companies needed major partners if either was to be involved in the project. The Canadian company needed a manufacturing partner because Northern Telecom's switches could not meet the Saudi specifications, which were based on European standards. ITT brought its manufacturing ability, as well as its extensive connections in Saudi Arabia, to the partnership. On the other hand, it required Bell Canada's extensive network management experience.

However, Harold Geneen, the ruthless chairman of what was then one of the world's wealthiest and most powerful conglomerates, was no pushover. He could not abide people who challenged his views or did not give him what he wanted, much less talked back to him. Employees that did were fired on the spot. Outsiders were usually sued. Although Geneen's reputation had been tarnished after the cover-up of the company's role in helping the U.S. Central Intelligence Agency overthrow the Chilean government of Salvadore Allende, he still wielded considerable clout.

Geneen may have needed Bell Canada's services but, in keeping with his style, he tried to browbeat de Grandpré and Scrivener into accepting an inferior deal. That made for some fairly tense discussions after the two Canadian executives arrived at the ITT Americas building in New York City on March 24, 1976, to negotiate a partnership deal with Geneen. It happened to be the day that Bell Canada's board of directors appointed de Grandpré chairman and chief executive of the utility. When the telephone call came to ITT's offices notifying the Bell Canada officers of their job changes, Scrivener was summoned back to Montreal. Although de Grandpré was left alone with Geneen, the Bell Canada chairman-to-be was able to negotiate as an equal and with the comfort of knowing he had his board's full support.

The first round of meetings simply considered the broad terms and conditions of a partnership between the two companies, but it quickly became clear that the positions of the two leaders on the role each

company would assume were markedly different. It was equally clear
that neither man was prepared to back down.

"Geneen wanted us to be a sub-contractor, instead of a full partner," de Grandpré says. "I told Harold Geneen, 'I am not going to be
subservient to ITT. We are there [in Saudi Arabia] because we believe
that our technology and experience are good, of international standards, and if you want us to be a sub-contractor, then we will not be
bidding jointly with you. I'll be a full partner, or nothing else.'"

De Grandpré says Geneen reacted very negatively to his position,
"because I was a small boy from the northern part of the continent.
But after a while, he accepted that we would become his partners. He
backed down."[33]

Scrivener's visit to Riyadh helped break the impasse with Geneen. He
obtained an invitation to make what amounted to an unsolicited bid. Bell
and ITT formed their partnership after his return and began to prepare a
joint bid for the largest telecommunications contract in history. Bell
Canada incorporated its international contract unit as a separate subsidiary, named Bell Canada International Management, Research and
Consulting Limited (BCI), in July 1976 and Kyles was named president.
But three months later, without inviting any tenders, the Saudi PTT
began direct negotiations with NV Philips Telecommunicatie Industrie
of the Netherlands.[34] De Grandpré says competitors criticized both the
idea of awarding the entire contract to a single player and Philips' lack of
experience in managing a telephone utility.[35] The U.S. ambassador to
Saudi Arabia, William Porter, attempted to intervene on behalf of potential U.S. bidders, and was rebuffed.[36] Shortly after the Philips negotiations
began that October, Bell Canada and ITT broke off their partnership.

There are many unwritten rules about doing business in Arab
countries. One is that it is acceptable to inflate a contract bid to
account for annual inflation rates. This is particularly important
when doing business in Saudi Arabia because standard contracts
there rarely include a clause to automatically account for inflation.
Philips, therefore, must have wanted a hell of a lot of money to have
had its bid rejected as inflationary, even by Arab standards, and to
have lost its chance of winning and keeping the entire $3.5-billion
contract for itself. Their bid was later estimated by industry sources
at a whopping $6 billion.[37] Unable to negotiate a compromise, the
Saudi PTT broke off their talks with Philips in April 1977. It was the
Dutch company's greed that provided Bell Canada with an unparal-
leled opportunity.

News traveled quickly through the telecommunications industry of the Saudi decision to break off talks with Philips. Kyles sent out letters to major equipment makers throughout the world looking for a possible joint venture partner, and de Grandpré called AT&T. Shortly after, Bell Canada was invited by telecommunications equipment maker L. M. Ericsson of Sweden to join a consortium it had formed with Philips. The invitation came about because of a trip de Grandpré had taken to Stockholm in early May for an international symposium on communications, held to celebrate the centennial of the company founded by Lars M. Ericsson in 1876. It was de Grandpré's first trip abroad after his appointment as chairman became effective. He met Hans Werthén, then president and CEO of L. M. Ericsson, and Bjorn Svedberg, soon to be promoted president, and the three talked about the Saudi contract and Bell Canada's previous venture with ITT. De Grandpré told them he was interested in exploring the prospect of joining their consortium, but that he would have to get back to them; he couldn't tell them that he was still waiting to hear back from AT&T.[38]

In the wake of the failed Philips bid, an alternative to the single tender had to be devised for the PTT by the Saudi Council of Ministers. It opted to split the telephone contract into three separate parts. Bell Canada was formally invited to bid on the three-part modernization contract and received the 630-page tender and specifications on May 28, 1977. The tender included two equipment contracts to provide and install the switching equipment and facilities needed to connect a total of 450,000 new lines — a reduction from the 476,000 lines budgeted for in the original five-year plan — as well as provide long-distance and international service. The third contract was an O&M contract to manage, operate and maintain the Saudi network. Bidders could tender on all three parts or on one of the equipment contracts plus the O&M contract.[39] Bell Canada and Northern Telecom formed a joint proposal team directed by BCI to see if Northern Telecom could satisfy the new tender. But Northern soon withdrew because its switches could not meet the specifications.[40] The Saudi tender called for switches with the miniature reed relays enclosed in a glass capsule to protect the sensitive contacts from desert dust and sand.

On July 27, 1977, Bell Canada announced that it was joining the Philips and Ericsson consortium. Its partners would bid jointly on parts one and two, and Bell would bid on part three of the tender. At around the same time, two other rival consortia emerged: AT&T's

Western Electric led a bid by a group of British equipment makers, including Plessey, BICC and Cable and Wireless PLC; and ITT and two of its European subsidiaries joined with United Telecommunications of Kansas. Nippon Electric of Japan also bid for part of the contract.[41] A little more than two weeks later, with no public fanfare, Bell Canada gained greater exposure in the kingdom when BCI signed one of several contracts it had received from the Arabian American Oil Company (Aramco). Aramco, which was fully nationalized by Saudi Arabia in March 1980, had extensive connections with business agents, government departments and the ruling Al Sa'uds. Under the terms of the contracts, BCI would provide mangement staff and operating systems for the giant oil company's communications department, which ran its own telecommunications networks throughout the vast Saudi oil fields.[42]

Thackray, Bell Canada's president, and Kyles were ultimately responsible for the PTT bid negotiations. Kyles appointed Doug Delaney, a vice-president of marketing at Bell Canada, as head of the fifty-member bid team made up of officials from all parts of Bell Canada. Delaney conducted most of the face-to-face negotiations in Saudi Arabia, and Len Lugsdin, an assistant general counsel at Bell Canada assigned to BCI, drafted the contracts. Kyles recalls flying back and forth every weekend to the frequent, and frantic, meetings in Stockholm and Amsterdam to put the bid together in the short time the consortium had to submit it — the deadline was September 27, only two months from when Bell Canada joined the consortium. Although Bell Canada's chairman and CEO wasn't involved in the actual negotiations, Kyles recalls receiving some precautionary advice from de Grandpré, but nothing that said "don't do it."[43]

De Grandpré briefly entered the fray before leaving for vacation. Bell Canada's partners came to Montreal on the Friday before the Labor Day weekend to finalize their proposals and to resolve a dispute over Bell Canada's contract bid. Along with Bell Canada's team, they took up suites on the thirty-fifth floor of the Hotel Chateau Champlain to hammer out an agreement. "These were very difficult negotiations because Ericsson and Philips were asking us to bid with tender conditions that were, in our view, unacceptable to us,"[44] de Grandpré says. He did not want Bell Canada to bid on some of the conditions that were spelled out in the documentation from the Saudi PTT. "And they were afraid that if one of the partners was not compliant, the whole bid would fall apart. And I was not prepared to give in on what we considered unacceptable conditions for the administration, operations and maintenance of the system."[45]

Bell Canada's contract tender for the O&M job called for it to fully staff a new telephone company that it was to organize. "But the key thing we decided was that we would not attempt to staff the whole operation ourselves; we would only provide managers," Thackray says. "We would leave the Saudi staff in place and we would provide advisors."[46]

It was a gamble because it deviated from the letter of the tender documents. However, over the course of the weekend, Kyles's team convinced the European partners that the idea made sense. It would allow Bell Canada to put in a far less costly bid, which meant the consortium's bid as a whole would have a better chance of being the lowest. And, de Grandpré figured, Bell Canada's proposal would appeal to the Saudis because it would have more Saudis in place to run their own telephone company than rival bidders had.[47]

Bell Canada and BCI spent $4 million just to put the bid together. BCI's proposal alone was six volumes and totalled 2,620 pages in length. Kyles and his team delivered the proposal to Philips in the Netherlands on September 20, where it was combined with the rest of the consortium's bid for a total of forty-six volumes. More than five copies of each volume were required, and they were packed in fiberglass crates about two meters long. Kyles accompanied the crates of documents in a chartered jet to Riyadh, where the ministry's office was located on the third floor of the PTT building. The elevators were broken, so he and Delaney had to find and hire some temporary workers to help carry the crates up the stairs in order to meet the deadline. He remembers Delaney telling him that you were not supposed to identify your bid in any way on the outside, yet AT&T had put their bid in blue and silver cases, "which, of course, are the company's colors."[48]

The Saudis opened the bids a week later on September 27, 1977. The PTT retained four international consultants to examine them — Arthur Little; Norconsult of Norway; PCR–Swedetel, a consulting joint venture between Swedetel of Sweden and Preece, Cardew and Rider of Great Britain; and the Geneva-based International Telecommunications Union.

As expected, there was a vast difference in the prices of the competing bids, with Bell's consortium coming in initially at $2.5 billion and ITT's at $4 billion. Bell Canada's contract bid on the third part of the project "turned out to be the key," Thackray says, "because the other people who were putting in bids had visualized that they would have all these people on their payroll, so that the price that they were charging for operating and maintenance was way the hell higher than ours. So when the bids were opened, of course, all the others screamed foul and said we cheated on the specs."[49]

Bell Canada's low bid on the O&M contract made their consortium the favored team over the rival AT&T and ITT bids. Geneen, for one, was furious. ITT had spent $20 million (U.S.), five times the amount of money Bell had spent in preparing its bid, ITT's chief negotiator, George Zoffinger, once told a New York businessman. And Zoffinger was supposedly authorized by his New York head office to pay a commission on the contract that was double the standard rate, according to a secondhand account of ITT's tactics.[50] As de Grandpré expected, the Saudis accepted his consortium's proposal. Acceptance, though, was only the beginning of the process. The real haggling was about to begin.

The contract called for numerous buildings to be built for the telephone company, including switching centers, repair facilities and administrative centers. BCI told the Saudis that it really did not want to be involved in the construction business and that it would undertake to find somebody who did. It was near the end of November and the contract was close to being signed off by the Council of Ministers, Thackray recalls, when Delaney called from Riyadh and said Bell Canada's consortium would be awarded the contract only if it undertook to build about $400-million worth of buildings for Saudi Telephone. The buildings had to be guaranteed for ten years and Bell Canada had to find and pay the contractors. Although the demand added "quite a bit to the size of the contract,"[51] Kyles says, the sticking point was that the Saudis wanted Bell Canada to offer its budgetary estimates for the buildings and other capital assets, such as vehicles and computers, as firm prices.[52] "So I had to go tell Jean and the board of directors that we were on the hook for another $500 million,"[53] Thackray says.

De Grandpré calculated it was not too great a risk for Bell Canada to provide the Saudis with that sweetener because they were allowed to budget for their costs in their bid. Their lower manpower costs would also help. But Bell Canada's board of directors wasn't convinced, and management's insistence that the Saudi proposal be accepted led to "an acrimonious session of the board," Dr. H. Rocke Robertson says. He adds that it was the only Bell Canada board meeting that he attended in twenty years at which he recalls any visible anger. "And I was the cause," the former director says. "Management

said, 'We have to act now,' but I was convinced it wasn't a good deal."
He argued it would be better to have no contract than a contract with
lousy terms.[54] At the end of the meeting, management was told to give
the negotiations one more shot before committing the company and to
come back to the board if the Saudis had no other position. De
Grandpré backed his team at the board meeting and, after another two
weeks of discussions with the Saudi PTT and other officials, Bell
Canada agreed to offer firm estimates of its prices.

What really steamed de Grandpré, though, was not his board's cau-
tion, but the lack of support from the Canadian government. "I never
had the impression that the federal government cared, or wanted to
help us," he says. "The Queen of England, the president of the United
States, the King of Sweden and the prime minister of Holland were all
pushing for their horse," he says, referring to the numerous state visits
to Saudi Arabia at the time. "And at one point, the Saudis [Dr. Kayal]
asked us, 'Do you have the support of your government?'"[55]

Kyles, however, says the Canadian government "supported us as
much as they could," adding that E. L. Bobinski, Canada's ambassador
to Saudi Arabia, shared his official residence with BCI executives.[56] De
Grandpré also elicited the services of Jean Chrétien, then the federal
minister of Industry, Trade and Commerce, after he ran into Chrétien
during one of his visits to Paris shortly after Dr. Kayal asked his
embarrassing question. He asked Chrétien whether he would stop off
in Riyadh to pay a courtesy call on some government ministers?
Chrétien agreed and told his Saudi counterparts that Bell Canada had
the support of the Government of Canada.

To de Grandpré, it was not the same as a state visit. Not that they
necessarily made a difference. That point was underscored when *Air
Force One*, the U.S. president's personal aircraft, landed in Riyadh late
on December 13, 1977. Jimmy Carter had sent his secretary of state,
Cyrus Vance, aloft several days earlier on a round of shuttle diplomacy
to drum up support for his latest Middle East peace proposal. Vance
had separate meetings on December 14 with King Khalid, Foreign
Minister Prince Saud Faisal and Crown Prince Fahad. The public part
of his mission, to report on his discussions in other Middle East capitals and to discuss oil prices, was widely reported.[57] But Vance's meetings were a day late to lend any last-minute support for an AT&T or
ITT bid. Bell Canada's consortium notified the Saudis on December 13
that it would make a fixed-price estimate on capital costs and, later that
day, the Ministry of PTT advised the consortium of their intention to

award it the contract with certain amendments.[58] The PTT made the selection public while Vance met with members of the ruling house.[59]

Bell Canada's skeleton bid team was hard at work on Christmas Day, 1977, in Riyadh. While Lugsdin was busy hammering out the final contracts with the Saudi PTT, a small group of pathfinders were sent to arrange accommodation, food, transportation and other services for Bell Canada's first employees to be posted to Saudi Arabia. The negotiations produced what seemed to be a mutually satisfactory contract about two days before the deadline for signing, Thackray recalls. Bell Canada translated it into Arabic and everything seemed set when the Saudis unexpectedly handed it back rewritten. Lugsdin had to sit down and bargain with the Saudis yet again for a straight thirty-six hours. By the time the contract was agreed upon, there was no time left to refine the English into polished legal prose.

The contract was signed in Riyadh on January 25, 1978. The Saudis were worried about signing such a large contract with BCI, a small subsidiary, and wanted Bell Canada's name on it. De Grandpré was busy in Paris and had a commitment back in Canada, so Thackray did the honors. Contrary to the Bell chairman's later view that federal officials had done nothing to help the company win the contract, the utility's news release announcing the contract-signing singled Chrétien out and thanked him for his "invaluable support."[60]

'Baksheesh! Baksheesh!'

The word reverberates throughout the Middle East, from the crowded streets of Cairo to the palace of the House of Sa'ud in Riyadh. Translated literally from Arabic, the word means "tip," but is more akin to a bribe exacted for services rendered — whether from an Egyptian street hawker or a Saudi Arabian prince. Bell Canada learned quickly about this sometimes exasperating Arab business custom.

For Bell Canada and its international contract arm, BCI, the extra cost of doing business in Saudi Arabia after the contract with the Saudi PTT was signed exceeded $100 million in commission payments to business agents. The Canadian utility and its subsidiaries either did business with, or were approached by, several major business families and agents in the Gulf States for a slice of their lucrative contract in return for helping to broker the deal. The list included: Sheikh Ahmed Alireza, head of Haji Abdullah Alireza & Company, the Jiddah-based trading and contracting giant; Sheikh Salem Binladen, the mercurial young head of Binladen Brothers for Contracting and Industry, Saudi Arabia's best-known construction company; Adnan Khashoggi, noted entrepreneur, agent and arms dealer; and Prince Mohammad ibn Fahd al Sa'ud, the flamboyant son of current Saudi King Fahad ibn Abdul Aziz al Sa'ud.

There is at least one pragmatic reason why the services of an agent are essential in Saudi Arabia: outsiders cannot conduct business or deal with the bureaucracy without one. The country's merchant and trading houses are part of an ancient tradition and are protected by Saudi commercial law, which requires all foreign companies doing business in the kingdom to retain the services of an agent. Certain contractors must have Saudi partners as well. A Saudi royal decree passed in late December 1977 further states that any foreign contractor that has signed a government contract must either have a Saudi partner or retain a Saudi agent.[61] These laws not only reflect long-standing Arab practice, but are also shrewdly designed by the House of Sa'ud to guarantee the political support of the powerful business and merchant families of the kingdom.

Choosing the right agent is crucial. All else being equal, a bidder will receive a contract if the king or a prince favors that contractor. Which, logic dictated, was all the more reason to select a business agent with princely connections to the House of Sa'ud. That is why Bell Canada's consortium opted to retain as their principal agent a company headed by Prince Mohammad.

There is a crucial distinction, often a fine line, between a commission and a bribe. Technically, a commission paid to a business agent is not a bribe; only if it is passed on to a government official is it considered a bribe. Khashoggi once defined the difference: "If one offers money to a government to influence it, that is corruption. But if someone receives money for services rendered afterwards, that is a commission."[62] There was no Canadian law or, at the time the contract was signed, American law, prohibiting the payment of commissions to business agents as required by Saudi law. But Opposition critics assailed Bell Canada in Parliament for the payments, in part because the Saudi contract was insured by a $430-million policy provided by the Export Development Corporation, the federal Crown corporation that provides services to Canadian exporters.[63]

Yet, for Bell Canada, there was no choice, de Grandpré says. "If you want to be in international contracts, you cannot export your way of life and your customs along with your technology and your know-how." As he perceived it, the alternative — "If you don't like the Moslem customs, you don't do business with Moslems" — was really no choice at all because "it seems to me this is contrary to international relationships."[64]

Critics would say that de Grandpré's remarks on baksheesh stem from a form of relativism, which argues it's the right thing to do because of where the business is being done. But others, such as trade groups and export associations, argue that commissions are not bribes and are simply a cost of doing business and of being internationally competitive. Bernard Lamarre, the head of the Montreal-based global consulting engineering firm, Lavalin Incorporated, once remarked to Peter Newman of *Maclean's Magazine* that when he does business in the Third World, he never hands out bribes without first demanding an invoice, and that he always pays by cheque, so that he can claim the expense on the company's income tax.[65]

According to public records filed with the U.S. Securities and Exchange Commission (SEC) in Washington, Bell Canada paid its agents an aggregrate amount that approximated 8 per cent of the gross value of its contract with the kingdom.[66] But details are scant. In shareholder prospectuses, for example, the company revealed no more than the percentage disclosed to the SEC and repeated the statement made in its SEC filings that it would "require extensive supporting services in Saudi Arabia in connection with its performance of this contract, and for this purpose has retained established business concerns in Saudi Arabia to provide such services."[67]

Saudi business practices have always remained highly secretive, reflecting traditional Arab abhorrence of publicly revealing business dealings. This penchant for secrecy is even stronger when powerful princes are involved. It is a way to shelter the al Sa'uds from embarrassment over their business dealings, which occasionally anger established merchant families. Bell Canada refused to reveal any specific details about its business agents because, like other contractors in the kingdom, it had learned a lesson from the experience of the Italian oil company, ENI. ENI had had its contract cancelled after it revealed details of its dealings with its business agent. I myself received a thinly veiled warning from a New York City lawyer in 1984 after *The Globe and Mail* published my story on one of Bell Canada's business agents. I was asked to cease any further inquiries so as not to antagonize the "powerful" interests he represented in Saudi Arabia.

Although Bell Canada never publicly revealed the identity of its agents and how much money they had received, enough information is now available from various documents and interviews to tell the complete story.

Bell Canada's first business agent was one of the largest and oldest trading companies in Saudi Arabia, Haji Abdullah Alireza & Company.[68] Owned by the powerful Alireza family, it emerged as a force in the 1860s and has since adopted a more modern name, The Golden Palm. It was one of the first Saudi trading companies to seek distributorships from U.S. and European-based multinational companies after the Second World War. Among others, it signed up Goodyear, Pepsi-Cola, Westinghouse, Ford Motor Company and ITT. Scrivener was introduced to the Alirezas through ITT, which had retained their services in the early 1950s after winning the contract to build the first broadcasting station in Saudi Arabia.[69] During his visit to the kingdom, Scrivener signed a contract with Haji Abdullah Alireza & Company (HAACO) on April 8, 1976.[70] It was a short-lived relationship. The contract was terminated by Bell Canada after the utility's equally brief partnership with ITT was disbanded in May 1977.[71] But HACCO felt that contract gave it a prior claim on Bell Canada's subsequent business when the utility's consortium won the kingdom's lucrative contract, which led the two companies into a contract dispute that was eventually settled out of court.

After BCI entered the Philips and Ericsson consortium, Bell Canada retained the services of Binladen Telecommunications Company of Jiddah as its commercial representative in Saudi Arabia on September 20, 1977.[72] Binladen Telecommunications was a joint venture that was majority owned by a wealthy family contracting business headed by a tempestuous young sheikh named Salem Bin Laden. Binladen Brothers for Contracting and Industry owned 61 per cent of the company, and Lansdowne Limited, a Bahamian-registered company owned by a group of businessmen based in New York and Paris, owned the other 39 per cent.[73] Binladen Telecommunications had been the agent for Northern Telecom's Middle East subsidiary, headquartered in Maidenhead, England, and "got on well" with Walter Light, according to Thackray.[74] It subsequently recruited a Northern Telecom employee to be its manager, according to company records.[75]

Binladen Brothers and Binladen Telecommunications both grew out of the family business founded by Salem Bin Laden's father, Mohammad Bin Laden. A small Hadhrami contractor from the sultanate on the southern coast of Arabia, he had created Saudi Arabia's largest construction firm in the early 1950s. According to Michael Field in his book, *The Merchants,* Mohammad was one of the first businessmen to directly lobby the king, which was how he gained the contract in 1951 to build the first major highway in the kingdom from Jiddah to Medina. Mohammad Bin Laden died in 1966 when he crashed his private aircraft, leaving one of the wealthiest modern Saudi business dynasties behind. King Faisal ibn Abdul Aziz al-Sa'ud took over the business in the name of the state and appointed the head of another construction company to run the Bin Laden concern until one of Mohammad's sons was old enough to manage the company. At the same time, he also banned any of the sons from flying for the next ten years.[76]

Sheikh Salem and six of his brothers took control of the family business and incorporated their partnership in January 1976. Sheikh Salem was named managing director. Like his father, he developed a passion for flying and earned his private pilot's license. Francis Hunnewell, president of Lansdowne Financial Services, once described how Sheikh Salem not only liked to throw his weight around, but also had the temerity to punch one of his American pilots in the spring of 1978 after the employee had spoken without permission to one of the sheikh's sisters. The pilot quit on the spot. That evening, during a private dinner at Sheikh Salem's residence in Riyadh, the pilot called on his former boss to ask for his pay cheque and exit visa. Hunnewell says: "Sheikh Salem Bin Laden, in my presence, launched a tirade against the man, saying that no one quit him, that he could fire people but no one could quit and finally telling the pilot that he would not allow him to leave the kingdom until he publicly apologized . . . and that if the pilot refused to work he would have him thrown in jail."[77]

The agreement between Bell Canada and Binladen Telecommunications stated that the Saudi company was retained as a commercial agent "distinct from Bell's capacity as a consortium member" with Ericsson and Philips.[78] Among its obligations, the agent would undertake to help resolve Bell Canada's contract dispute with Haji Abdullah Alireza. It would also help obtain work permits, as well as access to religious sites in Mecca and Medina[79] — slated to receive more telephone lines — even though those services were listed among the Saudi government's obligations in Bell Canada's contract with the PTT.[80]

Although Sheik Salem Bin Laden had a number of duties to perform, "It turned out he wasn't too helpful," Thackray says. "We kept him on merely because he was Northern's associate and we didn't want to make any more enemies than we had to."[81] Binladen Telecommunications was Northern Telecom's distributor in the region, and Bell Canada wouldn't let Bin Laden go for fear of harming that relationship. The young sheikh was already upset that Northern Telecom couldn't bid on the equipment portion of the Saudi contract, Thackray says. Yet, for not being too helpful, Binladen Telecommunications was entitled under the agreement to commissions of 1.5 per cent of the gross value of Bell Canada's first contract with the kingdom, or an amount equal to about $21 million against the final value of the contract.[82]

In addition to being a major expense, Sheikh Salem soon became a problem for Bell Canada when his partnership with Lansdowne Financial Services soured. Lansdowne claims to have assisted in the formation of Bell Canada's consortium, as well as to have brokered the agreement between the utility and Binladen Telecommunications. It had also signed an advisory service contract with Binladen Telecommunications under which it was to be paid a monthly retainer fee.[83] However, Sheikh Salem wanted to get rid of the middleman and, in 1980, brought in a new shareholder, the Saudi Economic and Development Company (SEDCO). The bank was represented on Binladen Telecommunications' board by its managing director, Sheikh Saleh bin Mahfooz, a member of a powerful Saudi banking family that owns the National Commercial Bank, the largest bank in the Arab world.[84] In March 1981, a special shareholders' meeting was held that terminated Lansdowne's contract.

In a suit filed against Binladen Telecommunications in a Montreal court on July 29, 1981, Lansdowne alleged that its partnership agreement had been wrongfully breached and that its Saudi partner had failed to pay it its share, more than $3.9 million, of the commissions paid by Bell Canada. The dispute was made even more complicated by the assignment of Binladen Telecommunications' contract with Bell Canada to the National Commercial Bank, which also intervened in the proceeding. It took until 1984 for a Quebec Superior Court judge to rule that his court had the jurisdiction to hear the case because the agreement between Binladen Telecommunications and Bell Canada was signed in Montreal.[85] Although Bell Canada terminated its business with Binladen after its first contract with the kingdom ended in 1982, and although Sheikh Salem died in 1987, when a Lear jet he was piloting flew into a set of power lines,[86] the lawsuit has yet to be resolved.

Not all business agents are retained at a company's choosing. It was early in the contract negotiations when Bell Canada's negotiators were approached by the extravagant Saudi arms dealer and middleman Adnan Khashoggi, who offered his services to help Bell Canada deal with "a fine point,"[87] Kyles recalls. Khashoggi was one of the princes' *wakeels*, or messengers, who are really high-priced go-betweens. For a fee from a contractor, a *wakeel* irons out minor problems that can suddenly crop up at the level of the Council of Ministers — problems whose precise nature are only known by the *wakeel* and a royal family member.

Kyles met Khashoggi at the Ambassador's Club in London for lunch in September 1977. The Saudi businessman arrived by private helicopter after having just landed in the city aboard his lavishly appointed green-and-white Boeing 727. The aircraft, which became his home and office, was the most vivid symbol of the newfound Arab wealth. Khashoggi told Kyles that he had flown to London after a breakfast meeting in Rome and was off to a dinner engagement in Madrid after their meeting. Yet it was the middle of the holy month of Ramadan, when Moslems are supposed to fast every day from dawn until dusk. The portly Khashoggi told Kyles that "the fasting only applies if you're less than one camel's day from home. If you're more than one camel's day from home, you can eat."[88]

Khashoggi made his fortune by using his connections to broker trade between oil-rich Saudi Arabia and foreign companies. He was a longtime supporter of the powerful princes in the House of Sa'ud. One way Khashoggi maintained his close ties was through a coterie of glamorous women. One of them, the Indian prostitute and former beauty queen Pamella Bordes, claims she was dispatched on a flight from Geneva to Riyadh as a present for a prince.[89] Although Khashoggi's uninhibited lifestyle and quick rise to wealth inspired Harold Robbins's novel *The Pirate*, he fell just as far and as fast as he had risen. He was implicated in the Iran-Contra foreign policy scandal that rocked the Reagan administration in 1986, and in the wake of the criminal investigations of his affairs, his overextended financial empire disintegrated.

Although Khashoggi was unsuccessful in garnering commissions for himself as an agent for the lucrative telecommunications contract, he did pick up some business as a go-between for his friend Prince Mohammad. The lion's share of Bell Canada's commission payments went to the bright, California-educated son of Crown Prince (and now King) Fahad. Few princes obtained more lucrative contracts than Mohammad, who did business through a company named Al Bilad, Arabic for "the country." Mohammad's name first surfaced in the Western press during a congressional investigation in 1975 into a controversial deal with two aircraft makers, Lockheed Corporation and Northrop Corporation. Khashoggi has always denied the charges that he was paid to bribe Saudi officials and, in the wake of the scandal, a U.S. law was passed forbidding payments of commissions abroad by U.S. businesses. After that law was passed, Mohammad gained the lucrative business of Bechtel Corporation after the San Francisco-based engineering consulting firm had a dispute with Suliman Olayan, the head of a powerful merchant business.[90] The twenty-five-year-old prince acquired a 10 per cent stake in Arabian Bechtel Company, which gave him both a share of Bechtel's massive $30-billion contract to build the industrial city of Jubail on the Red Sea and also a share of their $3.4-billion contract to build the Riyadh International Airport, the largest in the world.

According to an investigation of Saudi corruption by the *New York Times* in 1980, about a dozen princes were said to have profited from questionable commission practices.[91] During the height of the Saudi oil boom, which coincided with the telephone contract award, it was almost inevitable that profiteering by the princes would get out of hand. The avarice of Mohammad, however, was in a class of its own. By the late 1970s, he had almost single-handedly threatened the royal family's hold on power by antagonizing the established merchant families of the kingdom. Ironically, his grandfather and the House of Sa'ud's founder, Abdul Aziz, believed politics and business did not mix. He had maintained a strict ban prohibiting his family from taking part in business, a stricture that remained in force until his death in 1953. His rationale for the ban proved prophetic for his sons in the 1970s: Don't compete with the merchants for business and they won't compete with the Al Sa'uds for political power.[92]

In 1977, in part because Mohammad's activities had caused such anger among the merchant class and in part because of the embarrassment from the Lockheed scandal, King Khalid was forced to regulate the activities of the middlemen, including his nephew. His decree set a maximum amount for their fees, limited their number of clients to ten each and barred the payment of commissions to middlemen on military contracts.[93] While the new regulation did not go as far as Abdul Aziz's ban, it declared that agents were barred from exploiting influence.

Prince Mohammad first became involved with the Saudi telecommunications contract when he was retained to represent Philips in its abortive bid to win the entire contract in late 1976.[94] When Bell Canada's consortium was formed, the Dutch company convinced its two partners that Al Bilad should be retained as the consortium's agent.[95] Under their agreement, the young prince's company was entitled to receive a percentage of the entire contract price, which is what the prince originally stood to gain when Philips was in the bidding alone. The Saudi royal decree, promulgated three weeks before the contracts were signed, limited an agent's fees to 5 per cent of the contract, which would have grossed Mohammad at least $175 million on the first five-year telephone contracts alone. However, those restrictions did not apply to the prince in the case of the telephone contracts. Although they were signed after the new regulation was passed, they took effect on December 14 — the day after the letter of intent was signed, but three weeks before the restrictions took effect.[96]

As with the Bechtel contract, Mohammad also acquired a stake in a joint venture with BCI to handle the construction contracts that Bell Canada was responsible for. According to federal corporate records filed in Ottawa, he owned 60 per cent of Al Bilad-BCI Telecommunications.[97]

Although de Grandpré met Prince Mohammad, as well as several other members of the House of Sa'ud, BCI and Bell Canada's officers rarely had any direct dealings with the Saudi royal family. On a few occasions, Bell Canada paid tribute to King Khalid and his successor, King Fahad, Mohammad's grossly overweight father. In June 1980, for example, de Grandpré and Thackray attended a ceremony to mark the progress in Saudi telecommunications and presented King Khalid with a commemorative telephone set and a gold-embossed scroll. Ordinarily, Thackray and Kyles, as well as Delaney, dealt with Al Bilad's nonroyal employees.

The Al Bilad–BCI venture was dissolved at the end of BCI's second contract. Whether the prince shut down Al Bilad entirely is not known. He now has important duties of state as governor of the oil-laden Eastern Province, which borders the Iraq–Saudi Arabia neutral zone and Kuwait, and which bore the brunt of the military buildup and ground fighting in the Allied war against Iraq. In any event, the baksheesh paid to Al Bilad during the oil boom was enough to set the prince up for years to come.

Mohammad's commission on Bell Canada's contract, at 5 per cent of the gross of the revised contract value of $1.4 billion, meant the prince received $70-million worth of commissions from the utility alone. That sum was equivalent to an average of $11.66 per Bell Canada subscriber, which worked out to a rough average of nineteen cents added to the monthly bill of each of Bell Canada's 6 million telephone subscribers over the five-year period of the contract. The commission payments impacted upon the rates paid by Bell Canada's subcribers because, under the CRTC's method of regulating the utility's profit from the Saudi contract, the benefits were deemed to flow to subscribers — after deducting expenses, including the baksheesh paid to the agents. The subsequent renewal of the contract for a second five years, valued at $1.6 billion, pushed the total value of the prince's commissions from Bell Canada and BCI to $150 million — or an average of $25 per subscriber.

ITT more than just called foul after Bell Canada won the Saudi contract. Less than two months after it was signed, Geneen sought revenge at the long-running Restrictive Trade Practices Commission inquiry into Bell Canada's relationship with Northern Telecom. At a hearing in Vancouver on March 14, 1978, ITT's Canadian subsidiary charged that it had been effectively shut out of the Canadian market whenever Bell Canada was buying new equipment. Exhibiting visible frustration with the utility's practices, ITT officials also argued that Bell Canada continually blocked technological improvements. They concluded by calling for more competition in the Canadian market.[98] Little came of that particular intervention, but the point was made.

Meanwhile, shock at the scale of the Saudi victory began to sink in at BCI. Kyles recalls a meeting shortly before the contract award at which a Bell Canada official declared, "The very worst thing that could happen would be that we'd win. Then all of a sudden we'd have to produce seven hundred managers. Like pow. And they all have to be male and they can't be Jewish, and some have to be Arabic-speaking."[99]

The signing of the contract led to what Bell Canada termed a "massive, urgent recruitment effort."[100] BCI was able to draw on a pool of names collected by Bell Canada's human resources department over the preceding decade since the international consulting unit had been formed in the late Sixties. Managers were screened during their annual performance reviews and asked whether they were interested in off-shore assignments. "That was one of the real problems; always having to go back to Bell with your hat in your hand for people," Kyles says. But, he added, in the end, BCI solved its manpower problems for the Saudi contract because "the music had come down on high on that one."[101]

At about the time that the Saudi negotiations began, Bell Canada asked BCI for a forecast of its staff requirements. BCI's corporate parent told Kyles that the utility would take on as much extra manpower as the consulting unit wanted, but that BCI would have to pay for all the overhead — a proposal that BCI couldn't afford, Kyles says. He credits Scrivener with coming up with a solution. Scrivener told him that a similar consulting business set up by U.S. Steel had resolved the manpower issue with its corporate parent. Kyles visited Pittsburgh and was told that the solution was simple. The consulting arm submits manpower forecasts to the parent, just as salespeople give sales forecasts to a factory. The parent produces the staff levels required for the consulting business, and the consulting arm handles salaries but does not have to pay any additional staffing costs. BCI is required under its contract with Bell Canada[102] to pay the salaries and benefits of each employee it borrows from the utility, plus a slight administrative fee, but it does not have to pay Bell Canada's costs of staffing since they are considered to be an expense against the consulting income, which flows back to the parent company.

Life in Saudi Arabia for Bell Canada employees was both exciting and harsh. All aspects of life are regulated by the rigid Islamic shari'ah laws that govern conduct. Saudi Arabia is still one of the most conservative Moslem countries in the world. The strictures of the shari'ah forbid alcohol consumption, which precludes bars and Western-style nightlife; require women to be segregated; forbid women to drive or be seen unescorted in public; and restrict freedom of assembly and speech. In the Seventies, penalties for infractions were severe: caning, imprisonment or expulsion. And the mutawain, a religious police force directed by local Committees for the Protection of Virtue and Prevention of Vice, worked undercover throughout the kingdom to find transgressors. That still did not deter expatriates in some residential compounds from building rudimentary stills to make their own moonshine, says one former Bell Canada employee who lived there.

The hardest time for employees, in terms of their living conditions, was in the first two years when Bell Canada's residential compounds had not yet been built. Employees were billeted in sparsely furnished, rented compounds, apartments and hotel rooms throughout Riyadh, which made socializing difficult. Not all employees suffered, though. One of the homes leased for BCI's executives was a beautiful house in Riyadh owned by the Saudi ambassador to Canada. Kyles remembers telling his deputy, who was put up in the house, not to tell anybody about what he was going to show him in the bathroom: all the fixtures were gold.[103]

There was no shortage of applicants at Bell Canada who volunteered for a one- or two-year term. Despite the lifestyle changes, Saudi Arabia was considered a once-in-a-lifetime plum, because expatriates posted there paid no income tax. Basic living expenses were covered, so employees banked most of their gross annual salary.

Bell Canada hired an industrial psychologist to develop the procedures for selecting candidates. The company advertised internally for the positions it needed, then selected candidates that met the qualifications. Next, the candidates and their spouses completed a questionnaire, prepared and later analyzed by the psychologist to determine their motivations for wanting to go. Candidates that made it through that stage were briefed extensively on the country and, if they still wanted to go, were given the green light. Generally, there were two types of applicants, de Grandpré says: "Either the very young ones, with young children or no children, who wanted to take advantage of the nontaxable aspect of their income to build a pool of equity . . . or a group of employees who were close to retirement and who saw this as an opportunity to increase their income after retirement."[104]

Aided by an army of almost seven thousand laborers subcontracted from Dong Ah of South Korea, Bell Canada's partners provided and installed the equipment to add almost half a million telephone lines to the Saudi network and to expand service to an additional seventy cities and towns within three years. They also installed five thousand coin-operated pay phones. It was Bell Canada's job to help the Saudis develop a modern telephone company to run their network. Bell Canada's contract called for it to manage a new utility arm of the Saudi PTT, which was created on June 1, 1978, and named Saudi Telephone. Doug Delaney was appointed its first general manager. He worked in the PTT headquarters and reported to Faisal Zaidan, the deputy minister for telephone affairs.

Saudi Telephone inherited the 1,900 employees that were already employed by the Saudi PTT, and Bell Canada expected to increase that number to more than eight thousand Saudi employees by the end of the contract. Within two months of signing the contract, BCI had sent 150 Bell Canada managers to the kingdom; that number had almost doubled by June when Saudi Telephone was established.[105] Within a year, the Canadian utility had more than one thousand employees and their dependents in Saudi Arabia, or "Saudi," as the Canadians called it. The new phone company also hired more than 1,500 workers and laborers — 875 overseas laborers from Pakistan, Egypt and Sudan and 760 local workers.[106]

But Bell Canada walked into a mess.

Thackray recalls going into the basement of one Saudi PTT building in Riyadh to look for some records and finding them tossed in the corner on the floor near some goats. "The buildings they had were falling down. The equipment they had was hiccuping all over the damn place and they had no systems. They didn't know what the hell was going on. People were clamoring for service. Every office was just wall-to-wall people demanding telephones. The minister's office, the lobby outside his office, was choked with people. And every day he was getting complaints. And he wanted us to get all of this off his back." Kayal served at the pleasure of the king. Thackray's wife, Marie, once asked the minister how long he expected to serve in the cabinet, to which Kayal replied, "I will stay as long as the king smiles at me."[107]

Although faced with chaos, Saudi Telephone's advisors were keen to develop a modern utility. One of the first things they acquired for the new telephone company was a Madison Avenue-style visual identity, complete with logo: Saudi Telephone, in Arabic and English, printed around a green circle on a white background emblazoned in the center with a yellow desert palm tree, two crossed swords — and two telephone receivers.

The most important acquisition, though, was the Bell system's way of doing things. BCI's contract staff was the telephone company equivalent of the Peace Corps. In addition to developing the organizational structure for Saudi Telephone, Bell Canada's contract called on it to export its management practices. Access to Western methods and resources for operations, training, maintenance, research, engineering, finance, billing and planning was expected to bring order to the Saudi's backward system. Idealistic and confident in their state-of-the-art computerized methods, Canadian managers set out to convert the Saudi employees to the virtues of the Bell system, which they believed would bring progress and efficiency to the new telephone company.

The Canadian advisers quickly discovered, however, that Saudi values and customs had a profound effect on running a telephone company using North American methods. For example, because Saudi women are not allowed to be seen by strange men without an escort present, telephone installers and repairmen had to make visits when husbands were at home, which meant most residential installation work had to be done in the evenings.[108] The advisors also found that Bell Canada's methods took many of the most simple tasks, such as mailing a monthly phone bill or delivering a telephone directory, for granted. Performing those tasks alone posed a giant headache for Saudi Telephone because Saudi Arabia had no street numbering plan. Telephone subscribers had to pick up and pay their bills in person by the maturity date, or their service was automatically cut off. To solve these problems, BCI managers invented a country-wide address system by using the number designations of the line feeds and spray painting the number on the outside wall of each building served by the utility. When BCI began its work in mid-1978, less than 60 per cent of all telephone bills were actually paid. That number rose to 80 per cent within a year using the new address plan. As a result, the number of disconnections was cut in half.[109]

Bell Canada often exported North American attitudes, along with its methods and commodities. It had contracted to provide Saudi Telephone's new fleet of 2,300 green-and-white trucks and repair vans. Soon after the vehicles arrived — the bulk were for installation crews — Bell Canada organized a safe-driving clinic in Riyadh for telephone installers. The clinic, begun by Bell Canada in Ottawa, was prompted, in part, by the traffic problems caused by construction in Riyadh, but also by the Saudi drivers' well-known disregard for the basic rules of the road.

With Riyadh a vast sea of construction cranes in 1978, Saudi Telephone also launched a "call-before-you-dig" program to encourage excavation contractors throughout the kingdom to call a free service to locate underground telephone cables before they began excavating — assuming that contractors either had access to a phone or could be heard through the din.

Saudi Telephone soon added to the dust that choked most Saudi cities at the time as Bell Canada launched its $450-million construction project. Supervised by Bell Canada's contract administration, contractors hired by Al Bilad–BCI built eight Saudi Telephone compounds in Riyadh, Jiddah, Dammam, Mecca, Medina, Taif, Buraydah and Abha. They were actually entire communities, which contained all facilities and buildings for employees to work, live and shop in. The commercial center of each compound included administration buildings, work centers, maintenance buildings and training centers. Bell Canada constructed a total of eight administration buildings, twelve work centers and eight repair centers. The living area of each compound consisted of villas, townhouses and apartment buildings, plus recreation centers, medical facilities and a mosque. Bell Canada built 287 three-bedroom villas, 15 three-story apartment buildings and 24 townhouse-type residence buildings to house three thousand people. The compounds had all the basic services, such as electricity and water, as well as roads.[110]

In addition, three large training centers were erected — two large schools, one in Riyadh and one in Jiddah, and a third, smaller school in Dammam. Saudi Telephone had two mobile classrooms as well. A total of thirty-two instructors had taught sixty-one different technical courses in Arabic and English by the end of June 1979.[111] Saudi Telephone's district managers were also sent to Canada for management and other specialized training courses.

Two large IBM 3031 mainframe computers, part of IBM's powerful System 370, were shipped by Bell Canada from Europe to Saudi Telephone's data-processing center, making it the second-largest computer facility in Saudi Arabia and one of the most advanced civilian facilities in the Middle East. Saudi Telephone's systems operations center also received Bell Canada's proprietary data-processing software. The computer center in Riyadh was the first building finished in the construction project. It was opened by Kayal and de Grandpré, who visited Saudi Arabia frequently, in June 1979.

In late 1978, Saudi Telephone set up a special elite unit, named Royal Telecom, to provide communications support to the king, members of the Al Sa'ud royal family and the king's entourage wherever the king was in residence.[112] George Restivo, a Bell Canada advisor to the seventy-six-member unit during three years in the 1980s, recalls that King Fahad once took up residence with about two thousand people in his entourage for a couple of months in a huge desert camp about two hundred kilometers northeast of Riyadh. Royal Telecom had to string up lines in the compound and provide communication links to government facilities and family palaces in Riyadh, as well as to strategic military installations throughout the kingdom. Royal Telecom was also called upon to provide communications support to the king during the annual camel race, one of the largest cultural events in the kingdom.[113]

BCI and Royal Telecom were both enlisted to help provide services during the royal visit of Queen Elizabeth II and Prince Philip in February 1979. Bell Canada managed the communications services for the international press corps that accompanied the Queen and, in recognition of their company's help, de Grandpré and Thackray were

invited to meet the royal couple at a reception aboard the royal yacht
Britannia, which was docked in the Red Sea.

The Queen was allegedly decreed by the Saudis to be a man on that
tour. The visit of a reigning female monarch reportedly posed a serious
protocol problem for the Islamic House of Sa'ud, specifically, how the
Queen could be seen unveiled and unescorted by her husband in public, much less at a private audience with King Khalid. The British press
reported that the solution rested with a Saudi royal decree that
declared the British monarch was to be afforded the treatment and
privileges under Islamic law normally accorded to a man. When the
Arab News published a story denying the reports, a group of Bell
Canada employees seconded to the tour came to believe the rumor
that the decree had been passed. Said one, "Saudi newspapers were
worse than *Pravda* and only denied rumors when they were true."[114]

The Queen's visit came shortly after the first anniversary of the
contract, which was celebrated by simultaneously connecting service
to seventeen new switching centers in nine cities. In that one night,
another 80,000 telephone lines and 9,000 long-distance circuits were
connected to Saudi Telephone's network.[115] As well, by the end of the
first year, installers had worked ahead of the network crews and
installed 61,000 "silent" telephone lines and jacks in new houses and
buildings that had not yet been connected to a telephone company
switch. Saudi Telephone also opened sixteen new subscription offices
and twenty-five payment offices and, through Bell Canada's Tele-
Direct directory publishing arm, began printing eight regional telephone books for the kingdom. Meanwhile, Bell Canada's advisors
focused on improving both the quality of telephone service and repair
services. When Saudi Telephone was formed, one in every five subscribers reported a problem with their phone — a rate five times
greater than in North America. By the end of the first year, the "trouble rate" had dropped 25 per cent to fifteen faults per one hundred
telephones.[116]

One year into the Saudi contract, a Bell Canada advisor to Saudi
Telephone told the *Arab News* that the additional half-million lines
financed by the contract "won't even make a dent in the backlog of
demand." Kayal then stunned the consortium that June when he
announced King Khalid had approved a $1-billion supplement to
expand several of the kingdom's telecommunications projects. This
raised the prospect of an addition to the consortium's contract, which
was already escalating in value, due to a favorable exchange rate

against the U.S. dollar, to add another one million telephone lines to be installed by Ericsson and Philips.[117]

Yet the Saudis were rarely prompt in making their payments for the services they had bought, Robertson recalls. "There was a constant worry about being paid by the Saudis. We [the board of directors] were told at meeting after meeting that they were behind in their payments."[118] Management, on the other hand, wasn't overly concerned. "We took the heat off Dr. Kayal, and the king smiled at him more frequently, and so he smiled at me more frequently,"[119] Thackray says.

What did concern the company's leadership was the fact that Ottawa was not smiling.

"When we got the Saudi contract, I thought we would be the heroes. Instead, we were the bums,"[120] de Grandpré says. That sentiment was shared by other senior Bell Canada officers, including Thackray, who remarked, "We had more difficulty dealing with the governments in Ottawa than in Riyadh."[121] Public interest groups, including the Consumers' Association, criticized the utility's tacit support of the Arab embargo of Israel and the human rights consequences of doing business in Saudi Arabia.

One of the first people to bring the various restrictions in the Saudi contract to the public's attention was Herb Gray, the Liberal member of Parliament for Windsor West. In a question to Trudeau on April 24, 1978, Gray revealed Bell Canada's declaration to Saudi Arabia that it conducted no business with Israel and asked the prime minister whether the government would review its policy on international boycotts in view of the contract. The government's guidelines on the Arab boycott were coincidentally released on January 21, less than one week before Bell Canada signed its contract. Gray also asked Trudeau to consider the "possible existence of restrictions" under Saudi law barring entry into the country based on an applicant's religion.[122] Trudeau replied that the matter had been drawn to his attention a week earlier and that he had already asked that the government's policy be reviewed.

Gray then filed a request with the Canadian Human Rights Commission to investigate Bell Canada's employment practices under the contract. An investigator was appointed to determine whether Bell Canada unlawfully barred Jewish and female employees from working in Saudi Arabia and to prepare a report for the commission.[123] That investigation later led to two complaints, one filed by Gray and one by the commission.

Ontario premier Bill Davis also pressured Trudeau to act in the wake of the contract. He appealed to Trudeau directly to pass federal legislation to "discourage compliance with the Arab boycott," and told him that existing guidelines had been judged ineffective by the Commission on Economic Coercion and Discrimination.[124]

Thackray denied the charge that Bell had supported the Arab boycott of Israel and said the company was free to do business with "whomever we choose." In an employee newspaper in 1979, he stated that the Saudis had asked when Bell submitted its bid whether the utility or any of its subsidiaries involved with the bid "were doing business with Israel." Bell Canada declared they were not and "that was a statement of fact at the time and it still is today." He dismissed the accusation that Bell Canada discriminated against Jews and women under the terms of the contract and said that the company did not discriminate in its employment policy. "As far as Saudi Arabia is concerned, they have certain requirements for work permits, as does Canada and every other sovereign country."[125] He was even more adamant when he dismissed charges that Bell Canada was delinquent because it abided by Saudi laws and regulations: "Surely that is the ultimate in fatuous comment. Would we for one minute tolerate a foreign organization coming to Canada and saying they would not obey our laws?"[126]

"If the Saudis don't like women in the workplace, what can I do?" de Grandpré asked in a 1990 interview.[127] When discussing the employment terms of the Saudi contract, he likened the company's position to its obligation to provide telephone service and recalled an incident where Bell Canada had been asked to terminate service to a group that was making hate calls. "I said we cannot be the censor of calls. Don't ask the telephone company to be judge and jury and the party at the same time."[128]

The Human Rights Commission subsequently agreed with de Grandpré's and Thackray's logic. In its 1984 decision to drop both complaints, the commission stated: "Saudi law and custom dictate that men and women cannot work in the same place. Women may work away from home only as teachers and nurses to other women. This makes being a male a bona-fide occupational requirement for work on the contract, beyond Bell's control."[129]

Gray created a further stir when, in a highly irregular act, he showed up on his own behalf at the CRTC's public hearing room on the opening day of Bell Canada's rate increase hearings on May 2, 1978. The utility applied to the CRTC on February 1 to raise residential subscriber rates by 20 per cent, which would add an estimated $171.5 million to Bell Canada's revenue in 1978 and almost $400 million in 1979. Gray wanted to make sure that a resolution of the Windsor City Council was considered by the commission. His appearance was challenged by Ernie Saunders, Bell Canada's vice-president of law, who stated it "contravenes almost all the rules that have been laid down. . . ."[130]

After Gray had left the hearing room, lawyers for the commission and various intervenors began an energetic debate with Bell Canada's lawyers about the need to examine the Saudi contract. The CRTC's desire to regulate the profit from the contract had brought de Grandpré's already strained relations with the federal regulator to a boil. In a notice setting the date for the rate hearing, the CRTC stated that it would determine whether it had the authority to regulate Bell Canada's contract activities abroad. The legal opinion provided by the utility's lawyers during the negotiations was that the Saudi contract was not subject to CRTC approval because it was outside the commission's jurisdiction. They argued that the clause in the Railway Act that required Bell Canada's contracts and agreements to be approved by the CRTC did not apply in the Saudi case, because Parliament never intended its laws to apply outside the country's boundaries. Furthermore, the extension of that clause to the Saudi contract would constitute interference in the affairs of the Saudi Arabian government, Shaul Ezer, a Bell Canada lawyer, stated in the opinion he provided one week after the utility was told it had won the contract.[131]

At the hearing, Thackray gave six reasons why revealing the contract, even in a special closed hearing, would harm the phone company's business. They included the possibility of damaging the company's relationship with Saudi Arabia, which had asked that the contract be kept confidential; and the possibility that Bell Canada might receive fewer invitations to bid on major contracts in the future.[132] But company officials revealed under cross-examination by Andrew Roman, the lawyer for the NAPO anti-poverty group, that the utility had either filed the contract on a confidential basis with, or had shown it to, at least eleven other agencies and outside parties, including the U.S. Securities and Exchange Commission. This information enabled the lawyer to effectively challenge the company's reasons for not affording the CRTC the same treatment as a U.S. agency.[133]

Against Bell Canada's wishes, Charles Dalfen, vice-chairman of the CRTC, convened a secret two-day session on May 15 to enable the commission and intervenors' lawyers to study the contract and to question Bell Canada officials on its contents. Lawyers could not keep a copy, had only three hours to read it and were not allowed to take notes. The next day an in-camera session to cross-examine the Bell Canada witnesses was convened.[134]

"It was a fascinating exercise," says Gregory Kane, who attended the lock-up for the Consumers' Association. "It was helpful, but it wasn't a blockbuster. The frustration was we couldn't even talk to our clients about the contract! And we couldn't use anything from the contract or the in-camera portion of the hearing in our final argument, so it was interesting but not useful to us."[135]

Bell Canada argued in the public proceeding that all its profit from the Saudi contract, which it estimated to be $132 million over five years, belonged to its 228,000 registered shareholders, and that none of the revenue and expenses from the contract should be included in Bell Canada's regulated finances. In other words, the utility did not want any of the income from the Saudi venture to be used to offset subscriber telephone rates.[136] It did concede that the money it was paid by BCI for each employee loaned to Saudi Telephone (the employee's base salary, plus a small premium, plus a 15 per cent contribution to Bell Canada's overhead) should accrue to subscribers. That amount was estimated at $33 million over the life of the Saudi contract. The company also proposed to keep the books of its contract operations separate from its regulated accounts.[137]

Public interest groups and the CRTC believed that Bell Canada's employees should work for domestic subscribers alone. "What business does BCI have offering the services, if you will, of our employees to the whole world?"[138] Roman pointedly asked Thackray a year earlier at the CRTC's 1977 rate hearing. He picked up the theme again at the 1978 hearing when he asked the Bell president whether, given the contract's scale and demands on the company, senior management was convinced that none of the employees assigned to Saudi Arabia would be missed at home? Thackray replied that "you have to look at it against the base of more than 13,000 managers in Bell Canada" and answered that the company's abilities would not be impaired by sending any of the five hundred employees transferred to Saudi Arabia.[139]

One of the most crucial exchanges on the contract occurred when the NAPO lawyer asked the Bell Canada president what would happen if Bell Canada had a financial loss from its Saudi contract:

Roman: . . . where would Bell Canada propose to obtain that loss, even if we agree that it is an unlikely event? Where is that money to come from?

Thackray: I think the money will come from the shareholders, Mr. Roman.

Roman: Are you saying you would go to the market and say, 'We have now lost a whole lot of money and would you shareholders please cough up and make good the loss?'

Thackray: It would be entered in the accounts in a way that it would come from the earnings of the company . . . .

Roman: And if there were a loss or if the earnings of the company did not look terribly good, wouldn't you come back to the commission for a rate increase?

Thackray: I don't think we would, Mr. Roman, because I don't think it would be reasonable on our part to make such an approach.[140]

Roman then asked him for an assurance that subscribers would be insulated from any such loss:

Roman: Do we then have your undertaking, Mr. Thackray, that if the anticipated profit that you are going to be making on this contract one way or the other does not turn out that that amount will not be requested in the form of any rate increase?

Thackray: Yes, sir, I have no trouble with that.[141]

Lawyers also assailed the Bell Canada president for de Grandpré's remarks to shareholders at the utility's annual meeting in Hamilton, Ontario, at which the Bell Canada chairman remarked, "The regulatory process is an expensive one — for Bell, for the CRTC and for the intervenors. It would be a good thing for everybody if we could give it a rest."[142] Asked by Neil Burtnick, a lawyer for the Ontario government, whether that statement was "some sort of unreal hope on the part of Mr. de Grandpré," Thackray responded, "I think this is for us at least a very highly desirable objective and I am sure for our customers as well."[143]

The commission tossed aside Bell Canada's legal arguments that its international contract business was outside CRTC jurisdiction. Although it agreed that it had no authority to regulate the utility abroad, it stated that was not at issue. On the crucial question of how to regulate BCI, the commission pointed out that the contract was signed with Bell Canada and that BCI was exporting Bell Canada employees who had been trained at the company's expense. "What they bring to Saudi Arabia is their knowledge of how to run a telephone company, which they have gained at Bell Canada," the CRTC stated in what proved to be a momentous ruling on August 10, 1978.[144]

The CRTC then declared that, "For these reasons the Commission considers the Saudi Arabian operations to be integrally related to Bell Canada's telephone business."[145] The "integrality principle" meant that all revenue from a related business deemed to be integral to Bell Canada, "whatever form of organization the company may choose to employ for those activities,"[146] must be treated as regulated income. And that meant the profit from an integral venture, such as the Saudi contract, could not be accounted for separately and paid to the shareholders.

Dalfen's hearing panel rejected Bell Canada's contention that the risks under the contract were borne by investors. The CRTC noted that there was no capital being invested by Bell Canada in Saudi Arabia, that all its expenses would be reimbursed and that it had received a $211-million advance payment from the Saudi PTT. "The profits to be realized under the contract really amount to a 'return' on personnel employed, the risk of which is borne by both the shareholders and the subscribers,"[147] the CRTC stated in its decision. The commission also picked up on Roman's questioning of Thackray and, siding with the NAPO lawyer, found another reason to defend its controversial decision: "On the other hand, and in the event of a loss on the contract for some unforeseen reason, there seems little doubt that such loss would, indirectly but inevitably, lead to the need for further rate relief."[148]

The CRTC ordered Bell Canada to adjust its estimated revenue, used to calculate subscriber telephone rates, by adding the gross revenue from the Saudi contract.[149] The effect of that ruling was to increase the utility's total regulatory income, thereby reducing the amount of additional money the company could be allowed in a rate increase.

Dalfen was the architect of the decision and the integrality principle. The rationale for applying the principle to the Saudi contract, as he explained in a 1989 interview, was simple: "If you have the manpower to spare, then you probably have too much manpower. And the subscriber paid for that manpower, so they should get the benefits."[150]

De Grandpré called it expropriation: "It was unprecedented. The question of cross-subsidization by the shareholder to the subscriber was never an issue. It was the first time the question was placed before the regulators. I could not see how the providing of consulting services in a foreign country could be 'integral' to our operations in Canada. That I could not understand, and I still cannot understand the reason behind it."[151]

He also believed that the CRTC decision was contrary to the national interest, a point he stressed when Bell Canada requested a formal review of the decision in a twenty-nine-page application submitted to the CRTC on October 6, 1978. The ruling on the Saudi contract "continues to have an adverse effect not only on the common shareholders of Bell Canada, but on every potential investor in Bell Canada, and on every Canadian-regulated enterprise carrying on or contemplating ventures in foreign markets,"[152] Bell Canada argued. Although its common shares were trading at more than $60 a share, the utility had plans to raise an additional $150 million from another common share issue in early 1979, which made the company particularly sensitive to shareholder perceptions.

In less than three weeks, the CRTC announced the appointment of three commissioners from its executive committee who had not heard the original rate case to consider whether there were any grounds for a review of the decision.[153] The review panel unanimously rejected Bell Canada's request in a report released on February 2, 1979. Although the committee stated that it "recognizes the vital importance to Canada of a healthy Bell Canada," it found that the utility had failed to demonstrate that the CRTC's decision met either one of the four grounds that legally warranted a review.[154]

Again, de Grandpré invoked the national interest in his response: "It is a decision which, in my opinion, is shortsighted and is not in the best interests of our country. In the circumstances, we have no alternative but to appeal to the federal cabinet for a review of the CRTC position."[155] The petition, filed on March 2, 1979, stated that the long-term effect of the CRTC ruling, if allowed to stand, would be to "crush Bell Canada initiatives" abroad.[156] But the request was soon caught up in rapidly changing political events and a far greater crisis surrounding the Saudi contract.

An ill-conceived policy of the short-lived Progressive Conservative government of Joe Clark almost scuttled Bell Canada's business in Saudi Arabia. The problem began when Clark, as Opposition leader, declared during a trip to Jerusalem on January 15, 1979, that a Conservative government would move Canada's embassy in Israel from Tel Aviv to Jerusalem. The pronouncement was labeled by reporters accompanying Clark on his twelve-day world tour as a pitch for Jewish votes at home,[157] but it inflamed the Arab world because Jerusalem is also one of Islam's holiest cities.

Clark's statement took Bell Canada officials by surprise. As soon as he learned of Clark's remarks, Thackray telephoned Allan Gotlieb, who was then Under Secretary of State for External Affairs, and told him, "God, you've got a real problem."

No sooner were Clark's comments reported, however, than a representative of Al Bilad, the consortium's business agent, called Thackray to tell him that Bell Canada had a problem, and Clark's remarks "were very disturbing."

"I said, 'Well, it's not to worry too much' and, as I also told Dr. Kayal, 'it's just an election campaign thing,'"[158] Thackray recalls.

Clark repeated his promise to recognize Jerusalem as Israel's capital in late April,[159] just before the federal election campaign began. Shortly after his minority government was elected on June 4, 1979, Bell Canada publicly stated that it had been told that its contract was in jeopardy, and Thackray met with Robert René de Cotret, Minister of Economic Development and Trade, on June 13.[160] Dr. Kayal had called Thackray after Clark's election and told him, if Canada's embassy was moved, "we're going to have to throw you out of the country."[161] Bell Canada's advisors to Saudi Telephone had also begun to run into obstacles.

The Jerusalem embassy affair became a major political issue and led to intensive lobbying by numerous businesses. They selected de Grandpré as their chief spokesman to ward off the prospect of a threatened Arab boycott of all Canadian business. "I had the largest stake in the resolution of the issue," he says. "It was a very difficult period for me because I had to explain the situation, not just to Saudi [Arabia], but to the other countries in the region that Bell did business with," adding he had discussions with Iraq and Syria. To help get business's point of view across to the Clark government, de Grandpré also spearheaded the formation of the Canada-Arab Business Council.[162]

The fight dragged on for several more months as Clark ducked the issue and passed it on to former Conservative leader Robert Stanfield, who was asked by the prime minister that summer to study the matter and make a report to the government. A former advisor to the deputy minister of the federal industry minister says Clark tried to duck de Grandpré, too, during one of the Bell Canada chairman's whirlwind visits to the nation's capital to lobby the prime minister. The official says that Ottawa was abuzz after de Grandpré showed up unannounced at the prime minister's office, and was supposedly sent away because Clark wouldn't come out of his office to meet with him.[163] De Grandpré says he does not recall the incident. "Remember, when you're at the receiving end of the prime minister, you don't know whether he just doesn't want to see you or whether he is truly busy at the time when you want to see him. The people inside know the story. I couldn't pass judgment on that."[164]

Several Arab states and many businesses and government officials viewed the Stanfield mission as an inadequate response. One advisor urged Mickey Cohen, the deputy minister of the Department of Industry, Trade and Commerce, to convince de Cotret to put an end to "the Stanfield poultice." The aide stated that many businesses questioned the minister's intelligence and wondered how "someone so supposedly bright and plugged-in with the PM could have let this happen much less persist."[165]

De Grandpré and Thackray kept up the pressure and lobbied every cabinet minister for support. In October a Bell Canada vice-president attending an international telecommunications trade fair in Geneva sent a telegram to Clark that warned "doors are being slammed in the faces of Canadians and . . . Arab delegates are telling Canadians that contracts worth literally billions of dollars and thousands of jobs will not be negotiated or renewed if Canada does not quickly resolve its position on the possible removal of our embassy from Tel Aviv."[166] When the federal Opposition Communications critic, Jeanne Sauvé, asked the prime minister about the telegram in Parliament on October 12, he told her he had not yet seen it. Shortly before his minority government was defeated on a motion of nonconfidence over its federal budget on December 13, 1979, Clark accepted Stanfield's report and shelved the idea of an embassy move.

De Grandpré says his strong stance against Clark's aborted policy later gave Bell Canada an advantage in its negotiations with the kingdom for its second five-year contract. He also boasted that Dr. Kayal had singled him out for praise in a speech in October 1979 at the same Geneva fair, held once every four years: "He was pleased to note that it was a Canadian that came to the defense of the rights of the Moslem community. So I turned this difficulty, if you wish, into an opportunity."[167]

Thackray recalls that his standing with Dr. Kayal and other Saudi officials also improved, because "they had imagined that I had managed by my lobbying to get the thing turned around and get the move cancelled. Which was absolutely untrue. I was totally frustrated in all my dealings with the government."[168]

However, BCI executives in the kingdom said the chill in Canadian-Saudi relations did not thaw as quickly in the field. In fact, a year later, a Canadian embassy official in Riyadh compared the Saudi reaction to Clark's proposal to the anger later shown toward the British after the controversial film *Death of a Princess* was aired on commercial television in April 1980. The Saudis could not understand why Canada even considered the Jerusalem embassy policy. The proposal altered the Saudis' perception of Canada and Canadian companies as neutral players, Brian Tickle, director-general of Saudi Telecom (formerly Saudi Telephone), told a Canadian reporter during Trudeau's visit to Saudi Arabia in November 1980.[169]

Trudeau's visit, his first to Saudi Arabia, was widely perceived as an attempt to help repair Canadian relations with the Arab world in the wake of the embassy fiasco. It was on that trip that Trudeau developed a close friendship with Saudi oil minister Sheikh Yamani. The Petromin boss escorted him throughout the country and was his guide to Madain Saleh, an oasis on the Great Incense Route of Arabia, where Trudeau learned the traditional *moozmaad* dance.[170] Trudeau and Yamani continue to meet regularly and serve together on the international advisory council of Montreal-based Power Corporation.

De Grandpré still looks at the visit with disdain, however, and says it had no effect on the company, or its bid for a second contract. "I had told the prime minister, Trudeau, that we had a lot of people in Saudi Arabia representing Canada extremely well and [they] were an asset to the country. I think he thought it was an exaggeration on my part."[171]

Trudeau was genuinely touched by the outpouring of affection shown him by the Canadians in Riyadh, who held a rally for him on the tennis court of Bell Canada's Mursalat compound on November 16, 1980.[172] After that visit, de Grandpré says Trudeau told him, "'I never thought you had so many people in Saudi Arabia.' And I said, 'Well, I told you.'" At the time, more than seven hundred Bell Canada employees were posted throughout the kingdom on assignment to Saudi Telecom. The company had 584 employees in Riyadh alone and housed 1,200 employees and dependents. "I don't think he ever fully understood the importance of what we were doing in Saudi Arabia."[173]

Bell Canada's appeal to the federal cabinet to change the CRTC's rulings on integrality was put off yet again soon after the Trudeau government was re-elected in February 1980. That time, the delay was prompted by de Grandpré's compromise proposal to share the profit from the Saudi contract equally between shareholders and subscribers. The proposal was submitted to the CRTC along with the utility's 1980 rate increase application on February 19.[174]

De Grandpré had a far more compelling reason than retaining the profit from the venture to fight the integrality principle. Northern Telecom had just launched its much-touted DMS family of digital switches, which put flesh on the marketing promises of Digital World and gave the company a two- or three-year lead over its rivals. Although it was a relative newcomer to the international marketplace, Northern Telecom's management was intent on pursuing its ambitious goal of becoming the world's leading supplier of digital switches. That meant it had to spend enormous sums of money, particularly in the U.S. market, to expand its manufacturing capabilities. But it also had to be free to reinvest its earnings. Bell Canada had allowed Northern Telecom to do that, explains J. Derek Davies, executive vice-president of corporate strategy at Northern Telecom, by refusing to take a dividend on its majority shareholding and, instead, reinvesting that money in more of the manufacturer's stock each year.[175]

It was little wonder that Northern Telecom executives looked upon the CRTC's Saudi decision with great apprehension. "Bell was being hamstrung with its investments," says Walter Benger, then executive vice-president of marketing at Northern Telecom. "We couldn't afford to be limited and we feared that."[176]

The extension of the integrality principle to Bell Canada's investment in Northern Telecom would, management believed, destroy the economic benefits of vertical integration and create severe financial difficulties when the manufacturer could least afford it.

The first fully computerized switch made by Northern Telecom to be installed by a telephone company, a DMS-100, was put into service by Bell Canada in an Ottawa switching station in December 1978, ahead of its target for late 1979. The phone company also hooked up a much larger version for long-distance networks, a DMS-200, in January 1979. Having seen his pet project to fruition, Scrivener announced his much-deserved retirement from the manufacturer that summer, and de Grandpré was appointed chairman of Northern Telecom on November 29, 1979.[177]

De Grandpré's appointment as chairman of both parent and subsidiary was seen as linked to the strategic importance of regulatory developments to Northern Telecom's future. One of his first acts as Northern
Telecom chairman was to hire Bill Brennan, a New York headhunter, to
look for a new U.S. boss. Although operational control of Northern
Telecom was safe in the hands of Walter Light, the U.S. subsidiary,
Northern Telecom Incorporated (NTI), was in need of some firm direction. After John Lobb had retired from his post as chairman and CEO of
the Nashville-based unit in late 1977, Scrivener went on an acquisition
binge. He had bought several high-tech companies and had split the U.S.
operation into three distinct companies.

A few months after he had been retained, Brennan phoned back to say
he had found the perfect candidate to head NTI — an accomplished U.S.
businessman who had run a high-tech manufacturing and defense-con-
tract company and who had solid Republican connections to the White
House. In de Grandpré's words, "a wealthy man who didn't need to
work."[178]

The candidate was Edmund Bacon Fitzgerald, a mechanical engineer
by training. "Retired" for less than a year, he kept busy by serving as
managing director of Hampshire Associates, a small consulting company
that helped an investors' syndicate set up some coal research projects in
Wyoming. Fitzgerald had left his family-run business, Cutler-Hammer
Incorporated, in the summer of 1979. The company was founded in his
hometown of Milwaukee by his grandfather in the 1890s, and he himself
had joined it in 1946. He was appointed chairman and CEO of the manufacturing equipment maker in 1969.

A former Marine who had served in Korea, Fitzgerald had strong links
to the defense industry. After the war, he spearheaded Cutler-Hammer's
diversification into defense electronics and cultivated extensive contacts in
Washington. By holding positions such as vice-chairman of the
Pentagon's industrial advisory council, he helped his company garner
major defense contracts, including the electronics systems subcontracts
for the F-111B sweep-wing fighter bomber and B-1 strategic bomber.[179]

In fact, he was too successful.

Cutler-Hammer's performance led to several unwanted takeover bids
in 1978. That June, Eaton Corporation of Cleveland jumped into the fray
with a $378.5-million offer for Cutler-Hammer's 6.6 million shares.[180]
Although Fitzgerald did not want to sell, Eaton's extravagant offer was
endorsed unanimously by his board and accepted by the shareholders.
"We weren't happy, but the shareholders made a lot of money."[181]

Fitzgerald says he did not care for Eaton's business, which was in automobile and truck parts manufacturing. And he cared less for their plans to restructure the family business. He promised to stay on for a year as vice-chairman of Eaton's new division to manage the transition. Then, in the summer of 1979, he found himself unemployed. A baseball fanatic, he was also forced to sell his interest in the Milwaukee Brewers major league baseball team after the financially disastrous 1980 players' strike.

The worn-out former executive was finally settling into a more relaxed lifestyle when Brennan called to ask if he would be willing to travel to Montreal to talk to the head of a large company that needed a president for its U.S. subsidiary. Fitzgerald said he wasn't interested.[182] But it's a small world. When de Grandpré told Light about his interest in Fitzgerald, Light replied, "What a coincidence; we're on the board of Inco together."[183] Fitzgerald and he had only just been appointed to the board of the Toronto-based mining and metal giant that February.[184] Fitzgerald's fate was sealed after Light called him to ask if he would have dinner with de Grandpré. "I did it as a favor for Walter, but I'd pretty much made up my mind not to take the job," Fitzgerald says. He and his wife would enjoy a nice weekend in Montreal, and that would be the end of it.[185]

Fitzgerald talked the offer over with de Grandpré for about three hours. He knew a great deal more about Northern Telecom's accomplishments and plans than he perhaps let on. The company not only had a manufacturing license from Cutler-Hammer, but Fitzgerald had also known Lobb, who was a fellow Wisconsinite, for many years. After their meeting, the Bell Canada chairman took the American couple out to dinner at the prestigous Mount Royal Club. "Jean de Grandpré charmed me," Fitzgerald says. As the dinner went on, he became more impressed with de Grandpré's vision of Northern Telecom's future and says he found the challenge too tempting to resist. He accepted the offer to move to Nashville as president of NTI in May 1980. However, he did not want a repeat of the Eaton situation, so Fitzgerald made it a condition of employment that he would only work for a majority owner. "I said I'd quit if Bell Canada's holding fell below 52 per cent."[186]

De Grandpré called Fitzgerald's hiring "a shot in the dark," but says it could not have worked out better. The two men have developed a close friendship.[187] Fitzgerald, called "Fitz" by his friends, like de Grandpré, is tall, solidly built and small-c conservative. His intense shyness disguises his warmth and down-to-earth sense of humor, which borders on folksiness and tends to mask his prodigious intellect.

Looking back on his decision to join NTI, Fitzgerald says, "The nice thing about Northern was all the right decisions had been made in the Seventies."[188] He credits Scrivener with preparing much of that groundwork and says he was one of the finest corporate strategists he had ever come across.[189] There was one instance, however, where his strategy failed.

NTI had entered the hotly competitive business of designing and making computer terminals. In the late Seventies, Scrivener had acquired a bevy of companies in the field including Sycor Incorporated of Ann Arbor, Michigan, and Data 100 Corporation of Minneapolis. They were organized under Northern Telecom's Systems Corporation to manage the company's entrance into the electronic office equipment market. Although that unit accounted for 11 per cent of NTI's sales in 1979, the electronic office systems business suffered a $90-million drop in sales in 1980 to $259 million and reported a loss of almost $87 million. The division's future did not look promising. After all, the short-lived word-processor terminal was about to become a high-tech Edsel, replaced by the personal computer. And that was a market for giants. NTI's entry in that market drained precious resources and detracted from what Fitzgerald saw as the company's prime strategic mission — expanding the company's switching business.

"Everyone got skinned in that business,"[190] Fitzgerald says, referring to the office automation business. He made an immediate decision to get out of the field entirely and to take a $164-million write-off on the Sycor and Data 100 investments in the 1980 year-end. At least the timing was good; Northern Telecom was already having a horrible year financially, due to component delays and production problems with DMS manufacturing. Both conspired with the beginning of the 1980 economic recession to cause a drastic drop in profit. All of these factors led to a loss in 1980 of $185 million, despite revenue of slightly more than $2 billion, compared with a profit of $113.5 million on sales of $1.9 billion a year earlier.[191] The urgent need to concentrate on the U.S. market led Fitzgerald to side with the long-standing argument put forward by Walter Benger, executive vice-president of marketing at Northern Telecom, not to pursue every international prospect that came the company's way.

Northern Telecom's expansion to serve the burgeoning U.S. market had been both fast and furious, requiring major corporate financings and more loans. A mark of Scrivener and Lobb's success was the increased contribution from the company's growth in sales, which have doubled every five years since 1970. In order to sustain that growth, the company's capital spending had increased almost six-fold in the same five-year period, to more than $225 million in 1980 from less than $40 million in 1976. Northern Telecom's R&D budget also rose to almost 7 per cent of sales, or $140 million, in 1980.[192]

The bulk of that money was spent on adding millions of square feet of manufacturing and research space in both Canada and the United States for the DMS switches. Because the Canadian subsidiary already had a huge plant in Bramalea, Ontario, which had been making telephone company switches since 1962, there was not the same need to spend as heavily at home as in the United States. In 1980, the subsidiary opened its sprawling 250,000-square-foot manufacturing plant and R&D facility in Research Triangle Park near Raleigh and Durham in North Carolina. A measure of NTI's meteoric growth was the need to double the corporation's flagship U.S. facility within three years of its opening. Plans were made in 1984 to add more than one million square feet to the company's facilities there.[193] In addition to this plant, the company had opened its business communications plant in Santa Clara, California; started construction on BNR's associated R&D lab in Mountain View, California; announced plans to build a huge, state-of-the-art electronics chip-making plant in Rancho Bernardo in the famed Silicon Valley; and opened a large transmission manufacturing plant in Atlanta.[194]

At the same time, however, Northern Telecom smarted from the criticism of economic nationalists, who accused the manufacturer of abandoning the Canadian market by moving its investments and jobs to the United States. Yet the statistics show an increase in both the number of jobs and capital investment at home during this period. Northern Telecom Canada still employed more workers in 1980 than the U.S. subsidiary (15,736 versus 12,359).[195]

"It was terribly important for us to put up manufacturing operations down there and, to a lesser extent, some R&D," Benger says. "You have to put up bricks and mortar wherever you want to get substantial sales."[196] Fitzgerald was even blunter: "You have to do everything to appear Canadian in Canada and American in the United States. And that includes the added pleasure of paying corporate income tax in both countries."[197] Their comments are underscored by the spectacular growth of sales in the United States, which had increased to more than 35 per cent of the corporation's total revenue in 1980 from less than 15 per cent in 1976. U.S. sales increased to more than $807 million in 1980, compared with almost $92 million in 1975.[198] Propelled by sales of its digital business phone system, the SL-1 and the DMS switches, Northern Telecom's sales in the United States would more than double to $2.3 billion in less than five years.

Fitzgerald's credo has always been to take advantage of market "discontinuities." If a company isn't experiencing a major upheaval in its markets, then it's that business's job to create and exploit change to throw its competitors into disarray. The digital decision and related DMS switch introduction was one such discontinuity that "gave Northern Telecom a card it played very well,"[199] he says. Fitzgerald also planned to take advantage of a far greater discontinuity in the U.S. market as AT&T faced another consent decree — this one to break rival Western Electric's monopoly grip on the Bell system.

However, although his prime mission was to build on the company's opportunities in the United States, Fitzgerald also brought his own vision to the job. He dreamed of conquering the coveted Japanese market. And that required even greater financial clout and flexibility, which was only possible, Fitzgerald maintained, if the Bell Canada–Northern Telecom ownership alliance was preserved. "If anyone could write a book on the management of subsidiaries, the Bell Canada–Northern Telecom arrangement would be an excellent example," Fitzgerald says. "The growth at Northern Telecom has been quality growth because the financial resources of Bell Canada and now BCE have been plowed back into the future."[200]

Yet it was Bell Canada's ability to do just that that was in question at the CRTC rate hearing in early 1980. Although the Restrictive Trade Practice Commission's (RTPC) long-standing inquiry to study Bell Canada's relationship with its equipment-manufacturing arm had just about exhausted itself, the CRTC's principle of integrality was now perceived by both Bell Canada's and Northern Telecom's management as a far greater threat than the RTPC's vertical integration proceeding. It was seen as a potential impediment to Northern Telecom's need to reinvest large sums in its expansion.

At the hearing, Bell Canada's senior lawyer was rebuffed when he tried to argue that the federal government had indicated its support for the company's positions in an exchange of letters on the Saudi contract between de Grandpré and Communications Minister Francis Fox. Ernie Saunders boldly suggested that Fox and his cabinet colleagues wanted the CRTC to reconsider the principle of integrality:

I would be reluctant to propose any interpretation of the Minister's statements made in his letter, but I would suggest . . . there is this clear implication that for what it is worth the Minister and his colleagues believe that there should be reconsideration by this commission of the question.[201]

In fact, the minister's letter wasn't unsolicited, nor was it an expression of the cabinet's policy on the CRTC principle. Rather, Fox wrote the Bell Canada chairman on May 26 in response to de Grandpré's query about the status of the phone company's March 2, 1979 petition to cabinet.[202] In his letter, Fox stated that the cabinet decision had been deferred because of the rate hearing and, in response to de Grandpré's request for the minister's views on the international contract, wrote that cabinet viewed "trade initiatives such as the one taken by Bell Canada as of vital importance to Canada's economic growth." Fox concluded by stating that it was preferable for the regulatory agency to consider "the interests of your subscribers" and that he hoped "the rules of the game will be clarified."[203]

Although no one challenged Bell Canada when it filed the letters on the opening day of the hearings on June 6, lawyers for NAPO immediately cried foul after Saunders uttered his statement two weeks later. And Bell Canada had egg on its face after NAPO cross-examined Thackray on the letters. NAPO had, in fact, asked the company before the hearings started, in a written question dated April 3, whether Bell Canada was "given any assurances" by anyone in the federal government that de Grandpré's compromise proposal to the CRTC would be looked upon favorably. The exchange was particularly illuminating:

Schultz: You will note that the answer is: 'Bell Canada submits that the information requested is irrelevant, immaterial and unnecessary to a decision in this case. Bell Canada's application dated February 19, 1980, is before the commission and not before any other body of government.'. . .

You have already stated in your response . . . that any meetings with officials of government in connection with the status of that petition are irrelevant?

Thackray: Okay, that is what we said.

Schultz: May I ask then, sir, why you filed Exhibit Bell-80-626?

Thackray: Well, we thought that it was a useful contribution. . .

Schultz: Why is the attitude of the government relevant?

> Thackray: Well, to the extent that it reflects what you
> would call public policy. . . .
> Schultz: You have already agreed with me, Mr. Thackray, that
> the proposal that Bell Canada sets out in this correspondence is
> found in your evidence. I assume, therefore, that the only thing
> new in this correspondence is Mr. de Grandpré's impression that
> the government is supporting your position on this?
> Thackray: Yes, okay.[204]

Bell Canada had to eat its words, and NAPO moved that the CRTC
strike the letters from the record on the grounds that they were "irrelevant, immaterial and prejudicial" to the proceeding.[205] The CRTC
ruled in its decision that it could not withdraw the letters, particularly
after they had been admitted without contention and been the subject
of cross-examination. However, the commission "categorically"
rejected Saunders's interpretation of Fox's letter and, in a 147-page
decision released on August 12, 1980, lambasted the Bell Canada
lawyer for making a declaration "inconsistent" with the response to
NAPO's written question.[206]

Bell Canada fared even worse on the integrality issue in the decision.
De Grandpré's proposal to split the profit from the Saudi venture was
rejected and the CRTC upheld its integrality principle. Although the
commission termed Thackray's evidence and testimony "a sincere expression of concern on behalf of the company's directors and management," it
stated that Bell had "simply failed" to make a case for any different treatment of the Saudi income.[207] It ruled that the full amount of the 1980 profit from the contract, which Bell Canada estimated at $185 million, had to
be treated as regulated income. The CRTC also said it would hold a separate hearing in the future on how to treat such ventures.

"The point we tried to make, and we were not successful, was that
the customer did not invest in Bell; the customer pays for service,"[208]
Thackray says.

Management's worse fears about the extension of the integrality
principle to Northern Telecom also came to pass in that decision. The
CRTC declared that it "will require a return on Bell's investment in
Northern Telecom at a rate it deems to be commensurate with the
inherent risk involved."[209] The CRTC then pegged the rate of return on
Bell Canada's equity investment in Northern Telecom at 15 per cent a
year — 1 per cent higher than Bell Canada's allowed rate of return. As a
justification for its decision, the commission stated it wanted to protect

Bell Canada's subscribers "from the potential burden of having to cross-subsidize continuing capital expenditures of this type."[210]

The decision put a ceiling on the dividends from Bell Canada's investment in Northern Telecom that the utility could expense for regulatory purposes. If Northern Telecom had a banner year, for example, and paid twice the rate of return Bell Canada was allowed to earn, Bell Canada could not claim the full amount of dividends it reinvested. "It didn't matter what Northern Telecom earned or not, they were deemed to earn 1 per cent more than Bell," Thackray says, adding it was a "really stupid" decision.[211]

It was also the first Bell Canada decision released by the CRTC since Dr. John Meisel, a former professor of political science at Queen's University, had been appointed chairman of the agency by Clark in late 1979. Meisel remembers his first day on the job, when Charles Dalfen took him around. "Dalfen said, 'See that tower next door? That's where Bell's regulatory people are.' I asked him if they could listen in to our offices."[212]

Although Meisel had a chauffeur and car at his call, he preferred to bicycle to work most days. His relaxed and informal style, typified by his usual attire of turtleneck and jacket, was alien to most of the telephone company lawyers and executives who were used to formality, deference and hierarchy. Meisel came under harsh criticism from the telephone companies because of his lack of experience in utility economics. On the other hand, as a distinguished academic, he possessed an intensely curious and critical mind. He was not easily swayed by Bell Canada's self-serving positions or threats. "Bell kept arguing that if we, the CRTC, didn't change our integrality decision, then they would lose business abroad or not pursue further contracts. I always thought it was a fraudulent argument to try to blackmail us,"[213] Meisel says.

Fox and the federal cabinet seemed to have agreed with that assessment when they rejected Bell Canada's appeal of the 1980 decision on February 3, 1981. The minister noted in a statement that the cabinet had given "particular attention" to the utility's argument that the CRTC decision removed a significant incentive for Bell Canada to seek offshore business. Although the company was commended for "opening up new international business," the CRTC decisions on integrality were upheld because, Fox stated, the cabinet had "no practical alternative."[214]

"As a matter of policy we wanted to see Canadian enterprise go out into the international marketplace," Fox said in a 1990 interview. But, he added, cabinet was "very wary" of changing complex CRTC decisions. And it was also mindful of the consumer groups' arguments.

"The point was also made that if Bell shareholders would somehow benefit from the Saudi experience in profits, what if there were losses? Should subscribers be the ones to bear the losses? As I recall, there was no clear answer at that stage, and the CRTC was not able to find a solution that Bell would have found adequate,"[215] Fox says.

After the Saudi and integrality rulings were upheld, "de Grandpré probably perceived that the CRTC had it in for Bell Canada,"[216] John Lawrence, vice-chairman of the commission during Meisel's tenure and now a lawyer at Blake, Cassels & Graydon, says. Fox concurs, adding, "I think it was really the Saudi Arabia thing and his desire to find a solution to it which really pushed him to do the BCE reorganization. There really didn't seem to be any way out of the setup where Bell, as a monopoly operation, had to go to the CRTC for rate increases. There didn't seem any way out of it, any solution."[217]

But de Grandpré soon showed Ottawa that Bell Canada could engineer more than technological solutions.

Chapter 6

The Reorganization: De Grandpré's Revenge

We tend to meet any new situation by reorganization and attribute to this the illusion that progress is being made.
— PETRONIUS ARBITER (A.D. 66)

De Grandpré's long-simmering anger at the treatment his various ventures had received at the hands of Bell Canada's federal regulator boiled over on September 28, 1981. That was the day that the CRTC, in what had become an annual ritual, issued its decision on yet another request from Bell Canada for a general rate increase. The 1981 rate decision seemed bland at the time and, at only forty-nine pages, was one of the shortest on record. Yet it contained a ruling that set in motion de Grandpré's sweeping and controversial corporate reorganization of the 101-year-old utility.

He even knows the minute his quest to insulate the utility from further CRTC interference began in earnest. It was 10:15 p.m. Greenwich Mean Time. De Grandpré received a telephone call in Bell's apartment in London, England, from a Bell Canada official, who was calling from a pay phone in the lobby of CRTC's headquarters to brief the chairman on the decision. It was 4:15 p.m. in Hull, Quebec, and the official had just emerged from what federal officials call a "lock-up," in which reporters and interested parties get an advance look at the decision before its release time. To the reporters and lawyers who read the decision that afternoon, there did not seem to be any surprises or ground-breaking regulations in it. But when the Bell official told de Grandpré that the CRTC had ruled that it would now treat the profit from Bell's unregulated publishing subsidiary, Tele-Direct (Publications), as an integral part of the regulated telephone company,[1] de Grandpré went into a rage. His voice could be heard over the phone by a reporter at the neighboring pay phone. Years later, his voice still boomed and his fists still clenched when he recounted his version of the phone call:

"Some people said to me, 'Jean, you're so damn mad.'

"But I said, 'Never again! That will be the last decision that goes that way!'"[2]

The CRTC viewed Bell Canada's growing list of subsidiaries and affiliates, which numbered ninety-two the following year,[3] as highly problematic. "It is apparent that Bell Canada's investments have now diversified to an extent where the present regulatory treatment of subsidiaries and associated companies is no longer an appropriate method of dealing with the many problems that arise in this area,"[4] the CRTC stated in its decision. The commission also remained unconvinced that the risks of Bell Canada's investments in those ventures "should be considered to be assumed entirely by the shareholders."[5]

Using that rationale, the CRTC rejected the phone company's proposed method to account for the subsidiaries. Instead, it ordered the profit from Tele-Direct, which printed telephone directories and the lucrative Yellow Pages, to be added to Bell Canada's "other income" category on the utility's own financial statements, thereby further reducing any requirements for a rate increase by that amount. The commission then set an allowed rate of return on equity for those investments, too, which meant the subsidiaries could earn a profit of only 1 per cent above the limit set for Bell Canada. Any extra profit would have to be paid to subscribers.

The CRTC even went so far as to cast its regulatory net over all of Bell Canada's other investments that were not considered integral to the telephone company. The commission lumped all the subsidiaries together and decided to treat them as a single entity on Bell Canada's financial statements for regulatory purposes.

In hindsight, that decision "was probably enough to piss off de Grandpré forever,"[6] says John Lawrence, then vice-chairman of the CRTC and head of the hearing panel that wrote the decision.

After de Grandpré hung up the phone at Bell Canada's apartment in London, he jotted down a couple of ideas about how Bell Canada might reorganize its business to get its subsidiaries away from the CRTC. Those ideas written on the proverbial "back of an envelope" formed the basis of the actions he took when he returned to Montreal the following week. He and two of his most-trusted officials — Orland Tropea, his regulatory matters advisor and vice-chairman, and J. Stuart Spalding, vice-president of finance and treasurer — toyed with the rough idea for most of the month and came back to him with the opinion that a reorganization might be possible. The Bell Canada chairman created a highly secret task force to study Bell Canada's options. Seven of his in-house lawyers and advisors were appointed to the group. In addition to Tropea and Spalding, there were five senior lawyers on the select task force: Ernest E. Saunders, vice-president of law and corporate affairs, who was the supervisor; Guy Houle, corporate secretary and former general counsel; Josef Fridman, assistant general counsel; Richard Marchand and Marc J. Ryan, both general solicitors.

The existence of the task force was not only kept hidden from the outside world, but was not even revealed to Bell Canada's most senior executives. De Grandpré described its mandate in a memorandum that instructed the group to analyze the advantages and disadvantages of a possible corporate reorganization and to explore the question of whether, and how, Bell Canada could organize its corporate holdings under less restrictive structures.[7]

Asked about the memorandum during a 1990 interview, de Grandpré got up from behind his desk in his Montreal office and walked over to a shelf above the clothes closet in his private bathroom, where he pulled out the first of a four-volume set of leather-bound, gold-embossed books: "It's in here." The books are his personal chronicle of what he considers the greatest achievement of his life's work, the corporate reorganization that made Bell Canada a subsidiary of Bell Canada Enterprises Inc. (BCE). They contain original copies of every important document of the reorganization.[8]

The issues and questions faced by de Grandpré's task force, while not alien to its members, were novel when applied to Bell Canada. The task force quickly concluded that it needed expert advice on what a corporate reorganization of a regulated utility entailed. De Grandpré engaged the services of a large, venerable Montreal-based firm with close ties to Bell Canada's board: Ogilvy, Renault. As one of the country's leading corporate law firms, Ogilvy, Renault specialized in corporate reorganizations, which ordinarily are brought about by mergers and acquisitions.

"The preliminary questions originated with Bell Canada's lawyers, and they needed to know what they could do and how to proceed,"[9] says Kenneth Howard, the senior partner at Ogilvy, Renault who handled the Bell Canada reorganization account. Yet the initial questions from the task force were of such a general nature that the client's identity was not even revealed when the idea of a proposed corporate reorganization was first broached with the firm's chief corporate litigator, Arthur Campeau, in November 1981. The boyish-looking, blond-haired lawyer had done the court work for all of the major mergers and acquisitions brought to the firm since 1967, including Power Corporation's complicated reverse takeover, which transformed the holding company into a diversified operating company in 1972. He was asked for his opinion because, ultimately, a litigator had to be involved with any proposed reorganization of a company under the terms of the Canada Business Corporations Act (CBCA). All such schemes must be vetted by a court. The firm also involved a litigator at the planning stage because, according to Campeau, "a litigator knows the jurisprudence and is sensitive to any flaws in the plan, as well as to the court's thinking. So I'd look at a corporate lawyer's proposals and dissect them from the standpoint of the shareholders and the court."[10] He told his colleagues that the plan from the unnamed client required court approval and majority approvals from every different class of shareholder. Campeau did not hear anything more about the account for several months, until Bell Canada came back to the law firm in the late winter of 1982.

The Bell Canada task force reported back to de Grandpré on January 15, 1982. Based on the advice from Ogilvy, Renault, their study concluded a corporate reorganization under the CBCA was both possible and desirable because it would solve the problem created by the extension of the CRTC's integrality ruling to Bell's subsidiaries.

The only public clue that Bell Canada might be up to something was found in the utility's 1981 annual report, which was mailed to shareholders in early 1982. A passage where the company discusses the Tele-Direct ruling reads:

> We believe that such a treatment is unique in the history of regulation and is against the long-term interests of the subscribers and shareholders alike. We are assiduously exploring means to correct the consequences of this aspect of the decision.[11]

The means being explored was a legal "flip" to be carried out under a provision of the CBCA, the federal government's recently revamped law for federally incorporated companies. The corporate reorganization scheme adopted by Bell Canada was later described by Campeau in court as a "flip-flop."[12] Under the scheme, an obscure, existing subsidiary would become the new corporate parent of the phone company. Bell Canada, in turn, would move to a lower level of the organization chart and become a subsidiary of its new parent. Shareholders of Bell Canada would simply exchange their shares in the utility for identical shares in the new parent company. De Grandpré likened it to a "purification," because only the utility would remain regulated.[13] All other subsidiaries would be moved under the new parent company — and out of reach of the ambit of "regulatory treatment" under the CRTC's integrality principle.

Yet few outsiders at the time considered such a move even possible for Bell Canada. In order to reorganize, Bell Canada first had to be brought under the CBCA. That was considered an impossibility because, in order for it to do so, Bell Canada's legislative charter, the 1880 Special Act, would have to be dumped. And no one thought Parliament would agree to that.

The company had, in fact, tried to do something similar during Scrivener's tenure as chairman in the mid-Seventies, but to no avail.[14] The catch-22 was that Bell Canada had to ask Parliament to pass a private member's bill to amend its Special Act. That is what Scrivener and de Grandpré had hoped to do in November 1976 when a bill to amend Bell's Act was introduced, which included a provision taken from the CBCA that would have given the utility the power to create holding companies. The main purpose of the amendment, de Grandpré testified, was to put the phone company "in the same position as companies incorporated under the Canada Corporations Act or the Canada Business Corporations Act."[15] The bill died in the House, but was reintroduced in the next session by Martin O'Connell, the Liberal MP for Scarborough East, as Bill C-1001 on November 3, 1977. Both the Trudeau government and the Opposition, however, refused to support the bill.[16]

After the second bill died, de Grandpré decided to take another approach, never made public, which was to lobby to change the CBCA.

The new Corporations Act, which had received royal assent on March 24, 1975, was another law of the Trudeau era that reflected the prime minister's love of Cartesian logic. It was aimed at consolidating the confusing plethora of federal corporate laws into a single statute. It achieved that goal through a simple provision that required almost every federally incorporated company to continue its existing articles of incorporation under the CBCA. That simple move was termed "continuance."[17]

Before Bell Canada could take advantage of the CBCA's extensive provisions on corporate reorganization, it had to be granted a certificate of continuance from the federal Department of Consumer and Corporate Affairs, which administered the CBCA. There was the rub.

To uphold the doctrine of parliamentary supremacy, which has stated since the Magna Carta that only Parliament can pass a law, the CBCA prohibited any company incorporated by a Special Act from being granted a certificate of continuance. And the CBCA had made doubly sure that Bell Canada could not even try. The act included an additional clause that expressly prohibited any business "that carries on the business of constructing or working telegraph or telephone lines in Canada" from being granted a continuance under the CBCA. The inclusion of that restriction fostered the widely held belief that Bell Canada could not reorganize to escape the CRTC's tentacles of integrality.

But Bell Canada's reorganization task force and the lawyers at Ogilvy, Renault knew otherwise. De Grandpré had a trump card up his sleeve. "Jean found the solution and the way to get the changes made in a quiet and unspectacular way, which are the good, and best, ways," Scrivener says.[18]

Few people realized — no one at the CRTC, none of the public interest groups and no one in the media — that de Grandpré had succeeded in lobbying Ottawa to remove the CBCA's ban on Bell Canada's continuance. He had achieved that coup when the CBCA was extensively amended and updated in December 1978.[19] Bell Canada's role in the repeal of the clause was never revealed during the entire debate on Bell Canada's reorganization, although some observers suspected it. No evidence exists on paper to prove it.[20] Both de Grandpré and Thackray admitted in separate interviews for this book, however, that the restriction was repealed because of Bell Canada's direct intervention. Thackray says Bell Canada's lobby effort was linked to the company's desire to protect Northern Telecom:

> There was beginning to be evidence that the CRTC, their approach and attitude, was beginning to affect Northern's ability to be competitive. We had a feeling that they [the CRTC] were going to try and reach through Bell in to Northern and anything else that Bell wanted to undertake.

I think, in fact, I know, that Jean felt very, very strongly on this, and when the Canada Business Corporations Act was passed, it had that little stinger in the tail that corporations could restructure and all that kind of stuff, excepting federally regulated telecommunications enterprises — which, of course, was Bell.

It was removed by virtue of a lot of work done by Jean and Ernie Saunders. It was a very unfair thing to put there in the first place. . . .[21]

De Grandpré credits Guy Houle, then general counsel of Bell Canada, with the victory. Houle had previously served as de Grandpré's deputy when he was brought into the company's law department as assistant vice-president in 1967. He specialized in corporate law, which is why his boss dispatched him to Ottawa in 1977 to lobby the draftsmen at Consumer and Corporate Affairs to repeal the "stinger." De Grandpré says he wanted the change, dubbed the "Bell amendment," because: "I was very conscious of that impediment at that time. I'd tried in 1976 to get this provision removed, but I couldn't get anywhere. . . . I wanted it removed to have more flexibility and to be competitive. But the reorganization could not have been done with that restriction in place."[22]

Houle was later rewarded with joint appointments as corporate secretary of both BCE and Bell Canada.

The amendments to the CBCA were as lengthy and, in many ways, more confusing than the actual statute they changed. Bell Canada's clause was embodied in a section of the amendments that was, according to John Howard, the assistant deputy minister of the corporations branch at Consumer and Corporate Affairs, "one of the longest and one of the most complicated provisions in the statute."[23] The intent of that section was to bring all special-act federal corporations under the CBCA, Howard explained. Only one person in Parliament questioned Howard and his draftsman, Miles Pepper, about the change. Even then, Senator John Connolly was only concerned about the effects of continuance on the charter powers of special-act railway companies.

"I suppose that in a situation like that you would need to be careful,"[24] he stated during Howard's appearance before the Senate Standing Committee on Banking, Trade and Commerce on November 23, 1977.

That prompted Howard to volunteer that there was "one other thing" worth noting in the section — the repeal of the restriction on federally incorporated phone companies. Without mentioning Bell

Canada by name, he said, "Of course, we do this on the understanding that its business activities will be regulated by the appropriate commission, either the CTC or the CRTC."[25]

No one interjected when, a few moments later, Howard described the consequence that such a change would have by enabling "a special-act corporation that proposes to be continued, in effect, to amend its special act."[26] Although federal officials knew that Bell Canada's Special Act would legally be superceded by the CBCA, Howard's testimony indicated they operated under the misguided belief that Bell Canada's continuance would have to be "authorized by the CRTC."[27]

There were no further questions, no specific mention of Bell Canada and no headlines after the change carried.

Four years later, the task-force report approved by de Grandpré proposed a reorganization that was predicated upon Bell Canada gaining a continuance under the CBCA. De Grandpré gave the go-ahead in mid-January 1982, one week after AT&T and the U.S. Department of Justice reached their historic agreement to reorganize the U.S. telephone giant through a consent decree divesting it of its local telephone companies.[28] The task force immediately prepared the extensive paperwork necessary to obtain the company's continuance. Although the team remained the same, it soon involved about thirty more tax experts, accountants and corporate finance analysts who were brought in to research the issues for the more complicated reorganization that followed.[29] Only then were the company's senior executives told of the plans. At this time, de Grandpré also created a special committee of the board of directors, which was struck to act on behalf of the board to oversee and approve the technical details. De Grandpré chaired the committee, which also included Tropea and one outside director, John Henderson Moore, the former chairman of the executive committee of London Life Insurance Company and retired chairman of Brascan Limited, the investment holding company owned by Peter and Edward Bronfman.[30]

In early March, the Bell Canada lawyers went back to Ogilvy, Renault. "They came to us with volumes of questions and issues,"[31] Howard says, referring to the working paper that detailed each proposed step of the plan and the dozens of questions that Bell Canada had to resolve before its plan could be launched.

The first formal step was taken when Bell Canada notified the corporations branch at Consumer and Corporate Affairs that it was considering applying for a certificate of continuance. The phone company submitted its proposed filings in confidence and requested that officials check the material to ensure there was no impediment. The documents were delivered by hand on March 17, 1982.[32]

The process of being continued under the new legislation was relatively simple. Under another CBCA amendment, a company's board of directors could act alone to continue a company — without shareholder or court approval — if the company's articles of incorporation were unchanged. Therefore, Bell Canada declared in its proposed certificate that all provisions of its Special Act, which would become its new articles of incorporation after continuance, would be included unchanged in its articles of continuance. But Bell Canada noted that, after continuance, shareholders could make any amendment to the new articles of incorporation.[33]

Asked by reporters why Bell Canada's application was handled perfunctorily, Frederick Sparling, director of the corporations branch, told the *Wall Street Journal*, "We don't ask questions about a company's intent when they apply." However, the reporters incorrectly inferred from that statement that the department did not realize the implications of what the journalists termed a "seemingly innocuous move."[34]

Although the corporations branch no longer reviewed articles of continuance for their legality, Ogilvy, Renault advised Bell Canada to have the documents checked in advance by officials at Consumer and Corporate Affairs to ensure that the utility would receive the same treatment as other corporations.[35] That proved to be sound advice. In a memo to the head of the corporations branch, Jean Turner, then chief examiner of records, asked for advice on whether the continued application of Bell Canada's Special Act "should be permitted simply because we do not ensure that articles conform to law."[36]

In a confidential letter accompanying the filings, Richard Marchand, Bell Canada's general solicitor, stated that the planned continuance had "not been discussed with any other government agency or department" and requested that it not be mentioned to either the DOC or the CRTC, "so as to give us an opportunity to contact them first."[37] Neither Francis Fox nor John Meisel were ever advised by Bell Canada in advance about the continuance, or what it meant. Although Ottawa was a sieve in the latter years of the Trudeau reign, word of Bell Canada's plans somehow never leaked out of Consumer and Corporate Affairs that spring.

According to de Grandpré, all the transactions had to be secret because of fears of insider trading and to avoid the prospect of some investors being able to preferentially deal in Bell Canada shares because of their knowledge.[38] But Bell Canada had at least two other reasons for the intense secrecy. Although it was acting within the law, the company's lawyers no doubt feared that premature disclosure prior to approval might have led to a challenge from either the CRTC or opponents. And, because continuance was only the prelude to the reorganization, early disclosure could also have led opponents, or the government, to take steps to foreclose that option to Bell Canada.

Consumer and Corporate Affairs agreed on April 7 to issue a certificate of continuance if the articles submitted for approval were similar to the draft application.[39] Bell Canada submitted a completed application and associated documents on April 16. Marchand asked the department to hold the application "in escrow," pending its approval by the company's board of directors at a special board meeting scheduled for April 20 in London, Ontario. The board met on Tuesday morning, several hours before the company's annual shareholders' meeting at the Holiday Inn City Centre. That gave Marchand enough time to advise the department of the vote, so that "we can be issued a certificate of continuance within a few minutes after such approval."[40]

The board passed Resolution No. 3, authorizing the company to apply for continuance. Marchand telephoned Consumer and Corporate Affairs and a copy of the resolution was faxed to Ottawa. The certificate was granted and dated to take effect the next day.[41]

De Grandpré stood at the podium in the Commonwealth Ballroom after the routine business was transacted and began his remarks to shareholders with the ringing declaration, "Your company has transformed itself from what was essentially a local telephone company into what is now a world-renowned multinational operation." Then, near the end of the speech, he provided the rationale for Bell Canada's pending plans without mentioning the reorganization by name:

> How well we do in the years ahead will depend on our ability to stay abreast of, and indeed ahead of, the rapidly advancing fields of technology which are driving our various business enterprises. It will also require us to maintain a flexible and aggressive stance in the marketplace, and to order our business affairs in ways which will enable us to make the best possible use of all of our corporate resources. We intend to do just that.[42]

His promise was lost on the phalanx of reporters attending the meeting, who dismissed it as public-relations hype. They also snoozed through the next, and final, remark of the chairman's speech, which should have sent off alarm bells:

> . . . the company, acting on instructions from the board of directors, has applied for and been issued a certificate of continuance by the federal department of Consumer and Corporate Affairs, with effect from April 21, 1982. This will bring Bell Canada under the same corporation law, the CBCA, as is applicable to virtually all other federal business corporations.[43]

No one asked de Grandpré to elaborate on what that statement, and what continuance, meant. Although shareholders, the media and subscribers were oblivious to the significance of his remark, they had been publicly warned.

Secrecy again shrouded the next moves of Bell Canada's task force and the Ogilvy, Renault team as the lawyers prepared to consummate a corporate reorganization of the newly continued utility under section 185.1 of the CBCA.

On paper, the "flip-flop" scheme appeared deceptively simple. In reality, it was a complex undertaking, with an added element of irony in how it was effected. Bell Canada's lawyers did not simply switch Bell Canada and Bell Canada Enterprises Incorporated on the organization chart. Rather, Tele-Direct Limited, which was a unit of Tele-Direct Publications, was first renamed Bell Canada Enterprises.[44] The irony was that Tele-Direct was the company affected by the CRTC's rate decision in September 1981, the decision that had so angered de Grandpré and prompted the reorganization.

Yet de Grandpré says that was "purely coincidental."[45] He explained that the elaborate, intermediary steps involving Tele-Direct were required, because by then he had conceded that the publishing company was integral to the utility. That meant Tele-Direct Publications stayed behind, still under Bell Canada's ownership and the CRTC's regulatory scrutiny. "It was a bone that needed to be thrown out," de Grandpré says of the concession. "Maybe it was at a moment of weakness that I decided to do that."[46]

De Grandpré says he takes no credit for the format of the reorganization. It was certainly not a novel scheme, having been embraced more than a century earlier by Forbes and Vail as the corporate structure favored both for AT&T's precursor, National Bell Telephone Company of Boston, and the Bell Telephone Company of Canada. And, four months before de Grandpré formed his reorganization task force, another Canadian-owned and regulated utility with a variety of unregulated interests in natural resource exploration and distilling, Hiram Walker-Consumers Home Limited of Toronto, had performed a corporate flip that was almost identical to the proposed Bell Canada reorganization.

Hiram Walker-Consumers Home was created from the amalgamation of Consumers' Gas with Hiram Walker-Gooderham & Worts Limited in April 1980. The following year, the merged company changed its name back to Consumers' Gas Company Limited and, in a "flip-flop," became a subsidiary of one of its previous units — Hiram Walker Resources Limited.[47] Henry Clifford Hatch, chairman, president and CEO of Hiram Walker Resources and the brains behind those transactions, had been a director of Bell Canada since his appointment to the phone company's board by Scrivener in 1974. He reciprocated and appointed Scrivener to Hiram Walker's board in 1975. Scrivener was also invited to sit on the board of the amalgamated company in 1980.[48] The bond between the two men was strengthened when Scrivener's daughter, Katherine Ann, married Hatch's son, Richard.[49]

But de Grandpré dismisses any presumption that the Hiram Walker-Consumers' flip inspired his thinking or was a useful model for Bell Canada. He says his reorganization was different: "We were pioneers in this area and we were unique because we were a statutory company." He added that no other company incorporated under a special act had been continued and reorganized under the CBCA until Bell Canada.[50]

In any event, the corporate reorganization was far more involved than simply renaming a company and shuffling a couple of subsidiaries and a parent company around on a corporate organization chart. De Grandpré was not involved with the "clutter type of meetings" that dealt with the specific details worked out by the lawyers and tax accountants, Campeau says. Rather, in keeping with the Bell Canada chairman's management style, he was concerned with plotting strategy and resolving problems. Campeau met with de Grandpré several times between March and September and says the Bell Canada boss was available at any time to offer any assistance — especially in dealing with hurdles.[51] For example, he took a keen interest in a provision in the federal budget handed down by Allan MacEachen on November 12, 1981. A rather obscure income tax change in that budget actually played a key role in the financial aspects of the reorganization and dictated the sense of urgency de Grandpré attached to the approval processes. The provision waived the tax previously paid on the conversion of securities for a corporate reorganization; however, the exemption was only temporary, set to expire at the end of 1982. The deadline did not leave Bell Canada much room for delays and, according to Francis Fox, de Grandpré ultimately lobbied ministers directly for an extension.[52]

Much of the work between March and June was done by the lawyers at Ogilvy, Renault as they waded through the detailed planning documents provided by Bell Canada's task force. Each question was answered, legal opinions were provided, stacks of documents readied and financial analyses provided. Those tasks were made more onerous given Bell Canada's complex capital structure. It was not enough to merely swap common shares or gain a simple majority approval of all shareholders to complete the reorganization. The CBCA stipulated that each and every class of shareholder had to approve of the deal. Aimed at protecting the often-overlooked rights of minority shareholders, that provision in effect gave each class of shareholder a potential veto on the reorganization. "The problem was that there were so many classes of shares and all sorts of obscure, old issues. Then there were the people who forgot they even had the shares or had lost their certificates. We had to try to find them all,"[53] Donald Cruickshank, a former vice-president at Bell Canada recalls.

Howard's group at Ogilvy, Renault drafted the information circular that would be mailed to shareholders and worked out the legal aspects of the financial transactions involved. In addition to exchanging Bell Canada's shares for BCE shares, the new parent company had to put together a deal to acquire Bell Canada's subsidiary investments. Before those shareholdings could be transferred to BCE, they had to be valuated. The transfer was considered an intercorporate transaction, so the accountants calculated the price for Bell Canada's sale of those investments at book value, or historic cost. The other alternative, based on current stock market prices for the investments, was four times costlier than the historic cost. Much of the behind-the-scenes analysis was aimed at building a detailed defense of that transaction, because it was expected to come under intense scrutiny.

It was equally important for the lawyers to ensure that the transaction preserved Bell Canada's capital structure after the reorganization. The deal had to be painstakingly worked out to protect the interests of the utility, the shareholders and the subscribers,[54] Campeau says. If the transaction caused a revenue shortfall, for example, Bell Canada's critical financial ratios could be affected, which would raise the cost of its debt. And that would have an impact upon subscribers by putting additional pressure on the utility to ask for higher rates. It was also important to ensure that Bell Canada shareholders held the same proportionate stake in the new holding company and that holders of debt securities had the same call on BCE assets.

The proposed transfer was actually structured in such a way that Bell Canada subscribers stood to benefit in two ways.

First, the transaction would produce a slight capital gain for Bell Canada. This extra income would offset the phone company's revenue requirement by an equal amount. Bell Canada based the price of its investments on the CRTC's calculations of their historic cost, or $444 million. That price compared with the $1.9 billion it would cost to buy all of those investments on the open stock market at share prices on September 30, 1982.[55] BCE proposed to pay for the subsidiaries with an equal amount of preferred shares, which are nonvoting debt securities that are either convertible into cash or common shares after a fixed term. Unlike a straight cash transaction, however, those shares paid a 15 per cent annual dividend, which meant Bell Canada would collect an extra $66.6 million a year from BCE for the five-year term of the preferred shares.

The second benefit would be realized at the end of the five years when the preferred shares would be redeemed for cash. Bell Canada stated it planned to use most of the proceeds to help finance its capital spending projects and to reduce some of its external debt.[56]

Offsetting those gains, however, was the reduction in profit resulting from the transfer of assets to BCE's control. Circumventing the principle of integrality meant the profits from numerous investments, including Northern Telecom, would no longer benefit Bell Canada subscribers.

Part of Campeau's job was to prove Bell Canada's claims in court. He was involved in drafting memorandums on every issue that both he and Bell Canada's team anticipated might arise in court. The CBCA requires a company to obtain approval for a reorganization from the court of highest jurisdiction in the province in which it is headquartered. Bell Canada's plan had to be brought before the Quebec Superior Court. To expedite court approval, Campeau wanted to file the memorandums as evidence in advance. The timetable was dictated by the need to take advantage of the Income Tax Act provision, which necessitated court approval by late fall.

Bell Canada's lawyers took a preliminary step toward the reorganization when the first document to change the share structure of Tele-Direct Limited was filed at Consumer and Corporate Affairs on June 10, 1982. The Bell Canada board of directors met on June 15 and approved the circular for the forthcoming shareholders' meeting that would be called to approve the reorganization. A second document to change the name of the Tele-Direct unit to Bell Canada Enterprises Incorporated was filed on June 17 to take effect on June 22. The day after the board meeting, senior Bell Canada executives went to New York City to give secret briefings to representatives of the major bond-rating agencies, Standard & Poors and Moody's Investment. Bell Canada obtained their assurances that the plan, as structured, would not alter the utility's debt ratings.[57]

While Spalding and his financial experts made the rounds on Wall Street, Tropea quietly briefed Finance Minister Marc Lalonde; Consumer and Corporate Affairs Minister André Ouellet; Ottawa's most senior civil servant, Michael Pitfield, the clerk of the Privy Council; and both Opposition leaders to gauge political reaction to the proposal. His briefings were evidently couched in generalities and no word leaked out.

De Grandpré recalls taking over the old boardroom for a couple of days as the deadline to file the necessary documents approached. He spent an entire day just signing his name to the stacks of documents — articles of incorporation, amendments, shareholder circulars, board resolutions and court affadavits. After the reorganization committee of the board was briefed, he took Resolution No. 7 to a full meeting of the board of directors the following week. On June 23, the board approved the resolution, which authorized the company to take all steps to effect a reorganization of the Bell group of companies.[58]

The secret maneuvering fostered the illusion that the reorganization was a sudden act and a startling *fait accompli*. The lack of public understanding of the nuances of the complex steps and lobbying that led up to the continuance caused even favorable commentators to erroneously equate the reorganization with deregulation "through an act of absent-mindedness."[59] Those perceptions were spawned by de Grandpré's surprise announcement of the creation of Bell Canada

Enterprises Incorporated at a press conference at the Queen Elizabeth
Hotel in Montreal on June 23, 1982.

De Grandpré described three pressures that led to the reorganization: the increased complexity of Bell Canada's businesses; competition; and the need to "purify" the utility's regulation. In elaborating on
the first reason, he revealed a more personal rationale for the reorganization. "The chief executive officer of Bell Canada could not concentrate
his efforts as much as he would have liked to on the total corporate picture, caught as he is as CEO of an operating company as well as a holding company,"[60] de Grandpré told reporters in response to a question.
Several months later, he told the CRTC under cross-examination that:

> The new organization will permit me, as chief executive offi-
> cer, with my colleagues who will join me in BCE, not to have
> any executive responsibility of a telephone company. That, in
> my view, is a very significant aspect of the reorganization, an
> aspect that, for a number of years, had troubled the board of
> directors of Bell Canada.[61]

A decade later de Grandpré remarked, "You cannot compare the
responsibilities of a CEO at Bell Canada today and prior to 1983." In
his opinion, the utility is more "focused" today. The management
responsibilities of the telephone company's boss were far more complex "when it was [also] a holding company from 1973-1983" — a
period that coincided with his tenure as president and chairman.[62]

Journalists labeled the news of the reorganization as the most dramatic announcement in Bell Canada's history. Although de Grandpré
says the financial community was "dancing in the streets,"[63] the reaction of the financial analysts who attended a second press conference
at the Royal York Hotel in Toronto the following day was characteris-
tically understated. They termed it "positive."

The reaction of the CRTC, concerned government lawyers and
public interest groups, on the other hand, was one of wonder and
bewilderment. NAPO's Andrew Roman called the reorganization an
"end run" of the CRTC. It soon sunk in at the DOC and CRTC that
they had been left in the dark for more than two months, since Bell
Canada obtained its continuance. Although Tropea had briefed select
officials and ministers the week before the reorganization was
announced, other ministers and senior officials who were closer to Bell
Canada's affairs had been excluded. Communications Minister Francis

Fox says he was given no advance notice. "It took me by surprise and it took my deputy, Bob Rabinovitch, by surprise. I think it took most of the department by surprise."[64] The Department of Communications was told of the reorganization in a letter from J. Albany Moore, a Bell Canada assistant vice-president, to Alain Gourd, senior assistant deputy minister, only a few hours before the press conference.[65] Meisel says the CRTC was briefed the day before.[66]

After the initial shock had worn off, the regulators and Bell Canada's opponents began to ask: How did Bell Canada do it? Could it get away with it? Was the phone company still regulated? The federal government bookstore in Ottawa sold out of its copies of the CBCA as lawyers struggled with those questions.

"These grand things always resolve to very simple principles that have to be addressed directly,"[67] Campeau says. Foremost among everyone's concerns was how the reorganization would affect Bell Canada's Special Act. De Grandpré stated in his introductory remarks at his press conference, almost as an aside, that, "incidentally, the federal charter of Bell Canada, as approved by Parliament, will not be affected by the reorganization."[68] However, Fox and Meisel, as well as public interest group representatives, were not convinced. Fox recalls having two meetings with Bell Canada officials within a week after the reorganization was announced and telling the company that "we didn't have the necessary expertise or bodies in the department to be able to evaluate the reorganization that was being put forward. We were not against it, but we weren't for it because we didn't have enough information."[69] Although Fox left it as an open question whether the reorganization should have been such a surprise, Meisel says de Grandpré and Bell Canada made a strategic blunder by leaving the commission in the dark and "not putting themselves in the shoes of others."[70]

De Grandpré sought to alleviate those concerns by stating that even though BCE would not be regulated, Bell Canada would continue to be regulated by the CRTC. Pressed further, he maintained that the provisions of the utility's Special Act remained enshrined in Bell Canada's articles of incorporation after continuance.

Such simple assurances failed to satisfy Bell Canada's regulators. From Fox's and Meisel's perspectives, the reorganization itself was not at issue. What they grappled with was the legal impact of continuance, and whether it had impaired the CRTC's ability to regulate the country's largest telephone company.

The fact that the provisions previously enshrined in Bell Canada's Special Act were now part of Bell Canada's articles of continuance was particularly worrisome. The CBCA stated that these could be changed at any time by the company, provided a majority of the board of directors agreed and two-thirds of the company's shareholders concurred. The continuance of Bell Canada created a conflict between the CBCA and Bell Canada's Special Act. The CBCA declared that its provisions prevailed over any other act of Parliament, which meant that the Special Act did not have the same force as it did before continuance.

Hudson Janisch, an expert on regulation at the Faculty of Law at the University of Toronto, recognized a further complication in one of the first independent legal analyses of the reorganization, published later that fall by the Canadian Law Information Council's *Regulatory Reporter*. Janisch noted that Article 6 of Bell Canada's certificate of continuance stated, "there are no restrictions in these articles of continuance on the businesses which the corporation may carry on." Yet Article 8 reaffirmed all the provisions of the Special Act. Which provision prevailed? Janisch asked.[71]

Bell Canada created that confusion itself after it changed the wording of Article 6 from the original draft of the certificate first submitted to the corporations branch. The draft version, obtained under a federal Access to Information request in 1989, stated that "the corporation may not carry on any business which it is expressly prohibited from carrying on by the provisions of the Bell Canada Act of Incorporation."[72] To avoid limiting the company, and because the provisions of the Special Act were already included in its articles of incorporation, Bell Canada's lawyers dropped that statement from the final version that was approved by the Department of Consumer and Corporate Affairs.

Within six weeks of Bell Canada's public announcement, the Department of Communications, the Canadian Radio-television and Telecommunications Commission and the Restrictive Trade Practices Commission had each initiated separate inquiries into the reorganization. As the obstacles from those inquiries presented themselves, de Grandpré began a period of intense crisis management.

Shepherding the reorganization required a different kind of leadership, Dr. H. Rocke Robertson says. "De Grandpré had to get through all those hurdles, and to do it in one year was amazing." But the retired director was equally impressed by de Grandpré's ability to lead the company's own bureaucracy through such a momentous change. He described the special traits revealed by the Bell chairman during the reorganization as, "the leadership quality itself that emanates from a leader; a gift or ability to know what has to be done, who can do it and the ability to persuade them to do it."[73]

The leadership that was needed went beyond the daily activity of making and defending decisions and analyzing objectives director Louise Brais Vaillancourt says. It required someone to inspire allegiance to a radical objective and, above all, she says, it required a leader who possessed an intense drive and single-mindedness.[74] Campeau put it more bluntly: "This thing had real balls and needed someone with conviction to see it through."[75]

De Grandpré is what John Nicholls, a U.S. management consultant, has called a "transforming autocrat." A transforming autocrat must possess more than the perception to see what has to be done and absolute conviction in articulating his vision or goals. Such a leader must "lead by inspiration" and be able to "empathize with everyone involved," says Nicholls. Although the leader effects the transformation largely through autocratic decisions, he or she possesses a rare ability to work closely with subordinates.[76] The person who best embodies the traits of such a leader, Nicholls says,[77] is Lee Iacocca, the charismatic chairman of Chrysler Corporation, who rescued the troubled U.S. automaker from bankruptcy in 1979–80. De Grandpré and Iacocca have much in common and are, in fact, personal friends.

De Grandpré certainly inspired his board, his senior executives and the members of the Bell Canada team working on the reorganization. Control of all the crucial decisions required during the reorganization were placed "very much in de Grandpré's hands" by the board, Vaillancourt says.[78] Yet he lacked the ability to empathize with the many people outside the corporation who had a public interest in the reorganization. When it came to having to deal with those groups, particularly government bodies, the autocratic side of his leadership persona overpowered the transformer.

As an outsider, Campeau had a unique perspective on the reorganization, which enabled him to shed some light on de Grandpré's intransigence: "In the end, this case was just one of more than fifty files I had on the go at the same time, and was just one more reorganization. To the client, however, it is the only one and probably the only shot they have. If I looked at it from the client's perspective, I would be intimidated; it would make the task look almost too formidable or daunting."[79]

The Bell chairman took a special interest in helping Campeau plot the strategy for the court case, Campeau says, and it was very early on during those discussions that de Grandpré rejected the idea of involving the CRTC. They considered two options — going to the CRTC first, then seeking court approval as required, or bypassing the commission altogether. De Grandpré dismissed the alternative of voluntarily laying Bell Canada's proposal before the CRTC: "We opted to go to the court first under the CBCA and to argue that the CRTC had no authority over the proposal. We would let the court ratify the reorganization. It was being holier than thou, but the court afforded every intervenor any chance to appear."[80]

In fact, Justice Charles Gonthier broadened the definition of interested person at a preliminary hearing to approve the documents to be sent to shareholders on June 25.[81] Campeau says he was thrilled when he learned that Bell Canada's case would be heard by Judge Gonthier, who is now a Justice of the Supreme Court of Canada. "He is very fair and I knew if we had a judge who was fair and knew the law, we'd win."[82]

The Restrictive Trade Practices Commission was the first government agency to intercede in the reorganization. The RTPC wrote Bell Canada's lawyers on July 8, 1982, to ask a number of questions about the proposed change under the rubric of its existing sixteen-year-long inquiry into the Bell Canada-Northern Telecom relationship.[83] The previous fall, it had issued its first report on the telecommunications equipment industry, which recommended that Bell Canada's telephone equipment monopoly be ended.[84]

De Grandpré gave his concluding testimony before the RTPC inquiry on April 13, 1981. Walter Benger, then an executive vice-president at Northern Telecom, flew to Ottawa with de Grandpré and Alexander John MacIntosh. MacIntosh, who was the senior partner at Blake Cassels & Graydon, the Toronto law firm that acted for Bell at the RTPC hearings, was along to help brief the Bell Canada chairman on his testimony. Even Scrivener deferred to MacIntosh, whom he characterized as one of the most well-connected members of the Canadian Establishment. Benger, an engineer by training, was chewed out by the lawyer on board the Falcon when he second-guessed MacIntosh and told de Grandpré not to confuse the panel with the level of detail that the lawyer's instructions entailed. Benger said it was "irresponsible for de Grandpré to deal at such a miniscule level of detail."[85]

The Bell chairman scored a major victory in the inquiry when Lawson Hunter, the DIR, declared a stunning reversal of the combine branch's original position on vertical integration in his final argument submitted to the RTPC on July 17, 1981. The Director, who had originally recommended that Northern Telecom be divested, argued that vertical integration was in the public interest. In return for keeping Bell Canada, Northern Telecom and BNR intact, Hunter recommended that the telephone company be obligated to buy its telephone equipment through competitive bidding.[86]

"The DIR broke up Bell's monopoly on pole attachment, interconnection and telephone ownership, but we failed to break up Northern Telecom," Gordon Henderson says. Ironically, Henderson added that he, too, had been converted and become a proponent of vertical integration for the cable television business. "We need the economies of scale in this country because we don't have the pools of capital. It's the very opposite of what the DIR argued initially."[87] That view was reflected the following year in the RTPC's third and final report on vertical integration released on January 7, 1983, in which the concerns that had launched the inquiry seventeen years earlier were finally laid to rest.[88]

It is worth noting that nowhere in the RTPC report is the term "vertical integration" defined. Vertical integration has been used, particularly by government, to characterize a trait of the telephone business, rather than being seen as a broad term that describes a generic organizational form. That is because the term is synonomous with the Bell system's triad of utility, manufacturing arm and research and development subsidiary. The concept was assailed by regulators in North America in this century because of the stigma that had been attached to the extension of the utility's monopoly to the supply of equipment through a preferred, wholly owned supplier. The RTPC's affirmation of vertical integration and Bell Canada's retention of its corporate structure more than upheld the status quo, however, because it provided Northern Telecom with a strong competitive advantage in light of the impending breakup of AT&T the following year.

The RTPC released its slim, seventeen-page report on the proposed reorganization of Bell much more quickly. Too quickly, in fact. Bell Canada responded to the RTPC's written questions on July 21. MacIntosh's colleague, Warren Grover, stated that the CRTC had no jurisdiction over the reorganization and that Bell Canada therefore believed no legislative changes from Parliament were required.[89] Less than a week later, on July 26, without any hearing or request for formal evidence from Bell Canada, the RTPC sent its report on the reorganization to André Ouellet, the minister of consumer and corporate affairs. The RTPC concluded that the reorganization had "a number of positive aspects," but that "the public interest requires that a reorganization should not take place unless there has been full public consideration of the probable effects of the proposal, with respect to both subscribers and the telecommunications industry."[90]

MacIntosh provided Bell Canada's legal department with an opinion that the RTPC had improperly failed to give the company a fair hearing on the reorganization. By breaching its legal obligations on procedure, the report was a "nullity." MacIntosh agreed with the utility's lawyers, though, that "no immediate reaction on the part of Bell would be useful."[91] However, his opinion provided powerful ammunition if the government attempted to use the RTPC report either in court or at the CRTC.

The report did support the position of the DIR, who told Fox in mid-July that a public inquiry was needed because he believed the reorganization "may limit regulatory oversight of Bell's operations." Hunter was the only federal government official outside the DOC or CRTC who called for a hearing, according to a briefing document prepared for Fox on July 16, 1982.[92]

Meisel also met with Fox and lobbied the communications minister intensely for cabinet support for a hearing. A briefing document prepared by DOC officials after the meeting spelled out three alternate methods of holding a hearing. The CRTC could hold a general fact-finding inquiry on its own, either as an in-house review or in conjunction with the public hearing on Bell Canada's pending rate application. The third alternative, preferred by Meisel, was for an order-in-council from the federal cabinet, calling for a hearing and ordering the CRTC to report back to it. He believed such a reference would "add some weight" to an inquiry.[93]

Fox was also told by DOC's lawyers that the CRTC had no authority to block the reorganization, "nor could it, solely as a result of the special hearing, make any decision affecting rates charged to subscribers." Meisel had two chief concerns that he wanted addressed by a hearing: first, that the new holding company, and not the CRTC, might decide what services would be offered by Bell Canada on a regulated basis; and, second, that the reorganization might adversely affect subscriber rates. Although the DOC officials agreed with Meisel's concerns about subscriber rates, they rejected his first concern about the reorganization's impact on regulation. They argued that the CRTC's ability to regulate the telephone company remained unchanged and that any proposal related to the offering of new or existing services still had to be approved by the commission.[94]

The officials advised Fox to adopt the neutral course of urging the CRTC to examine the reorganization during the upcoming hearing on Bell Canada's June 28 rate application scheduled to begin on October 26.[95] The CRTC proceeded, instead, with a separate general inquiry that began with a public notice issued on August 12 asking interested parties to comment on several broad issues, including the effects of the reorganization on regulation, the question of whether prior CRTC approval of Bell's corporate reorganization was needed and the CRTC's jurisdiction. Meisel left it open as to how the CRTC would proceed, with the notice simply asking for comments by September 27, "after which the commission will determine what further action may be necessary."[96]

Publicly, Bell Canada had little comment on the CRTC notice. In a statement released the same afternoon as the notice, the company stated it was "studying" the questions and reiterated its position that the reorganization "will not impact subscriber rates."[97] Privately, however, de Grandpré vehemently objected to any investigation by the CRTC, no matter how perfunctory. At a meeting with Bell Canada's lawyers on August 20, he argued that the commission could not regulate Bell Canada's corporate affairs. He also stated his position that none of the laws governing the CRTC gave it authority to approve of the reorganization, and he also stressed that the CBCA had made no mention of the CRTC having any jurisdiction over the continuance or reorganization of a special-act company.[98]

But de Grandpré's views were both tempered and, to a degree, rejected by Ogilvy, Renault's lawyers, who stated in a memo on legal strategy written the same week that: "It would be nice to be able to say that the CRTC has no business investigating the corporate structure of the company. However, [the Railway Act] makes it clear that the CRTC is authorized to get into the company's affairs to a significant degree." Campeau cited several provisions, which, he concluded, "would appear to make it difficult to argue too strenuously that the present reorganization is none of the CRTC's business."[99]

That issue was at the heart of the company's pending court case to approve the reorganization. The formal application was filed in Quebec Superior Court in Montreal immediately after the special shareholders' meeting on August 18. Six separate votes for each class of shareholders were conducted and at least two-thirds of the shareholders voting in each class had to approve of the proposed change. More than 98 per cent of the votes cast (an average of all groups) approved the reorganization and more than 68 per cent of the company's eligible shareholders, from all classes of shares, were represented at the meeting.[100] De Grandpré optimistically told shareholders that he expected to have the reorganization "in place by October 2"[101] — a forecast that was contingent upon no opposition at the court hearing.

Although Campeau expected the CRTC to eventually "take a look at the reorganization," he says he was not sure whether it would intervene in court. However, he was "absolutely convinced that this thing was right and I never had any doubt about what we were going to achieve."[102] His opinion no doubt bolstered de Grandpré's belief in the propriety of the reorganization. The stakes were enormous for Bell Canada, underscored by the resources, including staff and money, which de Grandpré brought to bear on the initiative. There were high stakes for de Grandpré, too, who would face a leadership crisis if, for any reason, his bid failed. Campeau says he never had a corporate client that was so thoroughly prepared and that drilled him so intently in preparation for a court hearing.

Bell Canada's hearing, originally scheduled for the day after the shareholders' meeting, was delayed until August 23. Campeau took about four minutes to dispense with his opening formalities, then Pierre Bourque, a lawyer with the Montreal firm of Desjardins, Ducharme, Desjardins & Bourque, rose to address the court and asked for standing on behalf of the Attorney General of Canada.[103]

Lawson Hunter said the court intervention was his idea. "I think my shop even paid for the attorney general's intervention," he added. The DIR could not directly intervene in the proceeding because his interest in the matter would be perceived as too narrow, so he went directly to Ottawa's most powerful civil servant, Michael Pitfield, for approval to intervene in the reorganization proceeding through the attorney general's office.[104] After consulting with officials at the Department of Justice, Hunter selected Bourque for the case.

Bourque was a prominent Montreal litigator who had a reputation for his unorthodox, flamboyant style. Unlike most lawyers, who deferred to authority by appearing in court in traditional barrister's robes, Bourque eschewed the lawyer's garb and showed up in court that day wearing a sapphire-blue suit.

Campeau argued the government had no business to come before the court. Bourque responded by arguing that the court was obligated to hear the attorney general's argument that the court lacked the jurisdiction to approve of the reorganization because the law required the

The Reorganization: De Grandpré's Revenge

CRTC to give prior approval. In addition to an overriding public interest, Bourque added that the Crown reserved a traditional right to intervene in any action. "It was the Queen's court and she can come here whenever she wants,"[105] Campeau said in a 1989 interview. Judge Gonthier granted the attorney general's request for standing and adjourned the hearing until September 14 to give Bourque time to prepare his arguments.

De Grandpré huddled with his lawyers and asked how they should proceed. Campeau told him that he could preclude any arguments by the attorney general "with respect to anything other than the question of CRTC approval."[106] He had the advantage of acting for the petitioner. Because he had filed most of his evidence in advance, he could confine the oral arguments in court to the points raised by Bourque's intervention. Campeau had anticipated the attorney-general's intervention earlier that summer. Aided by Bell Canada's legal team, he had already prepared a detailed rebuttal to the key legal points that the attorney general would attempt to make his case on, which relied on a similar opinion prepared by Bell Canada's legal staff more than five years earlier.[107]

In the end, there was no contest.

Bourque's attempt to halt the progress of the reorganization was based on a single argument: that the prior approval of the CRTC was required because of Section 18 of Bell Canada's Special Act. His case hinged on the meaning of a single word, "disposition." The section stated:

> The company shall not have power to make any issue, sale or other disposition of its capital stock or any part thereof, without first obtaining the approval of the Canadian Radio-television and Telecommunications Commission of the amount, terms and conditions of such issue, sale or other disposition of such capital stock.[108]

The irony of Bourque's case was not lost on Campeau, who realized that de Grandpré had tried to throw Section 18 out of the act in 1968 during the hearing into the extensive amendments to Bell's charter. De Grandpré had agreed at the last minute to keep the provision in to alleviate some concerns that committee members had had.[109]

Bourque argued that the exchange of Bell Canada shares for BCE shares to effect the reorganization was a disposition of the utility's capital stock and therefore required the CRTC's approval. He spent most of the afternoon of September 14 citing numerous definitions of "disposition" and "stock" to support his case. Carlyle Gilmour, the disgruntled rooster farmer and shareholder, also intervened and attempted to argue that the reorganization was inimical to subscriber interests.

Campeau's rebuttal to Bourque was eloquently simple: Bell Canada was not the party disposing of shares in the transaction. Bell Canada's shareholders, not the utility, were the ones who were going to swap shares for BCE stock.[110] Judge Gonthier agreed, and rejected the attorney general's intervention as "unfounded" in his judgment of September 24. He declared that there were no grounds for Bell Canada to obtain the prior approval of the CRTC and approved of the utility's plan.[111]

The judgment created an urgent problem for the CRTC, Lawrence says. "I had argued that we had the power under our acts to at least look at the reorganization. But we lost that power in court."[112] And, by a strange twist, the federal government's "6 & 5" price-control program to fight inflation led Bell Canada to repeal its request for a rate hearing, which removed the sole option available to the commission to study the reorganization on its own.

The company withdrew its request for a 15.1 per cent rate increase on August 11, 1983, a week after it had announced its compliance with the restraint program. It had filed its rate increase on June 28 — just five hours before Finance Minister Allan MacEachen unveiled the "6 & 5" program in his second budget speech. The economic recession, a spiraling inflation rate soaring above 12 per cent a year, and more than 1.25 million people out of work had led MacEachen to announce a program to limit all federally controlled wages and prices to increases of 6 per cent in the year after the budget and 5 per cent in the following period in 1983-1984.[113] He advised the House on July 6 that instructions were being prepared for regulatory agencies, but that "due process ought to be observed in such a case as Bell Canada, for example."[114]

Fox echoed that view two days later when he told Parliament that "under its statutory obligation the CRTC must receive rate applications from Bell Canada, must hold hearings on those applications and must come to a decision on what rate of return ought to be allowed." Nevertheless, he stated that he expected the CRTC to follow the "6 & 5" guidelines. He also hinted that the cabinet might overturn any CRTC decision that breached those guidelines: "If the cabinet feels that the rate increase awarded by the CRTC is not justified, it can return the matter to the CRTC or indicate which way the decision ought to go."[115]

Meisel was prompted by those statements to write a secret letter to Trudeau the following week in which he told the prime minister that his agency could not override its statutory obligations in order to enforce the guidelines. He bluntly stated that, without legislation or a cabinet order, he could not be bound by MacEachen's statement and would have to judge each rate application on its merits.[116] Trudeau was equally blunt. He sent each of his cabinet ministers a letter on July 27, made public the same day, saying that he expected strict adherence to the guidelines and that exceptions would only be considered in exceptional circumstances. Trudeau also stated that any exception would have to be approved by a special sub-committee of the Priorities and Planning Committee of cabinet.[117]

Bell Canada had issued a statement a few days earlier in which de Grandpré both denied reports that the company had defied the restraint measures and indicated that the utility was prepared to discuss alternatives to its rate application, "which could be envisaged by the government, the CRTC and Bell."[118]

Those discussions led to an unprecedented action by the federal cabinet on August 5 when it approved an order-in-council that changed the previous CRTC rate decision in 1981 and awarded Bell Canada a 6 per cent rate increase in 1982 and a 5 per cent increase in 1983.[119] The action was in direct response to Meisel's concerns and eliminated the need for his agency to deal with Bell's application in the face of MacEachen's guidelines. Fox defended the government's action on the grounds that Bell Canada had requested a "clarification" on the application of the "6 & 5" guidelines for its corporate planning. He also said that the utility agreed to accept the guidelines "in view of the government's determination to grant exceptions to its administered prices policy only in exceptional circumstances."[120] Bell Canada stated publicly that it agreed to comply with the guidelines because it "recognizes the urgent national need for stringent measures to control inflation."[121] Yet "6 & 5" was not that stringent and, in the case of some of its services, gave the telephone company more money than it was requesting from the CRTC. That realization led de Grandpré to strike a bargain with the government.

Fox says he was approached after the government announced the "6 & 5" program by de Grandpré and J. V. Raymond Cyr, executive vice-president of Bell Canada, to discuss the implications of the program on the company's operations and financial performance. De Grandpré told the minister that compliance with the guidelines would result in some layoffs, a conclusion that DOC officials supported in their analysis of the telephone company's finances.

"That wasn't really the type of result the government was interested in," Fox says. After what he termed "frank and open" discussions, he and de Grandpré "came to, I don't want to say a deal, because that has a pejorative connotation, but an entente." Bell Canada agreed not only to withdraw its rate increase application, but also not to lay anybody off during the two-year life of the program. In return, the federal cabinet "allowed them to increase their rates immediately, thereby allowing them to pick up some additional funds between September and the time they would have had a rate increase," Fox says. "There are some persons in the House that felt that we made a sweetheart deal with Bell, but, on the whole, I think everybody's objectives were met."[122]

The entente between Ottawa and Bell was short-lived, however, and the cooperative spirit engendered by the "6 & 5" agreement was quickly shattered when the federal government found another way for the CRTC to probe Bell Canada's corporate reorganization.

Francis Fox rushed into the auditorium of the National Press building in Ottawa and took his place behind the microphone shortly after 4:00 p.m. on Monday, October 25, 1982. Tall, debonair and confident, the minister of communications began a short speech that was guaranteed to anger de Grandpré. He announced two moves by the Trudeau government that stalled Bell Canada's proposed reorganization. First, the attorney general of Canada had decided to appeal the ruling of the Quebec Superior Court approving the reorganization; and, second, the federal cabinet had directed the CRTC to conduct a public hearing into the corporate flip.[123] Any influence de Grandpré might have thought he had with Fox, who had articled in his law firm, was dispelled after he made his dramatic announcement.

The government's court appeal, lodged that moment at the Quebec Superior Court of Appeal in Montreal, came in just under the wire; the deadline was 5:00 p.m. The federal cabinet's order-in-council, dated three days earlier, directed the CRTC to hold a public hearing on the reorganization. The order also referred four questions to the CRTC for study and gave the commission until March 31, 1983, to report its findings back to cabinet.[124]

Ottawa's actions came on the heels of a stern request by the Quebec minister of communications, Jean-Francois Bertrand, who asked Bell

Canada to explain its proposed reorganization before a committee of the Quebec National Assembly on October 7. Arguing that Bell Canada had not yet revealed enough details of the plan, Bertrand stated, "The merits of this restructuring are still to be proven to the Quebec government and the consumers at large," whose financial contributions, Bertrand added, allowed Bell Canada to become a multinational corporation.[125]

"Those who dare to fly a little higher always draw flak,"[126] was Campeau's reaction to the government's interventions. But they were perceived by de Grandpré as hostile acts. "He felt Bell Canada would get through the reorganization without a public hearing and that the CRTC had no business to examine it," Donald Cruickshank, a former Bell Canada vice-president, recalls. "I've never seen Jean so mad. He was climbing the walls for three weeks."[127] De Grandpré's deep outrage stemmed from his belief that the government refused to "abide by the decision of a majority of the owners of the company in a democracy,"[128] pointing to the overwhelming vote in favor of the proposed reorganization at the special shareholders' meeting.

Contrary to de Grandpré's opinion, however, Fox echoed the sentiments of his senior advisors, who advised him that "the government does not in principle oppose the reorganization proposal."[129] Both the minister and the DOC stressed that the government had not attempted to block the reorganization, but was "merely providing an opportunity for all of the issues to be examined and resolved fairly. To proceed in any other manner would be irresponsible."[130]

"The issue was the autonomy of the State," Meisel told Fox. "I said, 'Will the government of Canada tell the telephone companies how to operate, or will the telephone companies tell the government?'"[131] He went back to Fox for a cabinet order empowering the CRTC to hold a hearing because the commission had neither the power nor the opportunity to investigate the reorganization after the Quebec court decision and Bell Canada's withdrawal of its rate application.

Both Fox and Meisel strongly believed that the public policy implications of Bell's reorganization, as well as the statutory obligations of the CRTC, provided compelling grounds for a public hearing. "The proposed reorganization can offer a number of potential benefits on the industrial side. However, it is important to find out whether and to what extent Bell subscribers would be affected, and whether the CRTC would continue to have regulatory control over Bell's telecommunications services,"[132] Fox stated in his announcement of the cabinet order under a special provision of the National Transportation Act.

Although a draft of that announcement had been leaked a week earlier, only those sworn to a privy councillor's oath knew how narrowly Fox's request had come to being squelched by his cabinet colleagues.

The order was first presented to the powerful Priorities and Planning Committee, which was normally chaired by Trudeau. When the reorganization hearing made the agenda early that October, the meeting was chaired by MacEachen, who was now the deputy prime minister and secretary of state for External Affairs. MacEachen mustered enough votes in the powerful inner cabinet to refer the request to the full cabinet. "His view was that a CRTC hearing would defuse the controversy surrounding the reorganization,"[133] a former aide to another cabinet minister says.

When the order came up for discussion at the weekly cabinet meeting on October 22, which was held that week on a Friday instead of the usual Thursday, however, it was clear that Fox had only a handful of supporters for a CRTC hearing. The ministers were divided over whether a CRTC public hearing was necessary or was an unwarranted intrusion into Bell Canada's corporate affairs and legal prerogatives. Yet when the vote was taken, Fox managed to find enough supporters to produce a split vote. One of Fox's aides had researched which ministers Bell Canada had lobbied on the reorganization and discovered a consistent pattern to the utility's methods over the preceding couple of years. The telephone company consistently targeted senior ministers and those in several important portfolios, while overlooking others. Armed with those findings, Fox went around the cabinet table and pointed to which of his colleagues Bell Canada had deemed important enough to be lobbied. There were enough bruised egos among his colleagues for Fox to sway a 50-50 tie.

"In those situations, the prime minister's way of doing things was very traditional," Fox says. Trudeau told him, "'In these circumstances, I always feel it is up to the responsible minister to decide on the course.'" He told Fox to think about it over the weekend and to call him back on Monday with a decision on "which way you want to go, and we'll go that way."[134] Hence the Monday press conference for an order-in-council dated Friday.

Two days later, Marc Lalonde, who had replaced MacEachen as finance minister in a September cabinet shuffle, granted Bell Canada its much-needed extension of the income tax exemption for convertible share roll-overs. That mitigated any financial penalty from the delay caused by the CRTC hearing.

The income tax exemption was crucial in order to consummate the reorganization, as Alain Gourd, DOC's senior assistant deputy minister, had stated in a briefing memo to Fox one day before the crucial cabinet meeting on the order-in-council. Prior to the proposed tax change, shareholders were allowed to defer any accrued capital gains tax that resulted when they exchanged their shares during a corporate reorganization. The proposed tax change would have imposed a capital gains tax on every Bell Canada shareholder at the time of the reorganization. Every shareholder who opted to retain new BCE shares would have had to pay a tax equal to one-half of the difference between the selling price of their Bell Canada shares on the day of the reorganization and the purchase price. "The proposed reorganization would be less attractive to many Bell shareholders if it would result in them being required to pay taxes now rather than when they ultimately dispose of their investment,"[135] Gourd told the minister of communications.

The only way for Bell Canada to obtain a delay of the new provision was for Fox to directly approach Lalonde, Gourd stated. Having gained de Grandpré's compliance with Finance's "6 & 5" program, as a quid pro quo, it was not too big a favor for Fox to ask. "And that's the way it went. There was no animosity with Bell Canada or anybody,"[136] Fox says.

According to Meisel, however, de Grandpré did harbor resentment toward Fox.

Less than a month after Fox's announcement, the CRTC chairman light-heartedly asked de Grandpré, "Have you forgiven me?" at the annual bash thrown by Thomas d'Aquino, the lawyer and head of the Business Council on National Issues. He says the Bell chairman lashed out in reply, saying he thought Fox has lied to him and "betrayed him." Meisel told him that Fox could not have misled him, because the Minister of Communications had not known whether a hearing would be held until the last minute. "But de Grandpré believed Fox had double-crossed him."[137]

De Grandpré admits he was angry, but says "betrayal" is too strong a word. "Disappointed, yes. Frustrated, yes. But I could not feel that I was betrayed because I had never received any assurances that there wouldn't be a hearing."[138]

Both Meisel and his telecommunications deputy, John Lawrence, believe de Grandpré could have defused the crisis atmosphere, and avoided delay, if he had simply volunteered to cooperate with the CRTC early on.

"Why didn't you come to the CRTC for a look at the reorganization to address and air the concerns?" Lawrence asked de Grandpré when the two men ran into each other at Telecom Canada's annual dinner that November. The two lawyers sat in a corner and discussed the recently issued order-in-council and the CRTC's pending inquiry for at least an hour and a half. Lawrence says he told de Grandpré that the reorganization not only raised a "valid public policy question," but that the CRTC "had no prior views." De Grandpré, however, explained how frustrated he was by the CRTC's rulings on the Saudi contract and integrality and told Lawrence that he had not wanted to get hammered again at what he called a "kangaroo court." Although Lawrence held the view "that Bell didn't owe it to us to brief us in advance," he says he found it hard to believe that de Grandpré would take "such a naive view that the government couldn't look at the reorganization."[139]

The CRTC issued its public notice setting out the procedures for the inquiry on November 2 and scheduled a public hearing to begin on February 1, 1983. As instructed by the cabinet, the CRTC asked parties to examine four key issues: whether the reorganization would result in increased rates for subscribers; whether it would impair the CRTC's mandate; what responses were needed; and, if the proposed reorganization were implemented, whether limitations should be imposed on Bell Canada.[140]

Although the CRTC kept an open mind, some DOC officials leaned toward the prospect of legislative amendments to uphold the restrictions in Bell Canada's Special Act. A DOC briefing paper prepared for government caucus members before the CRTC hearing began suggested that the need for legislative action would be more acute if the attorney general lost his pending appeal.[141]

Perhaps that is why Lawrence's assurances failed to calm de Grandpré.

In late November, the Bell Canada chairman made an astonishing decision to take his fight with the government directly to the company's shareholders. In a handwritten appeal reproduced and mailed to each of the company's 291,500 shareholders,[142] he moved to organize the small army of voters by eliciting their support for the company's lobby efforts. De Grandpré wrote that the government's actions were ironic, in view of Trudeau's televised appeal for greater efforts to improve productivity and exports: "The government's intervention in October, directed as it was against one of the country's few very successful, internationally competitive, high-technology exporters, was in stark contradiction to the prime minister's call for action." He urged each shareholder to write their MP in support of the proposed reorganization.[143]

"They were voters, too. It was a dangerous decision; it could have backfired," de Grandpré says, adding that the initiative was his own idea.[144] Although Donald Cruickshank, then Bell's vice-president of communications, takes credit for suggesting a handwritten note, the inspirational seed for the lobby effort seemed to have been planted by B. A. Redman, a Bell Canada shareholder, who asked de Grandpré at the special shareholders' meeting why he had not lobbied the company's shareholders to "bring pressure upon our government." De Grandpré replied, "If you succeed, sir, be my guest, and I hope that you will be the focal point to organize a petition to make sure that the harassment ceases."[145]

De Grandpré says MPs were "flabbergasted" by the response.[146] Many of the shareholders' letters to their MPs were relayed to the CRTC, where they were placed on the public record of the reorganization inquiry and were answered by the commission's secretary general. The letter campaign succeeded in heightening the sense of urgency attached to the proceeding by the government, states a DOC briefing document prepared for government MPs.[147]

Campeau says de Grandpré was "very eager" to get on with the reorganization and was equally anxious to dispense with the government's appeal of the Quebec Superior Court decision.[148] The Bell Canada chairman showed his impatience in mid-November, at about the same time that he launched the shareholder campaign, when he suggested a new legal tactic to Campeau. He wanted the lawyer the file a motion to ask the Quebec Court of Appeal for "an expeditious appeal on grounds of the urgency."[149] But Campeau was reluctant to do so. If granted, a petition for a preferential hearing boosts a case to the top of a court's crowded docket. Campeau believed the request would be an intrusion on the court and was not really necessary, because "I never had any doubt about what we were going to achieve."[150]

Not to be deterred, de Grandpré went over the heads of the lawyers at Ogilvy, Renault and retained a second outside law firm to prepare the motion. What was unreported at the time was that the lawyer who worked on the motion was de Grandpré's older brother, Louis-Philippe, senior counsel at Lafleur, Brown. The motion was filed with the Quebec Superior Court of Appeal on December 2[151] and stated that an urgent resolution of the appeal, which had not yet been scheduled by the court, was required because Bell Canada was hampered by the uncertainty created by the delay.[152] Bell had hoped to have gained all the necessary approvals for the reorganization by the end of the year — an objective that would have been met without the attorney general's appeal. Although Bell had removed the most pressing reason for seeking an expeditious hearing by gaining an extension to the tax law change, the CBCA imposes a time limit on how long the requisite approvals are valid. Chief Justice Marcel Crête granted Bell's request after a brief hearing on December 6, 1982. He extended the period for which the shareholders' approvals were valid and scheduled the appeal to be heard after the CRTC hearings during the week of February 21, 1983.[153]

All of the delays, however, did not stop BCE from making its first acquisition. The holding-company-to-be bought Comac Communications Limited in November 1982, to complement Tele-Direct's publishing business.

"No one is a good witness if they think the judge is biased," John Lawrence says. Although the CRTC vice-chairman failed to convince de Grandpré of the need for a public hearing at their chance meeting in early November, he believed he had allayed de Grandpré's concerns that the CRTC was either biased or opposed to the reorganization. De Grandpré, he says, "wasn't the least bit arrogant" when he took the stand at the hearing.[154]

Unknown to Lawrence, however, de Grandpré's smooth performance was due to professional coaching. "There was no way that Jean could take the stand at the CRTC hearing in the combative mood he was in," Cruickshank says. "He couldn't say the word 'government' without putting an expletive in front of it!" So Cruickshank locked up the Bell Canada chairman in a Toronto hotel room with some New York consultants hired to spend an entire day firing questions at de Grandpré to help him rehearse for the hearing. The practice session "took the edge off his hostility toward the regulators" and the difference "was like night and day,"[155] Cruickshank says.

As Bell Canada's chief witnesses, de Grandpré and J. Stuart Spalding, vice-president and treasurer of Bell Canada, spent five days on the stand in the CRTC's spartan hearing room at the federal government's sprawling Place du Portage complex in Hull. The reorganization proceeding was the first telecommunications hearing that Meisel had chaired in the three years since he had been appointed chairman. "I felt the chairman should chair a meeting of that importance,"[156] he says. The hearing also marked the debut of Jean-Pierre Mongeau, one of Fox's aides, who was appointed a CRTC commissioner on April 6, 1982.

Although Meisel says he found the reorganization hearing a fascinating experience, he did not enjoy the extent to which the lawyers dominated the proceedings, which is the main difference between a legalistic telecommunications hearing and a more informal broadcast proceeding. He was also struck by "how much the commissioners were at the mercy of the staff briefings. Our whole perceptions of the case developed from the briefings."[157] But the CRTC chairman's most vivid recollection was de Grandpré's repeated argument that telecommunications was one of the few areas of the Canadian economy that was on the cutting edge. "Given Bell's international track record, he argued we had to allow Bell to set up this corporate structure to make BCE a world-class holding company in high tech."[158]

Before his cross-examination commenced, de Grandpré delivered a short opening statement in which he immediately linked Bell Canada's reorganization to Canada's future role in the burgeoning field of high technology. He pointed to BNR's labs in Ottawa as "the jewel of Canadian research" and, to underscore his point, boasted to the CRTC commissioners that Egypt's president Hosni Mubarak was visiting the labs that same day.[159]

"High technology" was a highly evocative phrase in 1983. The Information Age had spawned a new class of "knowledge workers," whose prowess with silicon and software was viewed with a profound optimism. High tech, it was then believed, would not only lift Canada out of the doldrums of recession, but also eliminate the country's traditional dependency on the resource industry and manufacturing branch plants. De Grandpré made deliberate use of those perceptions to justify Bell Canada's corporate shuffle. As he remarked:

> This reorganization that is now before you is essential to continue what we have tried to do for the last ten years. . . there is no doubt in my mind that this reorganization is something that needs to be approved, because it is carrying so much importance on its shoulders, as far as Canada is concerned. It is one way of flying the Canadian flag, if you wish, all around the world.[160]

Bell Canada's legal defense of its reorganization at the Quebec Superior Court provided the utility's lawyers with the gist of the evidence it put before the CRTC. That eleven-page document, filed on November 15 along with two attachments and the circular to shareholders, was exceptional for its brevity. Bell Canada boiled the cabinet's questions down to two basic issues: the impact on the CRTC's ability to regulate the telephone company and the impact on subscriber rates.[161]

The first point was relatively easy to argue. The argument developed by Campeau earlier that summer stated simply, "no matter what form Bell Canada's corporate structure may take, the CRTC is in total and exclusive control over the company's rates."[162] The company went even further in its evidence and stated that the "purification" of Bell Canada's investments would actually facilitate "the exercise of the CRTC's mandate."[163] The national interest was invoked yet again to bolster the company's claim that no legal measures were needed to limit its business:

> Bell Canada believes that in principle, limitations on the scope of the activities of the Bell group should be considered detrimental to the public interest, as well as detrimental to the companies in question, their employees and investors.[164]

However, consumer groups held a conviction, and the CRTC a reasonable fear, that the corporate reorganization could possible lead the utility or some of its "integral" operations to escape regulatory scrutiny. Bell Canada vehemently opposed those claims in its written arguments:

> The atmosphere created by those persons who claimed that Bell Canada was seeking to escape from regulation appears to have blinded many to the fact that the proposed reorganization is a normal restructuring of a very large business group and that, absent regulation, it is unlikely that the reorganization would have attracted any interest whatsoever from anyone other than the directors and shareholders of the company.[165]

But Max Wolpert, counsel for NAPO, asked, "normal for what?" He said Bell Canada's "circular and self-congratulatory" arguments posed a conundrum. "It is precisely because in a monopoly the subscriber has no choice of supplier that it must be fully and effectively regulated."[166]

The second issue defined by Bell Canada, which concerned subscriber rates, was far more contentious. De Grandpré testified that the
corporate reorganization was "neutral as far as the subscribers are
concerned. The reorganization will not cause the subscribers to pay
one nickel less or more for their telephone service."[167]

A detailed analysis of the reorganization by Ogilvy, Renault and
Wood Gundy Limited on the impact of the reorganization calculated
that the transfer of Bell Canada's investments to BCE would remove
$65.7 million from the utility's rate base — a sum calculated by sub-
tracting the preferred share dividends from BCE from the lost return
on those investments. That cost to subscribers, however, was reduced
by an additional $55.2 million, which represented the carrying costs of
those investments that would be borne by BCE after the reorganization. That left subscribers about $10.5 million short of breaking even.
To compensate for that shortfall, de Grandpré proposed to reduce Bell
Canada's allowed rate of return by between one-quarter to one-half a
percentage point, which would result in rate savings of between $9.1
and $18.3 million and leave subscribers unaffected.[168]

The telephone company, on the other hand, calculated its treasury
would earn a one-time windfall from the transfer of its subsidiaries to
BCE due to a capital gain of about $560 million — the difference
between the $440-million historical value of the investments trans-
ferred to BCE and the $1-billion market value as of June 1982. The
fact that subscribers had borne a portion of the carrying charges of
those investments, however, led some consumers groups to argue
throughout the proceeding that the telephone company's customers
were therefore entitled to receive a portion of the capital gain.

The battle line between the competing interests of subscribers and
shareholders had already been drawn when Bell Canada announced the
reorganization. The utility declared in its initial shareholder circular on
June 25 that the utility's old corporate structure was inadequate because
the profit from its investments had to pass through the rate base before
reaching shareholders. The issue, as de Grandpré had maintained all
along, was "the proper sharing between subscribers and shareholders of
the risks and rewards associated with these investments."[169]

The arguments over the capital gain prompted a major clash between
subscriber and shareholder interests. De Grandpré was incensed when
the Consumers' Association of Canada (CAC) presented evidence from
Myron Gordon, a professor of finance at the University of Toronto,
who thought that $200 million "would be a reasonable estimate" of the

gain that should accrue to subscribers through lower telephone rates.[170] The Ontario government's support for that position led Ernest Saunders to bitterly complain in Bell Canada's concluding argument that "the commission has been bombarded with arguments suggesting that for some reason the shareholders of Bell Canada have fewer rights than the shareholders of so-called ordinary corporations."[171]

To de Grandpré, Gordon's argument was akin to advocating the confiscation of shareholders' profits. "It is Bell Canada's submission that the subscribers have no rights whatsoever to any share in the investments,"[172] Saunders stated in the company's reply argument. Besides, he argued, the telephone rates set by the CRTC were judged "just and reasonable" when initially set.

Telephone subscribers would have paid lower monthly rates had Bell Canada not invested in all its subsidiary companies, Kenneth MacDonald, then counsel for the CAC, countered in his supplementary argument on the issue filed on March 16. MacDonald did not dispute the fact that the rates paid by subscribers were considered just; rather, he stated that "Bell Canada received not only a just and reasonable price for the service provided to the subscriber, but also it received the carrying cost of its investments."[173] De Grandpré used an analogy between Bell Canada and Heinz ketchup when Ken Rubin, a consumer watchdog and head of a group called Action Bell Canada, tried to argue that subscribers, as well as shareholders, had contributed to Bell Canada's capital base:

> De Grandpré: . . . Furthermore, when a subscriber gets telephone service, and when he buys a car from General Motors, or from Ford, he is exactly in the same position; he gets value for his money. And nobody would suggest to me that when I buy a bottle of ketchup I have become suddenly the owner of part of Heinz. I have paid for my bottle of ketchup. I have paid for my car. I have paid for my truck. I have paid for my telephone service. . . .
> Rubin: I don't know if Heinz has the monopoly on ketchup.
> De Grandpré: Well, I don't know; but, do you think that you have a portion of the assets of Ontario Hydro or Québec Hydro? And they have a monopoly position.[174]

The utility assailed Lawson Hunter, the federal combines chief, when he cited a statement of the RTPC that the reorganization would benefit shareholders at the expense of subscribers. Hunter has stated in his September submission to the CRTC that:

> A strong case can be made that most of the increase in the value of Northern's shares held by Bell should accrue to the subscribers who, in a rate of return regulatory environment, were the real, if unknowing, risk-takers in Bell's reliance on Northern as its principal supplier.[175]

"This claim maligns the regulatory process and ignores the facts,"[176] Saunders stated in Bell Canada's final written argument, which was the utility's rejoinder to the previous filings and statements by other interested parties.

Brian Finlay, the DIR's counsel at the hearings, tried during his cross-examination to get de Grandpré to admit that BCE's board would have difficulty responding to the interests of its shareholders while at the same time making the tough decisions of determining how to allocate equity among its numerous subsidiaries. Finlay presumed it would be more difficult after the reorganization for Bell Canada to raise the same amount of capital it once had access to and that it might have difficulty defending its needs before BCE's board. The DIR then suggested that BCE should be ordered to sell a minority share issue in Bell Canada's stock to the public after the reorganization.

De Grandpré took great exception to this suggestion. As proposed by Bell Canada, the reorganization envisaged all of Bell Canada's equity being held by one shareholder — BCE. A minority shareholding, however, would give rise to a second class of shareholder. Purdy Crawford, a corporate lawyer and senior partner at the Toronto law firm Osler, Hoskin & Harcourt, stated in his evidence for the DIR that a minority holding "would impose in an imperfect world a rather significant discipline" on both BCE's management and its board of directors to safeguard the interests of Bell Canada's minority shareholders in any transactions between the phone company and its new parent.[177] Bell Canada would have to "conduct its affairs in its own best interests," rather than solely in BCE's interest, Hunter claimed.[178]

De Grandpré rejected any minority position in Bell Canada, except as "a last resort" and testified that the company would "exhaust all possible means to avoid that."[179] He explained that a minority interest would make it more difficult for BCE to guide and direct the group of companies under its control. It would also cause BCE's stock to be underpriced.[180] He gave several examples of holding companies whose market values had been discounted from the value of their holdings because of minority interests in their subsidiaries. They included McIntyre Mines, Canadian Pacific Enterprises, Power Corporation, Angus Corporation and Investors Group.

The Bell chairman firmly rejected Finlay's charge that Bell would have to compete more aggressively against other BCE subsidiaries for equity following the reorganization: "This is not new. We have faced these decisions for decades and decades. . . . I think you should place some confidence in the good judgment of management and the board of directors."[181]

Finlay persisted and tried to get de Grandpré to admit that the creation of BCE could further complicate life for the subsidiary if, under a hypothetical scenario he posed to the Bell Canada chairman, the utility faced a profit squeeze because of low telephone rates.

De Grandpré refused to be tricked and retorted:

> The regulators are not there to kill the company; the regulators are there to make sure that we have access to financial markets permitting us to give service when service is needed by customers. This is the interplay of regulations and management and the board of directors. . . .You are assuming, in your question, that Bell Canada is going to be shafted by the regulators, and I am not prepared to accept that.[182]

Peter Oliphant, a Toronto-based management consultant and chief witness for NAPO, dismissed one of de Grandpré's central rationales for the reorganization. He stated that the Bell Canada chief's claim that "managerial effectiveness can be improved by changing corporate structures is without foundation."[183] NAPO infuriated Bell Canada's chief lawyer when it challenged de Grandpré's testimony and told the CRTC it should ask the federal cabinet to nullify the reorganization for at least five years and that Parliament should pass a law to undo the utility's continuance under the CBCA.[184]

"Certain intervenors sought to denigrate Mr. de Grandpré's reasoning. Nevertheless Mr. de Grandpré is right and those who seek to denigrate his reasoning are wrong," Saunders wrote in Bell Canada's reply argument on February 28, 1983.[185]

One week earlier, Saunders had sat as an observer in a Montreal courtroom at the two-day "expeditious" hearing at the Quebec Court of

Appeal. Bourque repeated his arguments of the preceding fall in a futile attempt to convince the court that the CRTC had a prior right to approve the reorganization. When it was Campeau's turn to rebut Bourque's arguments shortly before noon on the second day of the hearing, Justice Georges Montgomery asked Bell Canada's litigator whether he thought he had enough time or needed to plead through the lunch hour. "I told him that we could deal with Bourque's argument in about ten minutes. Ernie [Saunders] just about shit his pants,"[186] Campeau says.

The three-member appeal court panel upheld Judge Gonthier's finding that Bell Canada was not disposing of its shares. "I would say that we are being invited by the attorney general to usurp the powers of Parliament to legislate by providing that the commission's approval shall be required,"[187] Justice Montgomery wrote in the appeal court's judgment on March 24, 1983. He also rejected Bourque's arguments that the arrangement should be rejected on the grounds that it was not in the public interest.[188]

But the court judgment was only half a loaf for Bell Canada. Although Justice Montgomery's decision neutralized the CRTC's ability to act independently to approve or alter the reorganization, it did not hamper the federal cabinet's or Parliament's powers to act on the CRTC's forthcoming report. More lobbying was needed.

On February 22, as Campeau was concluding his argument in court, de Grandpré was on Parliament Hill briefing the entire federal Liberal caucus from Ontario on the reorganization at a private meeting. He had met with Quebec caucus members a week earlier and was scheduled to meet the Opposition caucus on February 24,[189] according to a federal government memo obtained under the Access to Information Act. Fox says there was nothing unusual about the briefings. "It was all above board."[190] Yet few other corporate leaders have ever been granted such access. When asked whether corporate briefings to caucus are a regular occurrence, a former principal secretary to the prime minister, who has worked in Ottawa for the past twenty years, says he cannot recall any other instance.[191]

When Meisel reported to cabinet on the need for legislation, de Grandpré wanted to make sure that the parliamentarians understood the issues and Bell Canada's position. However, although charged to report back to cabinet, which considers any information furnished to it as a confidence of Her Majesty's Privy Council, Meisel ruffled a few feathers at DOC when he made a unilateral decision to make his report public on the grounds that it followed a public hearing. Despite the concern of

DOC officials that release of the report would affect the government's handling of the issue, as well as set a precedent for future inquiries ordered by the cabinet, Fox did not interfere with Meisel's decision.[192]

The CRTC released its report on April 18, a day before its extended deadline expired.[193] Out of courtesy, the CRTC agreed to give an advance copy of the report to de Grandpré early that morning before its public release at 4:00 p.m., says Hank Intven, former head of the telecommunications branch at the CRTC. To his surprise, though, Bell officials asked him to deliver the copy immediately to the Ottawa airport, where de Grandpré was waiting aboard his private jet after taking a detour from his planned flight from Montreal to New York City.[194]

The report concluded that subscribers would not be adversely affected financially by the reorganization. Although it stated there were numerous problems in trying to value Bell Canada's investments, it accepted de Grandpré's proposal to adjust the allowed rate of return to offset any loss of income from Bell Canada's rate base. As most observers expected, the commission asked the government to introduce legislation to reaffirm several special-act restrictions on Bell Canada. The report stated that the commission's ability to regulate the utility would not be impaired if legislation was passed that affirmed those provisions.[195]

The CRTC was particularly concerned about the effect that the reorganization would have on the restriction that limited Bell Canada to being a telecommunications carrier. It was worried about the prospect of the phone company entering the broadcast business and doing competitive battle with the other clients regulated by the CRTC. Although de Grandpré testified that Bell Canada had no interest in competing in that industry, the CRTC stated it was persuaded by the arguments of Peter Grant, a lawyer at McCarthy & McCarthy in Toronto who represented a large association of business telephone customers, that the previous limitation no longer had any legal effect.[196]

When Meisel turned to the issue of the capital gain at the executive committee meeting that determined the CRTC's conclusions, several commissioners voiced their support for the consumer-group arguments that the subscribers should receive it. But Lawrence's objections held sway — "He who bears the risks and owns the shares should share in the rewards."[197]

Likewise, the CRTC vice-chairman concluded that any value from the use of the Bell name by BCE was the shareholders' property. The issue had been raised by several parties who argued that subscribers should also be compensated for the use of the name by BCE and its subsidiaries. However, Lawrence also made sure that the CRTC report rejected de Grandpré's claim that the Bell name is not a major asset. De Grandpré had testified:

> At times I feel it is a substantial liability. But the Bell name, by itself, is not that substantial an asset because what is important is the interlinkage between Bell, Northern, BNR and the contributors to the flow of revenues. . . .
>
> For some reason we are still perceived in many quarters as a subsidiary of a U.S. corporation. Maybe the name 'Bell' is the stigma that we carry. . . .[198]

"They must have been kidding,"[199] Lawrence says in an interview. The CRTC report concluded that "the name 'Bell Canada' is, despite the disclaimers made by the company, an asset having very substantial value."[200] But the commission was at a loss to answer the question, What's in a name? It concluded it would be all but impossible to attach a monetary value on Bell Canada's goodwill, underscored by Bell Canada's statement that it has never carried its name on its books as an asset.[201]

The commission also rejected the DIR's call for a minority shareholding in Bell Canada and stated it was "not persuaded" that such an interest was needed "at this time."[202] Hunter says he was greatly disappointed by that rejection. A minority interest could have led the government to lessen its regulatory grip on Bell Canada — an aim that had underlaid his position since the outset, he says. But, as he perceived events, "the CRTC took a very bureaucratic approach to their own turf."[203]

Hunter added the federal government adopted a similar minority shareholding proposal recommended by the DIR when the government deregulated the financial services market in August 1989. Ironically, Crawford, who was then at Imasco, sought to escape the restrictions announced by Ottawa that would have affected Imascos' CT Financial Services Incorporated.

The report also contained two recommendations to beef up the CRTC's powers over the reorganized Bell Canada, which gives credence to Hunter's view of the report. The CRTC asked the government for a legislated provision allowing it to acquire information from BCE or any other affiliate that the CRTC deemed essential to conducting its job. And, second, two officials devised a proposed provision, dubbed the "in-and-out" clause, that would give the CRTC the authority to rule that a business activity entered into by Bell Canada could either be declared as integral and regulated, or as non-essential and to be run as an arm's-length subsidiary.[204]

"The report was very reasonable and did not unduly favor either side,"[205] Lawrence says. Fox's senior officials viewed the report as a "complete and balanced assessment." Although several of the proposals for legislative amendments required further consideration, Fox was urged by his deputy minister, Robert Rabinovitch, to permit Bell Canada to proceed with the reorganization and to ask cabinet for authority to draft new legislation. Rabinovitch recommended that both decisions be announced together after the weekly Thursday-morning cabinet meeting on April 21 (the report was released April 18).[206]

De Grandpré was in New York attending a board meeting when the CRTC made the report public. Later that day, he told reporters he was "gratified" with the report and was obviously pleased that the commission had sided with his view that subscribers would not be harmed.[207] Although he called the request for legislation "unnecessary and superfluous," he says in a 1990 interview that he did not lose any sleep over it.[208]

But he was sufficiently concerned at the time to show up at Fox's fifth-floor office in the Confederation Building on the western edge of Parliament Hill shortly after 9:00 a.m. on April 21. He made a last-ditch lobby effort just forty-five minutes before Fox was scheduled to present the report to the full cabinet meeting at 10:00 a.m. As advised, Fox was going to ask cabinet for authority to draft a new Bell Canada Act. De Grandpré told the minister he did not have much support in cabinet for any new legislation and that Bell Canada had been very thorough in its lobby efforts. Fox walked up to the Centre Block a few minutes later, wondering if de Grandpré really had lined up nineteen of his cabinet colleagues against him.[209]

The threat was a bluff. Fox issued a statement later that day saying the reorganization could proceed and the government would undertake a review of the CRTC's recommendations in order to prepare legislation to ensure the CRTC could properly regulate Bell Canada.[210]

Although Bell Canada had cleared its major obstacles, a Toronto newspaper reporter recalls that de Grandpré was still sensitive about the government's response to the reorganization at an interview the following Wednesday on the eve of Bell Canada's 103rd, and last, annual shareholders' meeting. Following Fox's approval, the final step was up to shareholders at the April 28 annual meeting when they would be asked to approve new corporate bylaws for BCE. Prior to his interview, de Grandpré had asked Jonathan Chevreau of *The Globe and Mail* for his tacit agreement not to ask any questions about the reorganization. Chevreau's reporting on the reorganization, which was the most comprehensive in the daily press, had antagonized Bell Canada by being the first to analyze the legal issues of the reorganization and to focus on consumer group criticisms. Once seated in the Bell Canada chairman's suite at the Harbour Castle Hotel, however, Chevreau could not avoid the topic. Although he ignored the prohibition a couple of times, he did not get past "About the reorganization . . ."

Finally, the Bell Canada chairman thundered, "That's it, Mr. Chevreau. This interview is over!"

De Grandpré got up from his chair, stormed out of the room and slammed the set of double doors behind him. A few seconds later, there was a knock on the door; instead of opening the doors to his private room, de Grandpré had walked out into the hallway and locked himself out of the suite.[211]

The following morning, de Grandpré faced the opposite problem when too many shareholders were admitted to Roy Thomson Hall. "Legally, shareholders can't be turned away from a meeting," de Grandpré says. Yet fire regulations would not permit everyone into the hall. By the time the scrutineers and organizers had worked out that problem, the much-awaited call from Consumer and Corporate Affairs in Ottawa and been delayed. That call was the one to say the reorganization certificate was approved and that a vote could be taken on the special resolution. "It was a real problem because we had people standing up in the hallways and lobby," de Grandpré recalls. It turned out the phone call was delayed by a fire drill that had evacuated the government building in Hull, Quebec.[212]

The bylaw passed and BCE was officially created.

The legislation was delayed much longer. Following completion of the DOC review, the draft legislation got bogged down because of its unanticipated complexity and a lengthy review of the draft bill by Department of Justice lawyers.[213] When it was finally tabled in Parliament in February 1984, the bill, which took almost four years to enact, left two issues unresolved.

The first intrigued Lawrence. It did not arise during the hearing, he says, because "it's intangible. And it's this: If you shift all the bright lights from Bell Canada to BCE, what happens?"[214]

But the second was of far-reaching, long-term significance and stemmed directly from Meisel's observation that the reorganization was promoted as being essential to Canada's economic future and high-tech development. "But I haven't seen that," Meisel says, looking back on the first six years of BCE's evolution in a 1989 interview. "I think they welched on that promise because they gobbled up companies that were already around and that were nontechnology companies."[215]

A former CRTC commissioner who requested anonymity says, "We made a mistake. We granted the reorganization with the expectation that some conditions, which have never been met, would be. We didn't have the quo with the quid. We didn't have the Act in place at the same time."[216]

To those who know de Grandpré and his aims, the reorganization was a masterstroke; a "brilliant solution to his problems,"[217] as Fox termed it. But his critics were less charitable, particularly those who looked disparagingly on BCE's soiled acquisition record with hindsight several years later. "It was a con job,"[218] Andrew Roman says.

Chapter 7
Back Rooms and Boardrooms

I am concerned with maintaining our leadership position in global communications. At the same time, I plead for all our industries.
— A. JEAN DE GRANDPRÉ, SPEECH TO THE CANADIAN CLUB, TORONTO,
DECEMBER 9, 1985

As head of Canada's wealthiest and largest corporation, de Grandpré enjoyed a significant degree of influence in both the government and business affairs of the nation. His clout in the Canadian Establishment grew as BCE's corporate empire expanded in the mid-1980s. At the peak of his career about two years before his retirement, de Grandpré was regarded as the ambassador of Canadian business abroad and as its chief spokesperson at home.

Yet many observers of Canadian business in the media outside Quebec failed to chart his career or appreciate his stature even after the reorganization. De Grandpré merited only a single entry in the text of Peter Newman's 1975 book, *The Canadian Establishment,* as one of a dozen French Canadians who qualified as members of the "national business Establishment."[1] He did not even make the index of John Sawatsky's 1987 book on lobbying, *The Insiders: Government, Business, and the Lobbyists.*[2] Soon after it appeared, however, de Grandpré received a flurry of attention centered around BCE's activities as the first Canadian corporation to top profit after taxes of $1 billion. He then came to be portrayed by some business writers as a corporate leader whose influence was of almost mythical proportions.

As if to atone for his previous omission, Sawatsky described the BCE chairman in a 1989 article for *Vista* as the "businessperson with the biggest clout" and wrote that de Grandpré's colleagues "viewed him with awe and envy for virtually controlling the federal cabinet over the years."[3] A similarly exaggerated perception of Bell Canada's political influence, as typified by one headline in 1978 that proclaimed, "Bell Canada usually gets what it wants,"[4] was fostered by the record number of rate applications to the federal regulator during de Grandpré's tenure.

When people say Bell Canada has power, however, they generally mean to say it has access. There is a crucial distinction. Neither de Grandpré nor the company he headed always got their way in the back rooms of the nation's capital, as the preceding chapters have shown. But few Canadian corporations have had the degree of access that Bell Canada has enjoyed in Ottawa through its century-long history. And fewer still have mastered the art of lobbying to the extent that the country's largest telephone company has.

Bell Canada's involvement in the nation's business stemmed in large measure from its position as a regulated monopoly and the belief of its corporate leadership that its business was inextricably linked to the national interest and the public good. One of the paradoxes of de Grandpré's leadership was his dependency upon, yet contempt for, politics and regulation. His abrasive and sometimes confrontational style led to some stunning defeats and made him a far less effective political lobbyist for his corporate causes than, say, Ian Sinclair of Canadian Pacific or Paul Desmarais of Power Corporation.

Like his grandfather, father and older brother, de Grandpré was a Liberal. He had been actively involved behind the scenes in both the provincial and national wings of the Liberal party since university days, and his personal connections to those in political power in both Ottawa and Quebec City ran deep. In 1962, he was invited to the famed Liberal policy conference at Queen's University in Kingston, Ontario, which marked the debut of future party leaders John Turner and Jean Chrétien. The policies on health care insurance, the building of a national gas pipeline and the Canada Pension Plan that were adopted at that conference "put the Liberals in power for a long time,"[5] de Grandpré says.

Not long after the Kingston conference, de Grandpré was asked to run in the 1963 federal election. Along with fellow Montreal lawyer Guy Favreau, he was approached by a Liberal organizer at a breakfast meeting at the Château Frontenac in Quebec City. "No way," de Grandpré told the organizer when asked if he would consider seeking a seat. Favreau, on the other hand, said "maybe" at breakfast, agreed later and went on to win a seat in the riding of Montreal-Papineau and an appointment to the Pearson cabinet.[6]

In explaining his decision not to enter politics, de Grandpré says, "Politicians very seldom lead; they follow public opinion."[7] Basically, he was too strong-willed to enter politics. Descriptions of him as "a doer" and a man who could not tolerate procrastination dovetail with his deeply rooted impatience with politics, which, after all, is loosely defined as "the art of compromise." By the time de Grandpré moved into the presidential suite at Bell, his antipathy had grown into contempt.

There was another reason he shunned politics as a career, de Grandpré says. "People go into politics to exercise power. I don't like power that much. It's not in my nature. That's not it at all. [It's] to do things right, to produce something that will be lasting. In the exercise of that ambition, then, you have to exercise power. Power is the process; it's the means, not the objective."[8] Yet some who dealt with de Grandpré believe he did enjoy the exercise of power. As Francis Fox says, "I think he was an autocrat by nature."[9]

As chairman, de Grandpré assumed the role of chief lobbyist for the company. John Armstrong, the former and pioneering director of the lobbyists registration branch at the federal Department of Consumer and Corporate Affairs, defines lobbying as "changing the rules of the game for pay."[10] A lobbyist, in turn, is defined by the federal Lobbyists Registration Act as someone who is paid to arrange a meeting with or to communicate with anyone who holds a public office in an attempt to influence either a legislative proposal, a law, regulation, policy or the award of either a grant or contract.[11] "Lobbying is not a dirty word. It's part of the democratic process," de Grandpré maintains. "And if you don't do it, then the legislators are not aware of your position and the reasons, or the rationale, for your position. I feel quite sincerely that lobbying is part of a way to educate the legislators about your problems and the advantages of what you are requesting."[12]

Prior to the mid-Seventies, Bell's lobbying was on an informal basis. Scrivener says he used to spend "quite a bit of time" in Ottawa, where he "wandered around the Centre Block." De Grandpré's predecessor lobbied as if he were campaigning for a seat in Parliament: "If a door was open to an office, I would just poke my head in — I didn't care who it was — and I said, 'I saw your door open, let me introduce myself, I'm Bob Scrivener, the chairman of Bell Canada.' This guy might be from Saskatchewan, so we'd talk about the Saskatchewan Telephone Company."[13]

As unsophisticated as Scrivener's method seems, it allowed him to meet most of the federal cabinet ministers. Yet his informality and charm disguised the contempt he also felt toward politicians. "Governments are the biggest enemy because they have so much power with so little intelligence."[14]

De Grandpré relied on his personal contacts even more than Scrivener did. Fox says de Grandpré was a very effective lobbyist, "in the sense that he knew who he had to see in Ottawa, he knew the players in the government very well, he knew the machinery of government and he knew the main players, like the prime minister and Marc Lalonde."[15] Fox, who has been a practicing lawyer with Martineau & Walker in Montreal since his defeat in 1984, is a registered lobbyist. He is on the board of Government Consultants International (GCI), the high-profile lobby firm owned by Frank Moores, the former premier of Newfoundland and a Mulroney crony. Despite the anger de Grandpré felt toward Fox during the corporate reorganization, the former federal communications minister now acts for the telephone company, which is a GCI client.[16]

"Anyone dealing with de Grandpré as a minister," says Fox, "had to know he was speaking to a lot of other people including the prime minister, which meant that you would want to be pretty sure of your facts and figures and your arguments before going to cabinet on some of these things."[17] That realization did not always work in de Grandpré's favor, however. Quebec cabinet ministers and government officials in the mid-Seventies, for example, resented his access to the premier and prime minister. In going over their heads, he alienated senior departmental officials responsible for drafting policies.

There was also the ever-present threat of appeal to the federal cabinet of any regulatory decision that was not in Bell Canada's favor, which led opponents to believe the utility was "exerting a lot of pressure at the Prime Minister's Office,"[18] Raynold Langlois, former counsel for the Quebec government, says. Even if the federal cabinet was reluctant to use that power, as Fox claims, it remained a source of contention with intervenors. In their opinion, the deck was stacked against regulatory fairness as long as federal politicians could retroactively alter a CRTC decision any way they saw fit. The order-in-council passed in 1982 that retroactively adjusted Bell Canada's rates to impose the "6 & 5" program on the utility is an example of how sweeping that power is.

Bell Canada solidified its ties with parliamentarians through events on the Hill, such as its annual dinner, held in the famous Railway Room in the Centre Block. "The MPs liked it very much, you know, having a senior officer of a major corporation like Bell to come and speak to them about his perception of the state of the nation and the problems of the country," Fox recalls. "It was very well done and it was good PR [public relations]. I suppose it's the way people would speak to senators and congressmen in the United States."[19]

Through more than a century of experience, the "Bell lobby," as it is called, came to be viewed as one of the most well-heeled and professional forces at work in Ottawa. "They had . . . become an institution,"[20] says Fox. However, the tendency of Bell Canada's leaders, from Sise onward, to equate the corporate interest with the national interest alienated some politicians and officials. The former DIR, Lawson Hunter, for example, likened de Grandpré's attitude to that of Charles Wilson, the former chairman of General Motors, who once remarked, "What's good for General Motors is good for the country."[21]

"Almost every institution looks at the public interest from a position of self-interest or from the position of what it conceives as its responsibilities," Northern Electric stated in its submission to the Royal Commission on Corporate Concentration in October 1975. "It is only on rare occasions that there is a consensus on what forms the public interest." Government, the company stated, interprets the public interest in terms of votes and maintenance of power while business must interpret it in terms of customers, shareholders and employees. Despite the difficulties of articulating the national interest on any matter, much less on getting society to agree to it, Northern Electric arrived at a broad definition:

> . . . the commission might wish to view the public interest as that which is to the advantage, profit or benefit of the majority of the nation; in other words, the national interest.[22]

That definition is nothing more than a restatement of the nineteenth-century doctrine of utilitarianism, which held that individual and corporate actions should be guided by the objective of producing the greatest happiness for the greatest number. But merely stating the theory, as great philosophers like John Stuart Mill learned more than a century ago, is of little use to society.

Many experienced policy makers and regulators who dealt with de Grandpré and Bell Canada take great pains to assert their neutrality and to stress that they were not Bell bashers. "Bell Canada was not the enemy," Allan Gotlieb, former deputy minister at DOC, says. "But we had to be able to say 'no' to get control of them. We could not be a rubber stamp."[23] Given that de Grandpré did not share that attitude and given his often conflicting views of the public interest, antagonism frequently characterized his relationship with government.

Although there was no public supervision of lobbying of federal government officials in Ottawa prior to 1989, Bell Canada's contacts with its federal regulator were strictly controlled because of the CRTC's function as a quasi court. When John Meisel was chairman of the CRTC, he tightened the rules relating to *ex parte,* or one-sided, contacts between the commission and officials in all companies regulated by the CRTC to avoid the perception of bias. Yet he recognized that some contact was needed, "because you can't regulate without knowing what the industry is doing."[24] Between hearings, his vice-chairman, John Lawrence, had regular, albeit formal, contact with a designated executive at Bell Canada. Lawrence recalls an attempt early in his tenure to break the ice with Bell Canada's chief lawyer, Ernie Saunders. He invited the fellow lawyer for a drink at the hotel next to the CRTC after a long day of hearings. "I wanted to try it, but I felt so uncomfortable, I had to leave."[25]

While Meisel questioned de Grandpré's strategies as a lobbyist and says he was perhaps too "gladiatorial," neither he nor Lawrence found Bell Canada's lobbying of the executive branch of government particularly worrisome. "I didn't think de Grandpré was bullying us in any way,"[26] Lawrence says.

Government relations became an increasingly important part of Bell Canada's business during de Grandpré's tenure. Faced with contending views of the "public" or "national" interest, both Bell Canada and Northern Telecom developed substantial public affairs programs. Lobbying, research, direct mail and advocacy advertising, grants to public policy organizations and political contributions were used to ensure that the utility's view of the national interest prevailed. De Grandpré gave those functions greater prominence through the creation of the regulatory and government affairs department. The advent of competition in the Seventies gave that department an even more critical mission — to preserve Bell Canada's monopolies.

The telephone company's efforts to influence government are backed by a small army of in-house regulatory and policy analysts, lawyers, lobbyists and pollsters whose functions are complemented by the professional economists and analysts in the economics department. By the time of the corporate reorganization, Bell Canada's regulatory and governmental affairs staff had grown to more than one hundred. The telephone company has never revealed how much it spends on lobbying or how much it budgets for its regulatory and governmental affairs operations. And the cost of those efforts has never been addressed by the CRTC, which does not require such expenses to be tracked.[27]

Bell Canada has augmented its influence by hiring several former cabinet ministers and top civil servants to senior corporate or board positions, a practice that is reminiscent of the U.S. business tradition of appointing "ins-and-outers" following a change of government. Among the appointments: Donald J. Johnston, former president of the Treasury Board, who joined BCE's board in May 1989; Gordon Osbaldeston, former clerk of the Privy Council Office, who joined Bell Canada's board in 1988; Roger Tassé, former federal deputy minister of the Department of Justice and a leading expert in constitutional law, who was named executive vice-president of governmental and environmental affairs in 1989; and Richard French, former Quebec minister of communications in the Bourassa government, who was vice-president of regulatory and governmental affairs. French is registered as a Tier II lobbyist for the telephone company. Full-time employees whose jobs require them to spend a significant amount of their time lobbying must register under the Lobbyists Registration Act, as must outside lobbyists hired on behalf of a client, who are classed as Tier I lobbyists.

Tassé joined Bell Canada from the Ottawa office of Lang Michener, where he had been retained by Mulroney to help resolve the fishing dispute with France, work on the Meech Lake Accord and provide legal advice to the Osbaldeston committee on security and intelligence. Bell Canada loaned Tassé's services to the Mulroney government during the last-ditch, but abortive, week-long marathon to rescue the Meech Lake Accord in June 1990. It was Tassé who advised both the government and a parliamentary committee that the provision recognizing Quebec as a distinct society did not compromise the Charter of Rights and Freedoms.[28]

De Grandpré, on the other hand, viewed the Meech Lake process as "horrendous," adding, "you don't legislate distinctiveness."[29] In opposing the distinct society clause, he parted company with much of the Quebec business community, including the influential policy committee of BCNI, who advocated acceptance of the accord. His opinion also diverged from that of his older brother. "The Meech Lake Accord is not perfect, but it's a step in the right direction," Louis-Philippe de Grandpré said in an interview before the accord was rejected. "Many are afraid of the term 'distinct society,' but we are all distinct. This is the only part of Canada built by the two founding groups, so it's normal to want to protect it and to promote the distinction."[30] It was de Grandpré's brother who advised the government of Robert Bourassa in 1989 to invoke the notwithstanding clause of the Constitution, which allows any legislative assembly to exempt a law from the Charter, after another controversial amendment to Quebec's language law was struck down by the Supreme Court of Canada because it violated the Charter of Rights and Freedoms. That measure, known as the "sign law," or Bill 54, stipulated that all public signs posted outside commercial establishments must be written in French. They could be bilingual only if posted inside. Bourassa's decision to invoke the notwithstanding clause precipitated the backlash against the Meech Accord.

Given de Grandpré's own strong views against the accord, one wonders whether he would have been predisposed to loan Tassé to help rescue it if he had still been the head of BCE. As chairman of the giant holding company, de Grandpré maintained a keen interest in Bell Canada's governmental and regulatory affairs and still found time to intervene on the utility's behalf. One such case landed federal Communications Minister Flora MacDonald in hot water in 1987.

Four interest groups, including the National Anti-Poverty Organization (NAPO) and the Consumers' Association of Canada (CAC), accused MacDonald of interfering with the CRTC after she wrote a letter to de Grandpré expressing her support for BCE's position on how much money its international contract arm should pay Bell Canada for borrowing employees for offshore assignments. At issue was a 25 per cent markup that the CRTC had ordered BCI to pay in a rate decision a year earlier, which, MacDonald stated in her letter dated July 14, 1987, "may indeed be inappropriate." The interest group lawyers did not object to MacDonald's meetings with de Grandpré and BCE officials on the issue or to her statements of policy on BCE's international business. What they did find offensive was her suggestion that Bell Canada file evidence on the matter at the company's upcoming rate increase hearing that fall, and her promise to review any decision of the CRTC that was at odds with what the utility felt was acceptable. "I would be prepared to recommend to the Governor in Council appropriate action to ensure that BCI can continue to compete effectively in international markets,"[31] MacDonald wrote in her letter to de Grandpré.

And that is what subsequently transpired.[32] On July 31, 1987, Bell Canada filed the letter along with the utility's evidence, which included a request that the CRTC reopen its previous decision on BCI's compensation to Bell Canada. However, the commission rejected Bell Canada's audit of the costs. It stated that Bell Canada's fees were about $4 million too low and it upheld the previous ruling in its rate decision on March 17, 1988.[33] A week later, BCE lodged one of its first petitions to cabinet to appeal the decision. As promised, MacDonald took the petition through the cabinet machinery and, in less than a month, BCE garnered a favorable order that set the fee at a lower amount.[34] But cabinet's refusal to consider a rebuttal to the petition led NAPO to appeal the order to the Federal Court of Canada. Although the trial court ruled that the cabinet had denied NAPO the right to a fair hearing, contrary to the Bill of Rights, that decision was quashed by the Federal Court of Appeal.[35]

MacDonald denied the charge that she was one-sided or that she had supported any position of Bell Canada. Referring to her letter to de Grandpré, she wrote in a letter sent to the CAC and NAPO lawyers on September 9, 1987, "I am surprised that you or Bell would feel that BCE could find much comfort in it."[36] The lawyers' charges that MacDonald's assurances to BCE were unethical and "contemptuous of the CRTC"[37] also prompted her to write to CRTC chairman André Bureau on October 9, 1987. She stated that her letter to de Grandpré did not imply "any particular view as to how the value of these resource transfers should be assessed."[38] Yet the subsequent actions of the government made a mockery of her claim.

Andrew Roman, counsel for NAPO, wrote that de Grandpré's lobbying of MacDonald went beyond displaying contempt for the CRTC:

> Had the letter from Ms MacDonald been solicited by an executive of Bell Canada itself, that might have been more normal. The fact that it was done by the chairman of Bell Canada Enterprises may reveal who is really running the show at the subsidiary as well. It may also be an attempt to place the protagonist beyond the CRTC's jurisdiction, to hide behind the corporate veil.[39]

Hardly, says de Grandpré, who argues neither BCE nor BCI, which were the companies affected by the CRTC ruling, had an application before the commission. He still has harsh words for the CRTC ruling, which, he claims, made BCI uncompetitive abroad and caused the company to lose its bid for a lucrative third contract in Saudi Arabia to an Australian competitor. "Don't export your know-how, that's the way I read it. I was not the only one to react that way; there were a lot of people who thought it was an unfair decision as far as Bell goes."[40] He added that the ruling was not only inimical to Bell Canada's interests, but "was contrary to the country's best interest."[41]

Prime Minister Brian Mulroney agreed.

The BCE petition to cabinet in March 1988 was actually put aside by DOC officials, who were incensed that the correspondence between de Grandpré and MacDonald had been placed on the public record at the CRTC by Bell Canada without their knowledge, said a former high-ranking official at DOC who asked not to be named. "There was no way we were going to take the petition to cabinet,"[42] the official says. But de Grandpré obtained a meeting with Mulroney, the official says, and the prime minister ordered MacDonald to bring BCE's petition against the CRTC's ruling before cabinet.

Lucien Bouchard, who was then Canada's ambassador to France and a close confidant of the prime minister,[43] played a key role in orchestrating BCE's fight against the CRTC decision, de Grandpré says. De Grandpré's personal access to Bouchard and other members of the Mulroney cabinet was guaranteed by his personal campaign contributions to the Progressive Conservative party. His cheque of $40,000 in 1988 was the largest single private donation to the PC party that year, according to the party's annual financial returns filed with Election Canada.[44] Just to be safe, he also donated $25,000 to the Liberal party of Canada the same year.[45] Prior to 1988, de Grandpré had wavered between the Liberals and the Conservatives. In 1984, the PC party was the sole beneficiary of his largesse, when he donated $25,000 to the party fund. The following year, he gave $26,000 to the PCs and $1,000 to the Liberals. He was alienated by the Tories in 1986 because of what he considered to be dismal corporate tax policies and mounting corruption scandals. In 1987, he donated $26,000 to the Liberals.[46]

Yet throughout his tenure at Bell Canada, de Grandpré followed Scrivener's strict policy that prohibited partisan corporate political donations or corporate sanction of political fund-raising activities. "Bell Canada never made any political contributions, ever, to anyone,"

Scrivener says.[47] It was that policy that led de Grandpré to shun a key business fund-raising dinner in Toronto for Mulroney when he was the PC's leadership hopeful in 1983. The event, hosted by cable television czar Edward S. Rogers, attracted about five hundred business people. Both de Grandpré and Tropea, who was then deputy chairman of BCE, refused to attend on the grounds that the holding company, like Bell Canada, could not be seen to be playing partisan politics. Although Bell Canada still abstains from party fund-raising activities because it is regulated, BCE has revised its policy and has since made occasional contributions to federal political parties, including $15,000 to the Liberals in 1988.[48]

The direct political contacts of Bell Canada's leaders were supplemented by the work of parliamentary agents, as paid lobbyists were then called, who were hired to help prepare the way for private members' bills to amend the company's Act. The utility now hires professional lobbyists, or "public affairs consultants," as they are called, to deal with single issues, such as long-distance service competition. The first such firm retained by the telephone company, JDE Consulting Services, was one of the first modern lobby firms in Ottawa. It was set up in 1970 by Duncan Edmonds, a former ministerial aide and assistant professor of political science. One of its first clients was the TransCanada Telephone System, which hired Edmonds to provide advice on dealing with the newly created Department of Communications. Shortly thereafter the company was retained by Bell Canada.[49]

The telephone company later did business, until 1989, with another pioneering and powerful firm, Executive Consultants, Limited.[50] ECL was formed in 1968 by Bill Lee — Trudeau's election campaign manager in 1968 and an advisor to Eric Kierans — and William Neville, former executive assistant to Finance Minister Edgar Benson. It was set up on the assumption that most businesspeople did not properly understand the workings of government or have enough political information to know how to properly lobby government on an issue. Unlike other public affairs firms, however, ECL refused to lobby on behalf of a client and confined its business to providing advice and guidance. This suited de Grandpré, who preferred to conduct important meetings with ministers himself.

Bell Canada has since retained Moores's Government Consultants International. In addition to Fox, the firm has registered Gary Ouellet, its president and a close friend of Mulroney and Deputy Prime Minister Don Mazankowski, as one of the phone company's lobbyists.

GCI has also represented American Express Canada, now Amex Bank of Canada, which won a cabinet order making it a Schedule B bank on election day, November 21, 1988, in part due to GCI's efforts; Nabisco Brands Canada, a unit of RJR Nabisco; and Messerschmitt-Bolkow-Blohm A.G. of Germany, which won a $1.8-billion order from Air Canada for its Airbus planes in 1989. The ensuing controversy over Moores's use of his connections to the Mulroney government to help clients land lucrative government contracts led him to appoint a prestigious bipartisan board of directors and to bring in high-profile Liberals, like Fox.

Bell Canada has long backed up its lobbying efforts and policy analysis with its own sophisticated public opinion polls and direct mail programs. The utility developed a polling unit in the 1950s. However, unlike large manufacturing companies that embraced survey research for marketing, Bell Canada initially used its polls to measure subscriber satisfaction and to help track its quality-of-service indicators. The growth of competition led to a significant expansion of the telephone company's polling operations for its competitive marketing operations. By the mid-1980s, the polling unit employed fifty full-time survey researchers in Toronto, Montreal and Ottawa.[51] Direct mail and advocacy advertising are also effective tools to garner support and to directly shape public opinion, as Bell Canada showed when it appealed to its shareholders during the reorganization. In addition to BCE's quarterly mailings to its 270,000-plus registered shareholders, Bell Canada's monthly bills sent to its almost seven million customers can be an effective medium when they include billing inserts on regulatory matters.

Although many of the methods employed by Bell Canada and BCE in their government relations emulate the techniques used by sophisticated U.S. lobbies, the differences in the political cultures of the two countries have a significant effect on the way lobbying is conducted. Lobbying in Washington, for example, is more open, yet more complex and time-consuming, simply because of the sheer number of people that have to be approached on any single issue; there are more than thirty thousand senators, congressmen, committee staffers and aides on Capitol Hill alone. Guided by Edmund Fitzgerald, Northern Telecom mastered the Washington scene in the early 1980s.

Fitzgerald had spent a great deal of time "working the Washington scene,"[52] first as head of Northern Telecom's U.S. subsidiary, then, after 1982, as president of the Canadian parent. With de Grandpré's support, he opened a Washington-based public affairs office in 1980 and appointed Norman Dobyns as its vice-president. The lobbying unit was modeled after a similar office that had been set up by Northern Telecom's Canadian subsidiary in Ottawa a decade earlier.

The initial purpose of the Washington office was to lobby the U.S. Trade Representative, the Department of Commerce and the White House for a special ruling from the U.S. government declaring that Northern Telecom's U.S. subsidiary was a U.S. company for purposes of exports from the United States.[53] Under American law, any company with more than 20 per cent foreign equity is treated as a foreign company. This meant that if NTI exported telecommunications equipment to Asia, for example, it would be ineligible for U.S. Export Bank financing.

In late 1981, however, Dobyns and Fitzgerald were forced to turn their attention to a crisis posed by two pieces of legislation before the U.S. Congress. Both bills were aimed at deregulating AT&T and restructuring the Bell system. After five years of work, the Senate version, Bill S.898, was passed by a 90-4 vote on October 7, 1981, and referred to the House of Representatives to be combined with the Congressional version of the bill.[54]

The Senate bill contained a dangerous "reciprocity provision" aimed at foreign telecommunications equipment makers. Although reciprocity in trade provisions had been banned under the international General Agreement on Trade and Tariffs after the Second World War, protectionist sentiments were sweeping through Congress because of mounting trade imbalances with Japan. Senator Barry Goldwater explained the intent of section 238, which proposed to give sweeping new powers to the U.S. Federal Communications Commission:

> The commission shall use this authority in cooperation with executive agencies to encourage foreign countries to open their domestic markets to U.S. suppliers. Should such efforts fail, however, this authority may be used to deny foreign corporations the advantage of access to our markets while having the benefit of protected domestic markets.[55]

The House version, Bill HR. 5158, introduced by Representative Timothy Wirth, chairman of the House Subcommittee on Consumer Affairs, Telecommunications and Finance, on December 10, 1981, contained a similar, but even stricter reciprocity clause.[56] It stated that foreign-made equipment could be purchased only from countries permitting U.S. equipment sales on a reciprocal basis.[57] Canadian telephone companies bought the bulk of their equipment from preferred domestic suppliers, a practice considered discriminatory by the proposed legislation.

Fearing the bill would entirely close the U.S. market to Northern Telecom, Fitzgerald decided the company would fight the reciprocity clause and not take a stand on the question of deregulation, which he viewed solely as a domestic political concern. First, he raised the company's profile on the issue by going public on the matter in the trade press, on Capitol Hill and in the Congressional districts where NTI did business. Meanwhile, Dobyns and his staff went to work lobbying each of the twenty congressmen on Wirth's subcommittee and the twenty professional staff members that served the subcommittee. Northern Telecom's lobbyists met regularly with the staff specialists and attended every subcommittee hearing for six months through the spring and summer. Although there were few members on the committee who came from constituencies with a Northern Telecom plant in their district, NTI's argument that Congress should consider the impact of the reciprocity clause on thirteen thousand U.S. workers was compelling.[58] In his testimony before the House subcommittee on March 3, 1982, Fitzgerald reminded the members that reciprocity legislation would not of itself bring about fair trade. He likened reciprocity to a hunting license "which does not in any way define or restrict the type of armament or method of the hunters or ensure that all hunters will enter all fields on equal terms."[59]

The fall brought the mid-term Congressional elections and, in keeping with a century-old historical trend, the Democrats stepped up their pressure for protectionist measures. However, the pro-free-trade Reaganites refused to budge on the issue, and that was Fitzgerald's trump card. He won high-level White House support for Northern Telecom's stand from at least two senior Reagan cabinet secretaries — Malcolm Baldridge, the Secretary of Commerce, and William Brock III, the U.S. Trade Representative. The White House, in turn, sent Brock to deliver a stern rebuke to the subcommittee and put further pressure on Congress to remove the reciprocity clause. His unspoken message was that any such provision would be defeated by a presidential veto. As it happened, the Wirth bill died after failing to clear a joint conference of the Senate and House.

Fitzgerald revisited Capitol Hill two years later to testify against a toned-down version of the reciprocity clause contained in a trade bill introduced by Senator John Danforth in May 1984. Although Senate officials said NTI was considered a U.S. company and not affected by the clause, Fitzgerald told Danforth's Senate Finance Subcommittee on Trade that such a law "would inevitably lead to retaliatory trade legislation in capitals around the world."[60] The bill was subsequently defeated. A study on international trade by Arthur D. Little & Associates in 1985 pointed out the fallacy of protectionist measures. It concluded that the greatest direct cause of the mounting U.S. trade deficit was the dramatic increase in value of the U.S. dollar, which made imports cheaper and exports more expensive.[61]

The Northern Telecom head had come to the same conclusion and not only decried protectionism, but also believed many of the problems facing U.S. industry were self-inflicted. It was an unpopular view on Capitol Hill. "I think they see me as a pain in the ass, but I believe we have to make ourselves competitive."[62] Although Fitzgerald may not have personally converted former protectionists to that view, Wirth echoed his sentiments when he told John Berry of *The Washington Post* in a 1985 interview for *Financier* that the pressures for protectionism "are coming, for the most part, from large enterprises who are having trouble keeping up."[63]

"The people trying to throw rocks at NTI have pretty much gone away now," Fitzgerald says. "I really think that NTI has established itself as a fully legitimate U.S. producer that has created high-paying, high-quality jobs in the United States." Equally important, he added, "our customers want us here."[64] It helps when one of those customers is the White House itself. Since 1985, all telephone calls to or from the U.S. president's mansion and offices at 1600 Pennsylvania Avenue have been routed through a Northern Telecom–made computerized switch leased to the secretive White House Communications Agency by the Chesapeake & Potomac Telephone Company.[65]

Northern Telecom's ties to the White House and Republican party leaders were strengthened further when the company elected Frank C. Carlucci III to its board of directors in October 1989. Carlucci's résumé "reads like an organizational chart of the federal government,"[66] James Bamford wrote in *The New York Times Magazine*. Carlucci returned to the private sector in January 1989 from the Pentagon, where he had been Reagan's secretary of defense since November 1987. Prior to that appointment, he had served as Reagan's top national security advisor; deputy secretary of defense during Reagan's first term; deputy director of the Central Intelligence Agency under President Carter; and deputy director of the powerful Office of Management and Budget under

President Nixon. The former bureaucrat has a reputation for his wise advice and was appointed by Fitzgerald to Northern Telecom's board because of his "comprehension of global issues."[67] His contacts and insights cannot hurt, either, in the quest for U.S. government contracts.

Although the BCE group did not require much assistance from de Grandpré to win government contracts in North America, he devoted considerable time and effort to raising the visibility of both Northern Telecom and BCI abroad and in lobbying to support their efforts to win international contracts throughout Europe, Asia and Africa. De Grandpré continued to tap offshore equity markets for both BCE and Northern Telecom, not simply to raise more capital, but to achieve the twin goals of raising the companies' profiles in foreign markets and ensuring access to financial and political elites abroad. In addition to BCE's four stock-market listings in North America, he listed the company's stock on nine European stock markets between 1976 and 1984 — London, Paris, Frankfurt am Main, Dusseldorf, Zurich, Basel, Geneva, Brussels and Amsterdam. "De Grandpré had a wide range of business contacts and friends throughout the world," Fitzgerald says, adding that several of those individuals were of help to Northern Telecom in its forays into Europe, particularly France.[68]

Both Bell Canada and Northern Telecom shunned federal government grants after the companies were stung by criticism following the Microsystems International Limited debacle in 1975. That aversion kept the BCE group relatively clear of patronage politics at home. But an obscure international development contract that BCI was asked to work on in the French-speaking nations of West Africa caught the company in the web of patronage that embraced Ottawa following the election of the Mulroney government in 1984.

The Pan-African Telecommunications, or Panaftel, project was conceived by the Geneva-based International Telecommunication Union (ITU) in 1973. Its goal was to use a cheap, but vast, microwave system to link five nations of western and central Africa on a communication path roughly equal to a link between Miami and Montreal. The proposed 3,000-kilometer-long network would join Senegal, Mali, Burkina Faso, Niger and Benin as part of the ITU's efforts to support telecommunications projects in the developing world.[69] Financing was the first obstacle. "There was no stampede to underwrite the cost of a project generally considered risky,"[70] Michel Hébert, a former general manager of the project for BCI, wrote in a 1986 trade journal article on the project.

The Canadian government responded to ITU's call in late 1976. After engaging Intel Consultants of Ottawa to help set the specifications for a contract, the Canadian International Development Agency (CIDA) called for tenders. The response was slow and doubts about the project led BCI to shun it altogether. Although Intel had initially estimated the construction costs of Panaftel at $30 million, the first five-year CIDA project was valued at $78 million, including contributions from the recipient countries.[71] CIDA announced the award of the contract in the fall of 1977 to Elinca Communications, a Montreal-based consortium that brought together Spar Aerospace of Montreal; Raytheon Canada of Waterloo, Ontario; Andrew Antenna of Whitby, Ontario; SNC Group, an engineering company based in Montreal; and DGB, a unit of SNC. The contract was signed in April 1978 and work began in early 1979 after crews and equipment arrived from Canada.[72]

CIDA soon discovered, however, that coordination of the project and crews in five countries posed a significant hurdle for the contractors. BCI was awarded a management contract on January 28, 1980, valued at $9.7 million.[73] By late 1982, the Elinca crews had completed the microwave route and built fifty-five working microwave relay stations. Although this was a significant achievement, the harsh environment led to major problems. There are no electricity grids in most of the region served by the Panaftel network, which meant the relay towers had to be powered by petroleum-fueled electrical generators. Getting enough fuel to these towers so that they could operate twenty-four hours a day year-round was virtually impossible because roads disappear during the four-month-long rainy season. That meant there were frequent and prolonged outages, which were exacerbated by the lack of any backup network.[74]

CIDA went back to the Treasury Board Secretariat (TBS) and won approval on October 20, 1983, for a second five-year $45-million grant to expand and improve the first network.[75] The extended program was named Devtelao, for Développement des télécommunications de l'Afrique de l'ouest. CIDA turned yet again to a reluctant BCI to supervise the project. BCI signed the second CIDA contract in a joint venture with a brother of Mulroney's senior adviser who had special expertise in building microwave systems. The scope of BCI's role was broadened beyond contract supervision to include acting as purchasing agent. The four-year contract, valued at $14.3 million, was awarded to Douserv-BCI Incorporated on January 4, 1985.[76] Groupe Douserv

Incorporated of Montreal operates a chain of consulting engineering companies and is owned by Raymond Doucet, a former Bell Canada engineer and brother of Jean Alfred "Fred" Doucet, then the powerful senior adviser to the newly elected Mulroney. Nicknamed "Doctor Death" because he always used his Ph.D. title and was known as the PMO's axe man, Fred Doucet had been a friend of the prime minister's since the two met at St. Francis Xavier University in Antigonish, Nova Scotia. Doucet's oldest brother, Gerry, is a Halifax lawyer and Moores's partner in the Ottawa-based GCI lobbying firm.

Doucet vigorously denied that he benefited from his brothers' political connections. The Devtelao contract had been launched by the previous Liberal government, which allowed Doucet to maintain that the timing of the contract was merely a coincidence. Officials applied to incorporate the Douserv-BCI venture on September 6, 1984 — two days after the 1984 federal election — and the new entity was officially incorporated on October 2.[77] The dates gave the impression that Doucet had wasted little time in seeking new contracts from the Mulroney government. Public records for the first term of the Conservative government show Douserv's business with the federal government soared after the 1984 election.[78]

The growing controversy surrounding patronage led the government to take numerous actions, some aimed at reining in Mulroney's friends. In March 1987, Monique Landry, the minister of external relations responsible for CIDA, released a public list of contracts to much fanfare as part of the government's efforts to improve its image. It is worth noting that CIDA hid the existence of the legally incorporated joint venture in that list.[79] Although the contract was awarded to the joint venture, CIDA's list attributed the full value of the Douserv-BCI contract to Douserv Group Incorporated. BCI showed up in a separate entry as the recipient of a lesser $2.5-million contract for Devtelao. The correct entry was recorded in unpublished contract documents obtained by the author.[80]

Why CIDA would hide the existence of the Douserv-BCI joint venture from public records is not known. But the agency had good reason to want to downplay the Devtelao contract, according to the minutes of a TBS meeting of January 24, 1985. That record confirmed the long-standing suspicion in Ottawa that CIDA, in contravention of TBS guidelines, did not award professional service contracts let under tender on the basis of price. CIDA's head of the Devtelao contract selection committee got the agency in hot water when he admitted that fact in a letter accompanying a set of bid documents returned to a competing bidder that had lost the tender. "Cost is not a factor in the award of this contract," the official declared.[81] TBS instructed CIDA's president, Margaret Catley-Carlson, in its decision awarding the contract to Douserv-BCI that "approval was granted on the understanding reached between our respective officials that CIDA has agreed to seek costed proposals in future."[82]

In addition to his companies being asked to help government agencies, de Grandpré himself was occasionally asked for assistance by the federal government. In April 1973, he was appointed to the five-member Executive Compensation Advisory Board headed by Allen Lambert, chairman of the Toronto-Dominion Bank, to advise the government on how much money senior civil servants should be paid.[83] And in June 1979 he was one of ten executives asked to brief Prime Minister Joe Clark prior to the Western economic summit meeting at Akasaka Palace in Tokyo.[84] Only his January 11, 1988, appointment as chairman of the Advisory Council on Adjustment prompted a flurry of criticism. The council's mandate was to advise the Mulroney government on what actions regarding employment and training it should take following the signing of the free trade agreement with the United States.[85] The NDP criticism, voiced almost a year later, was in part fueled by the layoffs of 630 Northern Telecom workers whose Aylmer, Quebec, plant was being shut down. Union representatives had told the NDP caucus that Northern Telecom's plant in Stoney Mountain, Georgia, outside Atlanta made a similar product, leading MP Dave Barrett to charge in the House of Commons that de Grandpré's "fox-in-the-henhouse appointment" was a conflict of interest. Mulroney defended the appointment by responding, "only an NDP'er would think that success in the private sector represents failure in Canada."[86] In turn, Northern Telecom reassured the government that the "rationalization" of the inefficient Aylmer plant was not related to free trade, a view accepted by the government.[87]

Paradoxically, de Grandpré had not been a strong proponent of bilateral free trade and had argued instead for global trade liberalization. One month after the 1984 federal election that brought Mulroney to power in part because of his party's opposition to free trade between Canada and the United States, de Grandpré told a conference in New Orleans that sectoral free trade agreements would invite retaliation from other trading partners.[88] Although business leaders later warmed up to the Canada–U.S. Free Trade Agreement, which was signed on

October 4, 1987, many, like Bell Canada's chairman, were initially lukewarm.[89] So, too, was Northern Telecom's U.S.-born chairman, who believed that a company had to compete globally to succeed.

"Business called the tune on free trade and now it has to pay the piper to give Canadian workers the training to compete head-to-head with the Americans," de Grandpré's Advisory Council on Adjustment concluded in its report, *Adjusting to Win*. Both business and government were told they would have to spend more on training and retraining, education and R&D if Canada was to remain competitive in the wake of the agreement, which took effect on January 1, 1989. Although he recommended that the federal government double its $250-million training programs, he urged the government to place the burden on the private sector through measures such as a special tax penalty on businesses that neglected to train employees, a tax that could be worth $3 billion a year.[90] De Grandpré took to the hustings to promote his recommendations amidst criticism from labor and business organizations alike. Union representatives called the recommendations inadequate, and manufacturers and small business groups opposed the training tax, which, they argued, was a "carrot-and-stick" that opened the way for coercion by government.

For the most part, however, de Grandpré spent far more time chastising the government for its economic policies than he did working with it. In the Seventies, his criticisms frequently earned banner headlines: "Bell Head Slams Fiscal Attitudes," proclaimed The Montreal *Gazette* in May 1973, and "Bell Chief Swipes at Ottawa," said *The Montreal Star* in April 1977. De Grandpré became a frequent and vocal basher of Trudeau's economic policies. On April 4, 1976, he joined forces with Alfred Powis, president of Noranda Mines, and Simon Reisman, who had resigned a year earlier as deputy minister of the federal Finance Department, on a business seminar panel at McGill University to denounce the government's policies. The three assailed Trudeau's reliance on Keynesian economics and policies that favored State intervention. After thirty years in the public service, Reisman's attack on Trudeau for "demonstrating economic illiteracy" guaranteed good coverage of the event.[91] De Grandpré's attacks on Trudeau's policies were unrelenting and continued unabated until his former classmate left politics.

De Grandpré did not speak only as the chairman of Bell Canada or BCE. He often intervened in federal affairs as the spokesman of big business, which was the role handed him by the Quebec Chamber of Commerce during talks between labor and the federal government on wage and price controls in August 1977. The Quebec business establishment called de Grandpré "the boss of business."[92]

"He was a very capable leader and spokesman for the business community and for national organizations,"[93] says Robert Bandeen, the former head of Canadian National Railway. De Grandpré often took charge at meetings with his peers on issues of mutual concern, recalled Ghislain Dufour, president of Quebec's largest employers' group, the Conseil du patronat du Québec.[94] Like others who dealt with him, Dufour was struck by de Grandpré's leadership and tenacity.

The interests of the business community and the companies de Grandpré led were often synonymous, particularly on issues such as industrial policy and research and development. A federal government proposal in the mid-1980s to eliminate the already inadequate tax deduction for expenses associated with R&D facilities led de Grandpré to vigorously lobby federal officials to oppose the idea. "Do you think you can do R&D without a roof over your head?" de Grandpré says he asked a deputy minister.[95] BCE's control of BNR and Northern Telecom gave him the power to back his words with a threat to halt the company's R&D expansion and to turn away new scientists if the deduction disappeared.

He also decried the federal government's dislike of big business, which he attributes to an "anti-bigness" attitude. "Thinking big is not a Canadian virtue," he told Northern Telecom shareholders in 1981. At that time, he slammed the federal government's R&D policy as a "national failure." BNR, for example, had not come about through a national industrial strategy, he maintained, but because of the decisions and determination of management.[96] In his eyes, the inquiry to break up Bell Canada and Northern Telecom was "a syndrome of anti-bigness" and ran counter to encouraging R&D. He maintains that large corporate organizations and R&D go hand-in-hand, particularly if Canada is to compete on a global scale. Like Vail and Forbes a century earlier, de Grandpré also believes that the federal government should pursue policies that preserve Bell Canada's long-distance monopoly to ensure the telephone company can "maintain a strong R&D presence for all of Canada."[97]

Both de Grandpré and other executives in the BCE group actively lobbied the government to live up to its previous promises to increase R&D spending. In addition to serving on the National Advisory Board on Science and Technology, which met regularly with Mulroney, de Grandpré backed groups such as the Corporate Higher Education

Forum (CHEF). A CHEF task force chaired by then Bell Canada chairman J. V. Raymond Cyr issued a detailed report on the state of Canada's research and development community in October 1985. Cyr directly lobbied Mulroney to strengthen the financing of post-secondary education and research and to approve the five-year plan of a key federal research granting council, the Natural Sciences and Engineering Research Council.[98] But his pleas, like those of his successors, along with the recommendations of the report, fell on deaf ears.

De Grandpré also believes that improving Canada's competitiveness requires reining in the federal government's burgeoning deficit, which stood at $300 billion in March 1987. Its rate of growth halfway through the Mulroney government's first term added an additional $40 million to the total debt every twelve hours. In a 1983 speech, de Grandpré likened the federal government to "a drunken sailor on a buying spree."[99] Servicing that debt consumed about one-third of all tax revenue in 1986, compared to one-tenth of each dollar collected in 1974. "The need for deficit reduction is not a narrow concern of business. It is a broad public issue. The deficit is a liability for all Canadians," he told the Canadian Construction Association in February 1986, adding, "Our government pensions and family allowance are at risk."[100] His views today are the same as he expressed a decade ago: "There comes a point when the disciplines of the international money market will come to bear. I believe the worst mistake governments made was to borrow for paying bills."[101]

De Grandpré was particularly active with the Business Council on National Issues (BCNI), which is both a policy think tank and Ottawa-based lobby for the CEOs of Canada's 150 largest corporations. He assumed a key role with BCNI after Scrivener, who co-founded the council in 1976, retired. BCNI hired Thomas D'Aquino, formerly a Toronto lawyer and public affairs consultant, as its first full-time president in 1981. Until 1986, de Grandpré was a member of BCNI's elite policy committee, which hammers out the group's agenda and drafts its positions on issues like free trade.

Many observers and academics have held BCNI up as a lobby group with unrivaled influence. Professor William Stanbury, professor of regulation and competition policy at the University of British Columbia, stated that BCNI has significant access because government and political decision makers attend meetings it has arranged to "discuss the items on its agenda."[102] Stanley Beck, now chairman of the Ontario Securities Commission, exaggerated BCNI's clout even further when he wrote in a 1985 article that BCNI "is a power group like no other in our society and is so recognized by government."[103]

Although it is certainly a unique organization, its ability to influence policy is not as considerable as those observers believe, according to de Grandpré. "I think BCNI has done some very important work and has permitted the business community to focus on national issues," he says, pointing to the council's policy papers on national defense spending, the deficit and job re-training issues as recent examples. "But I don't think it has the recognition in government it should have. My feeling is they've been taken too lightly by governments." He lamented the fact that the federal government "has not paid attention" to most of the policy recommendations drafted by BCNI and the advice it has given.[104] And he is chagrined that the Mulroney government did not give business greater tax rewards in return for BCNI's endorsement of the Goods and Services Tax.[105]

"The biggest mistake in the relationship between the federal government and business is that the government resides in Ottawa," de Grandpré says. He believes federal politicians and officials "live in an ivory tower" and that the national capital's geographic separation from the major centers of Canadian commerce "creates serious impediments which are compounded by the parliamentary system." As he put it, "politicians devote more time to politicking and not enough time to governing." Although he has no simple solution, he says it "would take some very tough decisions and a very strong leader," along with a long-term vision for Canada. "I don't think that anyone has a vision about where they want Canada to be in the year 2000 or 2010."[106]

Back rooms were places where de Grandpré, like his predecessors, cajoled the nation's powerbrokers and where he tried to have his way. Boardrooms, on the other hand, were where the titular figurehead of Canadian business presided and where his opinion carried the greatest weight. The power and prestige of his own position was enhanced by the network of corporate board appointments de Grandpré established during his business career, and which proved helpful to his company over the years. "It could have been a much quieter life if I had not been on these boards," he says. "But, on the other hand, I don't think I would have had the network of connections, of associates, who opened doors for me and who have helped me to participate in a number of informal discussions on a number of issues."[107] (Informal quarterly meetings are regularly held by a group of CEOs of Canada's largest corporations to discuss economic, political and social issues.) In his view, the profile he has in the business community is due to his participation on outside boards. Without that profile, he believes he would have been excluded from the upper ranks of the Establishment because "they would have said, 'We don't know him well enough. Can we trust him?'"[108]

His appointment to the boards of almost a dozen leading blue chip corporations greatly extended his influence beyond the telecommunications industry. That influence reached into the boardrooms of several leading corporations in the United States, including two companies that ranked among the top ten on the *Fortune 500* list. Such appointments have helped to make his name widely known throughout the world's financial capitals and have made him one of Canada's most sought-after corporate directors.

For the same reasons that he believed it important to hold directorships, de Grandpré maintains memberships in every key Establishment club in Montreal and Toronto: the Club Saint-Denis, the St. James Club, the Royal Club and the University Club — all of Montreal; and the Toronto Club and the York Club — both of Toronto. In addition, he golfs and fishes at the Mount Bruno Country Club and the Forest and Stream Club and is a member of the elite Montreal Racket Club, which boasts two indoor clay courts.

Clubs are generally perceived as proof that class exists. The Toronto Club, founded by the Family Compact of Upper Canada in 1835, is the second-oldest club in North America and has been considered the most difficult private club in Canada to join.[109] Only men with old money or who are CEOs or heirs apparent are invited to become members. The ultra-secretive and highly conservative Mount Royal has been called the most snobbish club in the country.[110] The purpose of these clubs has remained unchanged over the years. They act as sanctuaries where CEOs can dine in privacy at lunch, meet other members of the elite with similar interests or concerns and entertain fellow executives or directors from out of town. Clubs also broaden a member's network beyond the boards he or she serves on. An added benefit of belonging to a national club, such as the Mount Royal and the Toronto, is the guest privileges that members are given in clubs throughout the world.

One's contacts are particularly important when doing business abroad, especially in Europe. "When we decided to be aggressive in international markets, we had to be visible," de Grandpré says. "I strongly believe that in international markets the presence of the chief executive officer is extremely important to develop networks."[111] That is why he joined local business associations, such as the Montreal Chamber of Commerce, and similar international groups, including the Chambre de commerce France-Canada. "I knew most of the French elite in business because of that association. I'm not saying that it made our life easier," but it put the company's name "in the daily conversations of people with some influence either at the political or the financial or commercial level."

But his greatest opportunity for networking came from his plethora of corporate directorships. At the peak of his business career, he sat on the boards of thirteen corporations. He does not, however, have a gluttonous appetite for directorships, unlike Peter N. Thomson of Power Corporation, who held forty-three directorships in the early 1970s,[112] and Ian Sinclair, the former chairman of Canadian Pacific Limited, who once sat on twenty-three corporate boards. De Grandpré not only cultivated friendships with influential fellow directors — including Charles Bronfman, chairman of the Seagram Company; John Henderson "Jake" Moore, former chairman of London Life Insurance Company; and John L. Weinberg, chairman of Goldman, Sachs & Company — but strengthened those allegiances by appointing members of boards he served on to the boards of BCE, Bell Canada and Northern Telecom. He also rubbed shoulders with the world's financial and political leaders at the annual symposium for world leaders at the Davos ski resort in Switzerland, which is touted as the most important gathering of the international elite. In 1984, when Canada was featured at the symposium, he was a key member of high-powered Canadian delegation put together to sell Canada as a place to do business.

The role of corporate directors and boards had changed dramatically in the postwar period, particularly in the past twenty-five years. In days past, the "old boys" approved the minutes of the previous meeting, considered a management forecast or budget and then adjourned for lunch and an open bar.

"It's no longer a private club where views are not expressed and the board rubber stamps the decisions made by management. In that sense, their role has changed considerably,"[113] de Grandpré says. Gone, too, are the days when the board simply represented the interests of the company's shareholders. Boards now see themselves as representing a diverse group of "stakeholders" in the corporation: shareholders, employees, customers and society. Exclude one group from a decision and there will be "negative consequences on the corporation," de Grandpré believes, echoing the current thinking of leading business schools.[114]

Although corporate boards should consider issues of concern to stakeholders, such as the environment and employment practices, de Grandpré remains opposed to the appointment of environmentalists, labor representatives and other representatives of single-interest groups to a board because he believes their presence fosters confrontational attitudes. Boards must function by consensus for or against management decisions. The role of the board is to make decisions, and a good director "should at all times weigh all the factors affecting the company," he says.

Edmund Fitzgerald, the retired chairman of Northern Telecom who sat with de Grandpré on the boards of both Northern Telecom and BCE, says he knew of few directors as adept as de Grandpré at thinking strategically and focusing a board's attention on a company's objectives and goals.

"He's on enough boards that he understands what the role of the director is, which is not to manage the company day to day, Fitzgerald says. "A board should ask management the right questions. He's very good at that. I remember when he's asked me the right question, which has caused me to reconsider what I was doing — without being told. He's a powerful leader." At his own corporation, de Grandpré expected BCE's directors to ask him the right questions, too, Fitzgerald says.[115]

Many of the functions of large boards, like BCE's, are delegated to committees, which meet regularly with corporate executives. BCE has four committees — the audit committee, management resources and compensation committee, pension fund committee and the investment committee — in addition to an executive committee. De Grandpré calls the latter a "committee of convenience" because it exercises the powers of the board when it is not in session. He personally favors corporate governance through board committees and, as a matter of policy, involved each BCE director on at least one committee. This was often a difficult challenge because he wanted to "create specialists," while simultaneously avoiding cliques.[116]

The investment committee keeps track of the company's financial performance and sets the company's strategic goals and objectives. In the case of BCE, it also plans acquisition strategy. The audit commit-

tee, which consists solely of outside directors, is responsible for veri-
fying the company's financial results and producing quarterly and
annual statements for the board's approval. In addition to working
with the company's controllers, the committee helps to develop internal accounting controls and is responsible for ensuring that corporate
assets are properly accounted for and safeguarded.[117] The pension fund
committee is responsible for protecting the company's pension fund
and supervises its administration and investments.

The management resources and compensation committee is the
most important committee. It is responsible for selecting the CEO's
successor, for hiring and firing senior executives, for nominating directors and for setting the salaries of both officers and directors to attract
and promote top talent to the corporation. To distance BCE's directors from management, the committee is made up entirely of outside
directors. "They have to make sure the compensation is fair; that it is
competitive. We cannot, as a board, allow the deterioration of compensation for executives to fall substantially behind the trend in the
communities in which we operate or we won't keep our management," de Grandpré says. But the board should not overpay executives, "or be bribed by management," he says, adding, "I'm thinking
about the F. Ross Johnson situation."[118]

Johnson, the jocular Winnipeg-born president and CEO of RJR
Nabisco of Atlanta, personified the acquisition-frenzied, greed-rid-
den decade of the 1980s — an era when going into debt to take over a
company's stock was euphemistically termed a leveraged buyout (or
LBO). LBO deals used high-risk, but high-yield, junk bonds to
finance an acquisitor's purchase of common stock in a publicly traded corporation. Following the merger, the massive debt was unloaded onto the target company and left the acquisitor owning the stock
in the target for next to nothing. The targets were often valuable
companies, but were either inefficient or had management and
boards who were unresponsive to shareholders. The junk-bond
raiders essentially sought to exploit what William Fruhan of the
Harvard Business School calls the "value gap" — the difference
between what a company could be worth if restructured and what it
is currently worth.[119] An LBO, the theory went, was a good thing
because it would force management to restructure its assets to avoid
financial ruin and to maximize shareholder value.[120] In practice, however, LBOs fattened the pockets of the deal makers and lawyers, like
junk-bond king Michael Milken, who earned hundreds of millions of

dollars in fees, and corporate honchos who sold the original shareholders short.

The story of the leveraged buyout of RJR Nabisco is a case study of
what can go wrong when a public company is governed by an ineffective
board co-opted by selfish management. Johnson ran the company like a
gravy train. Senior executives and loyal directors benefited from opulent
compensation packages, lavish stock offerings and luxuriant perks,
including corporate junkets to luxury golf club retreats attended by big
name sports figures flown in at corporate expense. Although billed as
promotion, the extravagance was at the shareholders' expense.[121]

Johnson tried to take RJR Nabisco private through an LBO, which
would have personally netted him $100 million. He won the support
of top management and inside directors, who also stood to benefit
from the management bid that would have resulted in the executives
controlling the company through the exercise of stock options and
interest-free loans. But his takeover offer put the company in play and
launched a bidding war that led to RJR Nabisco's takeover in 1989 by
the world's largest leveraged buyout firm, Kohlberg Kravis Roberts &
Company of New York. The $24.7-billion deal was the largest merger
in history at the time. Johnson and his supporters were purged, but
Johnson's severance deal, or platinum parachute, cost $30 million.

If boards pay management well and link executive compensation to
increased shareholder value, the experts say, unwanted arbitrage can be
partly avoided. "Stock ownership at all levels can encourage people to
think like owners rather than like employees or managers," a 1990
report on compensation by the Conference Board of Canada stated.[122]
Although that sentiment is corporate gospel at companies like BCE, it
does not always describe reality, as the Johnson case made clear.

BCE pays its officers and the executives of its subsidiaries well,
by Canadian standards, and provides stock plans as bonus incen-
tives and highly competitive pension benefits. Although the chairman of BCE holds the most senior position in the group, Northern
Telecom's chairman and president, Paul Stern, continued to be
Canada's highest-paid executive in 1990, earning $1.68 million
(U.S.). He is paid more because of his global management responsibilities, which involve extensive travel. J. V. Raymond Cyr, the
chairman and CEO of BCE, was paid $1.3 million (Canadian) in
1990, and Bell Canada president and CEO Jean Monty earned
$582,500 that year.[123] As one of Canada's top-paid executives before
his retirement, de Grandpré's salary was the subject of yearly atten-

tion by the business press. He earned $1 million in 1987; $852,900 in 1988 after stepping down as CEO; and $320,200 as chairman in 1989.[124] Yet, he argued, his pay was modest by U.S. standards and given the scale of BCE's enterprise.

BCE's directors are also well paid, but not richly rewarded, by international standards, for their time. A director received $20,000 a year in 1991, plus $3,500 for serving on a committee ($4,000 as committee chairman) and an attendance fee of $1,000 per meeting, plus expenses.[125] A director who is in demand, like de Grandpré, used to be able to earn between $200,000 and $300,000 a year in base pay and meeting fees for holding ten seats. However, the extended responsibilities of directors make it increasingly difficult for them to sit on more than seven boards and still be effective. Given a shortage of qualified outside directors, experts argue that directors' compensation will likely start to rise more than management pay so that directors are also rewarded for acting as owners rather than mere custodians.[126]

Well before corporate raiders and LBOs launched the trend toward shareholder-sensitive boards, BCE — and, in the pre-reorganization period, Bell Canada — had been said to possess one of the most effective boards in Canada.[127] It's not hard to see why: with more than 270,000 registered shareholders in 1991, BCE is Canada's most widely held corporation. By consciously setting out to attract more shareholders than any other business in the country as a means of thwarting hostile takeovers — a century-old strategic goal inherited from Bell Canada — BCE must seek to place the financial interests of its shareholders uppermost in the collective minds of both management and the board, or face a massive shareholders' revolt. Although Bell Canada's management has historically defined its role as "guardians of shareholder interests,"[128] it has, at the same time, often had a condescending attitude to shareholder opinions.

BCE actually has many thousands more unregistered and indirect shareholders. For example, a trust company is a single registered shareholder, but may hold stock for five thousand separate accounts. "And there isn't a pension fund in the country that isn't dependent economically on Bell Canada's economic well-being," Dr. H. Rocke Robertson says.[129]

With twenty members on BCE's board, there simply is not enough time for each director to address every issue at a full board meeting. Hence the committee structure is critical to the board's functioning. The chairman's role, in turn, is to forge a consensus. But, de Grandpré says, "the chairman has to be very careful that he is not creating two levels of directors, the more informed and the less informed . . . ."[130] Some executives would argue that that desire is irreconcilable with governance by committee, which is why some corporations have gravitated toward smaller boards. William Dimma, deputy chairman of Royal LePage, for example, is opposed to board committees because too much of the full board meeting is spent considering committee recommendations. He stated in a *Business Quarterly* column that any board with more than twelve members is too large.[131]

Bell Canada and BCE board meetings open with each of the committee chairmen giving their reports, followed by discussion. (After the BCE corporate reorganization, the utility retained its own board, which continues to devote about one-quarter of its meetings to presentations on regulatory issues.) After reporting on financial results and other general matters, management then tries out various ideas or proposals on the board, which, Robertson says, was when the meetings were the most interesting — "Bell's management never moved without board approval."[132] However, the real haggling still takes place in the committees, where directors often meet with vice-presidents and other executives. Louise Brais Vaillancourt says, "It's not always easy because sometimes you have to make hard decisions," but she adds BCE's directors have always been comfortable dealing with the company's executives.[133]

After 1976, de Grandpré wore two hats on the board as both chairman and CEO. Although outsiders may have viewed him as a tough or autocratic manager, he did not resent his board or the fact that his company had to justify itself to the directors or their committees. De Grandpré was in firm control of his board, says one former outside director, and could "extract opinions from people who otherwise would say nothing," Robertson says. "But he was very fair in his chairmanship and he always indicated his views up front."[134] It is on that point that he differed markedly from other corporate titans of his day. Harold Geneen, for instance, could not abide outside directors and tried to find a way to pack ITT's board with inside directors — members who also held executive positions with the company.[135] De Grandpré, on the other hand, has strong views on the need for a board to have a majority of outside directors. "I feel there should not be too many inside directors on the board because it is very difficult for them to express views that are not the views of the chairman. It's just impossible . . . . When you have too many executive vice-presidents or vice-presidents on the board, I don't think you have an effective board."[136] Although BCE still has a large board, its members are better informed and sit on fewer boards than directors two decades ago, in keeping with the changing role of directors and a trend toward boards made up of outside directors. The net effect is that the old boy network "is justly on the decline,"[137] Richard Mimick, professor at the University of Western Ontario School of Business Administration, wrote in *Business Quarterly.*

As a young articling student, de Grandpré first learned the intricacies of the workings of corporate boards by watching a consummate "old boy," F. Philippe Brais. He obtained his first corporate directorship in the early 1960s, when he was appointed to an insurance company board by his father-in-law's best friend. Henri Ouimette appointed him to the board of Société Nationale d'Assurances because he knew of the insurance lawyer's work and had known his father.[138] De Grandpré later served as a director on the boards of several educational and community organizations, including McGill University, the École des Hautes Études Commerciales and Hôpital Marie Enfant.

It did not take long after de Grandpré was hired by the phone company for him to be offered a directorship on a major blue chip corporation. The invitation to join the board of the Toronto-Dominion Bank came in early 1969 from Herbert Lank, the founding president of Du Pont Canada and an internationalist who had worked in Argentina and France for the multinational chemical giant. Lank was one of the country's leading directors and knew de Grandpré because he sat on the boards of both Bell Canada and Northern Electric. He soon became de Grandpré's mentor in the corporate world by taking the executive under his wing and grooming him for blue chip board appointments.

Although T-D was then the smallest of the five banks chartered under the Bank Act of Canada, a seat on a bank board was a clear signal that an individual had arrived among the corporate elite of the Establishment. Bank directors have tremendous responsibility and influence. However, the position in some ways resembles a Senate seat in Canada: the appointment is largely honorific, is effective until the director reaches his or her seventies and requires the director to own property — the Bank Act states a director must buy 2,500 shares in the bank before being appointed. The bank boards are also probably too large to truly govern their institutions. With as many as fifty directors on the board, effective corporate decision making resides with management and board committees — not with the full board. Nevertheless, de Grandpré says there was a lot of discussion on the

T-D board — "I felt the directors had an influence, a major influence, on some of the decisions taken by the bank over the years."[139]

Like any board, the directors of the bank are entrusted with preserving the value of the corporation and shareholders' investments. They review the competence of the CEO and other senior officers and ensure that employees are properly trained and that key positions are filled. But unlike directors of other corporations, they are also responsible for protecting depositors whose funds are entrusted to the bank.[140] In the discharge of these fiduciary duties, they approve the bank's investment and lending policies.

Bank directors must also ensure that the bank complies with the Bank Act, which describes their unique responsibilities and authority in this regard. Among its provisions is the requirement that directors approve all loans over a certain size. To avoid favoritism or conflicts of interest, a bank director is required by law to leave a board meeting when a loan to any of his or her own companies is discussed, or risk severe penalty under the Bank Act.

Both the bank and its directors have self-interested reasons, however, for granting and seeking the appointments. Because there is significant competition in the provision of credit and a bank's customers can borrow from a wide range of alternate sources, Canadian chartered banks use their directorships to maintain their hold on key corporate accounts and to ensure that those CEOs appointed to the board continue to steer new business the bank's way. That made de Grandpré's appointment to the T-D Bank all the more noteworthy at the time because the Bell Canada heir-apparent's boss held a seat on the board of the Canadian Imperial Bank of Commerce. Although the CIBC was assured of Bell Canada's business during Scrivener's tenure, it must have dreaded the day that de Grandpré would assume power over the utility's purse and move its business from the Commerce to T-D. And, in fact, that's just what he did.

The outside directors appointed to a bank's board are usually leaders in their industries and provide the bank with crucial insights and opinions about the marketplace. They, in turn, gain economic intelligence from the board meetings and briefings by the chairman and officials that is unsurpassed by any other source. This includes knowledge of the workings of the Bank of Canada and first-hand information on capital and money markets, which may be useful in plotting the financing plans of their own companies. Although a bank director is not supposed to receive preferential treatment, he or she is personally well known to the bank's senior executives and "there's an interplay between your own activities and the activities of the bank,"[141] de Grandpré says.

As an example, he describes the time he telephoned Ernest Mercier, the T-D's executive vice-president of corporate banking, at 11:30 one morning in late 1983 to obtain a $1.2-billion line of credit. BCE needed the money to finance its offer for a minority block of the shares of TransCanada Pipelines, the Calgary-based gas pipeline company. "We got the approval before lunch, by 12:30, and the necessary credits. It was a classic case of credibility and integrity."[142]

He sat on T-D's board for twenty years and saw the bank, now ranked as fifth of the top six by assets, emerge as the most profitable bank in Canada under the management of Richard Thomson, chairman and CEO. In 1989, however, he was forced to leave the bank's board when BCE acquired a rival trust company, Montreal Trustco.

More directorships soon followed the T-D appointment. The United Group of Companies appointed de Grandpré to the board of United Accumulative Fund and its three associated mutual funds in December 1969. Also that year, he was asked to serve on the board of the Pulp and Paper Research Institute of Canada as McGill University's representative. Lank appointed him to the board of Du Pont Canada in April 1970.

De Grandpré's seat on the T-D Bank board led Charles Bronfman and Leo Kolber, president of Cemp Investments — the Bronfman's family trust — to ask the Bell Canada boss to join the board of Cemp's publicly traded real estate arm, the Fairview Corporation of Canada, in 1973. Fairview was a shopping center and office building developer with long-standing ties to the T-D Bank, and it often looked to the bank's board when considering appointments to its own board.[143] De Grandpré's short stint on the Fairview board led to his twenty-year-long friendship with Bronfman.

In 1975, Fairview became the largest property developer in Canada after it acquired Cadillac Development Corporation and merged with Canadian Equity and Development — originally a partnership between Cemp and E. P. Taylor. De Grandpré, however, resigned before the new board was appointed to avoid a potential conflict of interest. Bell Canada had begun to plan a new real estate development project in downtown Montreal on Lagauchetière across the street from its Beaver Hall Hill headquarters. "If Cadillac Fairview was invited to make a bid for the project, it would be perceived as being totally unfair if they won,"[144] he says.

In 1973, de Grandpré earned a seat on the board of TransCanada
Pipelines (TRAP for short). Jim Kerr, its chairman, asked him to fill a
vacancy created when Marcel Vincent, who had retired as Bell Canada
chairman that month, also left the board. "It was important for
TransCanada to have a member of Bell on the board," de Grandpré
says, "because they were very similar companies in many ways. Both
are extremely regulated and they have capital structures that are not
dissimilar."[145]

In addition to his directorships on the boards of banks, developers,
high-tech and regulated companies, de Grandpré held seats on several
major industrial corporations. Like his predecessor, who sat on the
board of U.S. Steel, he had ties to "Big Steel" and was appointed to the
Steel Company of Canada of Hamilton, Ontario — now Stelco
Incorporated — in February 1976. He subsequently reciprocated by
appointing Stelco's chairman, J. Peter Gordon, to BCE's board.

Although Stelco's influence over the industrial heartland of Canada
is formidable, it lags behind first-place Dofasco Incorporated. North
American steelmakers have been assaulted by aggressive, more modern
offshore competitors who export cheaper steel. Stelco's shaky position
has been further exacerbated by protracted strikes at its Hamilton
mills. De Grandpré left Stelco's board in 1989 because of time commitments. He says aggressive unions and a cyclical downturn in the auto
industry have brought Stelco "to its knees."[146]

De Grandpré witnessed the many changes affecting the steel industry's position as the auto industry's largest supplier through his membership on the board of Chrysler Corporation of Detroit, the marginal
producer of the "Big Three" automakers. He was named to Chrysler's
board by chairman John Riccardo in 1976 and says he is now the most
senior member of Chrysler's board, having joined two years "B.I. —
Before Iacocca." Chrysler "fell apart" financially within a couple of
years of his appointment. "We lost market share, our products weren't
good and we had lots of problems with the public. We were on the
verge of bankruptcy,"[147] de Grandpré says. The story of Chrysler's
turnaround is a twentieth-century business legend that made its architect, Lee Iacocca, one of the most popular businessmen in U.S. history.
A lesser-known story is that de Grandpré played a role in hiring the
man who was destined to become America's corporate hero.

Iacocca first achieved fame in the auto industry in the mid-1960s for
spearheading the development of the widely acclaimed Ford Mustang.
The success of the Mustang led to his appointment as president of Ford

Motor Company of Detroit in 1970. He was approached to join Chrysler after Henry Ford II abruptly fired him on July 13, 1978, following a feud between the two men. After a series of informal meetings with two powerful Chrysler directors and Riccardo, Iacocca decided he was interested in the challenge of running the troubled automaker, provided he had a shot at running the company as CEO in a year.

To begin the formal hiring process, a secret meeting with the compensation committee of Chrysler's board was held in November 1978 at Chrysler's suite on the thirty-sixth floor of the Waldorf Towers in New York.[148] The committee was chaired by de Grandpré, who met Iacocca along with fellow committee members Dick Dilworth, who ran the Rockefeller family's financial trusts, and Bill Hewlett, cofounder of computer maker Hewlett-Packard.

De Grandpré's committee agreed to "make him whole," as Iacocca phrased his package in his 1984 bestselling autobiography. He asked for the same salary he had made when he left Ford — $360,000 (U.S.) a year. Because he had forfeited a $1.5-million settlement package with Ford, which stipulated he would lose his severance pay and unpaid benefits and bonuses if he went to work for another car maker, Chrysler gave him a bonus on joining the company that in effect bought out that contract. Iacocca dismissed the press criticisms by saying that it was money earned at Ford.[149]

"I was publicly criticized for Iacocca's compensation package," de Grandpré says. "But the market pays Iacocca. And we have to be competitive." The new executive received stock options that entitled him to buy Chrysler stock at a later date at $6.25 a share. The option was not much of a bargain when Chrysler's stock was trading as low as $3.50 a share shortly after he joined the company, but Iacocca came under more criticism when he exercised the options after the stock shot up past $75 a few years later. De Grandpré argued Iacocca got a bum rap because "he had the right to exercise this option."[150]

Although Iacocca knew running Chrysler would be a challenge, he soon learned it was on the verge of financial collapse. He swallowed his pride and approached the administration of U.S. president Jimmy Carter in June 1979 for federal assistance through a $1-billion cash advance against proposed future tax credits. Although supported by the U.S. treasury secretary, G. William Miller, and Chrysler's bankers, Manufacturers Hanover Trust Company of New York, the administration procrastinated in making a decision by seeking further studies. As the days went by, Iacocca believed Chrysler would default on its bank loans and be unable to meet its payroll unless immediate aid was forthcoming. Supported by several board members, including de Grandpré, he decided to take Chrysler's request for help directly to the U.S. Congress that October. De Grandpré accompanied Iacocca to Washington several times during the Chrysler chief's frequent trips to testify to various committees that fall. If Mexico could receive a $1-billion loan without delay, argued Iacocca, why couldn't Chrysler, whose survival directly affected about 400,000 jobs in the auto industry.

Chrysler won a controversial $1.5-billion rescue package when the House of Representatives approved the Loan Guarantee Act on December 18, 1979.[151] In Canada, de Grandpré helped the company reach a compromise with the federal government, which demanded assurances of fixed employment levels in Canada in return for government guaranteed loans to Chrysler Canada. Chrysler agreed to a percentage of North American employment, rather than an absolute number.

Chrysler was fortunate that it had a major new model development project in the works, the K-car project begun in 1977. The K-cars got off to such a slow start when they rolled off the assembly line in the fall of 1980 that Chrysler had to draw another $400 million against its government loan guarantee in 1981 to get through the recession. Tough belt cutting and the later success of the fuel-efficient, front-wheel-drive K-cars led to a spectacular recovery and allowed Iacocca to announce the company's repayment of all its obligations to the U.S. government seven years early. On August 12, 1983, the loan agreement was terminated.[152]

But it was the launch of the T115 minivan in 1984 that put Chrysler squarely on its feet. The company was still living hand-to-mouth when Iacocca and Chrysler's chief designer, Hal Sperlich, took the radical new design to Chrysler's board (they had both pitched the idea of a minivan to Henry Ford in 1974).[153] De Grandpré recalls the decision of Chrysler's board to go ahead with the development of the minivan: "It was probably the worst period in our crisis. We had $1 million in the bank account. That was all. And we authorized $780 million to develop the minivan."

"You better be right," de Grandpré told Iacocca after the board meeting.

"No question, that's what the public wants," Iacocca said.

"If you're not right, we may both end up in jail,"[154] de Grandpré told him in jest.

Iacocca later termed Chrysler's turnaround the "miracle of Windsor," after the huge $400-million effort to gut and modernize

Chrysler Canada's aged manufacturing plant to produce the first line of minivans.[155]

De Grandpré was appointed to Chrysler Canada's board in 1976. His connections to Chrysler appeared to extend to Bell Canada's procurement practices for its corporate vehicle fleet. One of the utility's largest vehicle purchases was made in 1981 when it switched the bulk of its urban fleet to K-car station wagons from traditional utility vans. It bought 1,200 K-cars. According to Bell's purchasing department, about 81 per cent of its fleet vehicles in 1984 were Chrysler products, 9.7 per cent were made by Ford, 3.1 per cent by American Motors (Canada) and 2.9 per cent by General Motors of Canada.[156]

Those statistics seem to explain Bell Canada's loss that year of a major competitive bid to install a new telephone switching system for Ford Canada at its head office and auto plant in Oakville, Ontario, and parts center in Bramalea, Ontario. Officials for both Ford and Bell Canada denied that the automaker switched suppliers because the telephone company would not give Ford a bigger share of its fleet business. Nevertheless, what it did illustrate was that the phone company was no longer insulated from those sorts of tradeoffs and concessions as it had been before its terminal equipment market was opened to competition in 1982.

In 1986 Chrysler was reorganized. De Grandpré says Iacocca modeled the corporate reorganization of the automaker after BCE's reorganization. The corporate flip-flop made Chrysler the new publicly traded parent company. Chrysler Motors Corporation, which had been both the parent and the automaker since its incorporation by Walter Chrysler in 1925, became the manufacturing arm. The reorganization gave the new parent responsibility for Chrysler's financial subsidiaries and technology companies as well. "You can see there was an affinity between the organizations,"[157] de Grandpré says. There is also a bond between the two corporations' leaders, forged by their similar temperaments and views: "There's no doubt that we're on the same wavelength. We think alike on many, many issues."[158]

Chrysler is close to being on the ropes again as Iacocca nears retirement. Its market share has declined in the face of greater competition, its debt has mounted, it has $3.6 billion in underfunded pension liabilities and it had a negative cash flow of $2 billion in 1990. Although it ranked seventh on the 1990 *Fortune 500* list in sales, it ranked far behind — seventy-fourth — in profit.

Iacocca's failed and highly problematic takeovers after the turnaround have been viewed by industry analysts as the greatest source of Chrysler's woes as it entered the Nineties. His strategy caused the company to divert a substantial portion of its lucrative profits during the boom of the late Eighties away from new model development into ill-conceived acquisitions, including luxury Italian automaker Maserati, corporate jet maker Gulfstream Aerospace Corporation, and American Motors. Although Chrysler has shucked Maserati and Gulfstream, the company's problems and diversions have conspired to cause bankers to hike the cost of financing Chrysler's current $15-billion, five-year new model development program. Although Iacocca's competitors haven't written him off, some critics say his efforts now may be too late.[159]

One of the few companies to have adroitly divested assets and invested wisely in the Eighties is Seagram Company Limited, the world's largest wine maker and distiller. De Grandpré was appointed to the board in May 1979 by Charles Bronfman, who, along with brother Edgar Sr., Seagram's chairman, owns 40 per cent of the company. Control of the liquor empire, which was launched by their father in 1928, passed to the brothers when Samuel Bronfman died in 1971. The sons also inherited Mr. Sam's investment savvy. Despite all the millions of cases of rye, gin, vodka and trendy coolers sold by Seagram last year, almost three-quarters of the profit came from dividends paid to it as the largest shareholder of U.S. chemical giant E. I. du Pont de Nemours & Co. of Wilmington, Delaware.

Seagram's foray into chemicals wasn't planned, de Grandpré says. It began when the Bronfman brothers decided to sell the lucrative U.S.-based oil and gas interests their father had amassed in the Sixties, which they had hung onto through the Seventies as energy prices soared. The sale of those assets to Sun Company of Radnor, Pennsylvania, in April 1980 gave Seagram a $2.3-billion war chest.

Seagram set its sights a year later on St. Joe Minerals Corporation, the largest U.S. lead producer. De Grandpré and his fellow directors learned of the $2-billion proposal at a hastily called meeting of the Seagram board on the night of March 10. The secret meeting, held at the company's medieval fortress-style building in downtown Montreal, was prompted by an apparent leak, which had led to active trading in St. Joe's shares on the New York Stock Exchange earlier that day. The share price had been driven up $2.[160] Although the board approved an offer of $45 a share, the Seagram bid was foiled when

Fluor Corporation of Irvine, California, countered with a more attractive offer.

Seagram's U.S. subsidiary, Joseph E. Seagram and Sons Incorporated of New York, was rebuffed again that summer when it made a friendly offer for 40 per cent of Conoco Incorporated of Stamford, Connecticut. It then launched a $2.6-billion hostile takeover bid, which started a bidding war for the U.S. oil company that July.[161] Conoco executives decried the Canadian raid, which followed an earlier offer from Dome Petroleum of Calgary, as a threat to the U.S. petroleum industry. De Grandpré found himself on both sides of the battle after Du Pont stepped in as Conoco's white knight; he was an insider of both Seagram's and Du Pont's subsidiary. Du Pont's $7.3-billion (U.S.) acquisition of Conoco's shares was, at the time, the largest merger in history.

Although thwarted in its bid for Conoco, Seagram spent $2.6-billion to acquire a minority 20 per cent stake in Du Pont. At the time, Seagram's chairman was criticized by the financial community for going into debt to acquire the holding and for taking a $1-billion paper loss on the transaction. The Du Pont stock, which cost the Bronfmans $54 (U.S.) a share to buy, traded at $30 on the New York Stock Exchange later that summer.[162] Critics have since choked on their doubts as the profit contribution from Du Pont continues to rise.

The Bronfmans nominated de Grandpré to the board of the chemical giant. After his name was accepted, he stepped down from the Du Pont Canada board to take a seat on its U.S. parent's board. But in 1989 he resigned from that board because its meetings clashed frequently with BCE board meetings.

He continues to be active on Seagram's board. "It's a very stimulating board," he says, naming fellow board members Paul Desmarais, chairman of Power Corporation; John Weinberg, head of Goldman Sachs & Company; Marie-Josée Drouin, executive director of the Hudson Institute of Canada; and Bill Davis, former Ontario premier.[163]

Although the Bronfman family had debated who would become the third-generation successor to Mr. Sam for several years, outside directors are the ultimate kingmakers when it comes to succession. Edgar, Sr. surprised the Bronfman family and Seagram's board, de Grandpré says, when he put forward the name of his thirty-four-year-old son, Edgar, Jr., in the summer of 1989. The number-two son had never gone to college and had originally pursued a career as a movie producer.[164] In

1979, his father asked him to join the $6-billion family business, where he proved to be a sophisticated marketer and astute businessman.

"There was some reluctance to move him up so quick, particularly because the outside directors felt there was a risk for both his career and the company,"[165] de Grandpré says. However, based on his record in the preceding seven years, the board decided in February 1989 to name Edgar, Jr. as the successor to David Sacks as president effective July 1. His performance has been a "pleasant surprise," de Grandpré says. "He's a good manager and a chip off the grandfather's block — but more polished!"[166]

The creation of BCE brought greater recognition for de Grandpré, along with awards and offers of more board appointments.

One of the first was from Sun Life Assurance Company of Canada, which named de Grandpré to its board and its executive committee in May 1984. Having grown up with the insurance industry, he viewed the election to the board of Canada's largest insurance company with special pride. The primary reason why he was asked to join Sun Life's board, however, was because the insurance company had become one of the largest investors in BCE stock. That made de Grandpré extra cautious in what he told the Sun Life board about BCE's affairs. "You have to be careful not to make them insiders, which would place the board in a very difficult situation."[167] De Grandpré also left Sun Life's board in 1989 when BCE bought Montreal Trust.

His increased international stature landed him positions on the international advisory boards of two prestigious Wall Street financial institutions: the Chemical Bank of New York and Goldman, Sachs & Company. He was nominated to the Chemical Bank advisory board by former AT & T chairman Charlie Brown, who knew de Grandpré from the Du Pont board. Goldman, Sachs' highly influential chairman, John Weinberg, picked de Grandpré himself because of their contact on the Seagram board and his friendship with Northern Telecom's chairman, Edmund Fitzgerald. International advisory boards are like high-level political science seminars whose members advise, rather than decide. Those appointments are coveted because of the insights members gain into world political and economic developments. De Grandpré says they are the most intellectually rewarding boards he has ever served on.

For example, he recounted the briefing that Robert McNamara, former U.S. defense secretary to President Lyndon Johnson and former Ford Motor president, gave to the Chemical Bank's advisory board after his visit to the Soviet Union to meet newly appointed Soviet secretary-general Mikhail Gorbachev. "His big concern was that Gorbachev's economic policies were in big trouble." McNamara told the group, "'There was nothing on the shelves and Gorbachev would have some difficulty maintaining himself in power.' We learned firsthand that the pot was boiling and that the economic problems would bring social problems in the Soviet Union."[168] Although none of the members could predict the unraveling of *perestroika* since 1990 and the abortive coup to overthrow Gorbachev in August 1991, de Grandpré says McNamara's earlier briefing had made him more skeptical of Gorbachev's reforms — and less surprised when they failed.

Although de Grandpré viewed himself as an internationalist, his corporate fame made him an icon in the Quebec business community. His stature was reflected by his installation as the fifteenth chancellor of McGill University — the first francophone appointed to the position since Charles Dewey Day was named first chancellor of the university in 1864.[169]

He was simultaneously revered as a role model by the younger generation of Quebec business graduates, and vilified by the student body for his outspoken view that they should do more to pay their way.

In business, money talks. As BCE's coffers filled up after the reorganization, de Grandpré's influence took on another dimension following the transformation of the country's largest telephone monopoly into the Grand Acquisitor of Bay Street.

Chapter 8

The Empire Expands:
BCE's Diversification

An institution is the lengthened shadow of one man.
— R. W. EMERSON, *SELF-RELIANCE, 1841*

After the creation of what became Canada's largest corporate conglomerate, de Grandpré was guaranteed his place in the annals of
Canadian business history. It was only the beginning, though. Fueled
by BCE's rapid profit growth, he orchestrated a spectacular string of
acquisitions, the pace and scope of which has rarely been seen in
Canada. The wipeout was equally sensational.

"They're going to be a great big octopus," exclaimed *Business Week*
magazine in March 1985.[1] As BCE unleashed the enormous financial
potential of Bell Canada, the new holding company soon eclipsed the
size of all other Canadian corporations, including powerful Canadian
Pacific. "We're on a roll," de Grandpré proudly told reporters in May
1985 as BCE approached its zenith.[2] By the end of that year, BCE not
only ranked as the wealthiest corporation in the country, but also
achieved the simultaneous distinction of being the first Canadian corporation to earn more than $1 billion in after-tax profit — a goal de
Grandpré says the holding company achieved a year early. Although
he later claimed it was meant as a joke, he confidently boasted in a
March 1986 interview that BCE could, if the economy performed
well, double its profit to $2 billion in about seven years — "It's an
ambitious goal; no doubt about that. . . ."[3]

BCE's corporate structure served as a model for other utilities seeking to build their own unregulated empires, particularly the seven regional holding companies, or "Baby Bells," spun off from the divestiture of AT&T. Emulating that structure allowed U.S. West Incorporated of Denver to meet the terms of the breakup consent decree, company chairman Jack MacAlister told a 1984 meeting of the Toronto Society of Financial Analysts.[4] He later called Bell Canada the "leaders in a strategy that makes a lot of sense."[5] Other holding companies that adopted the BCE structure included Bell Atlantic of Philadelphia, and two Canadian telephone companies affiliated with BCE: New Brunswick Telephone Company, which created Saint John-based Bruncor Incorporated as its new parent; and Newfoundland Telephone Company, whose parent is Newtel Enterprises of St. John's. Union Gas also completed a corporate reorganization in December 1984 that created Toronto-based Union Enterprises.[6]

Accolades followed. BCE's meteoric rise was featured, along with a picture of the BCE chairman, on the front page of *The New York Times* business section in April 1986.[7] The Harvard Business School Club of Toronto bestowed de Grandpré with the Canadian Business Statesman Award in 1987. Two years later, he was inducted into the Canadian Business Hall of Fame and made an Honorary Associate of the Conference Board of Canada, the highest award granted by the non-profit economic think tank to individuals judged to have served both their companies and the country with distinction during their careers.

For the past decade, the left lapel of his suit jacket has sported a white "snowflake" pin that proclaims de Grandpré's membership in the prestigious Order of Canada. Each of the 2,055 living Canadians who have been appointed to the Order share the purpose expressed by the Latin motto inscribed around the tiny maple leaf at the center of the pin, *Desiderantes Meliorem Patriam* (They desire a better country). Many of de Grandpré's friends and colleagues, including Walter Light, Northern Telecom's former president; J. V. Raymond Cyr, BCE's current chairman; and Louise Brais Vaillancourt, BCE director; as well as his eldest brother, also sport the pin. The Order of Canada, which is a mystery to most Canadians, was created in 1967 and is headed by the Governor-General. It is Canada's equivalent to a British knighthood and is the highest honor bestowed on civilians. The Order has three levels, ascending from Member through Officer to Companion. De Grandpré was appointed an Officer "for achievement and merit of a high degree" on December 14, 1981, and was elevated to the rank of Companion in 1989 "for outstanding achievement and merit of the highest degree." One person who has not been named to the list, despite his accomplishments, is Robert Scrivener.

BCE's success changed the man running the corporation. Although the day-to-day concerns of managing a safe, secure and highly competent telephone company disappeared from de Grandpré's shoulders, in their place came the responsibilities of directing a highly visible conglomerate in search of a mission. His elder brother has seen a change in de Grandpré since the reorganization: "Human beings change, particularly due to the number of responsibilities they have. And if you have to make decisions, day in and day out, you change. He's possibly more assertive than he was in his younger days."[8]

Just as the combined economic muscle of the BCE group was unleashed by the reorganization, so, too, was de Grandpré's ambition. The corporate Darwinist brought the fiery instincts of a monopolist, whose first priority is to preserve the corporation and its prerogatives, to the corporate takeover jungle, as well as a desire to confer an evolutionary advantage to BCE. Once the reorganization was approved, he immersed himself in the field he truly loved: corporate law. Propelled by the heady growth of the Eighties and the removal of regulatory constraints on the use of BCE's profit, the chairman embarked on a major acquisition binge. With corporate assets exceeding $15 billion and more than $1 billion in the bank, he became the dean of "M&A" (mergers and acquisitions).

Although the financial community applauded the reorganization because it clearly separated BCE's regulated business from its riskier nonregulated affairs, at least one financial analyst cautioned investors not to expect any sudden windfall from the change. The reorganization produced minimal immediate benefits for shareholders, other than a higher share price as BCE's stock reached $33.75 on the Toronto Stock Exchange at the end of the first year for the reorganized company, compared with a high of $24.50 for Bell Canada's share price in 1982. "The benefits would only be derived when BCE's management makes prudent and profitable acquisitions in new business endeavors,"[9] Douglas Cunningham, a financial analyst in telecommunications, wrote in one of the first investment research reports following the approval of the reorganization in April 1983. But it is there that critics say de Grandpré's record falls flat.

As chairman and CEO, de Grandpré's first task should have been to plot BCE's strategy and chart his vision for the future of the new corporation. What would, or should, BCE become? What business fields should it enter? As chief executive officer, de Grandpré was answerable to BCE's board of directors on those questions, but, as the chairman of that board, he had significant latitude to chart the corporation's course. His credo, however, seemed to be simply "to act": "Bad decisions can be corrected. But missed opportunities are lost forever,"10 he said in a 1986 interview. Free to design and build a new empire, he resembled a corporate Caesar — he came, he saw — and he bought. But why?

Like most corporations, BCE had broad strategic goals, such as preserving the quality of its assets and enhancing the value of its stock for shareholders. But when de Grandpré is asked about BCE's particular acquisition strategy, and when the record of that period is reviewed, one is struck by the absence of a clearly articulated plan detailing what the corporation wanted to become and what fields it wanted to command. BCE seemed to move from deal to deal in an ad hoc and frenzied fashion.

In the preceding century, every attempt by Bell Canada to diversify could be related directly to its core business of providing plain, old-fashioned telephone service. It is not difficult to find the synergy between the phone company and Northern Telecom, BNR, Bell Canada International or Tele-Direct. De Grandpré did not abandon that core business; in fact, the record shows he greatly enhanced it. Yet, in seeking to break away from those ties, he departed from a strategic vision that gave his empire one single, clearly defined purpose. The newly created holding company struggled to find its identity as de Grandpré acquired businesses in fields as diverse as finance, energy transportation, newspaper publishing, real estate and computer systems integration.

BCE's first acquisition target in late 1983 was never made public at the time. It was the finance arm of Paul Desmarais' Montreal-based Power Corporation. "Maybe I was too greedy,"11 de Grandpré says, reflecting on his unsuccessful effort to negotiate either the purchase of or a deal with Power's subsidiary, Power Financial Corporation. Nevertheless, that quest did lead in March 1989 to BCE's $877-million acquisition of Montreal Trustco, the trust company arm of Power Financial and one of Canada's largest and oldest trust companies.[12]

The increased reliance of banks and financial service institutions on telecommunications networks to move funds electronically and to dispense cash from automated banking machines led de Grandpré to believe there could be a synergy between a trust company and BCE that would make an equity investment or joint venture attractive. He also admits he was induced by the prospect of combining the assets administered by Montreal Trustco with the $7 billion of pension fund assets managed by BCE's giant pension fund arm, Bimcor Incorporated. Besides, there were not many other trust companies that were willing to talk to BCE, de Grandpré says.[13] Royal Trustco was firmly within the orbit of Peter and Edward Bronfman (the Toronto wing of the family), Canada Trustco Mortgage Company was held by Imasco and National Trust Company was content to remain independently owned. But Power Financial would have been a good catch. In addition to Montreal Trustco, it controlled two Winnipeg-based companies whose businesses appealed to de Grandpré — Great-West Lifeco, which was in the insurance business, and Investors Group Incorporated, one of Canada's largest mutual funds.

De Grandpré began talks with Power Corporation's chairman and controlling shareholder, Paul Desmarais, in late 1983 and into 1984. Desmarais, who as a young man had dropped out of university to rescue his family's near-bankrupt bus company in Sudbury, Ontario, was the wealthiest and most influential businessman in Quebec. His reputation as a shrewd dealmaker was legendary. When de Grandpré approached him, Desmarais controlled almost two hundred companies that spanned industries as varied as forestry, finance, media and transportation. Their total assets exceeded $8 billion.

But Desmarais' growing international business and ties to offshore financial interests complicated any bid to acquire Power Financial. Power held a 19.4 per cent stake in Pargesa Holding S.A., a Geneva-based conglomerate and merchant bank that controls oil giant Petrofina S.A. and investment bank Drexel Burnham Lambert of New York. De Grandpré's bid was thwarted by Pargesa's holding in Paribas Suisse, the Swiss subsidiary of the French banking giant Compagnie Financière de Paribas, which, in turn, owned more than 15 per cent of Power Corporation.[14] The alliance between Power Financial and Paribas meant the sale of full control of Power Financial was not in the cards. When this became clear to de Grandpré, he broached the idea of a joint venture between Bell Canada and Montreal Trustco to Desmarais, but that idea, too, never got off the ground.

By the late Eighties, Montreal Trustco no longer fit into Desmarais' global financial plans, and he decided to sell it. De Grandpré was still interested in the venerable trust company; however, the Mulroney government's policies to deregulate the financial services sector threw a hurdle in BCE's path. The federal government's policy paper, released on December 18, 1986, stated that new commercial institutions would be prohibited from entering the financial sector. In other words, Ottawa would not issue a charter for a deposit-taking institution to any business that was not already in the financial services business.[15] The chief reason given for the prohibition was concern over concentration of ownership and a desire to limit existing conglomerates, like BCE, from expanding into the financial services area. Outside ownership of a financial institution was viewed suspiciously, because it was feared that the institution would be used by its corporate parent to issue loans on favorable terms to its other nonfinancial subsidiaries.

Fortunately for Desmarais and de Grandpré, by 1988 Ottawa still had no legislation in place that implemented the prohibition announced in its policy paper. The prospect that the restriction would become law led Power Financial to place Montreal Trustco's federally chartered subsidiary, Montreal Trust Company of Canada, under provincial jurisdiction. The way for the BCE acquisition was cleared in July 1988 when a private member's bill was passed by Parliament authorizing Montreal Trustco to place its financial subsidiary under the Quebec Trust Companies Act.[16] Unlike banks, trust companies may choose to be regulated provincially, rather than federally. The Quebec government, for its part, welcomed the move.

Once the details of the regulatory transition to provincial jurisdiction were worked out, the renewed talks between Desmarais and de Grandpré moved swiftly. Desmarais received $547 million in cash and BCE shares for his stake in the company. The transaction gave Power Corporation close to a 3 per cent stake in BCE, making it one of the largest single shareholders in the company — second only to Quebec's powerful pension fund, the Caisse de dépôt et placement du Québec, which owns 6 per cent. BCE's offer ended up costing $877 million, of which $336 million was spent on issuing new shares for its public offer to buy the rest of the trust company's stock for $23.50 a share.[17] BCE raised $680 million in short-term debt and issued another $200 million of preferred shares.

BCE summoned Robert Gratton, Montreal Trustco's CEO, to its headquarters only a couple of days before the deal was announced on

March 8, 1989.[18] The financial community was surprised when he stepped down as chairman, president and CEO of the trust company that July to move to Power Financial as its president. However, he denied the move occurred because of differences with BCE.[19] A career Bell Canada executive, J. V. Raymond Cyr, was appointed chairman of Montreal Trustco in Gratton's place, an appointment that underscored the importance the conglomerate attached to its new investment.

Some investment analysts initially questioned the investment because BCE revealed in its initial offering that Montreal Trustco would add only $7 million to BCE's 1988 profit of $846 million. Since 1988, however, Montreal Trustco's profit growth and contribution to BCE's earnings have been more impressive. The trust company contributed $37 million, or thirteen cents a share, to BCE's profit in 1989 (from April 24 when the acquisition was finalized to year-end); and $44 million, or fourteen cents a share, in 1990. Montreal Trustco also provided an $11-billion boost to BCE's assets and gave it an important position in one of the fastest-growing segments of the financial sector. Its profit potential was further improved by its withdrawal from the depressed real estate business in 1990.

How well Montreal Trustco performs as an investment for BCE will depend to a great extent on Lynton "Red" Wilson, who was appointed president and chief operating officer of BCE in November 1990. Wilson was persuaded to leave the Bank of Nova Scotia, where he had been vice-chairman since early 1989. He was first appointed to BCE's board in May 1985 by de Grandpré, but had to step down in 1989 because of the Montreal Trustco acquisition. "The Bank of Nova Scotia has had a problem with succession,"[20] de Grandpré says. Wilson's position at Scotiabank was created by Cedric Ritchie, chairman and CEO, to halt the ambitions of Peter Godsoe, also a vice-chairman and heir apparent to the mercurial chairman. However, Ritchie nixed his chance at succession when he downgraded Wilson's responsibilities and shifted him into a low-profile position.[21] Analysts believe he now must further expand Montreal Trustco's profitability, which is integral to justifying BCE's investment.

BCE's first major acquisition bid that de Grandpré made public at the time came in November 1983. The financial community was stunned when he made an offer for TransCanada Pipelines (TRAP) shares. The opportunity to buy a stake in TRAP, which operates one of the world's longest natural gas pipelines, arose when beleaguered Dome Petroleum put its almost 12 per cent stake in the company up for sale. Dome Pete's boss, John P. "Smilin' Jack" Gallagher, had himself acquired the block in 1978 from Canadian Pacific. Most analysts could not understand why de Grandpré wanted to buy into another regulated utility, much less one that was financially troubled. TRAP's profitability had been damaged by the Liberal government's much criticized National Energy Program, by falling energy prices and by Dome's rapid expansion.

Having served on TRAP's board for a decade, de Grandpré viewed Bell Canada and TRAP as being in the same business; both were heavily regulated and both moved things. Better yet, TRAP had set its sights on increased natural gas exports from Alberta to the energy-starved United States. Before he took a recommendation to make the offer to BCE's board, de Grandpré invited James R. Schlesinger, the former energy czar to U.S. president Jimmy Carter, to Montreal. The professorial Schlesinger, who had also been secretary of defense and is an expert on U.S. national security policies, briefed the BCE chairman for several hours on the world energy market, on price trends and on U.S. demand — the topic closest to de Grandpré's heart.

Because natural gas was both cheaper and safer than other energy sources, Schlesinger "was of the view the United States would be begging for gas in the next decade," de Grandpré says. The market in the U.S. northeast, in particular, was poised to grow in the wake of the Three Mile Island nuclear accident. The economics, in turn, dictated strong political support in Washington for authority to build an additional pipeline spur from TRAP's network deeper into the United States. Build it and there would probably be a market for as much gas as could be piped down. On the other hand, a failure to build a pipeline might lead the United States to tap competing sources in Mexico. Schlesinger helped convince de Grandpré of the efficacy of buying TRAP and of constructing a pipeline.[22]

BCE spent $167 million on Dome Canada's 5.3 million shares. It also used a $1.2-billion line of credit to finance a public offer to all other TRAP shareholders to buy their stock at $31.50 a share. Although TRAP's board initially rejected the offer, BCE went ahead and about 35 per cent of TRAP's shareholders tendered their shares under the offer, which expired on December 21. Through the public offer, BCE acquired an additional 14 million shares, which gave it a 42.3 per cent stake in the company for $606 million. It increased its shareholding to 47.3 per cent for a total of $675 million in the summer of 1984.

Bay Street money managers were in a serious quandary over the BCE offer, Avner Mandelman, a Toronto brokerage executive, stated in *Barron's*. Compared with many other acquisitions, the TRAP share price was considered a bargain for BCE; the offer price was only 9.2 per cent higher than the share price a day before the offer was made. In other words, why sell? TRAP earned a 19 per cent rate of return on its equity, its dividend had steadily increased and it had extensive reserves of natural gas, which would become more valuable with the passage of time. Those were good reasons to keep the stock. On the other hand, a pension fund manager, for example, would be in trouble if not enough shares were tendered to BCE and TRAP's share price plummeted.

Most managers tendered their shares. They then found out they had been astutely outwitted by BCE, Mandelman says. Although BCE bought the shares for less than a 10 per cent premium, TRAP's stock rose by more than 14 per cent shortly after. The deal was nicknamed the "Tender TRAP."[23]

The addition of TRAP to BCE's portfolio increased the conglomerate's already staggering dominance of the Toronto Stock Exchange. BCE's total share capitalization alone accounted for 8 per cent of the entire capital value of the TSE. Northern Telecom added 3 per cent to that total, and TRAP added another full percentage point. Together, BCE, Northern and TRAP accounted for 12.5 per cent of the TSE's entire market capitalization, or one dollar in every eight of the total value of all shares listed for trading.[24]

Although a BCE spokesman dismissed the possibility at the time, the acquisition also gave Bell Canada and the other monopoly telephone company members of the Telecom Canada consortium access to one of the world's longest continuous rights of way, stretching from the Alberta-Saskatchewan border to southern Ontario and Montreal.[25] Access to utility or railway rights of way had become a valuable commodity in the 1980s as telephone companies sought to lay new fiberoptic cables for national long-distance networks. As it happened, much of Telecom Canada's seven-thousand-kilometer fiberoptic network, which became operational in 1989 and 1990, was laid along TRAP's right of way.[26]

De Grandpré flexed his newfound muscle over TRAP's suppliers in the wake of BCE's acquisition when he delivered an ultimatum to one of Toronto's largest legal firms. McCarthy & McCarthy, which was

TRAP's law firm and had a senior lawyer on TRAP's board, also had a busy telecommunications law practice. The firm's lawyers acted for a large group representing business telephone subscribers, which de Grandpré viewed as a constant thorn in Bell Canada's side. Give up the Bell bashers, the BCE boss told the lawyers, or he would take the more lucrative energy account to another law firm. The Canadian Business Telecommunications Alliance hired an in-house counsel after BCE's ultimatum; a change, McCarthy's senior communications lawyer said, that the association had already been contemplating. "These kinds of things happen all the time, particularly after companies conclude transactions of that sort,"[27] a BCE spokesman said.

Radcliffe Latimer, TRAP's president and CEO, fared no better than Robert Gratton at Montreal Trust under BCE rule. In the summer of 1985, he fell victim to a poison-pen letter. The letter, which contained allegations about Latimer's lifestyle, was sent to TRAP's board and to Revenue Canada that April and led to an investigation by lawyers and auditors of his expense account.[28] Although the probe cleared him of any wrongdoing, de Grandpré summoned the high-flying, but successful, executive to BCE's offices and asked for his resignation on June 14. A company statement issued the next day simply stated that Latimer had left "for the purpose of pursuing private interests."

"Jean and Rad didn't see eye to eye," says Latimer's cousin, Jim Kyles, former president of Bell Canada International. "Jean is deeply conservative and Rad is flamboyant." Latimer, who was then fifty-one, enjoyed casinos, but what landed him in hot water was an allegation that he had a penchant for using the corporate jet in pursuit of his recreational interests. "Yeah, sure, he took the jet down to Las Vegas. But he took people down who were businesspeople,"[29] Kyles says, referring to a trip Latimer took with a group of oilmen, which combined discussions about a gas deal with sun and fun on TRAP's expense account.

"We could not tolerate that,"[30] de Grandpré says.

However, severance deals for CEOs do not come cheap. Latimer's personal shareholding in TRAP was valued at $5.36 million when he quit. He sold $2.3-million worth of shares back to the company and repaid a $1.1-million interest-free loan that key officers could take out to buy company stock. In addition to the profit on that deal, Latimer received a golden parachute that gave him an $842,000 retirement allowance to boost his pension. He opted to collect the retirement allowance in cash, which, with the profit from the stock sale, netted him about $2 million. Latimer also received a $291,200 annual pension, according to TRAP's annual meeting notice.[31]

Latimer left the company on the same day TRAP announced that it had signed a major gas export deal in the United States. Although spurned by the BCE chairman, Latimer played a key role in landing the big fish that de Grandpré had coveted for so long. Yet news of the three contracts with buyers in the northeastern United States was virtually lost amid the publicity surrounding Latimer's resignation. The potential exports of Alberta gas to a storage area in Pennsylvania would double TRAP's lucrative natural gas exports to the United States. Before this could happen, though, regulatory approval was required on both sides of the border, as well as the construction of a new $1.8-billion pipeline, dubbed the trans-Niagara project — a joint venture of Canadian and U.S. gas companies.[32] It was left to Latimer's successor, Gerald J. Maier, a petroleum engineer and former president of Bow Valley Industries, to push for those approvals.

Meanwhile, de Grandpré was trying to spend more money in the oil patch. He let it be known as early as the fall of 1984 that BCE was looking for an acquisition, and speculation briefly touched on Gulf Canada. It was put on the block after its parent, Gulf Oil — then the fifth largest oil company in the world — was acquired by Standard Oil of California in a $13.2-billion takeover in March 1984.

"Do you think we should buy it?" de Grandpré jokingly asked a reporter as he left a Toronto press conference on June 25, 1985.[33] In fact, he was not interested in the business of pumping gas and was content to let Toronto's Reichmann brothers place a $2.8-billion bid for 60 per cent of Gulf Canada. But he was not content to let Olympia and York Enterprises go unchallenged when the Reichmanns put in an unwanted takeover bid for Hiram Walker Resources of Toronto in April 1986. Although TRAP entered a $4-billion bid as a white knight, Maier soon surrendered to avoid a high-priced bidding war after the Reichmanns raised the stakes. Their 69 per cent piece of Hiram Walker cost the developers $3 billion.

"It was wise for us to withdraw. I wasn't there for my ego,"[34] de Grandpré said when asked at a press conference following BCE's annual meeting on May 1, 1986, why the bid for Hiram Walker had been withdrawn. He reminded reporters that BCE was sitting on a $475-million cash surplus and stated yet again that the company was looking for potential candidates for acquisition in the North American oil patch. There was more to de Grandpré's pitch than the desire to boost TRAP's flagging earnings, which had dragged BCE's profit down in 1986. The only way to improve TRAP's profit would be to deal with its money-losing oil and gas production investments. TRAP had been forced to acquire its 13 per cent interest in those properties by Dome Petroleum. Rather than selling those investments outright, BCE wanted to combine them with the productive assets of another company in a bid to make a profit from them, rather than suffer a loss. To that end, the conglomerate launched a complex $4.3-billion bid for Dome Petroleum the following year amidst much flag-waving.

For Maier, the bid held more than a touch of irony. He had ended up working for Dome Pete in 1981 after it acquired the company he headed, Hudson's Bay Oil and Gas Company, from Conoco. That particular deal had started Dome Pete on its downward path to financial ruin.

Although Maier was surprised to learn of a rival, higher-priced bid from the Canadian subsidiary of U.S. oil giant Amoco, de Grandpré was infuriated when TRAP's offer was rejected. At a press conference following BCE's annual meeting in Halifax on April 29, 1987, he accused Amoco Canada of not conforming to securities rules requiring disclosure of their $5.5-billion offer. He backed up his claim with a legal challenge mounted against Amoco.

BCE, through TRAP, finally made its acquisition in the oil patch later that year when it purchased Encor Energy Corporation (formerly Dome Canada) for $1.1 billion. But it was a company TRAP and BCE were destined not to keep. De Grandpré wanted to overhaul TRAP's production operations by consolidating it with Encor's interests in the North Sea oil field and properties in Australia, Indonesia, Turkey, Spain and Italy. The Encor deal separated TRAP's regulated and unregulated businesses. Referring to the parallel with the BCE corporate reorganization, de Grandpré quipped, "It seems I started a trend."[35]

However, the scheme to reorganize Encor was delayed because Amoco was a partner in TRAP's oil and gas properties. After the legal spat between Amoco and BCE was resolved, the properties were finally spun off into a separate, publicly traded subsidiary, Encor Incorporated, in January 1989. Each TRAP shareholder received an Encor share for each TRAP share they owned, which gave BCE a 49 per cent interest. BCE also held 30 million redeemable, convertible preferred shares, giving it a future option to hold an additional third of Encor's stock. The deal gave Encor $211 million in cash, which offset the $500 million it paid TRAP for the pipeline company's original investment in the lands.[36] The deal was good for TRAP in many respects: its share price rose two dollars on the TSE, to fifteen dollars from thirteen, on news of the restructuring; its profit increased and its credit improved, which was crucial to raising the funds needed for the proposed pipeline expansion to the United States; and Encor assumed $800 million in debt, which otherwise would have saddled TRAP's books.

Meanwhile, hearings to consider the proposed Iroquois pipeline project began before the National Energy Board (NEB) in Ottawa in March 1990 and before the U.S. Federal Energy Regulatory Commission in August 1990. The 580-kilometer underground pipeline from Iroquois, Ontario, to Long Island, New York, would supply gas to most of New England, New York and New Jersey. The Canadian hearing began on a sour note when NEB chairman Roland Priddle disqualified himself from chairing the proceedings because he had met privately with a senior TRAP official three days after the board lost a court case that limited the scope of the hearing.[37]

No one in the financial community was surprised when BCE announced an agreement to sell its 49 per cent stake in Encor to a group of investment dealers on June 27, 1990, because BCE had made known its intention to spin off TRAP's energy properties when it acquired the pipeline company.[38] There was some surprise, though, when BCE's new chairman, Raymond Cyr, quickly followed that transaction with a two-phased plan to sell all of BCE's stake in TRAP on September 10, 1990, to another group of investment dealers. That transaction could be worth up to $1.3 billion by December 1992.[39] Although de Grandpré did not openly challenge Cyr's decision to sell the TRAP stake, he argued it reflected the new CEO's preference for making what he called a "short-term" decision to cash in on BCE's investment in TRAP, rather than hanging in for the potentially more lucrative returns that de Grandpré is convinced will flow from TRAP's gas exports to the United States. All together, BCE stands to make a profit of between $600 million and $700 million from the sale of its TRAP shares,[40] de Grandpré says.

He vigorously defends his decision to buy the stake in the first place. "It was not a stupid decision to buy TRAP. And we made it profitable, plus its role is to be a leader. That's not a bad situation to be in."[41]

That is why some financial analysts were taken aback by Cyr's decision. Douglas Cunningham, a senior utilities analyst, asked, "Why sell it before TRAP knew whether it could conclude its U.S. gas export deals?"[42] That consternation was heightened when the U.S. energy commission approved the Iroquois project two months later.[43] Maier told

The Globe and Mail he was disappointed to lose BCE as a major share-holder, but that TRAP would be able to raise the debt and equity need-ed for the Iroquois project, thanks, in large measure, to the Encor deal.[44]

So why did Cyr sell BCE's stake? He says the decision was related to his strategy to focus on BCE's core telecommunications business.[45] De Grandpré says, "The new CEO wants to put his imprint on the business and he doesn't want to be in the pipeline business. I don't have any regrets because he will re-allocate those resources and he hopes to make a better deal with those resources.

"It's a different philosophy."[46]

In 1984, de Grandpré confounded some observers when he over-looked Cyr and selected an outsider as the first president of BCE. Then forty-nine years old, the short, round, cigar-chomping Cyr was the model of the loyal telephone company executive, having worked for Bell Canada since graduating as an engineer in 1958. Robert Richardson, a fifty-five-year-old executive vice-president of E. I. Du Pont de Nemours & Company of Wilmington and a Ph.D. in chemical engineering, was named president in March 1984. Maybe it was de Grandpré's fascination with TRAP's pipelines that had something to do with the appointment; Richardson had earned his doctorate by studying the mechanics of moving sludge through a pipeline. More likely, however, he landed the job because of his decade-long associa-tion with de Grandpré on the boards of several corporations. They had been appointed to the TD Bank board together in January 1973 through Herb Lank, who was Richardson's boss at Du Pont Canada. Richardson later replaced Lank as chairman and CEO of the Du Pont subsidiary and de Grandpré appointed Richardson to Bell Canada's board in 1978.

"We never really knew his philosophy for BCE, although he was ostensibly brought in to guide its acquisition strategy," says Donald Cruickshank, a former vice-president at Bell Canada. He recalls Richardson's avid affection for his pipe: "His pockets were always full of pipe stuff and his hand was always dirty from packing his pipe, which was anathema to de Grandpré. Richardson had a much different philosophy and style."[47]

During a 1985 interview at BCE's offices, which were then located at Victoria Square atop the Montreal Stock Exchange Tower, Richardson put his feet up on his desk and leaned so far back in his swivel chair that it seemed he was about to slide out the window of the forty-fourth-story office. Pointing to the amateur photos of his summer cottage and winter ski chalet that adorned one wall, he declared, "I'd rather be canoeing in the summer or skiing in the winter."[48]

He soon got his chance.

Richardson elected "to take early retirement," as the BCE press release stated, in November 1986.[49] But that did not mean his pension was based solely on his two years of pensionable earnings. Under BCE's compensation arrangements and benefit packages for officers, Richardson racked up thirteen years of credited service: three for the two years he had actually worked there (an officer gets an extra half-year of service for every year worked) and ten years under his employment contract, which gave him an extra five years of service on each anniversary date of his hiring for up to five years. The additional years of service ensured he would receive a pension of close to $100,000 a year. As a BCE officer, he was also paid a year's salary on retirement of $749,800 under a supplementary agreement.[50]

"He has a lot of good qualities and was a good manager of a subsidiary company, like Du Pont, and he was a good manager of R&D. But I think he was not a good manager of the parent company. You need someone to make tough decisions,"[51] de Grandpré says.

As a holding company, BCE is run and staffed almost entirely by accountants, finance experts and lawyers. It employed about two hundred people when Richardson was brought in. The majority of employees work for the accounting and treasury departments, a small team of financial analysts study enterprises in which BCE might be interested and the deals are put together by an elite group of in-house corporate lawyers, supported by an investor and public relations group. The high-level executive team is drawn largely from Bell Canada.

BCE executives divide their time between Montreal and Toronto. BCE's head office is now located on the top three floors of the state-of-the-art, gleaming black Industrial Life Tower at 2000 McGill College Avenue in Montreal. Its move there in 1986 helped transform the street that runs between McGill University and Place Ville into the financial heart of Quebec. BCE's Toronto offices are located on the sixty-seventh and sixty-ninth floors of First Canadian Place, the city's tallest office tower.

Security is stringent in both buildings. Visitors must first buzz an electronically locked door before they can enter either of the well-appointed reception areas. The foyers, in turn, are sealed off from the offices by more electronically controlled locked doors, placing the visitor in a secure area called a "man trap." For added security, de Grandpré's own office is also cipher-locked, creating a second trap outside his corner enclave.

That penchant for security and secrecy permeates the culture of most corporate deal makers, including BCE. The nature of the enterprise and concern about leaks of financially sensitive information dictates that ethos. But it has sometimes made for strained relations between BCE and employees of subsidiaries, including Bell Canada, who sometimes feel their corporate parent is arrogant and aloof. BCE executives, on the other hand, make no apologies to the utility. "BCE is not part of the telephone company," Gordon Inns, the former executive vice-president of planning at BCE, is quoted as telling a Bell Canada employee in the telephone company's employee newspaper, Bell News. Asked why BCE didn't provide more information to the phone company, Inns replied: "You see, I either don't know what we are going to do next, or what we are intending or I really don't know. If I don't know, then of course I have nothing to tell you. And if I do know something that we're going to do, then I can't tell you folks because it could dramatically affect the outcome. Can you imagine what would have happened to the price of TCPL stock if it were known what we were going to do? Heck, I can't even talk about what we are not thinking about doing, without it having an effect."[52]

Inns said BCE had no secrets to hide from Bell Canada — provided they were matters related to its business. BCE was forthcoming when required but, when it was given the discretion, it preferred silence. Typical of that attitude was BCE's submission to the Ontario Securities Commission in March 1984 asking the commission not to require companies to file a list of subsidiaries with an annual filing under the prompt offerings prospectus system.[53]

Viewed operationally, BCE is really in the financial services business, and its staff manage and plan financial investments on behalf of its shareholders. Its function differs only slightly from the traditional merchant banks, who make investments for their own gain. One other critical difference, however, was BCE's mission declared at its founding in 1983 to support its telecommunications triumvirate. The question that Cyr himself asked in a speech on BCE in 1986 was, "How does Bell Canada fit in? It is no longer at the top of the corporate pyramid. It has become part of a constellation of sister companies."[54]

The TRAP acquisition and wooing of Desmarais' Power Financial Corporation would have been enough to satiate the appetites of many acquisition-hungry corporations. But there were yet more deals in the works. It was like a year-round corporate Christmas for BCE in 1984. And BCE found "that one special gift for the company that seemed to own everything" when it bought a license to print money that October. The contract to print Canada's paper currency came with the acquisition of 117-year-old British American Bank Note Incorporated of Ottawa by BCE's publishing unit, Tele-Direct (Canada) Incorporated. The deal was part of a major expansion of Tele-Direct that year through a string of acquisitions that, de Grandpré believed, complemented the company's telephone directory publishing and printing businesses.

Yet few printing businesses were as lucrative as Tele-Direct's existing business of printing telephone directories and selling advertising space in the *Yellow Pages*. It had revenue of $417 million in 1984. Much of that came from two subsidiaries that sold *Yellow Pages* advertising in in the United States and Australia.[55] And that figure excluded Bell Canada's directory publishing and advertising sales, which was done through its own subsidiary, Tele-Direct (Publications) Incorporated.

The first deal of 1984 was the purchase of an outstanding 30 per cent minority interest in Ronalds-Federated, the Montreal-based printing and packaging unit acquired by Tele-Direct in 1980. The acquisition of Canada's largest printing company enabled BCE to vertically integrate its telephone directory business.

BCE followed that deal with the acquisition in June 1984 of Case-Hoyt Corporation of Rochester, New York, a leading high-quality printing and graphic design firm. The privately held company, founded in 1919, had revenue of more than $100 million and complemented Tele-Direct's magazine printing arm, Comac Communications.[56] Tele-Direct had acquired Comac — the largest publisher of controlled circulation publications in Canada — in November 1982. The purchase of British American, which specializes in engraving, gave Tele-Direct state-of-the-art labs in printing and engraving research and a significant share of the lottery ticket and credit card markets.

In late 1985, BCE attempted to bring all of its varied publishing and printing units together by reorganizing Tele-Direct. A new holding company, BCE PubliTech Incorporated of Toronto, replaced Tele-Direct (Canada) in early 1986. But the hoped-for synergy did not fully materialize.

BCE decided in late 1987 to sell some of PubliTech's smaller assets, which de Grandpré says were too small to effectively compete. They were swapped for a minority stake in Québecor, a large Montreal-based newspaper publishing, printing and forestry product company.

De Grandpré had his first conversation with Pierre Péladeau, Québecor's controversial majority owner, CEO and publisher of *Le Journal de Montreal,* Quebec's largest circulation newspaper, in December 1987 at the Club Saint-Denis. They agreed to negotiate a deal for BCE to buy a stake in Québecor in return for the sale of some of BCE PubliTech's assets.[57]

Péladeau agreed in June to buy Ronalds Printing and British American's securities, currency and cheque printing operations for $185 million.[58] The price tag was reduced to $161 million after Pélodeau dropped one of the PubliTech units before the deal closed in October. He paid for the companies with $80-million-worth of stock and ten-year notes bearing 9.5 per cent interest. BCE ended up with 21 per cent of Québecor and took a $69-million write-down on its printing operations. It also acquired a 60 per cent interest in Matthews Ingham and Lake, a large Toronto-based quality offset printer, in early 1988. While PubliTech disappeared from BCE's list of subsidiaries, its directory and printing units were grouped under a new holding company, BCE Information Services.

The Québecor acquisition signaled a change in BCE's policy by indicating its willingness to buy a noncontrolling interest in a company. Yet, de Grandpré says, BCE had significant clout over Québecor. He believed the company would face a succession problem in picking an heir to Péladeau, which could lead to a hostile takeover bid. Any potential suitor would need BCE's backing. In effect, the equity block made BCE a very important player in the company's management. "It's like a chess game. You put a pawn down there to have an advantage over a king," de Grandpré says.

He maintained the combination of BCE's assets with Québecor's management "could make something stronger and better than the sum of the parts."[59] So did Péladeau. As the strains of Beethoven's Ninth Symphony died down at the press conference to announce the deal with BCE, Péladeau proudly declared that the company was poised to become the second-largest printer in North America with revenue of $700 million for the year.[60] Only a year later, he and the late British media mogul Robert Maxwell were forced to shut down Québecor's troubled *Daily News* — which was to have been Maxwell's entrée into

North America — after twenty months of losses. Péladeau's widely reported slur that Jews "take up too much space" in his publications made him a liability.

No music played at Québecor when Cyr reached a $35-million deal to sell up to 12 per cent of BCE's holdings in April 1991. BCE sold 2 million units, each consisting of a Québecor subordinate, class-B share and a half warrant to buy an additional share, to Lévesque Beaubien Geoffrion Incorporated. The Montreal-based stockbroker placed the units privately with institutional investors. Six months later, Cyr sold the balance of BCE's 2.1 million shares in Québecor in a $36.4-million deal with Nesbitt Thomson Incorporated of Montreal.

Although BCE unloaded its interest in British American Bank Note to Pélodeau, few realized that it kept the currency printer's R&D arm. According to Lynton Wilson, president of BCE, British North America Technologies may become part of the perceived synergy between Montreal Trust and BCE's telecommunications units. The printing R&D company could play a role in the future development of "smart" credit cards to work with a less expensive automated teller machine. He said BCE is exploring whether Northern Telecom or another manufacturer could cut the cost of an ATM by moving the software out of each machine and putting it into the telecommunications network, which a bank would lease from Bell Canada.

BCE partly owes the creation of one of its most innovative and potentially lucrative business units, BCE Mobile Communications Incorporated of Montreal, to a decision by former federal communications minister Francis Fox. Although Fox's 1984 ruling regarding a new mobile telephone service was considered inauspicious by de Grandpré at the time, it spawned a chain of events that eventually launched BCE's most entrepreneurial subsidiary.

"BCE Mobile has been a tremendous investment," de Grandpré says. "If we were to sell our investment, we would make a billion dollars."[61] Yet the minister's decision to delay the start-up of cellular radio-telephone service led to what Fox says was one of his three biggest "run-ins" with the BCE boss.[62]

Cellular radio used new technology to make more efficient use of the scarce radio frequency allocated to it. The portion of the radio spectrum covering Toronto, for example, was split into smaller units, called "cells," which allowed more customers to be served than with existing mobile telephone services. It was viewed as a hot business because of the strong demand for mobile communications and because cellular phones ushered in the first generation of battery-powered cordless phones.

Fox licensed the new cellular radio service as a duopoly in December 1983. Cantel Cellular Radio Group, now Rogers Cantel (owned by cable television czar Edward S. Rogers), won the coveted national license to serve twenty-three cities. In these markets, it would compete against similar services provided by the local telephone companies. The trouble began when Bell Canada announced that its cellular radio unit would be ready to begin service that September. Cantel's, on the other hand, would not be ready. Cantel lobbied Fox for a delay of the start date, arguing that a head start by the phone company would gravely damage its prospects. Fox granted Cantel's request because, he said, a head start by the telephone company would likely put Cantel under within ten months.[63]

Fox wrote Cyr on March 14 and informed him that Bell Canada would not be licensed to begin its service until July 1, 1985. But what startled the utility was the addition of a new condition for the license, which ordered Bell Canada to transfer control of its cellular unit to BCE. Fox argued the requirement was necessary to prevent cross-subsidization and to ensure fair, unregulated competition.[64]

"I keep coming back to basics and the October 1982 policy decision to license a competitive system. My recent decisions are not anti-Bell but pro-competitive,"[65] Fox said.

Nevertheless, Bell Canada officials were incensed at the delay and the added costs that would be incurred because of it. They were also against forming a separate subsidiary, claiming that the utility had already agreed to share equipment with Cantel. But placing the cellular radio unit within the telephone company would allow it to draw upon the full resources of the utility, which was perceived as giving Bell Canada an unfair advantage.

"I was surprised at Bell's reaction to the decision because I thought that was what their reorganization was all about, to operate through other subsidiaries without regulation,"[66] Fox said in a 1984 *Globe and Mail* interview.

Although Bell Canada set Bell Cellular up as a separate subsidiary, as Fox ordered, it placed the company under its own competitive marketing arm, Bell Communications Systems — and not in BCE. Although it made some sense to put cellular radio in a business unit

responsible for selling competitive products, it also meant Bell
Cellular could share overhead costs with another business unit. De
Grandpré took a keen interest in the new service and presided over
Bell Cellular's inaugural in Toronto on June 25, 1985. His presence
underscored the importance BCE attached to the new business.[67] Bell
Cellular was transferred later in 1985 to the utility's own holding company, Bell Canada Management Corporation (BCMC). In the fall of
1986, the utility decided to reorganize its other mobile communications holdings under a new unit named Bell Mobile, which was also set
up under BCMC.[68] That move was a prelude to a major and important
deal a month later when BCMC merged its mobile communications
holdings, including cellular radio, with those of National Telesystem
Limited, a Quebec-based company that was owned by a dynamic
young entrepreneur, Charles Sirois.

The young financial whiz who ran BCMC and approached Sirois to
do the deal was Jean Monty — now chairman and CEO of Bell
Canada. Both companies were driven by a need to collaborate to
improve the efficiency of their respective services, as well as a desire to
combine each other's strengths. Monty could not have found a better-
suited individual to inject entrepreneurship into the new mobile communications business unit.

Sirois, whose family owned a radio paging business, Setelco, earned
his master's degree in finance from Laval University in 1979. His acu-
men with numbers helped garner backing from Quebec's mighty pension fund, the Caisse de dépôt et placement du Québec, and the
National Bank of Canada for his acquisition of his father's business in
Chicoutimi, Quebec, after he graduated. However, he wasn't content
to simply operate a beeper service. His ambition was to become the
country's largest radio common carrier — the mobile telephone equivalent of Bell Canada. With his uncanny knack for deal making, he parlayed his backing from the banks into greater sums of collateral, which
he then used to orchestrate even bigger acquisitions through sophisticated share exchanges. The sleepy radio carrier industry woke up
when he acquired well-known Pagette Airsignals in January 1986. At
the same time, he stunned the industry by beating out giants, including
Motorola, with a $30-million bid for another large paging company,
TAS Pagette, which was a unit of IU International Holdings (and,
despite its name, no relation to Pagette Airsignals). He merged all his
paging holdings under the name National Pagette Limited that June
and achieved his goal of building the country's largest paging company

when he was only thirty-two years old. Sirois took National Pagette public later that year.

Sirois found a kindred spirit at Bell Canada in Monty, and they launched National Mobile Radio Communications Incorporated on October 1. Bell Canada was the majority shareholder with 55 per cent of National Mobile's equity, while National Telesystems held 45 per cent. Sirois's National Pagette also acquired Bell Canada's paging business in return for 10 per cent of its common shares plus warrants to increase the holding to 20 per cent in 1988.[69] One of National Pagette's directors was Fox, who was appointed to the board after his defeat in the 1984 election.

Two weeks after the National Mobile deal was announced, the CRTC tossed a major hurdle in Bell Canada's path when it revisited its earlier ruling on the integrality of investments. The utility had wanted the CRTC to exclude all capital invested in subsidiaries that were not related to providing telephone service from the calculation of its rates. But the commission rejected that proposal in its October 14 decision and, instead, set a deemed rate of return for those investments only 2 per cent more than the midpoint of what Bell Canada itself was allowed to earn.[70]

"We will have to reconsider our role in establishing those new business ventures," Cyr said in a 1986 interview. "That has penalized us worse than the decision five years ago,"[71] he said in a reference to the 1981 CRTC decision that prompted the creation of BCE. Cyr was also greatly angered by the CRTC's unprecedented action to retroactively reduce Bell Canada's prior rate of return for 1985 and 1986. After the adjustment was made, the CRTC declared the phone company had made too much profit and ordered it to pay back $206 million to its six million subscribers.[72] Bell Canada appealed the refund, arguing the CRTC could not retroactively adjust rates it had allowed. But the Supreme Court of Canada ruled in June 1989 that Bell Canada's rate of return had only been temporary and that the CRTC could retroactively adjust any interim decision.

Supported by de Grandpré, Cyr pulled Bell Canada out of all of its nonintegral investments. The company's directors approved the transfer of BCMC — including Bell Cellular, Bell Mobile and National Mobile — to BCE at their board meeting on December 17, 1986.[73] Léonce Montambault, who took over as president and CEO of the utility on January 1 when Cyr moved to BCE as its president, disclosed that the transfer cost BCE about $160 million.[74] Montambault, who had had his heart set on taking early retirement to his Quebec farm, was cajoled by de Grandpré into assuming the presidency. Monty also moved to BCE as chairman and president of BCE Commcor, the renamed BCMC.

BCE then acquired all of National Pagette through a share exchange offer with Sirois in November 1987. Sirois swapped his interests in both National Radio and Pagette for a 10 per cent stake in BCE Mobile Communications, which was created in December to hold all of BCE's mobile interests, including Bell Cellular.[75] In addition to his other duties, Monty was named chairman, CEO and president of the new company and Sirois joined as vice-chairman.

The company went public on the Toronto Stock Exchange the same month. A $54-million issue of 3 million shares was sold to the general public in May 1988, which reduced BCE's holding in BCE Mobile to 81 per cent. BCE owned 70 per cent in 1990. Taking BCE Mobile public had the advantage of raising funds for capital expansion from outside sources. At the end of its first year in 1988, BCE Mobile's stock was valued at $1.5 billion — nine times the $176-million book value of BCE's original investment.[76]

The unbridled growth of Bell Cellular, which leapt from no subscribers in 1985 to 160,000 on its fifth anniversary, consumed massive amounts of capital. BCE Mobile invested $200 million in cellular expansion throughout Ontario and Quebec alone in 1989 — an amount equal to a comparable five-year program in British Columbia — to expand and convert Bell Cellular's network to digital, or computerized, cellular technology. The market-driven company also spent heavily on financing numerous entrepreneurial start-up ventures related to its core businesses. These included Skytel Communications, a joint venture with GTE Airfone to put telephones in regularly scheduled aircraft, and an investment in Telesat Canada's mobile satellite MSAT project.

To help market its services, BCE Mobile brought in one of Bell Canada's most aggressive marketing executives, Alan Walter, as president and CEO after Monty left the company in late 1988. Walter had presided over the response to the introduction of competition in Bell Canada's terminal equipment market in the early 1980s. He moved to Telecom Canada in April 1990, and Sirois brought in John McLennan, previously an executive vice-president at Mitel and, before that, president of Rogers Cantel.

Sirois was one of the few BCE subsidiary heads to draw on the full range of companies in the group. He did this when he launched BCE's international venture capital fund, BCE Venture International, to finance wireless communications projects offshore. With personnel from BCE Mobile, Northern Telecom and Bell Canada International, BCE Venture put together a successful bid for two cellular radio franchises in Mexico City.

Sirois then turned his attention to his personal merchant bank, Intermedia Financial Corporation, which he launched in May 1990 to explore mobile communications prospects abroad. The Caisse de dépôt et placement du Québec bought 25 per cent of his bank, which was capitalized with $100 million in equity. Although Sirois stepped down as chairman in December 1990 to spend more time on that business, he has kept his seat on BCE Mobile's board and executive committee. He has since launched a new telecommunications company, Optinet Telecommunications, to exploit the rapidly growing resale market, and has acquired the French-language rights to "Much Music," a rock video television program which he has launched in Quebec as "Musique Plus" and is taking to France.

Some financial analysts avoided BCE Mobile's stock because of the company's financial losses. Although it had a loss in 1990 of $20 million, its revenue was $329 million. Sirois denounced the sellers as "imbeciles."[77] De Grandpré shares Sirois's enthusiasm and says the company will probably turn a profit in "three or four years."[78] It may take even longer, but cellular radio is expected to be highly profitable once the network is completed.

One of the greatest coups for the BCE group came in 1985 when Northern Telecom became the first Western telecommunications equipment maker to crack the seemingly impenetrable Japanese market. Although de Grandpré and BCE may have appeared to have been ignoring their existing telecommunications businesses amidst all the other activities of the holding company, Fitzgerald says that was not the case, particularly when it came to pursuing Japanese business.

Japan and Britain were the two international markets targeted by Fitzgerald and Light in the early Eighties. Although Northern Telecom had enjoyed spectacular success in its assault on the U.S. market in the early 1980s, Fitzgerald believed a stronger international presence was essential to fuel continued growth of the company in the 1990s. And the test of whether the equipment maker could become a global player, or whether it would remain a North American equipment maker, would be determined in those two markets. Japan — which spent more than $30 billion (U.S.) a year on telecommunications equipment, making it the world's second largest market — was the oyster.

Light, in fact, had tried to sell switches to state-owned Nippon Telegraph and Telephone (NTT), the world's second-largest telephone company, beginning in 1976 — four years before Fitzgerald joined Northern Telecom's U.S. subsidiary. He made his first request to issue price quotes to NTT on June 24, 1976, after Japanese suppliers began to quote their products to Canadian telephone companies.[79] But NTT officials told Canada's ambassador to Japan, Bruce Rankin, that Northern Telecom could bid on contracts but would not get any business.[80] NTT's procurement practices were of concern in the United States as well. Its repudiation of Northern Telecom gave further ammunition to the Washington-based Electronic Industries Association, which had launched an intensive lobbying campaign to have the U.S. government address NTT's practices at the General Agreement on Tariffs and Trade (GATT) multilateral negotiations — the first new set of trade rules since the Second World War.

Trade relations between Washington and Tokyo worsened in 1979 when the Japanese government sought to exclude NTT from the Tokyo GATT talks. The United States insisted that NTT be added to the list of companies covered by GATT's government procurement code. Its addition would extend international competitive bidding to the state-run monopoly. But a delegation of parliamentarians from the Japanese Diet that visited Washington to discuss bilateral trade relations, including NTT's procurement practices, said the U.S. request was unreasonable. Their committee, which regulated NTT, could not change Japanese policy because:

> We are afraid that an easy compromise on this issue will make it difficult to maintain the support of the people for the incumbent government and [the] Liberal Democratic Party as well. It will be very difficult for the party to stay in power and [to] maintain conservative government. As members of the Parliament, this is the critical point of our anxiety.[81]

NTT's leadership maintained the issue was a diplomatic one. NTT president Tokuji Akikusa stated the company would open its doors to foreign firms gradually. However, he also said the company had to maintain its negotiated contract formula under which orders are issued to various firms,[82] a policy clearly aimed at protecting NTT's four major preferred suppliers — Hitachi, Fujitsu, NEC and Oki Electric Industry Company. The Japanese government's position led to an impasse at GATT and an extension of the talks, which stalled yet again in Brussels in early 1991 over farm subsidies.

It is against that context that protectionist legislators in Washington in late 1981 included the "reciprocity in trade" clause in the proposed telecommunications deregulation bills brought before the U.S. Congress that fall. The measure both threatened to begin a trade war and, if passed, would potentially damage Northern Telecom as much as the Japanese firms it was directed against. As chairman of the Washington-based Committee for Economic Development (CED), a nonprofit policy research group comprised of CEOs of large U.S. companies, Fitzgerald was in a position to try to do something about the impasse. The tall, beefy executive organized a joint meeting of the CED and its Japanese counterpart, Keizai Doyukai (Japan Economic Development Committee), in Hawaii in March 1982. The seeds of Northern Telecom's breakthrough in Japan were planted at that meeting.

The business leadership from both sides of the Pacific hashed out their views for several days to consider how to defuse the tensions. The U.S. executives explained that the protectionist policies on Capitol Hill did not reflect the consensus opinion of the country's leading businesses. The CED told their Japanese counterpart that protectionism was best fought as a domestic issue and that the U.S. organization would lobby to defeat the reciprocity clause. For their part, Japanese leaders tried to impart the message that their country was neither as closed nor as protectionist as it was made out to be. "But I told them that telecommunications was a market that the U.S. felt particularly abused in," says Fitzgerald. "I said my company, for one, would like to try to do business with NTT, but we believed we couldn't get access to it."[83]

Shortly after Fitzgerald returned to Nashville, he received a surprise telephone call from Yoshio Okawara, Japan's ambassador to the United States. He asked Fitzgerald if the two could meet to talk about the Japanese telecommunications market. They met in Washington, where Fitzgerald told the ambassador that "NTT is a lightning rod for trade frictions between the two countries" and reiterated his desire to do business with NTT.

Okawara surprised Fitzgerald again by telling him, "If you would go to Japan, I will get you a meeting with Hisashi Shinto, the head of NTT."

Fitzgerald did the diplomatic thing and responded with a polite "Yes," as opposed to a more heartfelt "hot damn." His meeting with Shinto, which was arranged in Tokyo for that May, was the entrée the entire U.S. high-technology community had sought for a decade since the Japanese government abolished its "Buy Japanese" national policy in 1972.

Like Fitzgerald, Shinto was an industrialist and a newcomer to the telecommunications industry. He had been the former boss of ship-builder Ishikawajima–Harima Heavy Industries Company before he was appointed by the Japanese government in 1981 to head the giant telephone company, nicknamed Denden. His appointment was in anticipation of a forthcoming proposal from the government to priva-tize Denden and to break up its monopoly.

Shinto told Fitzgerald, "If you want to succeed here, you'll have to come over here to do business; whatever you make or supply will have to be zero-defect; you have to assure us that you can meet NTT's stan-dards; you will have to have the integrity to meet your delivery dates; and you must assure us that you'll be here for the long term. You have to be patient and persistent and, if you're good, you'll get in."

"So I took him at his word,"[84] Fitzgerald says. After his meeting with the NTT head, he called his good friend and former classmate, Gordon Tagasaki, and asked if he would be available to set up Northern Telecom's office in Tokyo. Tagasaki, who worked for Fitzgerald as a consultant, reported to a new U.S.-based subsidiary, Northern Telecom Japan Incorporated.

Implicit in doing business with NTT was the need to be the lowest-cost supplier because Northern Telecom faced competition from four Japanese rivals. Unlike Bell Canada and AT&T, NTT did not own a stake in any equipment maker. That meant patent protection was not an issue. Contrary to the widespread Western perception that the Japanese are patent bandits, NTT has scrupulously protected the pro-prietary knowledge of all its suppliers, Fitzgerald says.

Quality was of far greater concern to the Japanese and became an obsession. One BNR employee recalled that a tin of spray paint destined for Japan was not sent from Northern Telecom's factory in Raleigh, North Carolina, because the can had a dent in it. Doing business in Japan has taught the North American manufacturer numerous lessons about Japanese manufacturing practices and methods. As early as 1985, Hugh Hamilton, the founding president of Northern Telecom Japan, recognized that the need to get the smallest details right the first time for the Japanese market would result in changes throughout Northern Telecom's manufacturing organization.[85]

Northern Telecom and NTT spent two years from 1984 to 1986 getting to know each other. Teams of Northern Telecom and BNR engineers were sent to Japan to learn about Japanese technical standards and the workings of NTT's telephone network. At the same time, NTT engineers pored over Northern Telecom's equipment specifications and product manuals. "We began putting big money into Japan in 1984," Fitzgerald says, and estimated the company spent between $50 million and $60 million in start-up costs alone.

Early that year, Northern Telecom was ready to take what Fitzgerald says was, from a price point of view, an unsatisfactory contract to provide a transportable switch to route phone calls in an emergency. The company modified its smallest computerized switch, the DMS–10, which was designed to serve sparsely populated areas in the Canadian North, so that it could be hooked up to NTT's network. The switch and its power supply were put in a tractor-trailer to make it portable. "The market is not large for emergency switches, but it offered us an inside view of how NTT's network works and was a strategic education,"[86] Fitzgerald says. He personally negotiated the letter of understanding for the three-year contract, which was signed on May 1, 1984.

Fitzgerald took a bold plunge later that year when he approached Shinto with an unsolicited bid to sell NTT 1.5 million telephone lines for use with DMS–10 switches. The bid stemmed from Northern Telecom's analysis of NTT's modernization plans. "NTT had made a commitment to all its subscribers that they would get modern, digital services in rural areas," Fitzgerald says. But on closer scrutiny, Northern Telecom's engineers realized that the plan's budget was predicated upon NTT buying all of the necessary switches from Japanese suppliers. However, they offered only two versions: a giant switch for local calls and another giant switch for routing long-distance calls — both designed for heavily populated urban areas.

"We developed a strategy to do the same thing, only with our smaller and cheaper DMS–10s organized in clusters around a few of the Japanese-made local switches to handle the heavy, or long-distance, traffic." The proposal drastically slashed NTT's proposed expenditure for the project and promised to give the utility additional features that were not available on the larger Japanese-made products. It also astutely appealed to the Japanese desire for consensus; Shinto's budget for the project would be shared between the foreign newcomer and the domestic suppliers.

Shinto accepted the bid proposal and Fitzgerald revealed that the two companies had signed a memorandum of understanding while testifying before a U.S. Senate subcommittee on international trade. NTT announced the award of the $250-million (U.S.), seven-year contract for the switches in late December 1985. Northern Telecom beat out a bid from AT&T and came in lower than Japanese suppliers because of the Japanese yen's rise against the U.S. dollar.

De Grandpré worked closely with Fitzgerald when Northern Telecom began to assess the Japanese market. And the two were often side by side at meetings and seminars in Tokyo, particularly during trips related to BCE's application to obtain a listing on the Tokyo Stock Exchange. BCE's first share issue in Japan was underwritten by Nomura Securities. "Our relationship with Nomura Securities and the share issue were all related to Northern Telecom's sales in Japan,"[87] de Grandpré says. On November 19, 1985, BCE became the first Canadian company to list its shares for trading on the Tokyo Stock Exchange after completing a $188-million preferred share offering through Nomura. Trading opened at ¥6,300 and more than 412,000 shares traded in Tokyo on the opening day.

The BCE listing on the Tokyo exchange served to remind NTT and other Japanese businesspeople that Northern Telecom had a larger corporate parent behind it. It also paved the way for Northern Telecom's own Tokyo share listing. "BCE's share issue was primarily to broaden its knowledge of the financial markets," Fitzgerald says. "Japan is a crucial market for financial capital and BCE has a voracious appetite for capital. So it made them known in that market." Northern Telecom's share offering, on the other hand, "was strictly marketing."[88] A stock listing was one way of showing NTT that the company was committed to the Japanese market. But Northern Telecom also used it to promote its identity in the Japanese business community. It began selling its business telephone systems in Japan in 1985.

Japan was also a crucial proving ground for Northern Telecom's "OPEN World" concept, unveiled on a lavish three-day cruise aboard the *SS Rotterdam* later that year. Light hired Henry Bowes, a consultant and friend of John Lobb, to plan the event. Bowes suggested they rent the luxury cruise ship for the marketing extravaganza, an idea that de Grandpré blessed. The company sponsored two trips, one for telephone company executives and one for business customers.[89]

OPEN World was Light's way of dealing with a rampant form of technological protectionism. Just as Northern Telecom embarked on its bid to go global, the telecommunications market was in chaos due to the large number of incompatible proprietary, or closed, systems. These prevented computers and telephones from talking to each other, thereby undermining the promise of Digital World, which Northern Telecom had unveiled in 1976 at Disney World. Northern Telecom's proposal with OPEN World was twofold: it would open its interfaces to all other computer equipment makers, and it would ensure its phone systems could work with any other system architecture. It was a stroke of marketing genius aimed at increasing its total market and setting Northern Telecom apart from its international rivals in the increasingly competitive business equipment market.[90]

The DMS–10 contract was signed in mid-May 1986 after NTT certified that the switch could be safely connected to its network. The first delivery was set for 1987. But two jealous U.S. rivals cried foul. AT&T and Rockwell International took their complaints to the White House after Fitzgerald asked Secretary of Commerce Malcolm Baldrige for access to financing from the U.S. Export–Import Bank. In his opinion, this was a reasonable request, because all of Northern Telecom's exports to Japan were from its U.S. factory in Raleigh, North Carolina. In addition, "NTT was quite anxious that it be so because they wanted their imports from us to lessen the trade imbalance with the United States." Lionel Olmer, the deputy secretary of international trade at the Commerce Department, wrote a letter to Fitzgerald and to NTT stating that Northern Telecom had the full support of the U.S. government, including the right to call upon the U.S. embassy in Japan. "We didn't use them, but it was important to have that support in principle," Fitzgerald says.[91] More importantly, the support of the Reagan administration quietly and effectively thwarted any attempt by Northern Telecom's rivals to derail its bid.

Appropriately, the first telephone call on the first of the seven hundred DMS–10 switches installed in Japan in early spring 1988 was placed by Fitzgerald to U.S. trade representative Clayton Yeutter at the State Department in Washington.[92]

It soon became clear that Japan could be the encore to Northern Telecom's performance in the United States. Since 1988, the company has racked up substantial sales of its business switches, won new switching contracts from NTT to install a large DMS–200 switch to provide backup service for NTT's congested Tokyo network and has won a major switching contract from International Digital Communications, a competitive carrier, to install its huge international gateway DMS–250 and DMS–300 switch. It also won a $20-million contract in early 1991 from Intec of Tokyo for eight switches for its data communications network.

"There's no doubt that [Fitzgerald] is probably the best known non-Japanese businessman in Japan today,"[93] de Grandpré says. He has developed close ties with NTT's current chairman, Haruo Yamaguchi, and the Japanese government paid tribute to him at a ceremony in Tokyo in June 1990 shortly after his retirement. Fitzgerald has "contributed greatly to the internationalization" of the Japanese telecommunications market and "assisted in educating the global business community about the Japanese telecommunications market," Japan's Minister of Post and Telecommunications, Takashi Fukaya, said at the ceremony.[94] The Washington-based CED also honored the retired Northern Telecom chairman when it created a new $500,000 program in international studies in his name to research U.S. global competitiveness.

"My hope would be that Walter [Light] and I made enough good decisions in the Eighties that will make the Nineties as fun for our successors as our years were,"[95] Fitzgerald says. He certainly left behind an ambitious dream for his successor, Paul Stern, to follow. Dubbed "Vision 2000," the corporate mission is to make Northern Telecom no less than the top global telecommunications equipment maker by the end of this decade.

Stern is a no-nonsense internationalist who is well suited to the task of leading a corporation that must revamp its management structures and manufacturing methods to compete in a single global market. Image means a lot to Stern, who wrote in his 1990 book, *Straight to the Top*: "I learned to enjoy the best in everything."[96] But one should not be misled by his highly polished image. Stern knows solid-state physics, engineering, manufacturing, management — and Washington. The son of a U.S. diplomat, he was raised in Mexico and the United States. After earning his doctorate in physics at the University of Manchester in England, he worked in the United States and Germany for E.I. Du Pont, IBM, Braun A.G., Rockwell and Burroughs Corporation. He speaks English, German and Spanish and understands French and Italian. And he speaks math.

He has not wasted any time in setting out to accomplish the vision handed to him. Like Fitzgerald and Lobb before him, when he assumed the presidency, Stern immediately launched a restructuring to get rid of unproductive assets and plants. He also knew how to say no to a pet R&D project to develop a computer network product that had

consumed hundreds of millions of dollars with little sign of an immediate payoff. And he launched a major reorganization of the company's subsidiaries and product groups early in 1991 to better assault
international markets as the company strives to earn at least 15 per
cent of its income off North America's shores. To help, he wooed
Edward Lucente away from IBM, where he had created a $30-billion
business unit, IBM World Trade Asia Corporation, to be his new
senior vice-president of marketing.

Stern and Lucente certainly have been left ample opportunity to
build on the foundations left by Fitzgerald and Light. In addition to the
two critical beachheads in Japan and Great Britain that will give its
Vision 2000 a shot, Stern's predecessor built a major plant in France,
launched joint ventures in Austria and Hungary and made substantial
gains in China and Australia. As potentially important as the NTT contracts was Fitzgerald's $985-million purchase from ITT of a 27 per cent
holding in British fiberoptic and computer maker STC PLC in October
1987. Stern seized the opportunity presented to Northern Telecom
when STC completed the sale of its computer arm, International
Computers Limited (ICL), to Fujitsu in 1990. Northern Telecom
bought the remaining 73 per cent of STC, which meant the profit from
the ICL sale could be used to help pay for the $2.6-billion (U.S.) acqui-
sition. In addition to strengthening Northern Telecom's fiberoptic business, it positions the company as a key supplier to giant British
Telecommunications PLC. More importantly, the deal moved Northern
Telecom into third place in the $80-billion world telecommunications
market — ahead of NEC of Japan, Siemens A. G. of Germany and L.
M. Ericsson of Sweden — and three steps closer toward realizing its
goal of taking the number-one spot away from AT&T.[97]

"The companies that lead have a dream," Stern says. But then he
repeats the message that both motivates the company and makes him
feared by employees: "Failure is not an alternative."[98]

Privatization began to sweep the country in the early Eighties as debt-
ridden governments opted to cash in on their successful state-run
enterprises. That offered some opportunities for de Grandpré, who
tried to pick up a few government-owned telecommunications companies to complement Bell Canada's business.

At the end of 1982, Ottawa owned 186 Crown corporations and
had investments in 129 other companies.[99] The Trudeau government

had begun the federal privatization drive in January of that year when Senator Jack Austin, one of Trudeau's most powerful, but unelected, cabinet ministers, was given responsibility for privatizing a group of Crown investments held by the Canada Development Corporation (CDC).[100] But after CDC's head, H. Anthony Hampson, balked at the plans, Ottawa created a new federally owned holding company, the Canada Development Investment Corporation (CDIC), to accomplish its aims. CDIC was formed by a cabinet order on May 27, 1982, at almost the same time that de Grandpré announced his corporate reorganization.

Austin quickly turned his privatization vehicle into a massive state-owned conglomerate. In November, his cabinet colleagues approved a small pile of orders-in-council that transferred authority for several of Ottawa's largest Crown investments to CDIC. These included Eldorado Nuclear Limited, Canadair, de Havilland Aircraft of Canada and the choicest plum in Ottawa's crop — Teleglobe Canada.[101] With profit of $41 million in 1983, Teleglobe was one of the most profitable companies acquired by the new holding company. On May 25, 1983, Austin tabled legislation that gave CDIC incredible scope, including the power to manage those investments transferred to it. CDIC "will help manage better what the government already owns. It will privatize,"[102] Austin said in a statement. Few were convinced, however, as the CDIC Act appeared to give Austin more power to keep companies than to divest them, as previously promised.

Teleglobe, which is Canada's sole overseas communications link and has important international treaty obligations, was of special concern. Few disputed Austin's rationale that Teleglobe could not be privatized until certain policy issues were dealt with and, even then, not until special legislation changing the existing Teleglobe Canada Act was passed. But some questioned whether Teleglobe should report to the boss of a holding company that also owned a number of money-losing investments, rather than the Minister of Communications, who had a keen interest in ensuring that Teleglobe could invest its capital in international facilities as required.

De Grandpré soon decided that he wanted Teleglobe. He first expressed his interest publicly in an October 1984 interview with the Dow Jones news agency[103] shortly after federal industry minister Sinclair Stevens said the Mulroney government wanted to dispose of CDIC's holdings. De Grandpré then registered a formal interest with the federal government that November on behalf of Bell Canada and the eight other telephone company members of Telecom Canada. The group, which had formed a shell company named Canadian Telecommunications Carriers International (CTCI) to submit the bid, proposed to buy Teleglobe at book value — or $350 million.[104] It also proposed to pay another $60 million for Ottawa's 50 per cent stake in Telesat Canada, the domestic satellite communications operator.

Telesat responded with its own bid for Teleglobe in December 1984 for an undisclosed price. Canadian National Railways and Canadian Pacific of Montreal also put in a joint bid in May 1985.[105] Stevens issued the policy guidelines for the sale and called for final proposals in August 1985. The names of other bidders were subsequently leaked and included: Power Corporation; Inter-City Gas Corporation of Winnipeg; First City Financial Corporation, a unit of Gordon Capital Corporation; and Memotec Data Incorporated, a small data communications equipment make from Montreal.

De Grandpré publicly defended his group's bid at a speech to the Canadian Club in Toronto in December 1985. He rejected the view that a carrier bid would pose a difficulty because of concentration of ownership in the telecommunications industry. Rather, he argued, Telecom Canada would enhance Teleglobe's technology and marketing. He also offered a sweetener, promising that the carriers would reduce international phone rates if their bid were accepted.[106]

De Grandpré turned up the heat at BCE's 1986 annual meeting in Toronto, which he used as a platform to lobby for CTCI's bid. He was visibly angry as he talked about the federal government restrictions that limited the telephone companies' ability to hold a stake in Teleglobe. "I cannot visualize the sale of Teleglobe in a way in which the carriers would not be given a role and a majority position in Teleglobe," de Grandpré said at a press conference before the annual meeting. He argued that BCE had the necessary expertise to run the company, as well as to support Teleglobe with R&D. "The people who know the business should run it."[107]

He also declared that the fear that Bell Canada would dominate Teleglobe was "a red herring" because of BCE's proposed equity structure, which envisaged all telecommunications carriers holding a stake in the company. And, for good measure, he added the old company line, "The position of the common carriers and the common good are one and the same — and the common good is in the national interest."[108] As he did during the Bell Canada reorganization, he appealed directly to his shareholders to lobby "those who will decide its future course."

However, de Grandpré believed the most compelling reason to sanction the telephone companies' bid was the need to forge international alliances in telecommunications. At stake, he said, was Canada's international position in communications and its ability to compete globally. He hinted that BCE had been approached by "some international carriers to circle the world. My answer to them is, 'I can't do anything.' Opportunities are being missed."[109]

Memotec, a tiny company run by William McKenzie and acquired by high-tech venture capitalist Eric Baker in 1977, emerged on February 11, 1987, as the surprising dark-horse winner with its successful $488.3-million bid for Teleglobe. It had arranged $345 million in financing from the National Bank of Canada, a large investment dealer and some pension funds to pay for its catch.[110] The company's shortfall provided a way for BCE to acquire an indirect stake in Teleglobe by buying a one-third equity position in Memotec for a total of $196 million.[111]

Although de Grandpré did not criticize Memotec itself, he remains intensely opposed to Ottawa's decision. "It was a very bad thing and very irrational. I have no words strong enough for that. Memotec does not have the resources they need."[112]

De Grandpré's concerns about Memotec's ability to run Teleglobe and the industry's ability to compete internationally soon proved prophetic. In mid-May 1991, BCE made its chagrin at Memotec's management public after the Toronto investment firm, Gordon Capital, launched a proxy fight to oust Baker and McKenzie from Memotec's board. In a circular mailed to Memotec shareholders a week before the company's annual shareholders' meeting on May 16, Gordon Capital charged that Memotec's financial performance was being adversely affected by the company's management.[113]

The brokerage blamed the founders for a costly $210-million string of acquisitions made by Memotec since it had purchased Teleglobe in 1987. Many of those acquisitions plagued Memotec with losses and, the brokerage stated, contributed to a cash deficit. Gordon Capital also pointed out that most of the money to finance Memotec's spree came from interest-free loans from Teleglobe. The CRTC, too, had serious concerns with the company's financial dealings with Teleglobe. Memotec tapped into Teleglobe's funds reserved for settling its accounts with overseas telephone companies under a cash management system it had set up immediately after Teleglobe was privatized. A month before the proxy fight, the CRTC issued a decision ordering Teleglobe to secure its parent's borrowings, requiring Memotec to seek collateral for its daily balances, which were as high as $133 million. The CRTC feared the consequences for telephone subscribers if Memotec could not afford to pay the loans back in time.[114]

Gordon Capital's concerns were shared by several other powerful minority shareholders, in addition to BCE, including two of the country's largest pension funds, the Caisse de dépôt et placement du Québec and the Ontario Municipal Employees Retirement Board. BCE was barred from voting for the new slate of directors put forward by Gordon Capital, other than its own four representatives, by the same law that limited BCE's ownership stake in Teleglobe. Yet Raymond Cyr, chairman of BCE, was not prevented from speaking out and was vociferous in his criticism of Memotec's investments. He was particularly angered by Memotec's $154-million purchase of ISI Systems, a small U.S. software company based near Boston. Memotec paid Charles Johnston, ISI's owner, four times the book value of ISI's share price, and Johnston is now the largest individual shareholder of Memotec.

However, on the eve of the shareholders' meeting, Cyr predicted in an interview that Gordon's bid would fail because Quebec's powerful pension fund, which owns 12.5 per cent of Memotec's shares, could not support a Toronto-based brokerage. "It's a Bay Street versus Quebec thing," Cyr said, referring to the proxy fight. "The caisse, which represents Quebec Incorporated, is at odds with Gordon over Steinberg's and some other things. And they do not want to vote against two Montreal-based companies."[115] Gordon Capital's bid was defeated. Cyr described the implication for BCE and the rest of the Memotec board: "The caisse will become the powerbrokers on the board. Management will need the caisse's support to run the company."[116] The irony was that Jean-Claude Delorme, the president of the giant $36-billion pension fund, had previously served as president and CEO of Teleglobe from 1971 until he joined the caisse in June 1990.

Memotec's public claim that Gordon acted for BCE was wrong. But the beleaguered company's president, William McKenzie, was closer to the truth when he proclaimed in a full-page newspaper ad placed throughout the country on May 13 that, "This is not an issue about Memotec management. The issue is control of Teleglobe Canada."[117]

Monty had admitted as much earlier that week. In an interview for *The Globe and Mail,* Monty said he had begun a lobby campaign on behalf of the country's largest telephone companies to remove the restriction that limits them to a one-third stake in the international carrier.[118] Cyr had stopped short of revealing the campaign when, at BCE's annual meeting in Ottawa earlier that month, he said the restriction made Canadian telephone companies uncompetitive, echoing the criticism voiced five years earlier by de Grandpré. In a later interview at the height of the proxy battle, Cyr said: "This whole situation has been good for forcing, or reopening, the policy issue of Teleglobe's ownership. I don't know if we'll be successful in having the policy rules re-examined or changed, but I think they could be re-examined along with the federal government's general telecommunications policy review."[119]

Looming underneath the back-room battle for control of Teleglobe, however, is a long-simmering fight with the domestic telephone companies over the rates for international telephone calls. Each side accuses the other of inflating the price of overseas calls to fatten its coffers. On the one hand, Bell Canada charges that Teleglobe is an uneconomic middleman that hampers business by keeping the price of outgoing calls too high; and, on the other hand, Teleglobe argues Bell Canada and its ilk are the culprits because they charge Teleglobe almost $500 million a year to handle international telephone calls. Efforts to reduce that settlement charge, which would allow Teleglobe to slash the price of its overseas calls, have been frustrated by the phone companies, Teleglobe officials say. BCE's motives for seeking control of Teleglobe, they argue, are aimed at protecting the telephone companies' annual half-a-billion-dollar share of overseas calling revenue.

That became clearer when Bell Canada filed details of its proposed ownership scheme with the CRTC on August 28 at the final day of a public hearing to consider how Teleglobe should be regulated in the future. Although the CRTC has no legal power to change Teleglobe's ownership structure, Bell Canada told the commission that joint ownership of the overseas carrier by the country's telephone companies would be "in the national interest." Monty's proposal also hearkens back to de Grandpré's and Scrivener's earlier scheme to co-opt Telesat Canada, envisaging Teleglobe Canada acting as a "carrier's carrier."[120] In resuming de Grandpré's quest, however, it remains to be seen whether Monty and Cyr will have better luck, or whether the attempt to relegate Teleglobe to a wholesaler will be repudiated, as was the 1976 Connecting Agreement between Telesat Canada and Telecom Canada.

BCE had better luck when it put in a bid for Northwestel Incorporated of Whitehorse, which it acquired from Canadian National in 1988. Northwestel provides service in the Yukon and part of northern

British Columbia, which complements Bell Canada's service area in the eastern Arctic. The $147-million deal followed Scrivener's tradition of acquiring other telephone companies that serve areas contiguous with Bell Canada.

A few years later, de Grandpré made a strongly worded pitch to Alberta government officials for a large stake in Telus Corporation, the publicly traded holding company created during the privatization of Alberta Government Telephones (AGT) in the summer of 1990. He actually recalled a clause in the 1908 agreement between the utility and the provincial government that in effect compelled the province to offer shares to Bell Canada if it ever contemplated selling its phone company. The agreement, which was signed when Bell Canada sold its Alberta phone business to the province on April 1, 1908, was one of those antiquated legal documents de Grandpré read in 1966 during his first months at the phone company.

After Alberta refused to sell any shares to the holding company, de Grandpré met with provincial officials. The officials said Alberta did not intend any specific offense toward BCE; rather, the government of Premier Donald Getty had decided to offer shares only to Albertans. The government would retain a 40 per cent stake (put up for sale in late 1991). But de Grandpré wanted a third.

"The terms are very clear," de Grandpré says. "If the province of Alberta ever sold its telephone company to a company that is not a municipal corporation, we have the right to overbuild in that province. It's a very strong right; it means we could come into Alberta and build a second, competing telephone company if they didn't offer to sell any equity to us."

His attempt to use that clause as leverage, however, did not sway the Alberta delegation. And de Grandpré says he does not know if Bell Canada would ever exercise its right. "But it's there."[121]

It is doubtful that current management at Bell Canada would move into Alberta with a second network because it would fly in the face of one of their key arguments mustered in their fight against long-distance competition. The company has stated that two networks are inefficient. Bell Canada is facing potential competition from Unitel Communications, which is 60 per cent owned by CP and 40 per cent by Rogers Communications. A second bid is also before the CRTC from a joint venture of two telecommunications resellers, Lightel of Toronto and B.C. Rail Telecommunications, a unit of Crown-owned B.C. Rail. Resellers lease wholesale facilities

from telephone companies and sell enhanced and discount services to
their customers.

"Some days, I'd like to be younger again," de Grandpré joked in a
1990 interview, referring to his desire to personally take on Unitel and
Rogers in the current long-distance competition battle. "They are trying to weaken us again and, as the victor of the first battle, I'd hoped
we wouldn't have to spend the time fighting this again."[122]

Unitel's bid is the second for the company, which used to known as
CNCP Telecommunications. De Grandpré was intensely active in the
first fight, which began in October 1983. A former Bell Canada vice-
president recalls telephoning de Grandpré and Cyr with the news of
the decision on August 29, 1985, after he was let out of the CRTC
lockup. "De Grandpré wanted to know how I would handle our reaction. I said, 'There's not much to handle because the decision seemed
to give us everything we had wanted.' Cyr then said, 'I don't care how
you handle it — I'm going to go and have a drink.'"[123]

The CRTC rejected CNCP's application because, according to its
decision, the company's business plan was too weak, leading the commission to conclude that the company would not be viable. In fact,
that was only part of the reason. John Lawrence, then vice-chairman
of the CRTC, says the commission turned CNCP down "because
there would still be regulation. And to set up regulated competition
would be insane. What does that do for anybody? It's just the CRTC
playing God. A regulated duopoly just shifts the gains from one company to another for the benefit of the new entrant's shareholders."[124]

De Grandpré says Bell Canada should be allowed to keep its long-
distance monopoly to maintain the strength of BNR and Northern
Telecom's research budget. "The government should not do anything
to reduce the role of Bell and Northern, which means they shouldn't
give a piece of the action to Unitel to line the pockets of Mr. Rogers."[125]

Rogers has also challenged Bell Canada's local telephone monopoly
by setting up a new telecommunications subsidiary that is building its
own fiberoptic lines in large cities. At stake is whether Bell Canada can
keep its local telephone service monopoly — and new video services
on demand — to the home. The Department of Communications is
currently studying the issue of convergence between cable and telephone companies and whether there should be competition in local
telephone service delivery.[126]

Marcel Masse, federal communications minister, announced the policy review at the annual convention of the Canadian Cable Television

Association (CCTA) in Toronto on May 8, 1989. He says two events the preceding week prompted the move.[127] The first was Rogers' acquisition of his stake in Unitel on May 1. The second was de Grandpré's pitch the following day, made to reporters after BCE's annual meeting in Vancouver, that Bell Canada should be able to own broadcast or cable television stations if Rogers could move into the telecommunications business.

"What we're talking about is competing in Rogers' territory [with] a duplicate cable system. That's what he wants to do in long distance — duplicate the long distance network,"[128] de Grandpré told reporters.

But this is not another squabble between the telephone company and a would-be rival. The outcome of the debate to develop a single set of rules governing both the telecommunications and broadcasting industries is profound. Both stand to lose their prized monopolies, and at stake is potential supremacy of the combined $20-billion-a-year communications industry. In what has become a landmark speech that launched the current policy debate, Masse laid out a sweeping scenario under which he envisaged the telephone company and cable monopolies competing with each other, and he indicated that his government "strongly favors competition" between the two industries. Communi-cations Minister Perrin Beatty has taken the review another step further by asking for recommendations from an industry committee.

Skeptics argue the industries may not fully converge and that telephone and cable companies may not go head-to-head for at least a decade. Just in case, though, BCE staked out a toehold in the cable industry less than three weeks after the DOC review was announced. It struck a deal to acquire a 23 per cent shareholding in Videotron Corporation, the British subsidiary of Quebec cable giant Groupe Videotron of Montreal, which is allowed to compete against British telephone companies.[129] The message was clear; although Bell Canada was not allowed to enter the cable industry, BCE would by investing in Rogers' largest competitor. For the current leadership at BCE, the key issue is not so much whether they will go into the cable business as it is whether they will be allowed to. As de Grandpré frequently declared, Bell Canada is willing to compete, "but not with my arms tied behind my back." Ironically, it was de Grandpré who, in 1968, consented to the legislative stricture keeping the utility out of the cable business.

"Why is Wood Gundy trading at BCE Place?" asked the billboard sign outside the then uncompleted second tower of the 2.5-million-square-foot monument to BCE in Toronto's financial district. In 1990, BCE's development arm erected the sign at the corner of Bay and Wellington streets in the hope that tenants of competing developers' office towers would phone to find the answer.

But that was not the question on the minds of people like Gerald Connolly, who had lost $50,000 of his retirement savings when he bought debentures in BCE's real estate subsidiary, BCE Development (BCED). "Why did BCE ever become a real estate developer in the first place?" he asked in January 1991 on the way to a meeting between BCED debenture holders and the company's new management partners in the Canada Trust Tower, the fifty-one-story first phase of BCE Place.[130] At the 1990 annual shareholders' meeting, many irate BCE and BCED shareholders had peppered the holding company's management with that and other questions about BCE's $440-million write-down on its BCED investment in January 1990. Connolly's anger mounted when he reached the twenty-third floor and found BCED had ushered the irate investors to an empty, cavernous floor. The debenture holders felt further humiliated as they sat on folding chairs on the naked concrete floor to be told by the developer's new bosses to be patient, and that the restructuring of BCED's assets, which left them holding paper with no equity value, was in their long-term best interest. Fight the restructuring proposal and they would hold shares in a bankrupt company, they were told.

The financial debacle at BCED was at odds with the image of elegance and sophistication that BCE Place's developers strove to impart at every turn, from the blue-lit spire atop the first tower — inspired by the Chrysler Building in New York — down to the six-story glass galleria connecting the twin towers of BCE Place with a set of twelve neighboring buildings restored by the developers. Their attention to promotional detail extended to the production of a lavish, hardcover book about the development, which adorns the coffee table in de Grandpré's office and which carries the same title as the project's slogan, "The Spirit of Place."

In answering why tenants like Wood Gundy chose BCE Place, the sales agent's carefully scripted message talked of the complex's amenities and features. But it did not tell the caller about Wood Gundy's trading relationship with BCE, particularly the fact that the large, institutional stockbroker made a good portion of its income as the lead underwriter for much of the holding company's share issues. Wood Gundy also holds the prestigious position of designated market-maker of BCE's stock on the Toronto Stock Exchange. The TSE appoints one brokerage to be the market-maker for every publicly traded company listed on the exchange to ensure that stock trading is always balanced.

Investors like Connolly had invested in BCED because of the blue-chip reputation of BCE's name. When things soured at BCED, leaving common shareholders with worthless stock and debt security holders with no dividends and no claims on any assets to recoup their losses, Raymond Cyr, BCE's chairman and CEO, told shareholders at the conglomerate's 1990 annual meeting that BCE had no responsibility to make good those losses. He defended BCE's original investment decision and argued that BCE had always informed investors that BCED was both a separate entity and a high-risk investment. That was true enough. But Cyr's message did not change investors' perceptions of the fact that BCE had attached its name and reputation to the developer, which had encouraged them to part with their savings. The subsequent reorganization of BCE's real estate assets into another joint venture company fueled the flames.

Ironically, it was a retired vice-chairman of Wood Gundy who headed a committee of preferred shareholders created to oppose the restructuring. "I've never seen a case where a public company has done something like this, and I believe it is damaging to BCE's reputation in the investment community," Ross LeMesurier said in putting a question to Cyr at the 1990 annual meeting.[131]

"It's not a scam," Cyr retorted to the charges of LeMesurier and other shareholders. He justified the complex transactions between BCE and an Edgar and Peter Bronfman subsidiary, Carena Developments of Toronto, to manage the holdings on the grounds that the joint venture and asset reorganization were made to avoid BCED's bankruptcy, which, Cyr says, would have left investors even worse off.[132]

Fingers can be pointed in all sorts of directions for the BCED fiasco: to de Grandpré, to the BCE board, to BCE money managers, to BCED's board and management, to high interest rates and to the trillion-dollar-plus Savings and Loan scandal in the United States. De Grandpré is repentant about the acquisition of BCE's 69 per cent stake in Vancouver-based developer Daon Development, which was one of British Columbia's fastest-growing property developers, in January 1985. "That's the only one I regret,"[133] he says in reference to BCE's costliest and most troublesome acquisition. However, de Grandpré blames John Wilson "Jack" Poole, Daon's founder and former chairman, for the costly mess.

"He welched on his promise," de Grandpré said as he flung a copy of BCED's 1988 annual report down on a table in his Toronto office. It was the day after BCED wiped $610 million from the value of its real estate portfolio and announced a staggering $709.8-million loss for the developer's 1989 fiscal year. BCE's exposure to BCED's losses would have been far less had Poole been true to his word to deliver a joint venture partner, de Grandpré says.

So why did de Grandpré invest in real estate in the first place, and what went wrong after he did?

BCE's foray into real estate development actually pre-dates Bell Canada's corporate reorganization. By the late Sixties, the telephone company had become one of the country's largest corporate property owners and corporate landlords. It had had to build or lease dozens of regional offices and hundreds of switching centers throughout its territory.

One of Bell Canada's first major urban development projects with an outside developer was in Ottawa. In 1967, the phone company had acquired a prime parcel of downtown real estate on Elgin Street for a future administration building. Scrivener struck a deal with the Reichmann brothers' Olympia & York Developments on November 1, 1969, to lease O&Y the land for a twenty-six-story building, which the developer would build and own. Bell Canada, in turn, would become a tenant in Place Bell Canada.

"We were Paul Reichmann's first major client and it was their first big project outside Toronto," Scrivener says. "He and I have been good friends ever since. O&Y also built our data centers."[134] Bell Canada moved into Place Bell in early 1972 and bought the building for $62.5 million when O&Y put it up for sale in November 1979 after changes in the Income Tax Act made leasebacks less financially attractive.[135] Although consumer group critics opposed the purchase, the CRTC sided with Bell Canada's analysis that it would be cheaper to buy the building and assume the mortgage than to lease space from another landlord.

The first major real estate project launched by Bell Canada under de Grandpré's tenure was the Bell-Banque Nationale project on Lagauchetière in Montreal. Bell Canada began amassing the properties on the block bounded by University and Vitre in 1974 for a two-tower project slated to be the largest development project in Montreal since the 3-million-square-foot Place Bonaventure was begun in 1963. But a rival developer, Trizec Corporation, threw a hurdle in de Grandpré's path when it bought a 5,700-square-foot piece of property located in the middle of the development site — a tactic called a "squeeze play."[136] Trizec hoped to use the property as leverage to gain the multi-million-dollar construction management project. Undeterred, de Grandpré went ahead and announced the $100-million development with Bell Canada's partner, the Banque Nationale, on December 8, 1978. Trizec be damned — de Grandpré told reporters that the bank "quite logically" wanted the contract to go to a Montreal engineering company. Furthermore, he asserted that Trizec would back down because it would not want to be known as "obstructionist."[137] The project went ahead as planned and the buildings were completed shortly before BCE's reorganization.

When BCE was a year old, de Grandpré decided the conglomerate should develop a major project in downtown Toronto that would be a monument to the holding company he had created. In keeping with BCE's role as a holding company, a separate subsidiary named BCE Realty was created in early 1984 to manage BCE's real estate interests and proposed property developments. The creation of BCE Realty shortly followed BCE's second annual meeting, where de Grandpré proudly told reporters, "I don't think there was one minute of BCE history in 1983-84 that was a disappointment."[138] In fact, the creation of BCE Realty, which soon became an unmanageable real estate empire, planted the seeds of his own, and his shareholders', subsequent disgruntlement.

The establishment of the subsidiary was a prelude to the launch of a joint venture with G. Donald Love, the reclusive chairman of Oxford Development Group of Edmonton, to build the BCE Place development project. Under that deal, announced on July 10, 1984, BCE agreed to arrange for the financing for BCE Place, while Oxford was to act as the developer and property manager.[139] Depending on what architectural proposal was chosen, the project was expected to cost between $250 million and $350 million, according to BCE. De Grandpré landed Canada Trust as BCE Place's principal tenant shortly afterward and signed a deal with Mervyn Lahn, chairman and CEO of Canada Trust, to share the cost of the first phase of BCE Place, to be named the Canada Trust Tower. On October 1, 1984, the trust company announced that it had bought its equity stake in BCE Realty through Truscan Realty.[140]

BCE Realty then bought an eighteen-story landmark heritage property in the heart of the Bay Street financial district, the head office of

The Permanent, a major trust company. "But experience gained with these transactions showed that a shortage of human resources was holding back the real estate subsidiary's growth," Cyr stated in a speech at the time, so "BCE decided to shop for expertise."[141] That quest led it to Daon Development.

BCE's problems began even before it had completed its offer, made on January 22, 1985, to acquire 68 per cent of Daon's shares. It bought into a company that was heavily in debt. Poole had led Daon as its president for twenty years when BCE made its offer. The civil engineer had cofounded Daon's precursor, Dawson Housing Developments of Vancouver, with Graham Dawson in 1964. Poole changed the company's name and expanded into Alberta in 1973 to cash in on the mid-Seventies oil boom. A high-roller, he lived in a $4-million home in West Vancouver, traveled in a private Hawker-Siddeley 700 jet and owned more than 4.5 million common shares of Daon, which were worth more than $13.5 million when BCE made its offer.[142]

In 1982, Poole's record was blemished when Daon defaulted on its public debt. Its extensive land development schemes had caused it to be overly exposed to soaring interest rates. The problems became acute when Daon was forced to pay off a partner in the Park Place office tower project in downtown Vancouver. Daon was rescued from the brink of bankruptcy by Leonard Rosenberg, the controlling shareholder of Crown Trust Company, who lent Daon $50 million for the transaction, plus a $10-million line of credit in return for a 50 per cent stake in the project.[143] But a few months later, Crown Trust was seized by the Ontario government after Rosenberg orchestrated an abortive $500-million flip of eleven thousand Toronto apartment units. That left the Ontario government holding the bulk of the mortgage on Park Place.

It was the efforts of the Ontario government and the Canada Deposit Insurance Corporation to negotiate the sale of their Park Place mortgages to a consortium of institutional pension funds that led to BCE's acquisition of Daon.

Selim Anter, president of BCE's giant pension fund, Bimcor Incorporated, was chief negotiator for the pension fund consortium in the Park Place deal. Anter began negotiations in the fall of 1984 at almost the same time de Grandpré let it be known he wanted to acquire a real estate developer. But Anter also held the simultaneous position of president of BCE Realty and it did not take long until he proposed the Daon acquisition to BCE.[144] He took the proposal to J. Stuart Spalding, BCE's executive vice-president of finance, who passed it on to de Grandpré. Anter subsequently was paid a bonus for arranging BCE's acquisition of Daon through an option to acquire one hundred thousand shares of Daon at $4.10 a share.[145]

Given Daon's loss of almost $91 million in 1983, BCE surprised analysts and investors with its offer to pay $150 million for the developer. BCE acquired almost 63 per cent of Daon's stock, then increased it to almost 69 per cent by exercising a share option, which gave BCE Realty control over assets of $1.2 billion. Although the book value of those properties was almost $930 million, only 19 per cent of them produced income.[146] Poole tendered 1.4 million of his shares, which gave him $4.2 million.

BCE proposed a share exchange offer to Daon's shareholders in October 1985 in a bid to both reorganize its real estate holdings under one roof and to get around the constraints imposed on Daon by its creditors' agreement.[147] The offer was quickly resisted, then challenged in court, by two trust companies on behalf of five thousand debentured shareholders. The creditors, who had received payment in Daon shares in lieu of cash, won a restraining order from the Supreme Court of British Columbia barring the share swap until their claim that the merger breached the previous financial restructuring agreement could be heard.[148] Daon reached an agreement with its creditors after National Trust Company of Toronto and Central Trust Company of Halifax lost their bid for an injunction.[149] Daon agreed to repay the lenders an undisclosed amount and, in return, the creditors agreed to remove the restrictions placed on Daon. The agreement left Daon free to invest in new properties and undertake financings.[150]

"For Daon, the agreement is a dusting off of the final ashes of its past, clearing the way for a rescue by its fairy godmother,"[151] a lawyer for one creditor told *The Globe and Mail.*

Shortly after BCE completed its Daon reorganization, BCED began discussions with the Love family, who had put Oxford's U.S. subsidiary up for sale. The company owned $915 million (U.S.) worth of commercial real estate in five U.S. cities. The package included a mix of office buildings, malls and other commercial developments located in Chicago, Denver, Minneapolis, St. Paul and Phoenix. That deal was also proposed to BCE by Anter and the Bimcor fund, de Grandpré says, and was recommended to him by Spalding.[152] Poole, in turn, made the presentation to the BCE board because he needed debt financing from BCE. He promised to bring in a joint venture partner and assured de Grandpré that he was close to signing a deal with a California pension fund.

"Poole came to the board and said he had a partner for the deal. It wasn't signed, but he said it would be within hours or days," de Grandpré says. "The board of directors approved the investment in Oxford Properties on the condition that we would get this partner."[153] Under that scenario, BCED would have assumed only half, or $390 million, of Oxford Properties' debt.

"The enthusiasm of developers is often shared by lenders," de Grandpré says in a 1990 interview. He explained how property developers became over-extended in the cash-flush Eighties, which, in turn, contributed to the subsequent collapse in real estate prices. But he, too, was caught up in the euphoria, as evident during an interview in his Toronto office on March 12, 1986, when he excitedly announced details of BCED's purchase of Oxford Properties in a $125-million cash-for-shares deal and the assumption of $790 million of debt.[154]

Not all of Oxford's U.S. properties were trophies, however, particularly in overbuilt downtown Denver, where one of its buildings, the fifty-six-story Republic Plaza, was nicknamed the "see-through building" because of its high vacancy rate.[155] The California pension fund that Poole promised to deliver backed away from the deal shortly after BCE acquired the properties and, as Poole and de Grandpré soon found out, no one else wanted to buy a stake in the properties. BCE found itself on the hook for the full amount of BCED's added debt. In financial documents issued at the time, BCED stated that the terms of the prospective partners were not as attractive as BCE's financing, which was provided under a private share placement.[156]

At the same time that the Oxford deal was announced, BCE was rocked by a real estate scandal involving the Bimcor pension funds, which raised serious concerns about Anter's judgment. The architect of the Daon and Oxford deals was summoned to BCE's boardroom high atop First Canadian Place on the morning of February 3, 1986, after he found his office padlocked shut. He was told he was being suspended on the basis of an audit that was conducted into a string of real estate transactions involving Bimcor's funds, and he was subsequently fired.[157] Christopher Morgan, vice-president of real estate at Bimcor, was also dismissed.

Coincidentally, news of the trouble broke in Toronto newspapers the same day that the Oxford deal was announced, after documents related to a lawsuit between a small developer, Maron Land Development of Toronto, and Bimcor and BCE were sent anonymously to several newsrooms.[158] Maron launched its lawsuit after de

Grandpré blocked a same-day, $63-million flip of a Toronto shopping center and warehouse by Maron to a numbered company owned by Bimcor. Maron would have netted a $27.5-million profit from the deal, and it sued BCE for the money on the grounds that it had a valid, signed agreement of sale.

Maron had been formed the previous November when a numbered company changed its name. It was headed by Henry Federer, a former vice-president of a shopping center development company run by Jack Burnett, brother of Toronto financier Joseph Burnett. Super Carnaval Food Stores Limited, a supermarket chain then controlled by Jack Burnett, helped finance the Maron deal with an $8.5-million second mortgage and intended to be a lead tenant.[159] According to Maron's statement of claim, the company began negotiating a deal with Bimcor in October 1985. BCE's lawyers had already been tipped to problems with the pension funds' real estate dealings a month earlier when Northern Telecom secretly dismissed Bruce Craig, the head of its pension fund, giving credence to the view that the Maron deal was a "sting" to snare Anter.[160]

The Bimcor affair took an even more sensational twist the day the story of the Maron lawsuit broke. That morning Metro Toronto Police homicide officers were called to the partially renovated mansion of Maron's lawyer, Garry Smith, where they found his body lying face down on the bottom of his empty swimming pool. He had fallen out of bed, the police said. His brass bed was on a concrete platform, much like a diving board, suspended over the deep end and below a thirty-meter skylight running the length of the house, which was camouflaged by its rundown storefront exterior and located in a seedy part of downtown Toronto.[161] His death was ruled an accidental fall. Described as brilliant, he was a real estate and constitutional law expert at the Toronto firm of Weir and Foulds and had previously acted for Burnett in the Crown Trust affair. The lawyer lived his forty-seven years "in record time," one of Smith's colleagues said in the eulogy.

When asked the day before his death who Maron was, and what his client's relationship was with the Burnetts, Smith had told *The Globe and Mail,* "I don't want to talk about, or get into, the personalities."[162]

BCE called the Maron flip, which was only one of seven that Anter and Craig had participated in with the Burnetts between August 1984 and January 1986,[163] "unconscionable" in its statement of defense. Among the documents BCE had obtained and was prepared to put before the court were Bell Canada long-distance records of calls made by the fired employees, according to its affadavit of documents.[164]

Despite a more comprehensive audit by Touche Ross & Company and the hiring of private investigators, BCE has never been able to trace a total of $12 million in missing commissions,[165] de Grandpré says. Both the Maron claim and Anter's suit against BCE alleging wrongful dismissal were settled out of court for undisclosed terms in 1988.

BCE Realty shelled out more money in early August 1986 for the remaining 50 per cent stake in BCE Place it did not already own. That deal, plus the Oxford acquisition, meant BCE had to turn down an offer to buy the Bronfmans' stake in Cadillac Fairview that fall. Leo Kolber approached de Grandpré to ask if he wanted to acquire Cemp Investment's 51 per cent share.

"But when we looked at the size of it and the debt we would have, I came to the conclusion that it was too rich for our blood. It was something we couldn't do."[166] The development company, whose properties included the Toronto Eaton Center, fetched $2.6 billion when it was sold to Chicago-based JMB Realty Corporation in 1987.

Meanwhile, BCED continued to announce new projects, despite a growing glut of unleased space throughout North America. These included a new building for Revenue Canada in Vancouver, the $1-billion MacArthur Place project in Orange County, California, and a forty-three-story office building at 1000 de La Gauchetière in Montreal, which BCED planned to develop with Teleglobe Canada and Group Lavalin, the giant Montreal-based consulting engineering firm.

Poole painted a bright picture in his report to shareholders accompanying the 1988 annual report, which, ironically, was dated on the Ides of March, 1989. Looking back, it's not hard to see why de Grandpré was so infuriated by its enthusiasm:

> We encourage shareholders not to measure the progress of BCED over the next few years by near-term earnings, but rather by the underlying growth in real market values. . . . It will be a year in which the emphasis will shift even further to the retention, rather than sale, of properties. It will also see the final step in our four-year transition from a land company to a company with a substantial investment portfolio of blue chip, income-producing properties, and one of the strongest balance sheets in the industry.[167]

Yet it had already become apparent by the end of 1988 that BCED's quest for cash flow by taking on more debt had only succeeded in building a house of cards. BCED's financial squeeze was made worse by higher interest rates and falling commercial real estate prices, a situation which, de Grandpré says, was precipitated by the Savings and Loan failures in the United States. These, in turn, had flooded the commercial real estate market with bargain-basement assets that further reduced prices.[168]

Nor did Poole's optimistic message mention the discussions that began in January 1989 to sell off a portion of BCE's stake in BCED.

BCE entered talks with Olympia & York to sell 49.9 per cent of a new subsidiary that would control BCED to the Reichmanns. The complex transaction would have injected $225 million of new equity into BCED and left BCE and O&Y each owning about 14 per cent of BCED. The deal was expected to lead to an offer to BCED shareholders in early March,[169] but it collapsed on March 17 — the very day of Poole's optimistic letter! The two companies merely said they were unable to reach a "mutually satisfactory" arrangement.[170]

Only after the deal's failure did Cyr and BCE move to replace Poole, who announced his plans to step down as BCED's boss at the company's annual meeting in Vancouver on May 1. BCE installed Spalding as chairman and Robert Naiman, the former executive vice-president of real estate at Bimcor, as president. An accountant and quintessential bureaucrat, Spalding seemed like an odd choice to head the troubled developer. Naiman had been senior partner of Touche Ross & Partners and was the auditor who scrutinized Bimcor after the pension fund scandal broke. He had left Touche when Bimcor named him to the real estate pose in June 1986. Two BCE lawyers, Josef Fridman and Shaul Ezer, were also named to BCED's board.

Poole told shareholders at the annual meeting that BCED would consider selling its U.S. properties, which represented about 80 per cent of its total portfolio.[171] Later that month, BCED announced that JMB Realty of Chicago was bidding on the assets. But the U.S. developer walked away, too.

The bid to sell BCE's stake in BCED became even more of a roller-coaster ride for investors as O&Y returned to the fold in late June with a reduced, $557-million bid for BCED's shares, or $2.80 a share. Its previous offer that spring was worth about $3.75 a share. Despite the reduction, Cyr opted to get out of real estate, in principle, and sell all of BCE's 67 per cent holding. But the bid, which was set to expire on August 15, required a minimum of 90 per cent of BCED's shares.[172]

There was the rub. Minority shareholders had been stung before during Daon's previous restructuring. Now, in this bid, they saw the

value of their stock devalued 25 per cent by the Reichmanns.
Although a large majority of shareholders agreed to tender their
shares, the terms of the offer meant that even one small, but alienated,
class of shareholders could easily scuttle the deal by refusing the offer.
The Reichmanns let the offer die on August 15 when they fell short of
their goal by less than 3 per cent after 87 per cent of the shares were
tendered.[173] BCE was again left holding the bag.

Minority shareholders did not like the offer and O&Y, it seemed,
did not believe BCED was worth more money. There was, however,
an allegation that the Reichmanns had walked away from the second
bid because of their eleventh-hour discovery of a confidential consulting report on BCED's U.S. problems.[174] De Grandpré denied reports
that BCE withheld any report in order to try to get a higher price
from O&Y. Besides, he says, the Reichmanns knew the condition of
the U.S. commercial real estate market, pointing to O&Y's withdrawal
from its $1-billion (U.S.) bid to buy the Sears Roebuck tower in
Chicago in November 1989 after it, too, was reappraised.

"I think they simply got cold feet," de Grandpré says. "I suspect they
walked away because of their own difficulties in Canary Wharf in London
and with their own U.S. investments, for the reasons I have given."[175]

In any event, Naiman publicly warned BCED shareholders in the
company's six-month results, released a week before the offer expired,
that BCED would have to review the worth of its properties if the
Reichmanns' bid failed. And, he said, that would likely cause the
developer to take a write-down in its financial statements. That reap-
praisal was required by federal corporate law, Naiman said a month
later. BCED hired Goldman Sachs & Company to prepare a study on
how to dispose of the U.S. assets, as well as review the carrying value
of the properties. Because of its losses, and in anticipation of a write-
down, Naiman announced that the company had suspended the payment of dividends on its preferred shares — a move that was
guaranteed to upset minority shareholders.[176]

On the verge of insolvency, BCED sold half its 80 per cent interest
in BCE Place to the Ontario Municipal Employees Retirement Board
in 1989. In return for its 40 per cent stake in the development, the
Ontario pension fund provided a fixed rate, ten-year mortgage for
BCED's share of the development costs. But paying the interest on its
$2.4 billion worth of debt was hardly enough to solve BCED's woes.
The company also required a $100-million line of credit to operate and
pay contractors at its job sites.[177]

As an ailing company that no one wanted, BCED was a candidate for a restructuring by a turnaround expert who would assume the management of the company and infuse the company with cash in return for equity. In October 1989, Gordon Arnell, the president of Carena Developments, the real estate arm of the Toronto Bronfman brothers, approached Cyr with a bid. Arnell had worked for Love and knew the BCED properties as well as anybody else. Carena proposed to pay up to $415 million for slightly less than 50 per cent of a joint venture company that would hold BCE's stake in BCED; it was the Reichmanns' first proposal revisited. Cyr had little choice but to agree.[178]

De Grandpré had stepped down as BCE chairman and was simply a bystander to the deal with Carena's parent, Hees International Bancorp — the merchant banking arm of Edward and Peter Bronfman's Edper Enterprises. Hees is recognized as Bay Street's pre-eminent financial restructuring experts. "They're so complex, I get lost in trying to figure their structures out,"[179] de Grandpré says, even though he is an expert in corporate law. Hees; also holds a 49 per cent stake in Edper's Brascan Holdings. The rescue of BCED is now in the hands of a Brascan-Hees-Carena triumvirate: Jack Cockwell, chief operating officer and executive vice-president of Brascan; Willard L'Heureaux, managing partner and president of Hees; and Arnell, who assumed the position of president and CEO of BCED after Naiman stepped down.

Carena increased its equity infusion in BCED to $500 million in January 1990 after BCED took its write-down. The reappraisal of BCED's properties led to a $550-million charge against its assets. When that was added to a $105-million loss for 1989 and a foreign currency loss of $61 million, BCED had a negative net worth, or shareholders' equity deficit, of $72 million.[180] BCED shareholders and debenture holders held worthless stock.

The new joint venture company, named Brookfield Development Corporation, became BCED's operating subsidiary. It holds the productive, or money-generating, properties, which have been transferred to it from BCED. BCE agreed to leave the partnership in four years, which will then give Carena and Hees full control of Brookfield.

To Hees, Brookfield is a cup half full. It is a recapitalized company whose assets have been rescued from bankruptcy. As L'Heureux told debenture holders in early 1990, at least Brookfield has a fighting chance. It has kept its assets from being seized by creditors, which would have left shareholders in an even worse position.[181]

Minority shareholders, on the other hand, look at the restructuring as a cup half empty. "A clever way to freeze out may creditors," J. A. Denton, a BCED debenture holder and committee member, wrote in a letter to bond holders in January 1991.[182] They are creditors holding notes on a corporate shell with no productive assets, he maintains, although Carena will offer BCED debenture holders options to buy stock in Brookfield on the same terms as BCE and Carena.

BCE left the sector as it entered it; out a bundle of money and with National Trust rallied around irate debenture holders who again believe they have been left at the bottom of the heap — as in 1985. De Grandpré blames Spalding, his former right-hand acquisitions man who, he says, failed to control BCED's debt. "Poole attracted too much debt and Spalding was supposed to keep an eye on Poole and his projects. He didn't, and we got too much debt, and so that's where I put the blame. On Spalding."[183]

But to blame Spalding for not cutting BCED's debt long after it had been incurred, and with the board's approval, is akin to shooting the messenger. And it glosses over questions that still linger, such as: Why was Poole kept on so long if, in de Grandpré's opinion, he was such a "son-of-a-bitch"? Why did the Oxford deal go through, even though its architect, Anter, had been fired a month earlier? The answer to Connolly's question — Why did BCE go into real estate? — helps to answer those questions, too: corporate ego, and BCE's desire to be a landlord, not a tenant, in a development named after the corporation.

Despite his role in a $440-million write-down, Spalding was considered untouchable by other BCE employees and analysts who watched the conglomerate.[184] However, one week before Christmas in 1990 he "elected to take early retirement" after nineteen years with BCE and Bell Canada.[185] Not surprisingly, no one thanked him for his efforts in the company news release; BCE had been forced to take a second write-down that year because an investment that he was even closer to than BCED had soured.

BCE took a $224-million write-down in the third quarter, ending September 30, 1990, against a total of $415 million of loans to Kinburn Technology Corporation of Ottawa and equity investment in its computer system integration subsidiary, SHL Systemhouse Incorporated.

Spalding had convinced BCE to loan the money to Kinburn and to acquire a stake in Systemhouse in 1988. Together, both Spalding-related write-downs totaled $644 million.

The Kinburn write-down was prompted after its owner and founder, Roderick Bryden, announced in April 1990 that he could not meet the deadline for payments on a total of $831 million in loans. BCE ranked second in line for the $415 million owed it after a syndicate of financial institutions led by the Royal Bank of Canada, which is owed $345 million. Other creditors are owed the balance.

"That was Mr. Cyr's and Mr. Spalding's deal," de Grandpré says. "That wasn't mine. I had nothing to do with that one."[186]

The first deal between BCE and Kinburn actually had its roots in Cyr's creation of Bell Canada Management Corporation in early 1985 — the small holding company for all of Bell Canada's mobile communications businesses. But there was more to BCMC than cellular radio. The corporate reorganization not only provoked envy at Bell Canada, because many of its investments were transferred out of its control, but also created a void. Cyr believed the phone company needed an entity to either acquire or spin off businesses. "I hope I've shown you that the day is not far off when people will say: 'You work at Bell? What do you do?' No one will automatically assume that you work with telephones," Cyr said in a 1986 speech.

Cyr spent $125 million and bought seven subsidiaries for his new holding company in 1985. One of those first deals was with real estate giant Campeau Corporation to buy Computer Innovations Distribution Incorporated (CIDI), the parent company of the ComputerLand personal computer retail chain, in July 1985. BCMC paid $35.1 million for almost 49 per cent of the franchise chain, which had thirty-five stores in Canada.[187] With personal computers becoming the hot high-tech item in the mid-1980s, Cyr believed the company would be a good investment for Bell Canada's new holding company. Computer Innovations ended up under BCE when BCMC was sold to Bell Canada's parent in 1987. As a general rule, however, de Grandpré and BCE were opposed to investments in cyclical businesses, which prompted the conglomerate to seek a buyer for the computer retail chain in 1988.

Meanwhile, Bryden was busy building a high-tech conglomerate of his own in the halcyon days of Silicon Valley North. The ex-lawyer, bureaucrat and political fund-raiser had built his empire around two major businesses: software and paper. He had also acquired Public Affairs International in 1980 and Decima Research in 1981.

Bryden wanted to become a $3-billion conglomerate by the end of the Nineties and, to that end, he ploughed cash into acquisitions. When he sought to increase his shareholdings in both Systemhouse and Paperboard Industries Corporation in 1988, he approached Spalding. He also eyed one of BCE's cardboard printing companies and had his sights on the ComputerLand franchises. "Here was a large, well-financed company in technology that also owned a paper business," Bryden says. "I though we had complementary interests."[188]

Bryden had known Spalding informally since 1983 when they met on the ski slopes at Mont Tremblant in the Laurentians, north of Montreal, where both businessmen had chalets. The two went skiing in the early winter of 1988, and Spalding agreed to discuss Bryden's proposal further at a later date and to take the proposal up with de Grandpré. The two companies struck a deal in late March 1988. BCE would inject $263 million in cash into Kinburn in return for a five-year option to acquire a 49 per cent interest in Bryden's holding company. Bryden also landed Computer Innovations and Rolph Clark Stone Packaging Corporation in the deal. The acquisition of CIDI in August 1988 ended up costing Systemhouse $160 million, and BCE invested an additional $319 million in notes in Kinburn. It was that deal that led Bryden to seek a $300-million loan with the bank syndicate to repay some short-term financing.

BCE invested another $51 million in Systemhouse, when it acquired a 12.4 per cent interest in the company in September 1989, because of a promise to purchase a major part of a $117-million rights offering.[189] Kinburn then owed BCE more than $400 million. But mounting losses and a problem in marketing an office automation system made by one of its high-technology businesses ate up another $100 million. That meant Bryden would not be able to meet his April 1990 loan payment deadlines unless he could sell a company or two. When a deal fell through, he announced he would default on the loans. Both de Grandpré and Bryden, in separate interviews, suggest that BCE could recover its investment and loans by acquiring a larger stake in Systemhouse, Kinburn's flagship subsidiary. However, BCE ranks behind the secured creditors' call on Bryden's assets.

Charles Sirois and the creation of BCE Mobile showed that entrepreneurship can indeed by nurtured by BCE. Yet the conglomerate has shown only a limited interest in the high-tech sector. Although it created a new venture capital unit in 1988, it confined its mandate to telecommunications-related areas. That meant BCE steered away from other leading high-tech companies, like laser-maker Lumonics

Incorporated, which was acquired by Sumitomo Heavy Industries of Japan in 1989. One could argue that laser technology would have been closer to the strengths of the BCE group than real estate.

But de Grandpré responds to criticisms, like Meisel's view that BCE should have done more in acquiring and developing high-tech companies, by saying that he was deterred by government. "The position of Bell and Northern in high tech and telecommunications is so big in Canadian terms that every time I was looking at a similar company in a similar field in Canada, I was always deterred by combines.

"I was coming out of sixteen years of hearings and investigation by combines and the Restrictive Trade Practices Commission. Once you're burnt, you're reluctant to do more in the area. If we didn't move in that direction more, the government has to take the blame." De Grandpré adds he cannot recall a single instance when Bell or BCE was approached by the federal government to support or encourage an acquisition that, from a standpoint of public policy, was believed to be in the public interest.

"They never, never talked to me, or us, in those terms. And then they tried to limit us, or said they had to limit us."[190]

Epilogue

De Grandpré boasted at BCE's zenith that the company he would leave behind would be completely different from the one he took over in 1973. It was. But the BCE since his retirement in 1989 has also been transformed as his successor, Raymond Cyr, endeavors to undo much of de Grandpré's expansion and refocuses the conglomerate on its core telecommunications business.

Many economists, financiers and academics now question the efficacy of holding companies after the acquisition binge of the 1980's. Given the checkered performances recorded by these companies, there has been an ongoing debate about whether mergers and sheer size are of value or are a quintessential myth of corporate culture, as distinguished economics professor Walter Adams believes.[1]

"In the Eighties, conglomerates were the name of the game," Pierre Laurin, senior vice-president of Merrill Lynch says. "In BCE's case, it was a pioneering move. Now, investors don't want to have the mix. They want to see what they get."[2]

In the face of the BCED disaster, which wiped out the profit from several well-performing companies, even Cyr said shareholders "may question the whole notion of a diversified holding company, or retain some nostalgia for earlier, more simple times."[3] In the rush to assess de Grandpré's record, however, there has been a tendency to both focus on the short-term performance of the corporation and to tar every acquisition with the same brush. A caricature in a leading Montreal newspaper illustrated BCE's below-average share price growth and diminished return on equity by portraying de Grandpré, who now holds the title chairman emeritus, riding a mule down the graph.[4] Yet another feature article termed all the moves into energy, real estate, publishing and computers "misbegotten." In Cyr's view, though, BCE's business has never been simple and the creation of BCE was absolutely essential just to cope with the profound changes in telecommunications. Robert Scrivener agrees. He believes it was essential to build upon the vision that Bell Canada's corporate leadership had for Northern Telecom.[5]

"You cannot have a time-frame of just one or two years,"[6] says de Grandpré. The difference of assessing corporate performance year-by-year, compared with looking at five-, or ten-year, periods is comparable to the difference between a snapshot and a panoramic view. BCE's year-by-year return on equity between 1980 and 1990 has had a roller-coaster ride with ups, as well as downs. Although shareholders earned a return on their equity of almost 11 per cent in 1990, compared with 9.5 per cent in 1980, profit has increased five-fold, to $1.1 billion from $273 million in the same period. And, despite the BCED wipeout and the sale of TRAP, BCE's assets have quadrupled, to almost $42 billion in 1990 from slightly more than $11 billion ten years earlier.[7]

"I can't take all the credit, but I shouldn't take all the blame,"[8] de Grandpré says, referring to both BCE's acquisitions and its performance.

There are critics, both shareholders and opponents of Bell Canada, who believe the reorganization produced nothing of intrinsic value. But to those on the inside and observers who believe that Bell Canada and Northern Telecom had to freely invest profit to take the companies abroad, it was a prerequisite to ensuring that BCE's core business can compete globally as telecommunications becomes a more strategic element in both the productivity and wealth of nations.

Charles Dalfen compared de Grandpré to William S. Paley, the legendary founder of the U.S. broadcasting conglomerate CBS Inc. "I did some work for Paley and I saw the same feelings of loathing and contempt for regulation held by the chairman of CBS that I saw in de Grandpré. But there is an incredible irony there because, for all his bluster and outbursts, Paley did really poorly in his forays into unregulated businesses. . . . The same is true for de Grandpré. His unregulated ventures have not performed as well for shareholders as his regulated ventures. It's a case of 'the Lady doth protest too much.' The ultimate irony, however, is that regulation may be responsible for their success. And that, said Caesar, is the unkindest cut of all."[9]

The new generation of leaders who will follow Cyr to the helm of BCE and are commanding Bell Canada, Northern Telecom and BNR, have been handed a legacy that has been enriched and strengthened by the decisions made by de Grandpré and his peers. Their styles of leadership will differ markedly, just as the circumstances in which they now, and will later, find themselves is radically different. But they probably face a challenge far more profound than that of de Grandpré's generation, for they must not only try to orchestrate an encore to Northern Telecom's recent successes, but they must try to keep the BCE group together in the face of accelerating competitive pressures that are tearing at the fabric of the telecommunications industry and its monopolies worldwide.

De Grandpré believes his own legacy was to build an organization with a vision to be, and to dream, big. "If you're a small architect, you'll always build a small building."[10]

Endnotes

INTRODUCTION

[1] See: Enchin, Harvey. "De Grandpré quits as CEO but remains BCE chairman," *The Globe and Mail, Report on Business.* May 4, 1988, B1.

[2] Clark, Gerald. *Montreal: The New Cité.* (Toronto: McClelland & Stewart, 1982), p. 115.

[3] de Grandpré, A. Jean. Interview with author. January 25, 1990.

[4] Fox, Francis. Hon. Interview with author. February 8, 1990.

[5] Ibid.

CHAPTER 1

[1] Mosca, Gaetano. "Elementi di Scienza Politica" (1896), trans. Hannah D. Kahn. In: *The Ruling Class*, ed. Arthur Livingston (New York: McGraw-Hill, 1939), p. 50.

[2] de Grandpré, A. Jean. Interview with author. January 25, 1990.

[3] de Grandpré, Louis-Philippe. Interview with author. February 7, 1990.

Details on family ancestry from notes in interviewee's personal files.

[4] de Grandpré, A. Jean. Interview with author. January 25, 1990.

[5] *Lovell's Montreal Directory*, 1912. National Library of Canada.

[6] de Grandpré, A. Jean. Interview with author. January 25, 1990.

[7] Dussault, Roger. Interview with author. December 3, 1990.

[8] de Lorimier, François Thomas Chevalier. "Testament politique de Lorimier écrit la veille de son exécution." Prison de Montréal, February 14, 1839. Original at Archives nationales du Québec, Quebec City. ANQ-Q, E17/37, no. 2971 (copies in files P1000-49-976 and 66-1317). See: Michel de Lorimier, "François Thomas Chevalier de Lorimier." In: *Dictionary of Canadian Biography.* Vol. 7 (Toronto: University of Toronto Press, 1988), pp. 512-15.

[9] See: Carr, Edward Hallett. *The Twenty Years' Crisis 1919-1939:*

An Introduction to the Study of International Relations (London: MacMillan, 1939).

[10] de Grandpré, Louis-Philippe. Interview with author. February 7, 1990.

[11] de Grandpré, A Jean. Interview with author. January 25, 1990.

[12] See: Johnson, J. K., ed., *The Canadian Directory of Parliament 1867-1967.* Public Archives of Canada (Ottawa: Queen's Printer, 1968), p. 560.

[13] de Grandpré, A. Jean. Interview with author. January 25, 1990.

[14] Ibid.

[15] Ibid.

[16] Ibid.

[17] Ibid.

[18] Ibid.

[19] de Grandpré, Louis-Philippe. Interview with author. February 7, 1990.

[20] de Grandpré, A. Jean. Interview with author. January 25, 1990.

[21] Cited in: Salter, Michael. "The Emperor of BCE: The Power and the Profit." *The Globe and Mail, Report on Business.* April 1988, p. 35.

[22] de Grandpré, Louis-Philippe. Interview with author. February 7, 1990.

[23] Dussault, Roger. Interview with author. December 3, 1990.

[24] de Grandpré, A. Jean. Interview with author. January 25, 1990.

[25] de Grandpré, Louis-Philippe. Interview with author. February 7, 1990.

[26] Dussault, Roger. Interview with author. December 3, 1990.

[27] de Grandpré, Louis-Philippe. Interview with author. February 7, 1990.

[28] de Grandpré, A. Jean. Interview with author. January 25, 1990.

[29] de Grandpré, Louis-Philippe. Interview with author. February 7, 1990.

[30] Rolland, Roger. "Remarques sur l'éducation et la culture canadienne-française." *Cité libre,* November 8, 1953, pp. 40-41. Cited in: Michael Behiels. *Prelude to Québec's Quiet Revolution: Liberalism versus Neo-nationalism 1945-1960* (Montreal: McGill-Queen's University Press, 1985), p. 166.

[31] Figures cited in: Behiels, p. 162.

[32] See: Briefs from the Confédération des travailleurs catholiques du Canada and Collège Jean-de-Brébeuf to Québec Commission royale d'enquête sur les problèmes constitutionnels, 1956. Cited in: Behiels, p. 163.

[33] The Canadian Press. "Shortage of Priests Puts Jesuit College in Laymen's Hands." In: *The Globe and Mail.* December 29, 1986.

[34] Clement, Wallace. *The Canadian Corporate Elite: An Analysis of Economic Power* (Toronto: McClelland & Stewart, 1975), p. 244.

[35] Collège Jean-de-Brébeuf. Alumni Association (1940). Class list.

[36] Off-the-record interview.

[37] de Grandpré, A. Jean. Interview with author. January 25, 1990.

[38] Ibid.

[39] Iglauer, Edith. "Profile of Pierre Trudeau." *The New Yorker.* July 5, 1969, p. 42.

[40] Bandeen, Robert. Interview with author. January 8, 1990.

[41] Dussault, Roger. Interview with author. December 3, 1990.

[42] de Grandpré, A. Jean. Interview with author. February 7, 1990.

[43] Cited in: Bantey, Bill. "Jean de Grandpré, le president de Bell Canada, un homme pas complique," *Actualité* October 1973, p. 20.

[44] See: Ibid.

[45] See: McGill University Archives. *Old McGill '43* (published by the undergraduates of McGill University, Montreal, 1943), p. 107.

[46] de Grandpré, A. Jean. Interview with author. February 7, 1990.

[47] de Grandpré, A. Jean. Interview with author. January 25, 1990.

[48] de Grandpré, A. Jean. Interview with author. February 7, 1990.

[49] Ibid.

[50] Ibid.

[51] Ibid.

[52] de Grandpré, Louis-Philippe. Interview with author. February 7, 1990.

[53] Ibid.

[54] See: Carroll, Campbell. "F. Philippe Brais Is Named MLC and Council Government Leader." *The Globe and Mail.* February 9, 1940; "Business Profile: F. Philippe Brais." *The Globe and Mail.* July 19, 1958.

[55] See: Thomson, Dale, C. *Louis St. Laurent : Canadian* (Toronto: Macmillan, 1967), p. 8.

[56] Vaillancourt, Louise Brais. Interview with author. February 20, 1990.

[57] See: Newman, Peter. *Bronfman Dynasty: The Rothschilds of the New World* (Toronto: McClelland & Stewart), 1978.

[58] de Grandpré, A. Jean. Interview with author. February 7, 1990.

[59] de Grandpré, Louis-Philippe. Interview with author. February 7, 1990.

[60] Vaillancourt, Louise Brais. Interview with author. February 20, 1990.

[61] Ibid.

[62] de Grandpré, A. Jean. Interview with author. February 7, 1990.

[63] de Grandpré, Louis-Philippe. Interview with author. February 7, 1990.

[64] de Grandpré, A. Jean. Interview with author. February 7, 1990.

[65] de Grandpré, Louis-Philippe. Interview with author. February 7, 1990.

[66] de Grandpré, A. Jean. Interview with author. February 7, 1990.

[67] See: McGill University Archives. *Old McGill '43* (published by the undergraduates of McGill University, Montreal, 1943), p. 107.

[68] de Grandpré, A. Jean. Interview with author. February 7, 1990.

[69] Beatty, Edward, Sir. In: *Old McGill '43* (February 25, 1943), p. 11.

[70] McGill University Annual Convocation Program. May 26, 1943. McGill University Archives.

[71] de Grandpré, A. Jean. Interview with author. February 7, 1990.

[72] Ibid.

[73] Vaillancourt, Louise Brais. Interview with author. February 20, 1990.

[74] de Grandpré, A. Jean. Interview with author. February 7, 1990.

[75] Ibid.

[76] Ibid.

[77] See: Pickersgill, J. W. *My Years with Louis St. Laurent: A Political Memoir* (Toronto: University of Toronto Press, 1975), p. 30.

[78] Canada. Privy Council Office. Order-In-Council P.C. 6444, October 6, 1945. In: Justice Robert Taschereau and Justice R. L. Kellock, Report (Ottawa : King's Printer, June 27, 1946).

[79] Cited in: Whitaker, Reg. "Spy Story: Lifting Gouzenko's Cloak." *The Globe and Mail.* November 6, 1984, p. 7. See also: Statement by

Mackenzie King. In: Canada.
Parliament. House of Commons.
Debates. 20th Parliament, 2nd
Session (10 George VI). Vol. 1,
1946 (Ottawa: King's Printer,
March 18, 1946), p. 49.

[80] Ibid.

[81] Pickersgill, J. W. *My Years with
Louis St. Laurent: A Political
Memoir* (Toronto: University of
Toronto Press, 1975), p. 31.

[82] Canada. Privy Council Office.
Order-in-Council PC 411-1946.
February 5, 1946.

[83] See: "Telegram 19, Secretary of
State for External Affairs to
Ambassador in Soviet Union,"
DEA/50242-40, Top Secret,
February 15, 1946. In: Canada.
Department External Affairs.
*Documents on Canadian External
Relations,* ed. Donald Page. Vol.
12 (1946). (Ottawa: Supply and
Services, 1977). Document 1246,
p. 2041. [Hereafter referred to as:
Department External Affairs.
Documents.]

[84] See: Letter from Commission
Counsel to the Minister of Justice.
February 23, 1946. In: Taschereau-
Kellock Report, p. 8.

[85] First Interim Report. March 2, 1946.
Appendix 1. In: Taschereau-
Kellock Report, pp. 693-96.

[86] See: Canada. House of Commons.
Debates. March 15, 1946, p. 5.

[87] Cited in: Ibid.

[88] Ibid., p.6.

[89] Ibid.

[90] See: "Text of Rose Warrant," *The
Globe and Mail.* March 16, 1946.

[91] See: "Commons Member Arrested."
The Globe and Mail. March 15,
1946.

[92] de Grandpre, A. Jean. Interview with
author. February 7, 1990.

[93] "Prima-Facie Case Against Rose;
Boyer Admits RDX Disclosure."
The Globe and Mail. March 27, 1946.

[94] "Non-Suit Is Refused Defense in
Rose Trial; Crown Case
Concluded." *The Globe and Mail.*
June 14, 1946.

[95] See: *The Globe and Mail.* June 17,
1946.

[96] For detailed descriptions of both the
treasures and the affair, see:
Lorenz, Stanislaw. "Canada
Refuses to Return Polish Cultural
Treasures." Polish Research and
Information Office. New York,
1950. Monograph. And Aloysius
Balawyder, *The Odyssey of the
Polish Treasures* (Antigonish, N.S.:
St. Francis Xavier University
Press, 1979).

[97] See: French, William. "Plodding
Style Spoils Tangled Tale of
Treasure." *The Globe and Mail.*
March 13, 1979.

[98] See: "Polish Art Treasures in
Canada." Memorandum from
Assistant Under-Secretary of State
for External Affairs to Acting
Under-Secretary, DEA/837-40
(May 16, 1946). In: DEA.
Documents, no. 1230, pp. 2021-22.

[99] See: Bethell, Nicholas. *Gomulka: His
Poland, His Communism* (New
York: Holt Rinehart & Winston,
1969), p. 105.

[100] Memorandum by Head. First
Political Division, DEA/837-40
(December 26, 1946). In: DEA.
Documents, no. 1239, pp. 2032-
33. Confidential.

[101] de Grandpre, A. Jean. Interview
with author. February 7, 1990.

[102] See: *The Globe and Mail.* October 31, 1947.

[103] See: "Ottawa Seeking 8 Missing Trunks of Polish Art." *The Globe and Mail.* March 8, 1948.

[104] See: *The Globe and Mail.* March 5, 1948.

[105] See: Baldwin, Warren. "Duplessis Would Smear Minister, St. Laurent Says." *The Globe and Mail.* March 5, 1948.

[106] de Grandpre, A. Jean. Interview with author. February 7, 1990.

[107] de Grandpré, A. Jean. Interview with author. January 25, 1990.

[108] Fitzgerald, Edmund. B. Interview with author. July 6, 1990.

[109] de Grandpré, A. Jean. Interview with author. February 7, 1990.

[110] See: Dame Marie Léontine Thériault v. H. Huctwith et al. In: Canada. Supreme Court. *Law Reports [1948 SCR].* (Ottawa: King's Printer, 1949), pp. 86-100.

[111] de Grandpré, Jean-François. Interview with author. December 3, 1990.

[112] de Grandpré, A. Jean. Interview with author. February 7, 1990.

[113] Ibid.

[114] de Grandpré, Louis-Philippe. Interview with author. February 7, 1990.

[115] de Grandpré, A. Jean. Interview with author. February 7, 1990.

[116] Ibid.

[117] Ibid.

[118] de Grandpré, Louis-Philippe. Interview with author. February 7, 1990.

[119] O'Donnell, J. V. Interview with author. April 10, 1990.

[120] de Grandpré, A. Jean. Interview with author. February 7, 1990.

[121] See: Canada. Privy Council Office. Order-In-Council P.C. 1955-1796 (December 2, 1955). In: Canada. Royal Commission on Broadcasting. *Report* (Ottawa: Queen's Printer, March 15, 1957), pp. 293-94. [Hereafter cited as Fowler Report.]

[122] de Grandpré, A. Jean. Interview with author. February 7, 1990.

[123] Ibid.

[124] See: Fowler Report, p. 22.

[125] de Grandpré, A. Jean. Interview with author. February 7, 1990.

[126] See: Canada. Statutes of Canada, 1958. c. 22. Broadcasting Act. Ottawa, Queen's Printer. For a discussion of the Fowler Commission, see: Marc Raboy, *Missed Opportunities: The Story of Canada's Broadcasting Policy* (Montreal and Kingston: McGill-Queen's University Press, 1990), pp. 119-36.

[127] de Grandpré, A. Jean. Interview with author. February 7, 1990.

[128] See: Young, Brian, and John Dickinson. *A Short History of Quebec: A Socio-Economic Perspective* (Toronto: Copp Clark Pitman, 1988).

[129] Guindon, Hubert. "The Modernization of Quebec and the Legitimacy of the Canadian State." In: Daniel Glenday et al., eds. *Modernization and the Canadian State* (Toronto: Macmillan, 1978), p. 214.

[130] Figures from: Posgate, Dale, and Kenneth McRoberts. *Quebec: Social Change and Political Crisis* (Toronto: McClelland & Stewart, 1976), p. 145.

[131] de Grandpré, A. Jean. Interview with author. February 7, 1990.

132 Ibid.

133 O'Donnell, J. V. Interview with author. April 10, 1990.

134 See: Canada. Royal Commission on Bilingualism and Biculturalism. Book III: The Work World (Ottawa: Queen's Printer, 1969). Report. And André Raynauld et al. "La répartition des revenus selon les groupes ethniques au Canada." Submitted to the Royal Commission on Bilingualism and Biculturalism. Research Report.

135 de Grandpré, A. Jean. Interview with author. February 7, 1990.

136 Porter, John. *The Vertical Mosaic: An Analysis of Social Class and Power in Canada* (Toronto: The University of Toronto Press, 1965), p. 171.

137 de Grandpré, A. Jean. Interview with author. February 7, 1990.

138 Ibid.

139 Canada. Royal Commission on Bilingualism and Biculturalism. A Preliminary Report of the Royal Commission on Bilingualism and Biculturalism (Ottawa: Queen's Printer, February 1, 1965), p. 79.

140 See: Canada. Report of an Industrial Inquiry Commission Concerning Matters relating to the Disruption of Shipping on the Great Lakes, the St. Lawrence River System and Connecting Waters (Ottawa: Queen's Printer, July 1963).

141 Fox, Francis, Hon. Interview with author. February 8, 1990.

142 Off-the-record interview.

143 O'Donnell, J. V. Interview with author. April 10, 1990.

CHAPTER 2

1 de Grandpré, A. Jean. Interview with author. February 7, 1990.

2 de Grandpré, Louis-Philippe. Interview with author. February 7, 1990.

3 de Grandpré, A. Jean. Interview with author. February 7, 1990.

4 O'Donnell, J. V. Interview with author. April 10, 1990.

5 See: *The Merck Manual of Diagnosis and Therapy.* 15th ed. (Rahway, N.J.: Merck & Co. Inc., 1987), pp. 1430-31.

6 de Grandpré, A. Jean. Interview with author. February 7, 1990.

7 See: Public Archives of Canada. RG 46. Vol. 846, case 1116, 1965 Bell Telephone Rate Case. The Board of Transport Commissioners for Canada. Notice of Hearing. Case 955.176 (September 22, 1964). Mimeo.

8 de Grandpré, A. Jean. Interview with author. February 7, 1990.

9 See: "'That's What It's For.' Reply to Bell's Stand Surplus Being Used Up." *The Globe and Mail.* March 9, 1950.

10 See: McManus, John. "Federal Regulation of Telecommunications in Canada." In: H. Edward English, ed. *Telecommunications for Canada* (Toronto: Methuen, 1973), pp. 389-428.

11 See: Public Archives of Canada. RG 46. Vol. 846, case 1114, part 1 and 2, 1965 Bell Telephone Employee Stock Plan. The Board of Transport Commissioners for Canada (March 23-24, 1965). Transcripts of Hearing.

12 See: "Bell, NE Scrutinized at Board Hearing." *The Globe and Mail,* May 14, 1965.

[13] Scrivener, R. C. Interview with author. March 11, 1990.

[14] For an historical analysis of the development of regulation in Canada see: Christopher Armstrong and H. V. Nelles. *Monopoly's Moment: The Organization and Regulation of Canadian Utilities,* 1830-1930 (Toronto: University of Toronto Press, 1988).

[15] See: Bell Canada. Evidence of Orland Tropea. "Telecommunications Regulation in Canada." Filed with the Restrictive Trade Practices Commission (RTPC). Exhibit T-1586 (Ottawa, May 1980).

[16] Southwestern Bell Telephone Company v. Public Service Commission of Missouri. In: United States. Supreme Court of the United States of America. 262 U.S. Reports (1922), p. 291.

[17] Bluefield Waterworks and Improvement Company v. Public Service Commission of the State of West Virginia. United States. Supreme Court of the United States of America. 262 U.S. Reports (1922), p. 692.

[18] de Grandpré, A. Jean. Interview with author. February 7, 1990.

[19] See: Bell Canada. Evidence of Orland Tropea. "Telecommunications Regulation in Canada." Filed with the Restrictive Trade Practices Commission (RTPC). Exhibit T-1586 (Ottawa, May 1980).

[20] de Grandpré, A. Jean. Interview with author. February 7, 1990.

[21] Cited in: Public Archives of Canada. RG 46. Vol. 847, case 1120, part 20. *1965 Bell Telephone Rate Review.* The Board of Transport Commissioners for Canada.

[22] Ibid, p. 3927.

[23] Scrivener, R. C. Interview with author. March 11, 1990.

[24] Figures cited in: Board of Transport Commissioners of Canada. *1966 Judgment,* p. 629.

[25] Cited in: Public Archives of Canada. RG 46. Vol. 847, case 1120, part 20. Transcripts. Vol. 20, p. 3923.

[26] Ibid., p. 3930.

[27] Cited in: Public Archives of Canada. RG 46. Vol. 847, case 1119, part 19. Transcripts. Vol. 19 (June 17, 1965), p. 3806.

[28] Cited in: Public Archives of Canada. RG 46. Vol. 847, case 1120, part 20. Transcripts. Vol. 20, p. 3898.

[29] Cited in: Canada. House of Commons. Debates. 27th Parliament, 2nd Session, 1967-68. Standing Committee on Transport and Communications. Minutes of Proceedings and Evidence No. 14. (Thursday, February 1, 1968). (Ottawa: Queen's Printer), p. 462.

[30] Board of Transport Commissioners. *1966 Judgment,* p. 731.

[31] de Grandpré, A. Jean. Interview with author. February 7, 1990.

[32] Ibid.

[33] Scott, Bruce. Interview with author. January 18, 1990.

[34] Robertson, H. Rocke, Dr. Interview with author. January 23, 1990.

[35] de Grandpré Jean-François. Interview with author. December 3, 1990.

[36] Ibid.

[37] de Grandpré, A. Jean. Interview with author. February 20, 1990.

[38] de Grandpré, A. Jean. Interview with author. February 7, 1990.

Transcripts of Hearings. Vol. 20 (June 18, 1965), pp. 3928-29.

[39] Robertson, H. Rocke, Dr. Interview with author. January 23, 1990.

[40] de Grandpré, A. Jean. Interview with author. February 20, 1990.

[41] de Grandpré, Louis-Philippe. Interview with author. February 7, 1990.

[42] Scrivener, R. C. Interview with author. March 11, 1990.

[43] Ibid.

[44] de Grandpré, A. Jean. Interview with author. February 7, 1990.

[45] Vaillancourt, Louise Brais. Interview with author. February 20, 1990.

[46] de Grandpré. Interview with author. February 7, 1990.

[47] See, for example: Bell Canada. "Material Relating to Section 320 (11) of the Railway Act." Bell Canada legal department. Mimeos.

CHAPTER 3

[1] Cited in: Abbott, Charles, ed, "Telegraphy and Telephony," *Great Inventions*, (New York: Smithsonian Institution Series Inc., 1944), The Smithsonian Series, Vol. 12, pp. 100-01. See also: Robert Bruce, *Bell: Alexander Graham Bell and the Conquest of Solitude* (Ithaca, N.Y.: Cornell University Press, 1990), pp. 139-41.

[2] See: Bell Canada, "Alexander Graham Bell," (Montreal: March , 1956), p. 18 (monograph). The most thorough description of Bell's work is in Bruce, Ibid., pp. 143-50.

[3] Bell Canada, Ibid., pp. 19-21 and: Bruce, Ibid., pp. 146-47.

[4] See: Bruce, Ibid., pp. 157-58.

[5] Ibid., p. 158. The story of Bell's dealings with Brown is told from Bell's perspective in Bruce, pp. 158-66, and from Brown's perspective in: J.M.S. Careless, *Brown of the Globe* (Toronto: The Macmillan Co., 1963), Vol. 2: Statesman of Confederation, 1860-1880, pp. 343-45.

[6] See: Bruce, Ibid., p. 160.

[7] Public Archives Canada. *George Brown Papers*, MG 24, B40. Vol. 10, Reel C-1603, p. 2398, George to Anne Brown, December 30, 1875.

[8] See: Careless, *Brown*, p. 344.

[9] Public Archives of Canada. *George Brown Papers*, MG 24, B40. Vol. 10, Reel C-1603, p. 2416, George to Anne Brown, February 16, 1876.

[10] Bruce, *Bell*, p. 166.

[11] Ibid., pp. 167-68.

[12] Bell Canada. "Bell," pp. 22-23.

[13] See: Bruce, *Bell*, p. 231.

[14] AT&T Canada Inc., "The History of AT&T in Canada" (mimeo), no date.

[15] See: Orland Tropea, Evidence.

[16] AT&T Canada, "History."

[17] See: Public Archives of Canada. *Sise Papers*. MG 30, D187, Vol. 2, File 22. "First Telephone Lease Made in Canada," October 18, 1877.

[18] See: Schlesinger Leonard, et al., *Chronicles of Corporate Change: Management Lessons from AT&T and its Offspring* (Lexington Books, 1987), p. 5.

[19] Armstrong and Nelles, *Monopoly's Moment*, pp. 66-67.

[20] See: AT&T Canada, "History," and: Robert Collins, *A Voice From Afar: The History of Telecommunications in Canada* (Toronto: McGraw-Hill Ryerson Limited, 1977), pp. 70-74.

[21] See: Bruce, *Bell*, pp. 271-77 and: American Bell Telephone

Company, *The Deposition of Alexander Graham Bell in the Suit Brought by the United States to Annul the Bell Patents [1892],* (New York: 1908). The deposition was published in full by the company because of its historical value. A copy can be found at Public Archives of Canada — National Library, Ottawa.

[22] See: Bruce, Ibid., pp. 258-60.

[23] See: Stone, Alan. *Wrong Number: The Breakup of AT&T* (New York: Basic Books Inc., 1989), p. 41. For a general biography on Vail, see: Albert B. Paine, *In One Man's Life* (New York: Harper & Row, 1921).

[24] Cited in: Josephson, Matthew. *The Robber Barons: The Great American Capitalists, 1861-1901* (New York: Harcourt, Brace, 1934), p. 87.

[25] See: Bruce, *Bell,* pp. 267-71.

[26] Alexander Melville Bell to Alexander Graham Bell, Nov. 14, 1877, cited in: Bruce, Ibid., p. 282.

[27] See: Armstrong and Nelles, *Monopoly's Moment,* pp. 67-68 and: Bruce, Ibid., p. 282.

[28] According to Confederate records, Sise fought at the Battle of Shiloh with Company B of the New Orleans Crescent regiment. See: U.S. National Archives. *Records of Confederate Soldiers.* Official Roll of the Crescent Regiment, microfilm (at New Orleans Public Library) and published in: Anthony Booth, *Records of Louisiana Confederate Soldiers* (New Orleans: 1920). See also: R. C. Fetherstonhaugh, *Charles Fleetford Sise 1834-1918* (Montreal: Gazette Printing Co. Ltd., 1944), private printing, pp. 54-55.

[29] See: Fetherstonhaugh, Ibid., pp. 58-72.

[30] See: Log of the Annie Sise, in: Public Archives of Canada. *Sise Papers.* MG 30, D187. Vol. 2, File 29.

[31] See: Fetherstonhaugh, *C. F. Sise,* p. 53.

[32] Canada. The Canada Gazette (Part I). Vol. XIII: No. 33 (Feb. 14, 1880), p. 1140.

[33] Canada. Statutes of Canada. 1880, Vol. I: Public Acts (43 Victoria), Second Session-Fourth Parliament, C. 67 (Ottawa: Queen's Printer, 1880), pp. 100-01. See also: Canada. Parliament. House of Commons. Debates. Second Session-Fourth Parliament (43 Victoria 1880), Vol. VIII (February 23, 2880) (Ottawa: Queen's Printer, 1880), p. 151.

[34] Bell Canada Archives, *Correspondance,* Box 2755-22, W.H. Forbes to C.F. Sise, March 8, 1880.

[35] Ibid.

[36] Cited in: Schlesinger, et al., *Chronicles of Corporate Change,* p. 6.

[37] Cited in: Ibid., p. 6.

[38] Canada. House of Commons Debates, 1880 - Vol. VIII, op. cit (March 12, 1880), pp. 624-25.

[39] Canada. Parliament. The Senate. Debates. Second Session- Fourth Parliament (43 Victoria 1880). April 8, 1880 (Ottawa: Queen's Printer, 1880), pp. 271-72.

[40] Senate Debates, Ibid., p. 270.

[41] Bell Canada Archives. Forbes to Sise, March 8, 1880.

[42] Bell Canada Archives. *C. F. Sise Letterbook 1880, No. 1.* This episode is also described briefly in Armstrong and Nelles, *Monopoly's Moment,* footnote 32, p. 342; and in: Feth-erstonhaugh, *C. F. Sise,* pp. 114-16.

[43] Bell Canada Archives. Ibid., Sise to W.H. Forbes, March 13, 1880.

[44] Ibid, Sise to Forbes, #25, pp. 441-46, May 11, 1880, and: Sise to Forbes, #28, pp. 459-61, May 15, 1880.

45 Bell Canada Archives, *Sise Correspondance,* Forbes to Sise, No. 1.

46 Bell Canada Archives, *Sise Letterbook 1880, No. 1,* Sise to Forbes, March 11, 1880.

47 Ibid, Sise to Forbes, #33, p. 498, June 2, 1880. Biographical information from: Armstrong and Nelles, *Monopoly's Moment,* p. 69 and: *Lovell's Montreal Directory,* 1879-80, 1880-81.

48 Ibid, Sise to Forbes.

49 Ibid., Sise to Forbes, #12, p. 398, March 31, 1880.

50 Ibid., Sise to Forbes, #28, p. 462.

51 Ibid., Sise to Forbes, Draft of letter dated July 4, 1880, pp. 3-4. [Written on stationary of Mt. Mansfield Hotel, Stowe, Vt.]

52 See: Bell Canada Archives, *Sise Correspondance,* Forbes to Sise, July 8, 1880 and July 13, 1880.

53 Sise was a staunch defender of the interests of his subsidiary and spoke out with characteristic aplomb and sarcastic wit at any perceived slight of the Canadian subsidiary by the Boston-based parent, as happened after he received the report from the American Bell Telephone Co.'s 1884 annual meeting that indicated "that you do not receive all the dividends which we remit to your company. The report says: 'The same is true for Canada. . .and 6% dividends are paid upon its capital of $1,000,000.' We sent you two dividends of 3-1/2% each; did 1% miscarry in the mails? If so, I will see the P.O. [post office] people about it. This is not quite so bad as last years' report, wherein you are quoted as stating that we 'would in all probability soon pay a dividend'

- and at that moment your treasury was overflowing with the $39,775.32 of dividends which we had sent you." From: Bell Canada Archives, *Sise Letterbook 1884,* Sise to W.R. Driver, March 26, 1884.

54 See: Public Archives of Canada. *Sise Papers,* MG 30. D 187, vol. 2, File 27, C.F. Sise Log Book, No. 1, 1880-1891, pp. 1-2.

55 Bell Canada. Annual Reports of the Bell Telephone Company of Canada, Dec. 31, 1880. reproduced in: Canada. Parliament. House of Commons. Select Committee on Telephone Systems. Vol. 1: Minutes of Proceedings, Exhibit No. 93 (Ottawa: King's Printer, 1905), pp. 404-05.

56 AT&T Canada, "History of AT&T in Canada."

57 See: Fetherstonhaugh, *C. F. Sise,* p. 160.

58 Canada, Statutes of Canada 1882, Vol. II: Local and Private Acts (44 Victoria), Fourth Session-Fourth Parliament, C. 95 (Ottawa: Queen's Printer, 1882), section 2.

59 Smith, *Anatomy of a Business Strategy,* p. 96.

60 Ibid., pp. 111-20. Text of Agreement reprinted as Appendix E, pp. 175-82.

61 Cited in: Ibid., p. 103.

62 Chandler, Alfred. *The Visible Hand: The Managerial Revolution in American Business* (Cambridge Harvard University Press, 1977).

63 Armstrong and Nelles, *Monopoly's Moment.* p. 73.

64 See: Ibid., footnote, p. 343 and: Bell Canada Archives, *Sise Correspondance,* 1885, Wood to Sise, Jan. 19, 20, 26, 1885; and: *Sise Letterbook 1882,* Sise to H.P. Dwight, April 1, 1882.

[65] Armstrong and Nelles, Ibid., pp. 107-8.

[66] Bell Canada Archives. People's Telephone, *Special Agency Letterbook 1885,* p. 2, Sise to Wainwright (Confidential), Nov. 10, 1885.

[67] Bell Canada Archives. *C.F. Sise Letterbook 1886,* p. 441, Sise to F.G. Walsh, April 20, 1886.

[68] Armstrong and Nelles, *Monopoly's Moment,* p. 111.

[69] Bell Canada Archives, *C.F. Sise Letterbook 1891,* Sise to J. E. Hudson, January 20, 1891.

[70] Ibid., p. 215, Sise to Hudson, June 19, 1891.

[71] Ibid., pp. 212-13, Sise to C.R. Hosmer, June 19, 1891.

[72] Ibid., pp. 208-09, Sise to Hudson, June 17, 1891.

[73] Armstrong and Nelles, *Monopoly's Moment,* p. 115.

[74] Bell Canada Archives, *Sise Letterbook 1891,* Sise to J. E. Hudson, p. 149, Feb. 10, 1891; and: Sise to Irving Evans, p. 179, March 26, 1891.

[75] Bell Canada. "Background Information: Bell Canada Ownership," 1983. (Mimeo).

[76] Canada. Parliament. Senate - Debates of the Senate 1892. (55-56 Victoria) Second Session - Seventh Parliament, May 10, 1892 (Ottawa: Queen's Printer, 1892), p. 206. See also: Canada. Statutes of Canada (55-56 Victoria), Second Session- Seventh Parliament, Vol. 1: Public General Acts, c. 67 (Ottawa: Queen's Printer, 1982), p. 101.

[77] Cited in: Schlesinger et. al., *Chronicles of Corporate Change,* p. 7.

[78] See: Chernow, Ron. *The House of Morgan: An American Banking Dynasty and the Rise of Modern Finance* (New York: Atlantic Monthly Press, 1990), pp. 82-86. Although the steel trust is still hotly debated by historians, Chernow writes that a subsequent U.S. government valuation of U.S. Steel at only half of its $1.4-billion selling price suggests "investors had purchased an enormous bag of hope, at least half it hot air," giving credence to the view that Morgan's biggest deal was a giant scam.

[79] Statistics from: Chernow, Ibid., p. 81.

[80] Cited in: Kennedy, David M. ed., *Progressivism: The Critical Issues* (Boston: Little, Brown and Co., 1971), p. viii.

[81] Armstrong and Nelles, *Monopoly's Moment,* p. 144.

[82] Ibid., pp. 165-66.

[83] Ibid., p. 166.

[84] See: Canada. Statutes of Canada 1902. (2 Edward VII), chapter 41 (Ottawa: King's Printer, 1902), pp. 13-14.

[85] Public Archives of Canada. *Sir Wilfrid Laurier Papers.* MG 26-G, Vol. 41, Reel C-748, pp. 13436-37, (words unclear) Laurier, March 28, 1897.

[86] Bell Canada Archives, *Sise Letterbook 1883,* p. 120, Sise to Francis Scott, April 14, 1883.

[87] Bell Canada Archives, *Sise Letterbook 1891,* p. 143, Sise to Hugh Baker February 4, 1891.

[88] See: Fetherstonhaugh, *C. F. Sise,* pp. 174, 176.

[89] Public Archives of Canada. *Sir Wilfrid Laurier Papers.* MG 26, G-2, Vol. 247, p. 68747 Minister of Justice, Memorandum in reference to the Telegraph and telephone Bill House of Commons during the session of 1903 (no date) p. 22.

[90] Great Britain, Treasury Minute, May 23, 1892, cited in: Public Archives of Canada, Ibid., p. 68747 (p. 22).

[91] Ibid., p. 68767 (p. 42).

[92] Public Archives of Canada, *Sir Wilfrid Laurier Papers*, MG 26, G. Vol. 323, reel C-813, pp. 86825-6. Sir William Mulock to Sir Wilfrid Laurier (Private), June 15, 1904.

[93] Ibid., MG 26, G, Vol. 337, Reel C-816, p. 90162. The Board of Trade of the City of Toronto to Wilfrid Laurier. September 29, 1904.

[94] Canada. Parliament. House of Commons. Parliament Debates, First Session-Tenth 1905 March 17, 1905, (Ottawa: King's Printer, 1905), p. 2681.

[95] For details on Belcourt, see: Armstrong and Nelles, *Monopoly's Moment*, p. 170. See also: Bell Canada Archives, *Sise Logbook No. 20*, March 14, 1905.

[96] Canada. Parliament. House of Commons. Select Committee, Vol. 1, p. 533

[97] See: Armstrong and Nelles, *Monopoly's Moment*, p. 170.

[98] Canada. *Select Committee*, p. 533; and cited in: Armstrong and Nelles, Ibid, p.171.

[99] Great Britain. Parliament. House of Lords - Judicial Committee of Privy Council, City of Toronto and Bell Telephone Company of Canada. [1905] A.C., pp. 52-65; See Peter Hogg *Constitutional Law of Canada* (Toronto; The Carswell Co. Ltd., 1977), p. 325.

[100] See: Garnet, R.W. *The Telephone Enterprise: The Evolution of the Bell System's Horizontal Structure, 1876-1909* (Baltimore: The Johns Hopkins University Press, 1985) and: R. Bornhotz and D.S. Evans, "The Early History of Competition in the Telephone Industry," In: D.S. Evans ed., *Breaking Up Bell: Essays on Industrial Organization and regulation* (New York: North Holland) and Amstrong and Nelles, *Monopoly's Moment*, pp. 171-72.

[101] Romaniuk, Bohdan and Hudson Janisch, "Competition in Telecommunications: Who Policies the Transition?' *Ottawa Law Review*, V. 18: No. 3, 1986, p. 571.

[102] Canada. Statutes of Canada 1906, Second Session - Tenth Parliament, Vol. 1, Public General Acts, chapter 61, (Ottawa: King's Printer, 1906), p. 21; Canada. Railway Act 1888, Chapter R-2 (Ottawa: Supply and Services, 1970).

[103] Bell Canada Archives, *AT&T Letterbook*, C.F. Sise to F.P. Fish, June 27, 1906

[104] For views on the capture theory, see: Kohlmeier Lewis, *The Regulators* (New York, Harper & Row 1969,) and: Thomas McCraw, "Regulation in America: A Review Article," *Business History Review* 49 (1975) pp. 159-83 and: Paul Sabatier, "Social Movements and Regulatory Agencies: Toward a More Adequate and Less Pessimistic — Theory of 'Clientele Capture'," *Policy Sciences*, vol. 6, 1975, p. 301-42; and: Q. Wilson, ed., *The Politics of Regulation* (New York: Basic Books, 1980).

[105] See: Chernow, *The House of Morgan*, p. 153.

[106] See: Armstrong and Nelles, *Monopoly's Moment*, p. 183.

[107] See: *Who was Who in America: A Companion Volume to Who's Who In America* (Chicago: The A.N. Marquis Co., 1943), p. 1266.

[108] Figures in: Chernow, *The House of Morgan*, p. 152.

[109] From: Canada. Department of Labour. *Report of the Royal Commission on Hours of Employment between the Bell Telephone Company of Canada Ltd. and Operators at Toronto Ont.* (Ottawa: 1907); cited in: Armstrong and Nelles, *Monopoly's Moment* pp. 217-18.

[110] Bell Canada Historical Collection (Ottawa). Letter from Katherine M. Schmit (no date).

[111] Armstrong and Nelles, *Monopoly's Moment*, p. 179.

[112] See: Canada, Parliament. House of Commons. Standing Committee on Railways, Canals and Telegraph Lines. 1947-48. Minutes of Proceeding and Evidence, Bill No. 8, An Act Respecting The Bell Telephone Co. of Canada, Appendix 1: Brief on Behalf of the Bell Telephone Company of Canada, p. 67.

[113] Cited in: Ramsay Cook, "The Triumph and Trials of Materialism 1900-1945," in: Craig Brown, ed., *The Illustrated History of Canada* (Toronto: Lester & Orpen Dennys Ltd., 1987), p. 393.

[114] Armstrong and Nelles, *Monopoly's Moment*, p. 182.

[115] Alberta. Legislative Assembly, Debates, 1907, Session, February 14, 1907 (Edmonton: King's Printer, 1907).

[116] Figures from: Armstrong and Nelles, *Monopoly's Moment*, p. 184. See also: Bell Canada Archives. *AT&T Letterbook*, Sise to T.N. Vail, Oct. 3 & 7, 1907; Manitoba. Sessional Papers, 1908, No. 4. pp. 356-64; Alberta. Sessional Papers, 1909. No. 12:

Bell Canada Archives, *Minute Book,* April 14, 1909; and J. Mavor, *Government Telephones: The Experience of Manitoba, Canada,* (Toronto: MacLean Publishing Co., 1917).

[117] AT&T Canada, "AT&T in Canada."

[118] Cited in: Romaniuk and Janisch, *Ottawa Law Review*, p. 568.

[119] H.M. Boettinger, *The Telephone Book: Bell, Watson and American Life, 1876-1983* (New York; Stern, 1983), pp. 167-68.

[120] Temin, Peter with Galambos, Louis *The Fall of the Bell System: A Study in Prices and Politics.* (Cambridge: Cambridge University Press, 1987), pp. 9-10.

[121] Von Auw Alvin, *Heritage & Destiny: Reflections on the Bell System in Transition* (New York: Praeger, 1983), note 16 at p. 68. Originally cited in: AT&T, *Annual Report, 1907,* p. 18.

[122] Von Auw, Ibid., p. 70 and Original citation in: AT&T, *Annual Report, 1910.* p. 32.

[123] Strictly speaking, a natural monopoly is a firm that is a sole supplier of a good or service for which there is no close substitute. Under this economic definition, there are few natural monopolies. While Bell obtained "monopoly power" by capturing such a large segment of the market in which it operates, it did not — and still does not — meet the test of being a "sole supplier."

A firm has "monopoly power" when it is insulated from competitive pressures and can therefore act with impunity. Economists define monopoly power as the ability to raise prices above competitive levels or to market inferior products while excluding competition. For an economic analysis of competition and monopoly, see: Franklin Fisher et. al., *Folded, Spindled and Mutilated: Economic Analysis and U.S. v. IBM* (Cambridge The MIT Press, 1983). pp. 19-41. Mabee cited in: Armstrong and Nelles, *Monopoly's Moment*, p. 203. Original citation in: Canada. Board of Railway Commissioners, Seventh Annual Report, Ottawa: 1912, pp. 289-93.

[124] Cited in: Armstrong and Nelles, Ibid., p. 203.

[125] See: Ibid., pp. 289-90.

[126] See: Bell Canada. *Evidence of Orland Tropea*, p. 6 and: Armstrong and Nelles, Ibid., pp. 272-76.

[127] Canada. Statutes of Canada 1920 (10-11 George V), Fourth Session-Thirteenth Parliament, Vol. 1: Public General Acts, chapter 100 (Ottawa: King's Printer, 1920), pp. 97-98.

[128] Armstrong and Nelles, *Monopoly's Moment*, pp. 276-77.

[129] Bell Canada. *Evidence of Orland Tropea*, pp. 6-8; Armstrong and Nelles; Ibid., p. 277; and The Board of Railway Commissioners for Canada. *Judgments, Orders, Regulations and Rulings*, Vol. XI, No. 2, Ottawa, April 1, 1921.

[130] Armstrong and Nelles Ibid., pp. 277-79; and: *Board of Railway Commissioners. Judgments. Orders, Regulations and Rulings*, Vol. XI, Ottawa: 1922, pp. 440-59.

[131] J. Derek Davies, "Lest We Forget: A paper dealing with the changes in the structure of the telecommunication industry in the United States and the associated

trade balance." Northern Telecom Ltd., (Unpublished paper), January 1986, p. 2.

[132] Details on amalgamation from: Canada. Department of Justice. Combines Investigation Commission. *Electrical Wire and Products, Report of H. Carl Goldenberg*, Special Commissioner, Ottawa: November 13, 1953, p. 2; and: Canada. The Board of Transport Commissioners for Canada. *Judgments, Orders, Regulations and Rulings*, Application of Industrial Wire and Cable Company, File No. 36730.4., Pamphlet No. 4, Ottawa: January 13, 1964, pp. 51 and 53. For a synopsis of the Bell-Northern relationship, see: Northern Telecom. *Submission to the Restrictive trade Practices Commission on behalf of Northern Telecom Ltd.*, Exhibit T-1990 B, A-2, Cassels. Brock. Toronto, July 17, 1981, pp. 82-85.

[133] Temin, *The Fall of the Bell System*, p. 13.

[134] Cited in: Canada Restrictive Trade Practices Commission, *Telecommunications in Canada, Part III: The Impact of Vertical Integration on the Equipment Industry* (Ottawa: Supply and Services, 1983). p. 85.

[135] See: Ogle, E.B. *Long Distance Please: The Story of the TransCanada Telephone System* (Toronto: Collins, 1979), pp. 44-92.

[136] Bell Canada. *Evidence of G.E. Inns*, Bell Canada Operations. Filed with the Restrictive Trade Practices Commission, Exhibit T-1422. *Ottawa Law Review*. p. 572. Ottawa: February, 1980.

137 See: Galbraith, J.K. *The New Industrial State* (3rd edition), (Boston: Houghton Mifflin, 1978), p. 390. Cited in: Romaniuk and Janisch.

138 Irwin, Manley R. *Telecommunications America: Markets Without Boundaries* (Westport, Conn.: Quorum Books, 1984), note 24 at p. 28. Cited in: Romaniuk and Janisch, Ibid., note 33, p. 571.

139 For details on the 1956 consent decree, see: Stone. *Wrong Number,* pp. 59-80; and *U. S. v. Western Electric Co. and American Telephone and Telegraph Co.* Civil Action No. 17-49. U. S. District Court, District of New Jersey. Final Judgment of Thomas Meaney, U.S. District Judge. January 24, 1956.

140 Davies, J. Derek. Interview with author. July 6, 1990.

141 See: Temin, *The Fall of the Bell System,* p. 15; and *U.S. v. Western Electric Co. and American Telephone & Telegraph Co.* Civil Action No. 17-49. U.S. District Court, District of New Jersey. Complaint. January 14, 1949.

142 See: Stone. *Wrong Number,* pp. 61-66 and 71-73; and, Temin. Ibid., p. 14.

143 Sandia Labs, located adjacent to the U.S. Air Force Special Weapons Center at Kirtland Air Force Base, New Mexico, is now operated under contract to the U.S. Department of Energy, which is responsible for all aspects of the U.S. nuclear weapons program. For a brief description of Sandia's role in the program, see: Samuel Day Jr. "The Nuclear Weapons Labs." *The Bulletin of the Atomic Scientists.* Vol. 33, no. 4 (April

1977); and, M. D. Fagen, ed. *National Service in War and in the Bell System.* Vol. 2 (New York: Bell Telephone Laboratories Inc., 1978), pp. 650-72.

144 See: *Temin. The Fall of the Bell System,* p. 15.

145 Ibid., p. 15; and, U.S. Attorney General Herbert Brownell, quoted in AT&T memo. March 3, 1954. In: United States. Congress. House of Representatives. Committee on the Judiciary. 86th Congress, 1st Session. *Report of Antitrust Subcommittee on Consent Decree Program of the Department of Justice.* January 30, 1959, pp. 53-54.

146 See: Stone. *Wrong Number,* p. 77; and, Franklin Fisher et al. *Folded, Spindled, and Mutilated: Economic Analysis and U.S. v. IBM.* A Charles River Associates Study. MIT Press Series on the Regulation of Economic Activity (Cambridge, Mass.: The MIT Press, 1983).

147 Northern Telecom. *Submission to RTPC.* Exhibit T-1990 B, pp. 88-89.

148 Davies, J. Derek. Interview with author. July 6, 1990.

149 See: U. S. Securities and Exchange Commission. *Bell Canada, Form 10-K,* 1985. Exhibit 10-A. "Memorandum of Agreement, November 3, 1939, between Northern Electric Co. Ltd. and The Bell Telephone Co. of Canada," pp. 431-58.

150 Scrivener, R. C. Interview with author. March 11, 1990.

151 See: Canada. Department of Justice. *Goldenberg Report;* and, Canadian Press. "Ten Wire, Cable Firms Accused of Price-Fixing." *The Globe and Mail.* November 25, 1953.

[152] Schlesinger et al. *Chronicles of Corporate Change*, p. 19.

[153] Bell Canada. Evidence of J. A. Harvey. Technology Development. Filed with the Restrictive Trade Practices Commission. Exhibit T-1421, p. 102 (Ottawa: January, 1980).

[154] Benger, Walter, C. Interview with author. March 20, 1990; and, John Elliott. "BNR: Past, Present, and Future." *Telesis*. 1989: Three.

[155] Parkhill, Douglas. Interview with author. May 26, 1987; and see: L. Surtees. "Bell -Northern Owes Prowess to Missile Race." *The Globe and Mail*, Focus. June 20, 1987, p. D2.

[156] See: Goodspeed, D. J., Capt. *A History of the Defence Research Board of Canada* (Ottawa: Queen's Printer, 1958), p. 203.

[157] Northern Telecom. *Submission to RTPC*. Exhibit T-1990 B, p. 90.

[158] Benger, Walter. Interview with author. March 20, 1990.

[159] Scrivener, R. C. Interview with author. March 11, 1990.

[160] Ibid.

[161] Ibid.

[162] Ibid.

[163] Ibid.

[164] MacPherson, W. C., and V. O. Marquez. "Report on Bell Canada-Northern Electric Relations." Cited in: Northern Telecom. *Submission to RTPC*. Appendix 1. "Analysis of Green Book Summary and Conclusions," p. 19. [Originally prepared August 8, 1977.]

[165] Northern Telecom. *Submission to RTPC*. Ibid. Appendix 1, p. 18.

[166] Scrivener, R. C. "Memo." May 30, 1966. Serial 4195. Cited in:

Northern Telecom. *Submission to RTPC*. Ibid., p. 22.

[167] Notes of discussion between W. MacPherson and Martin (Northern). Sept. 22, 1965. Serial 4570. Cited in: Northern Telecom. *Submission to RTPC*. Ibid., p. 22.

[168] Harvard Report. March 11, 1966. Serial 4211-4240. Cited in: Northern Telecom. *Submission to RTPC*. Ibid., p. 21.

[169] Marquez, Vernon, to Holly Keefler. June 21, 1965. Serial 2508-2509. Cited in: Northern Telecom. *Submission to RTPC*. Ibid., p. 20.

[170] Scrivener, R. C. Interview with author. March 11, 1990.

[171] See: Harvard Report. Cited in: Northern Telecom. *Submission to RTPC*, p. 19 (i) and (ii).

[172] Inns, G. E. Evidence to RTPC, p. 85.

[173] Northern Telecom. *Submission to RTPC*. Exhibit T-1990 B, p. 19.

CHAPTER 4

[1] de Grandpré, A. Jean. Interview with author. February 20, 1990.

[2] Canada. Supreme Court. *Law Reports [1966]*. SCR. Commission du Salaire Minimum v. Bell Telephone Co. of Canada (Ottawa: Queen's Printer, 1967), pp. 767-77.

[3] Ibid., p. 774.

[4] Hogg, Peter. *Constitutional Law of Canada*, p. 308. Toronto: The Carswell Company Ltd., 1977

[5] Canada. Department of Consumer and Corporate Affairs. Director of Investigation and Research. *Report and Statement of Evidence Submitted to the Restrictive Trade Practices Commission, in the Matter*

of a General Inquiry Under Section 47 of the Combines Investigation Act Relating to the manufacture, Production, Distribution, Purchase, Supply and Sale of Communication Systems, Communications Equipment and Related Products (Ottawa: December 20, 1976), p. 3. [Hereafter cited as "The Green Book."]

[6] de Grandpré, A. Jean. Interview with author. February 7, 1990.

[7] Scrivener, R. C. Interview with author. March 11, 1990.

[8] Memo to A. Jean de Grandpré. December 1966. Bell Canada Archives; and, S. Sykes. Communication with author. October 25, 1990.

[9] DIR. "The Green Book," pp. 3-4.

[10] Scrivener, R. C. Interview with author. March 11, 1990.

[11] For a description of the Laurentian, see: Peter Newman. *The Canadian Establishment.* Vol. 1 (Toronto: McClelland & Stewart-Bantam, 1977), p. 220.

[12] Scrivener, R. C. Interview with author. March 11, 1990.

[13] See: Collins. *History of Telecommunications,* p. 173; and George Linton. "Bell Plans Takeover of Northern Phone. Offers $14 a Share." *The Globe and Mail.* June 10, 1966.

[14] Scrivener, R. C. Interview with author. March 11, 1990.

[15] Cited in: Cruise, David, and Alison Griffiths. *Lords of the Line: The Men Who Built the CPR* (Toronto: Viking, 1988), p. 400.

[16] Northern Electric. Memorandum. June 10, 1966. Serial 3163-3164. Cited in: Northern Telecom.

Submission to RTPC. Appendix 1, p. 7, paragraph (i).

[17] See: Québec-Téléphone. Annual Report 1986 (Rimouski, Quebec: 1987), p. 2; and, Collins. *History of Telecommunications,* p. 177-78.

[18] Vincent, Marcel. Cited in: Langevin Cote. *The Globe and Mail.* August 19, 1966.

[19] Scrivener, R. C. Interview with author. March 11, 1990.

[20] For terms of offer for MT&T. See: Cote. *The Globe and Mail;* and for date of Northern Telephone acquisition, see: Letter from Gagnon (Northern Electric) to A. Lester (Bell). August 16, 1966. Serial 3128. Cited in: Northern Telecom. *Submission to RTPC.* Appendix 1, p. 6.

[21] Scrivener, R. C. Interview with author. March 11, 1990.

[22] Bell Canada. "News Release and Statement by Marcel Vincent, president." August 18, 1966. Bell Canada Archives. Montreal.

[23] RTPC. *Phase I Report,* pp. 2-3.

[24] Nova Scotia. Office of the Premier. "Statement by Hon. Robert Stanfield." August 22, 1966. Cited in: *The Globe and Mail.* August 23, 1966.

[25] As reported by: The Canadian Press. *The Globe and Mail.* September 2, 1966.

[26] de Grandpré, A. Jean. Interview with author. February 7, 1990.

[27] See: Nova Scotia. Statutes, 1966 (Second Session). 48th General Assembly, 4th Session (15 Elizabeth II), c. 5. "An Act to Amend Chapter 156 of the Acts of 1910, An Act to Incorporate the Maritime Telegraph and Telephone

Co. Ltd." (Halifax: Queen's Printer, 1967), pp. 6-10; and, Howard Windsor and Peter Aucoin. "The Regulation of Telephone Service in Nova Scotia." In: G. Bruce Doern, ed. *The Regulatory Process in Canada* (Toronto: Macmillan, 1978), p. 241.

[28] Reported by: The Canadian Press. In: *The Globe and Mail.* September 10, 1966.

[29] de Grandpré, A. Jean. Interview with author. February 7, 1990.

[30] See: The Canadian Press. In: *The Globe and Mail.* September 12, 1966.

[31] Cited in: The Canadian Press. "Hints Bell is Preparing For Fight." *The Globe and Mail.* September 13, 1966. Original transcripts of hearing at Public Archives of Canada. RG 46. Vol. 848. Case 1201. *Transcripts.* Vol. 1 (September 12, 1966).

[32] Canada. The Board of Transport Commissioners for Canada. *Judgements, Orders, Regulations and Rulings.* Order No. 122017 (September 14, 1966). File No. 36730.2. Pamphlet No. 27 (September, 1966). (Ottawa: Queen's Printer), pp. 1213-14.

[33] See: Gillan, Michael. "Commons to Investigate Bell." *The Globe and Mail.* June 14, 1967.

[34] Cited in: Canada. Parliament. House of Commons. 27th Parliament, 2nd Session. Standing Committee on Transport and Communications. Minutes of Proceedings and Evidence. No. 14 (February 1, 1968). (Ottawa: Queen's Printer), p. 459.

[35] For a clause-by-clause summary of Bill C-104, see: Appendix A to

Bell's testimony. In: Canada. Parliament. Senate. 27th Parliament, 2nd Session. Proceedings of the Standing Committee on Transport and Communications. No. 7. "Complete Proceedings on Bill C-104." March 6, 1968 (Ottawa: Queen's Printer), p. 70.

[36] See: Canada. *Statutes* 1929 (19-20 George V), c. 93. "An Act respecting the Bell Telephone Company of Canada" (Ottawa: King's Printer, 1929), p. 18. While Parliament maintained the power to set a statutory limit on Bell's share capital under the 1929 amendment, sub-section 1(2) declared Bell could not "make any issue, sale or other disposition of its capital stock, or any part thereof, without first obtaining the approval of the Board of Railway Commissioners."

[37] Canada. House Committee Proceedings (1968), p. 462.

[38] de Grandpré, A. Jean. Interview with author. February 7, 1990.

[39] Canada. Parliament. 27th Parliament, 2nd Session. Standing Committee on Transport and Communications. Minutes of Proceedings and Evidence. No. 12 (December 7, 1967). (Ottawa: Queen's Printer), p. 389.

[40] Vaillancourt, Louise Brais. Interview with author. February 20, 1990.

[41] Canada. House Committee Proceedings. Issue No. 10 (November 30, 1967).

[42] Canada. House Committee Proceedings (1968), p. 464.

[43] de Grandpré, A. Jean. Interview with author. February 7, 1990.

[44] For dates and details of that conference, see: Canada. Canadian Intergovernmental Conference Secretariat. *The Constitutional Review 1968-1971: Secretary's Report* (Ottawa: Information Canada, 1974); and, Canada. Canadian Intergovernmental Conference Secretariat. *Constitutional Conference Proceedings.* First Meeting. February 5-7, 1968 (Ottawa: Information Canada, 1968).

[45] Trudeau, Pierre E. Cited in: George Radwanski. *Trudeau* (Toronto: Signet, 1978), p. 92.

[46] See: Canada. House Committee Proceedings (1968), pp. 459, 491.

[47] See: Ibid. , pp. 466-67. Henry's testimony on that matter is in Issue No. 12. (Dec. 7, 1967), pp. 389, 416, 422.

[48] Ibid., p. 468.

[49] Ibid., p. 484.

[50] Ibid., p. 500.

[51] Ibid., pp. 501-2.

[52] Ibid., p. 483.

[53] See: Canada. *Statutes of Canada 1948,* 20th Parliament, 4th Session (11-12 George VI). Part I: Public General Acts, c. 81. "An Act respecting The Bell Telephone Company of Canada" (Ottawa: King's Printer, 1948), p. 13.

[54] Canada. Parliament. House of Commons. 12th Parliament, 4th Session. Standing Committee on Railways, Canals and Telegraph Lines. Minutes of Proceedings and Evidence. No. 3 (May 11, 1948). Bill No. 8: An Act Respecting The Bell Telephone Company of Canada (Ottawa: King's Printer, 1948), p. 61. See also: N. A. Munnoch, and D. K. MacTavish. "Brief on Behalf of the Bell Telephone Company of Canada." Appendix in: Ibid., p. 74.

[55] Canada. House Committee Proceedings (1968), p. 478.

[56] Ibid., p. 473.

[57] Canada. *Statutes of Canada 1968.* 27th Parliament, 2nd Session (16-17 Elizabeth II). Part I: Public General Acts, c.48. "An Act Respecting The Bell Telephone Company of Canada (Ottawa: Queen's Printer, 1968), p. 46.

[58] Canada. Parliament. Senate Committee Proceedings (1968), p. 65.

[59] Canada. House Committee Proceedings (1968), p. 490.

[60] Bell Canada . *Special Act, 1968.* Subsection 5(3).

[61] See: Letter to Standing Committee Re: Bill C-104, from B. Anthony Lawless, The Thorne Group and chairman, ratios committee of ADAPSO. December 1, 1967. Appendix A-11 in Proceedings (1968), pp. 503-4.

[62] Canada. House Committee Proceedings (1968), p. 462.

[63] See: "Memorandum on the History of Section 18." Prepared by Arthur Campeau, Ogilvy and Renault, counsel to Bell Canada. Re: "In the matter of a proposed arrangement concerning Bell Canada and certain of its shareholders under section 185.1 of the Canada Business Corporations Act, S.C. 1974-75, C.33." September 1982 [copy in author's files]; and, Quebec. Quebec Superior Court. District of Montreal. Bell Canada and Bell Canada Enterprises Inc. v. The Attorney-General of Canada.

Transcripts of Hearing. File No.
500-05-010452-827 (September 15,
1982), pp. 88-101.

[64] Cited in: Alderman, Tom. "How to
Earn a Quiet Scotch." *The
Canadian Magazine.* March 22,
1969, p.2.

[65] Kyles, James. Interview with author.
February 18, 1990.

[66] Light, Walter. Interview with author.
March 20, 1990.

[67] Kyles, James. Interview with author.
February 18, 1990.

[68] Scrivener, R. C. Interview with
author. March 11, 1990.

[69] See: Temin. *The Fall of the Bell
System.* Footnote, p. 44.

[70] Light, Walter. Interview with author.
March 20, 1990.

[71] Marquez, Vernon. Speech to the
Canadian Industrial Management
Association. Annual Convention.
Toronto, June 14, 1968. Cited in:
The Globe and Mail. June 15, 1968.

[72] Light, Walter. Interview with author.
March 20, 1990.

[73] Scrivener, R. C. Interview with
author. March 11, 1990.

[74] See: Harvey, J. A. *Evidence,* p. 104;
and, Northern Telecom.
Submission to RTPC, p. 90.

[75] Bell Canada. *Evidence of Alex Lester.*
History of Bell Canada and
Northern Telecom. Filed with
Restrictive Trade Practices
Commission. Exhibit T-1420.
January 1980, p. 59.

[76] Ibid., p. 44.

[77] Marquez, Vernon, to Holly Keefler,
president Northern Electric.
Memorandum. Filed with RTPC,
serial 2507 (June 21, 1965). Cited
in: Northern Telecom. *Submission
to RTPC.* Appendix 1, p. 20.

[78] Canada. Department of Consumer
and Corporate Affairs. Director of
Investigation and Research, "The
Green Book," p. 117; and,
Northern Telecom. Ibid., p. 18.

[79] See: George, D. A., and S. T.
Nichols. "Telecommunications
Technology." In: H. Edward
English, ed., *Telecommunications
for Canada* (Toronto: Methuen
Publications, 1973), p. 266; and,
Bell Canada. *Evidence of J. A.
Harvey,* pp. 32-33.

[80] Harvey. Ibid., p. 34.

[81] Ibid., p. 37.

[82] See: George and Nichols.
"Telecommunications
Technology," p. 266.

[83] Harvey. *Evidence,* p. 37.

[84] See: George and Nichols.
"Telecommunications
Technology," p. 267; and,
Northern Telecom. *Submission to
RTPC,* p. 44.

[85] Harvey. *Evidence,* p. 38.

[86] Benger, Walter. Interview with
author. March 20, 1990.

[87] Leenders, Michiel. "Study D: Cases
on the SP-1, SL-1 and DMS-100
Family of products." School of
Business Administration,
University of Western Ontario.
August 31, 1981. Attachment 3 to:
Bell Canada. Submission to the
Canadian Radio-television and
Telecommunications Commission.
*Bell Canada-Northern Telecom
Price Comparison.* CRTC Telecom
Public Notice 1981-18 (September
1, 1981), p. 66.

[88] Figures cited in: Leenders. Ibid.,
pp. 72-73.

[89] See: Ibid., p. 92; and Harvey. *Evidence,*
pp. 114-15.

[90] See: Millman, S., ed. *Communication Sciences (1925-1980): A History of Engineering and Science in the Bell System.* Vol. 5 (New York: AT&T Bell Laboratories, 1984), p. 359; and, Stone. *Wrong Number,* p. 201.

[91] For full details of "Project X," see: Fagen, M. D., ed. Service in War and Peace, pp. 296-317.

[92] See: "Companies in the News." *The Globe and Mail, Report on Business.* March 22, 1969.

[93] See: Webster, H. L. "The Digital World of Systems and Technology." Paper presented to Pacific Telecommunications Council. '83 Conference. Honolulu. January 19, 1983, in: Bell Canada. *The Digital World.* Ottawa: 1983.

[94] See: *The Globe and Mail, Report on Business.* June 29, 1971.

[95] Cited in: Harvey. *Evidence,* p. 132; and, Leenders. *Pricing Study,* p. 92.

[96] Cited in: Leenders. Ibid., p. 93.

[97] Ibid., p. 94.

[98] Canada. Department of Regional and Industrial Expansion. "Microsystems International: An Historical Sketch," 1985. Unpublished mimeo.

[99] Ibid., p. 2.

[100] Britton, John, and James Gilmour. *The Weakest Link: A Technological Perspective on Canadian Industrial Underdevelopment.* Science Council of Canada. Background Study 43 (Ottawa: Supply and Services, 1978), p. 101.

[101] Scrivener, R. C. Interview with author. March 11, 1990.

[102] Ibid.

[103] See: Wilensky, Harold L. *Organizational Intelligence: Knowledge and Policy in Government and Industry* (New York: Basic Books Inc., 1967), p. 45.

[104] Ibid., p. 45.

[105] de Grandpré, A. Jean. Interview with author. February 7, 1990.

[106] Off-the-record interview with author.

[107] See: Janisch, Hudson N. "The Canadian Transport Commission." In: G. Bruce Doern, ed. *The Regulatory Process in Canada* (Toronto: Macmillan, 1978), p. 168.

[108] Ibid., p. 200.

[109] See: Vandervort, Lucinda. *Political Control of Independent Administrative Agencies.* A Study Prepared for the Law Reform Commission of Canada. Administrative Law Series (Ottawa: 1979), pp. 33-43.

[110] Newman, Peter. "The Nobility of Sailor Jack's Last Career: Literary Chamberlain, History's Guardian." *The Globe and Mail.* January 17, 1976, p. A6.

[111] de Grandpré, A. Jean. Interview with author. February 7, 1990.

[112] See: Bell Canada. *Evidence of Orland Tropea,* p. 18.

[113] Canada. Parliament. House of Commons. Debates. 28th Parliament,1st Session (18 Elizabeth II). Vol. X (June 17, 1969), pp. 10244-45; June 18, 1969, p. 10311; and vol. XI (July 4, 1969), p. 10840 (Ottawa: Queen's Printer, 1969).

[114] See: Tropea. *Evidence,* p. 18; and, Canada Transport Commission. Railway Transport Committee. *Judgements, Orders, Regulations*

and Rulings. Vol. LIX (1969).
Decision, September 29, 1969
(Ottawa: Queen's Printer, 1970).

[115] CTC. Ibid., p. 33.

[116] de Grandpré, A. Jean. Interview with author. February 7, 1990.

[117] Scrivener, R. C. Interview with author. March 11, 1990.

[118] See: McManus, John. In: English, Telecommunications in Canada, pp. 410-11.

[119] Bell Canada. News Release. September 30, 1969.

[120] Kettle, John. Inside Bell, 1978, pp. 8-51. Montreal. Bell Canada Archives. Unpublished Manuscript.

[121] Ibid., pp. 8-53.

[122] Ibid., pp. 8-61.

[123] Ibid., pp. 8-58.

[124] Ibid., pp. 8-59.

[125] Ibid., pp. 8-59.

[126] Bell submitted rate increase applications to its federal regulator: 1970-73, 1975-76, 1978, 1980.

[127] Kettle. Inside Bell, pp. 8-61.

[128] See: Balfour, Clair. The Globe and Mail, Report on Business. March 21, 1969.

[129] Cited in: Braithwaite, Chris. The Globe and Mail, Report on Business. December 2, 1969.

[130] Ibid.

[131] Cited in: Kettle. Inside Bell, pp. 8-55.

[132] Gotlieb, Allan. Interview with author. January 12, 1990.

[133] Office of the Prime Minister. "Statement by the Prime Minister on Cabinet Committee Structure." April 30, 1968. Cited in: Gordon Robertson. "The Changing Role of the Privy Council Office." Canadian Public Administration. Vol. 14: No. 4 (Winter 1971), p. 6.

[134] Kierans, Eric. In: Debates, p. 6077.

[135] See: McLuhan, Marshall. Understanding Media: The Extensions of Man (New York: McGraw-Hill, 1964).

[136] Gotlieb, Allan. Interview with author. January 12, 1990.

[137] Pearson, L. B. Federalism for the Future: A Statement of Policy by the Government of Canada. The Constitutional Conference, 1968 (Ottawa: Queen's Printer, 1968), p. 38.

[138] See: Gwyn, Richard. The Northern Magus: Pierre Trudeau and Canadians (Toronto: McClelland & Stewart, 1980), p. 300.

[139] Ibid., p. 300.

[140] See: Swift, Jamie. Odd Man Out: The Life and Times of Eric Kierans (Toronto: Douglas & McIntyre, 1988), p. 209.

[141] Parkhill, Douglas. Interview with author. May 26, 1987.

[142] Gotlieb, Allan. Interview with author. January 12, 1990.

[143] Scrivener, R. C. Cited in: Alderman. "How to Earn a Quiet Scotch," p. 2.

[144] de Grandpré, A. Jean. Interview with author. February 7, 1990.

[145] See: Johnston, C. C. The Canadian Radio-television and Telecommunications Commission: A Study of Administrative Procedure in the CRTC. Study for the Law Reform Commission of Canada. Administrative Law Series (Ottawa: Supply and Services, 1980), p. 7.

[146] Gotlieb, Allan. Interview with author. January 12, 1990.

[147] Parkhill, Douglas. Interview with author. May 26, 1987.

[148] See: Radwanski. Trudeau, pp. 6-7.

[149] de Grandpré, Louis-Philippe. Interview with author. February 7, 1990.

[150] English, H. E. *Telecommunications for Canada,* p. viii.

[151] See: Canada. Department of Communications. *Instant World: A Report on Telecommunications in Canada* (Ottawa: Information Canada, 1971).

[152] Parkhill, Douglas. Interview with author. May 26, 1987; and see also: Organization for Economic Co-operation and Development. "Terms of Reference for OECD Panel on Policy Issues of Computer/Telecommunications Interaction." May 1970. In: OECD. Computers and Telecommunications. Vol. 3. OECD Informatics Studies (Paris: OECD, June 1973).

[153] See: Nora, Simon, and Alain Minc. *The Computerization of Society: A Report to the President of France* (Cambridge, Mass.: MIT Press, 1980).

[154] DOC. *Instant World,* p. 199.

[155] de Grandpré, A. Jean. Interview with author. February 7, 1990.

[156] DOC. *Instant World,* p. 233.

[157] Parkhill, Douglas. Interview with author. May 26, 1987.

[158] See: Canada. Department of Communications. *Branching Out: Report of the Canadian Computer/Communications Task Force.* 2 vols. (Ottawa: Information Canada, 1972).

[159] Canada. Department of Communications. *Computer/ Communications Policy: A Position Statement by the Government of Canada* (Ottawa:

Information Canada, 1973). See also: Douglas Parkhill, "The Necessary Structure." In: David Godfrey and Douglas Parkhill, eds. *Gutenberg Two: The New Electronics and Social Change* (Toronto: Porcépic, 1980), p. 81.

[160] Parkhill, Douglas. Interview with author. May 26, 1987.

[161] Kierans, Eric. In: Debates. November 10, 1969, p. 686.

[162] Canada. Privy Council Office. Order-in-Council. January 28, 1969. Cited in: Debates. Ibid., p. 688.

[163] Ibid., p. 688.

[164] Gotlieb expressed his concerns on those points in a May 4, 1969, letter to D. S. Maxwell, deputy attorney-general of Canada. Cited in: Robert E. Babe. *Telecommunications in Canada: Technology, Industry and Government* (Toronto: University of Toronto Press, 1990), fn. 8, p. 294.

[165] DOC. *Instant World,* p. 198.

[166] de Grandpré, A. Jean. Interview with author. February 7, 1990.

[167] de Grandpré, A. Jean. Interview with author. April 10, 1990.

[168] United States. Comptroller General. General Accounting Office. *Legislative and Regulatory Actions Needed to Deal with a Changing Domestic Telecommunications Industry.* Report to the Congress. CED-81-136 (September 24, 1981), p. 107.

[169] See: Kirton, John, ed. *Canada, the United States, and Space* (Toronto: Canadian Institute of International Affairs, 1986).

[170] Hartz, Theodore, and Irvine Paghis. *Spacebound* (Ottawa: Supply and Services, 1982), p. 16.

[171] See: Cox, Donald. *The Space Race: From Sputnik to Apollo...and Beyond* (New York: Chilton Books, 1962), pp. 19-30.

[172] Canada. Defence Research Board. Canadian Armament Research and Development Establishment. *An Outline of Some Characteristics of the Upper Atmosphere.* CARDE TR 470/64 (Valcartier, Quebec: March, 1964). Unclassified.

[173] For a comprehensive history of the program and texts of documents, see: Canada. Privy Council Office. Science Secretariat. *Upper Atmosphere and Space Programs in Canada* ("The Chapman Report"). Special Study No. 1, 3 vols. (Ottawa: Queen's Printer, 1967).

[174] Golden, David. Interview with author. May 28, 1987.

[175] See: Hartz and Paghis. *Spacebound*, p. 100.

[176] See: Canada. Interdepartmental Committee on Space. *Canada in Space* (Ottawa: n.d.), p. 100. Industry handbook.

[177] Hartz and Paghis. *Spacebound*, p. 18.

[178] U. S. National Aeronautics and Space Administration. "Orbits of Bodies in Space." Educational Brief, 1982. Mimeo.

[179] See: Northern Telecom Canada Ltd. "Belleville and NTC." Background Information, 1985, p. 4. Mimeo.

[180] "Chapman Report."

[181] Ibid. Vol. 1.

[182] See: Niagara Television Limited and Power Corporation of Canada Limited. "Proposal for a Canadian TV Network and a Domestic Space Satellite Distribution System."

Brief submitted to the Board of Broadcast Governors. Ottawa, October 25, 1966; and, Robert E. Babe. *Telecommunications in Canada*, p. 222.

[183] Salter, Liora. *Scientific Assessment and the Consideration of Introducing a New Technology: A Case Study of Satellite Policy.* Report prepared for the Science Council of Canada (Ottawa: Supply and Services, 1980), p. c-8.

[184] See: Parliament. House of Commons. 28th Parliament, 1st Session (1968-1969). Standing Committee on Broadcasting, Films and Assistance to the Arts. Minutes of Proceedings and Evidence. No. 36 (May 9, 1969). (Ottawa: Queen's Printer, 1969), pp. 1850-51.

[185] See: Salter, Liora, and Debra Slaco. *Public Inquiries in Canada.* Science Council of Canada. Background Study 47 (Ottawa: Supply and Services, 1981), pp. 135-36.

[186] Doern, G. Bruce, and James A. R. Brothers. "Telesat Canada." In: Allan Tupper and G. Bruce Doern, eds. *Public Corporations and Public Policy in Canada* (Montreal: The Institute for Research on Public Policy, 1981), p. 222; see also: Trans-Canada Telephone System and CN-CP Telecommunications, "Canadian Communications Satellite System." Brief submitted to the Minister of Transport" (Ottawa: May 1967).

[187] See: Doern and Brothers. Ibid., pp. 222-23.

[188] Babe. *Telecommunications in Canada*, p. 223; and, James

Brothers. "Telesat Canada: Pegasus or Trojan Horse? A Case Study of Mixed, Composite and Crown Enterprises." MA Thesis. Carleton University, Ottawa, 1979, pp. 3-32.

[189] Canada. Science Council of Canada. *A Space Program for Canada.* Report No. 1 (Ottawa: Queen's Printer, July 1967), pp. 5-10.

[190] Bobyn, E. J. Interview with author. May 27, 1987.

[191] Canada. Department of Industry. *White Paper on a Domestic Satellite Communication System for Canada* (Ottawa: Queen's Printer, 1968), p. 8.

[192] See: Doern and Brothers. "Telesat Canada," p. 224.

[193] *White Paper*, p. 46.

[194] See: "Three Companies Agree on Satellites Group." *The Globe and Mail,* January 19, 1968; and, *The Globe and Mail,* April 2, 1968.

[195] Gotlieb, Allan. Interview with author. January 12, 1990.

[196] Chapman, John. Cited in: Doern and Brothers. "Telesat Canada," p. 224.

[197] Canada. Parliament. House of Commons. Debates. 28th Parliament, 1st Session (1968-1969). April 14, 1969 (Ottawa: Queen's Printer, 1969), p. 7495.

[198] Telesat Canada. "Briefing Book" (Ottawa, August 1974), p. 9.

[199] Canada. Parliament. House of Commons. Standing Committee on Broadcasting, Films and Assistance to the Arts. Minutes of Proceedings and Evidence. No. 40 (May 27, 1969). (Ottawa: Queen's Printer, 1969), p. 2093; and Brothers. "Thesis," pp. 4-7.

[200] See: Doern and Brothers. "Telesat Canada," p. 225.

[201] Trans-Canada Telephone System and CN/CP Telecommunications. "Canadian Communications Satellite System: Considerations and Recommendations." Brief to the Minister of Transport (Ottawa: July 1968), p. 4.

[202] Canada. Parliament. House of Commons. Standing Committee on Broadcasting, Films and Assistance to the Arts. Minutes of Proceedings and Evidence. No. 38 (May 15, 1969). (Ottawa: Queen's Printer, 1969), p. 1987.

[203] Minutes of Proceedings. No. 37 (May 13, 1969). (Ottawa: Queen's Printer, 1969), p. 1912. Cited in: Babe. *Telecommunications in Canada,* p. 224.

[204] Gotlieb, Allan. Interview with author. January 12, 1990.

[205] See: Crane, David. "Telesat Canada is Threatened by Phone Companies." *The Globe and Mail.* May 7, 1969; and, Canada. Parliament. House of Commons. Standing Committee on Broadcasting, Films and Assistance to the Arts. Minutes of Proceedings and Evidence. No. 34 (May 6, 1969). (Ottawa: Queen's Printer, 1969).

[206] Swift. *Odd Man Out,* p. 231.

[207] Doern and Brothers. "Telesat Canada," p. 227.

[208] Canada. Parliament. House of Commons. Standing Committee on Broadcasting, Films and Assistance to the Arts. Minutes of Proceedings and Evidence. No. 41 (May 29, 1969). (Ottawa: Queen's Printer, 1969), p. 2122.

[209] See: Revised Statutes of Canada (17-18 Elizabeth II), (1968-1969), c. 51.

[210] See: Bain, George. "Former War POW Deputy Minister." *The Globe and Mail.* September 15, 1954.

[211] Telesat Canada. *Briefing Book,* p. 14; and, Doern and Brothers. "Telesat Canada," p. 228.

[212] Swift. *Odd Man Out,* p. 232.

[213] Gotlieb, Allan. Interview with author. January 12, 1990.

[214] Ibid; and, Swift. *Odd Man Out,* p. 232.

[215] Gotlieb, Allan. Interview with author. January 12, 1990.

[216] Figures from: Swift. *Odd Man Out,* p. 232.

[217] Ibid., p. 233.

[218] Telesat Canada. *Briefing Book,* p. 10.

[219] Telesat Canada. "Telesat Canada and Anik" (Ottawa: n.d.), p. 1. Brochure.

[220] Gotlieb, Allan. Interview with author. January 12, 1990.

[221] Northern Electric's first satellite study was commissioned by the federal Department of Transport. See: Northern Electric Co. Ltd. Research and Development Laboratories. *Satellite Communications in Canada,* 5 vols. (Ottawa: 1967); and the second study was for the Department of Industry. See: Northern Electric Co. Ltd. Aerospace Communications Laboratory. *Canadian Domestic Satellite Study.* Contract File IRA 9122-03-3 (Ottawa: 1968).

[222] See: Telesat Canada. "Spacecraft Program Requirement Specifications." Ottawa,

September 15, 1969. Mimeo; and: Brothers. "Thesis," pp. 8-19.

[223] Scrivener, R.C. Interview with author. March 11, 1990; and, Doern and Brothers. "Telesat Canada," p. 235.

[224] Fox, Francis, Hon. Minister of Communications. "A Governor in Council Review of Telecom Decision CRTC 81-13 with Respect to Telesat Canada." *Draft Memorandum to Cabinet* (Ottawa, October 30, 1981), p.17. Mimeo. Confidential.

[225] de Grandpré, A. Jean. "The Canadian Constitution and its Future." Address to the Insurance Bureau of Canada. Bell Canada Transcript (Toronto, September 15, 1977).

[226] de Grandpré, A. Jean. "Where are we going?" Address to the Chamber of Commerce of the District of Montreal. Bell Canada Transcript (February 15, 1977), pp. 11-12.

[227] de Grandpré, A. Jean. Interview with author. February 20, 1990.

[228] Canada. Federal Court of Canada. 13 Dominion Law Reports (3d), 1970. *Québec-Téléphone v. Bell Telephone Co. of Canada,* p. 200.

[229] de Grandpré, A. Jean. Interview with author. February 20, 1990.

[230] See: Canada. Canadian Transport Commission. *1972 Canadian Transport Cases.* File No. C955.181 (Ottawa: Queen's Printer, May 19, 1972), p. 207.

[231] Langlois, Raynold. Interview with author. January 9, 1990.

[232] Ibid.

[233] de Grandpré, A. Jean. Interview with author. February 20, 1990.

[234] Harvey. *Evidence to RTPC*, p. 133; and, Leenders. *Bell Canada Pricing Study*, p. 94.

[235] Northern Electric Research Division. Systems Engineering. *Prospectus.* October 1970. Cited in: Leenders. *Bell Canada Pricing Study*, p. 94.

[236] Benger, Wally. Interview with author. March 20, 1990.

[237] Davies, J. Derek. Interview with author. July 6, 1990.

[238] Light, Walter. Interview with author. March 20, 1990.

[239] Benger, Wally. Interview with author. March 20, 1990.

[240] Anderson, Hugh. "Man in the News: Very Model of Modern Master Scientist Given Tough Research Mandate by Bell." *The Globe and Mail.* December 11, 1970.

[241] For a discussion of Bellcomm's role, and Chisholm's stewardship, see: M. D. Fagen, ed. *National Service in War and Peace* (1925-1975), pp. 695-96.

[242] Scrivener, R. C. Interview with author. March 11, 1990.

[243] There are several histories of the Avro Arrow. For an engaging read, see: Greig Stewart. *Shutting Down the National Dream: A.V. Roe and the Tragedy of the Avro Arrow* (Toronto: McGraw-Hill Ryerson Ltd., 1988).

[244] Lake, Michael. Interview with author. November 17, 1990.

[245] Northern Telecom. *Submission to RTPC*, p. 113.

[246] Figures from: Northern Telecom. Ibid., p. 94; and, Anderson. "Man in the News."

[247] Anderson, Hugh. "Bell, Northern Firm to Get Staff of 1,800." *The Globe and Mail.* November 18, 1970.

[248] Chisholm, Donald. Testimony to Restrictive Trade Practices Commission. Inquiry under Section 47 of the Combines Investigation Act relating to the Manufacture, Production, Distribution, Purchase, Supply and Sale of Communication Systems, Communication Equipment and Related Products. *Transcript of Hearing.* Vol. 215 (Ottawa, November 12, 1980), p. 32,243.

[249] See: Schoenberg, Robert J. *Geneen* (New York: W. W. Norton & Co., 1985), p. 108.

[250] See: Thomas, David. *Knights of the New Technology: The Inside Story of Canada's Computer Elite* (Toronto: Key Porter Books, 1983), p. 24.

[251] Light, Walter. Interview with author. March 20, 1990.

[252] Benger, Wally. Interview with author. March 20, 1990.

[253] Kyles, Jim. Interview with author. February 18, 1990.

[254] Light, Walter. Interview with author. March 20, 1990.

[255] Scrivener, R. C. Interview with author. March 11, 1990.

[256] Ibid.

[257] Light, Walter. Interview with author. March 20, 1990.

[258] See, for example: Schmidt, Arno, assistant vice-president. Strategic marketing programs. Northern Telecom World Trade. Presentation on Global Marketing to Northern Telecom Canada Marketing Conference. October 18, 1990.

259 Fitzgerald, Edmund. Interview with author. July 6, 1990.
260 See: Shepherd, Harvey. "Northern Plans Major Quest for Markets." *The Globe and Mail.* June 29, 1971.
261 Benger, Wally. Interview with author. March 20, 1990.
262 Northern could not use its name in the United States, and the similar name, North Electric, was already taken by a competing U.S. manufacturer. Hence the genesis of the moniker, Northern Telecom.
263 Scrivener, R. C. Interview with author. March 11, 1990.
264 Cited in: Temin. *The Fall of the Bell System,* p. 71.
265 Scrivener, R. C. Interview with author. March 11, 1990.
266 Ibid.
267 AT&T marketing executive. Cited in: Schlesinger et al. *Chronicles of Corporate Change,* p. 19.
268 Leenders, Michiel. *Bell Canada Pricing Study,* p. 9.
269 Cited in: The Canadian Press. "Russian People Resist Change, Kosygin Admits." *The Globe and Mail.* October 19, 1971.
270 See: Northern Telecom. *Submission to RTPC,* pp. 131-32.
271 See: Intel Corp. "Some Background on Intel" (Santa Clara, Calif.: 1985). Mimeo; and, T.R. Reid, *The Chip: How Two Americans Invented the Microchip and Launched a Revolution,* (New York: Simon and Schuster, 1984), pp. 140-41.
272 See: DRIE. Mimeo; and, "Micro-systems Plans to Raise $93,392,000." *The Globe and Mail, Report on Business.* January 15, 1970.
273 See: Anderson, Hugh. *The Globe and Mail, Report on Business.* February

20, 1971; and, Harvey Shepherd. *The Globe and Mail, Report on Business.* April 30, 1971.
274 Shepherd, Ibid.
275 Cited in: Anderson. *The Globe and Mail, Report on Business.*
276 Cited in: Leenders. *Pricing Study,* p. 95.
277 Ibid., pp. 94-95; and, Harvey. *Evidence to RTPC,* pp. 133-34.
278 Leenders. Ibid., p. 73; and, see Northern Telecom Ltd. *Evidence of J. D. Davies, Nature of the Telecommunications Equipment Industry Throughout the World.* Reference Binders 1 & 2. Exhibits T-1566A, B. Filed with Restrictive Trade Practices Commission. Ottawa, May 1980.
279 Leenders. Ibid., p. 70.
280 Harvey. *Evidence to RTPC,* pp. 123-27.
281 See: Temin. *The Fall of the Bell System,* pp. 28-65, 72-73; and, J. D. Davies. "Lest We Forget." Mimeo.
282 Temin. Ibid., p. 74.
283 See: de Butts, John. "Closing Remarks." Presidents' Conference. Key Largo, Florida. May 12, 1972; and, Temin. Ibid., pp. 74-86.
284 de Grandpré, A. Jean. Interview with author. February 20, 1990.
285 Ibid.
286 Scrivener, R. C. Interview with author. March 11, 1990.
287 Inns, G. E. *Evidence to RTPC,* p. 83.
288 Leenders. *Pricing Study,* p. 115.
289 Benger, Wally. Interview with author. March 20, 1990.
290 Bell Canada. Minutes of Board of Directors Meeting. November 22, 1972; and, Stephanie Sykes. Chief historian, Bell Canada. Private

291 communication with author. March 16, 1990.

291 Fox, Francis, Hon. Interview with author. February 8, 1990.

292 Cited in: Murray, Don. "Bell Canada's New President." *The Montreal Gazette.* January 30, 1973.

293 Bandeen, Robert, Dr. Interview with author. January 8, 1990.

294 Cruickshank, Donald. Interview with author. January10, 1990.

295 See: Canada. Parliament. House of Commons. Debates. 13th Parliament, 2nd Session. 26 ER II (1977). Vol. IV (March 10, 1977). (Ottawa: Queen's Printer, 1977), p. 3861.

296 Details on Canada's corporate fleets and aircraft costs were compiled by Les Edwards of the Air Transport Association of Canada and published in: Peter Newman. *The Canadian Establishment,* pp. 455-56.

297 Vaillancourt, Louise Brais. Interview with author. February 20, 1990.

298 Robertson, H. Rocke, Dr. Interview with author. January 23, 1990.

299 Ibid.

300 Cited in: Surtees, Lawrence. "Profile: Now it's on to the Next Billion." *The Globe and Mail, Report on Business.* July 1986, p. 52.

301 Ibid.

302 de Grandpré, A. Jean. Interview with author. February 20, 1990.

303 Ibid.

304 See: Salter, Michael. *"The Emperor of BCE."* The Globe and Mail, Report on Business. April 1988, p. 37.

305 Light, Walter. Interview with author. March 20, 1990.

306 Thackray, James. Interview with author. February 23, 1990.

307 de Grandpré, A. Jean. Interview with author. February 20, 1990.

308 See: Murray, Don. "Bell Canada's New President." *The Montreal Gazette,* January 30, 1973.

309 Robertson, H. Rocke, Dr. Interview with author. January 23, 1990.

310 Figures provided from data in Dr. Robertson's files.

311 Cruickshank, Donald. Interview with author. January 10, 1990.

312 Cited in: Strong, Joanne. "The Informal Jean de Grandpré." *The Globe and Mail.* February 13, 1982.

313 Thackray, James. Interview with author. February 23, 1990.

314 de Grandpré, A. Jean. Interview with author. February 20, 1990.

315 For a discussion of the arrangements between Bell Canada and the cable companies, see: RTPC. *Phase I Report,* pp. 137-140.

316 Hoey, Eamon. Interview with author. February 21, 1990.

317 RTPC. *Phase I Report,* p. 138.

318 Bell Canada. "Integrated Telecommunications Plant." February 21, 1977. Mimeo.

319 Vail, Theodore. AT&T Annual Report, 1910. Cited in: Steve Coll, *The Deal of the Century: The Breakup of AT&T* (New York: Atheneum, 1986), p. iv.

320 Hoey, Eamon. Interview with author. February 21, 1990.

321 Ibid.

322 Henderson, Gordon. Interview with author. December 5, 1990.

323 Ibid.

324 Ibid.

325 RTPC. *Phase I Report,* pp. 4-5.

[326] Hunter, Lawson. Interview with author. January 30, 1990.

[327] See: Tropea, Orland. *Evidence*, p. 21.

[328] Cited in: Canada. Parliament. House of Commons. Debates. 29th Parliament, 1st Session. 22 ER II. Vol. III (1973), (April 2, 1973). (Ottawa: Queen's Printer, 1973), p. 2852.

[329] See: Slatter, Frans. *Parliament and Administrative Agencies.* Study Prepared for the Law Reform Commission of Canada. Administrative Law Series (Ottawa: Supply and Services, 1982), p. 49.

[330] Canada. Parliament. House of Commons. Debates. April 6, 1973, p. 3036.

[331] Benson, Edgar, Hon. Interview with author. January 23, 1990.

[332] Ibid.

[333] Kane, T. Gregory. *Consumers and the Regulators,* p. 15.

[334] Benson, Edgar, Hon. Interview with author. January 23, 1990.

[335] de Grandpré, A. Jean. Brief to the Royal Commission on Corporate Concentration. April 13, 1976; cited in: The Canadian Press. "Bell Chief Tells of Activism at Hearings." *The Globe and Mail.* April 14, 1976.

[336] de Grandpré, A. Jean. Speech to the Conference Board in Canada. Montreal, November 13, 1974. Cited in: "Bell Canada Hopes for Formula on Rates." *The Globe and Mail.* November 14, 1974.

[337] Tropea, Orland. *Evidence*, p. 22.

[338] de Grandpré, A. Jean. In: Canada. Restrictive Trade Practices Commission. *Transcripts of Hearings.* Vol. 226 (April 13, 1981), p. 33940.

[339] Canada. DRIE. "Microsystems International," p. 3.

[340] See: Rodger, Ian. "Wolff Leaves Post at Microsystems; 200 Others Fired." *The Globe and Mail, Report on Business.* November 20, 1973.

[341] See: Shepherd, Harvey. *The Globe and Mail, Report on Business.* November 9, 1974.

[342] See: Rodger, Ian. "Microsystems Idling 492, Closing Plants." *The Globe and Mail, Report on Business.* March 5, 1975.

[343] Cited in: Rodger, Ian. "Split in Bell Shares Viewed 'Advisable.'" *The Globe and Mail, Report on Business.* April 9, 1975.

[344] Palda, Kristian. *Industrial Innovation: Its Place in the Public Policy Agenda* (Toronto: The Fraser Institute, 1984), p. 114.

[345] Jewkes, John. *Government and High Technology.* Occasional Paper No. 37 (London: Institute of Economic Affairs, 1972). Cited in: Palda. Ibid., p. 120.

[346] Canada. Senate. A Science Policy for Canada: Report of the Special Committee of the Senate on Science Policy. Vol. 4: Progress and Unfinished Business (Ottawa: Supply and Services, 1977), p. 53.

[347] Britton, John, and James Gilmour. *The Weakest Link: A Technological Perspective on Canadian Industrial Underdevelopment. Background Study 43. Science Council of Canada* (Ottawa: Supply and Services, 1978), p. 101.

[348] Bell–Northern Research Laboratories Inc. "Microsystems

International Ltd. Spin-offs and their Founders." Ottawa, December 31, 1985. Mimeo.

[349] Cowpland, C. J. Michael, Dr. Interview with author. January 11, 1990.

[350] de Grandpré, A. Jean. Interview with author. February 20, 1990.

[351] Canada. Canadian Transport Commission. *Transcript of Public Hearing. File* 955.183.1. Case T-7/75. Vol. 2 (July 8, 1975), pp. 202, 204.

[352] See: CTC. *Transcripts. Ibid.* Case T-7/75. Vol. 1 (July 7, 1975); and, The Canadian Press. "Lawyers Walk Out on Bell Rate Hearing." *The Globe and Mail.* July 8, 1975.

[353] See: CTC. *Transcripts. Ibid.* Case T-7/75. Vol. 3 (July 9, 1975); and, The Canadian Press. "6,000 Face Layoffs if Bell Rates Vetoed." *The Montreal Gazette.* July 10, 1975.

[354] Bell Canada. News Release. "Bell says Rate Cut Will Mean Reduction in Capital Spending." July 29, 1975. Filed with CTC as Exhibit B-75-431; and see: Susan Wishart. *The Globe and Mail, Report on Business.* July 30, 1975.

[355] Quebec. Minister of Communications. "Le Quebec sera present aux audiences de la commission Canadienne des transports." July 29, 1975. Reprinted in: CTC. *Transcripts. Ibid.* T-14/75. Vol. 4 (October 30, 1975), pp. 608-9.

[356] See: Wishart, Susan. "More Cutting of Costs Implemented by Bell." *The Globe and Mail, Report on Business.* August 13, 1975.

[357] Kane, T. Gregory. Interview with author. October 14, 1989.

[358] See: "Brampton Plea." *The Globe and Mail.* September 10, 1975; and, "Rate Rise Backed." *The Globe and Mail.* September 19, 1975.

[359] See: Beaufoy, John. *The Globe and Mail.* August 22, 1975.

[360] The Canadian Press. *The Globe and Mail.* September 13, 1975.

[361] See: Johnson, William. *The Globe and Mail.* September 13, 1975; and, *The Globe and Mail.* August 23, 1975.

[362] CTC. *Transcripts.* Vol. 4, p. 603.

[363] de Grandpré, A. Jean. Interview with author. February 20, 1990.

[364] CTC. *Transcripts.* Vol. 26 (December 3, 1975), p. 3800.

[365] Johnson, William. *The Globe and Mail.*

[366] Latham, Roberta. Interview with author. June 21, 1990.

[367] de Grandpré, A. Jean. Interview with author. February 20, 1990.

[368] Kane, T. Gregory. Interview with author. October 14, 1989.

[369] Ibid.

[370] de Grandpré, A. Jean. Interview with author. February 20, 1990; and see: CTC. *Transcripts.* Vol. 4, p. 597.

[371] de Grandpré, A. Jean. Ibid.

[372] Roman, Andrew. Interview with author. January 9, 1990.

[373] Ibid.

[374] Langlois, Raymond. Interview with author. January 9, 1990.

[375] Ibid.

[376] Cited in: The Canadian Press. *The Globe and Mail.* October 29, 1975.

[377] Tropea, Orland. *Evidence*, p. 23.

[378] Cited in: Janigan, Mary. *Toronto Star.* December 23, 1975.

[379] See: Rodger, Ian. "Canadian to Get Bell Shares Held by AT&T." *The*

Globe and Mail, Report on Business. January 28, 1975.

[380] de Grandpré, A. Jean. In: RTPC. *Transcript of Hearings.* Vol. 19 (November 7, 1977), p. 2749.

[381] Thackray, James. Interview with author. February 23, 1990.

[382] Vaillancourt, Louise Brais. Interview with author. February 20, 1990.

[383] Robertson, H. Rocke, Dr. Interview with author. January 23, 1990.

[384] Harvey. *Evidence to RTPC,* p. 136.

[385] Northern Telecom. *RTPC Evidence,* p. 101.

[386] Light, Walter. Interview with author. March 20, 1990.

[387] Webster, H. L. "Digital World - Base for the Intelligent Universe." Presentation to Northern Telecom Executive Seminar. Geneva, Switzerland (September 21, 1979), p. 5.

[388] Benger, Walter. Interview with author. March 20, 1990.

[389] Scrivener, R. C. Interview with author. March 11, 1990.

[390] Benson, Edgar J., Hon. Interview with author. January 23, 1990.

[391] Gotlieb, Allan. Interview with author. January 12, 1990.

[392] de Grandpré, A. Jean. Interview with author. February 20, 1990.

[393] Roman, Andrew. Interview with author. January 9, 1990.

[394] de Grandpré, A. Jean. Interview with author. February 20, 1990.

[395] Cited in: Canada. Parliament. House of Commons. 13th Parliament, 2nd Session (1976-77). Minutes of Proceedings and Evidence of the Standing Committee on Finance, Trade and Economic Affairs. Respecting Bill C-42. Issue No. 67 (June 28, 1977). (Ottawa: Supply and Services, 1977), p. 67:76.

[396] Canada. Canadian Radio-television and Telecommunications Commission,Telecommunications Regulation - Procedures and Practices. *Transcript of Proceedings* (Ottawa, October 26, 1976), pp. 264-65. Cited in: Kane. *Consumers and the Regulators,* p. 20.

[397] Gotlieb, Allan. Interview with author. January 12, 1990.

[398] Parkhill, Douglas. Interview with author. May 26, 1987; and, Canada. Department of Communications. "Federally Regulated Carriers and Chartered Banks Participation in Commercial Data Processing." Joint Statement of the Ministers of Finance and Communications. Ottawa, January 16, 1975.

[399] Bell Canada. *Evidence of B. H. Tavner.* Exhibit T-1471. "Marketing and Services." Filed with the Restrictive Trade Practices Commission. Ottawa, March 1980, p. 116.

[400] Hunter, Lawson. Interview with author. January 30, 1990.

[401] Canada. Department of Consumer and Corporate Affairs. Director of Investigation & Research. *The Effects of Vertical Integration on the Telecommunication equipment Market in Canada. Statement of Evidence and Materials filed in the matter of an Inquiry under section 47 of the Combines Investigation Act relating to the Manufacture, Production, distribution, Purchase, Supply and Sale of Communications Systems,*

Communication Equipment and Related Products. (Ottawa: December 20, 1976), p. 11. (Hereafter Green Book).

[402] DIR. Green Book, p. 11; and, Jane Chudy. *The Globe and Mail.* February 4, 1977.

[403] Cited in: Chudy. Ibid.

[404] Canada. Department of Communications. National Telecommunications Branch. Canadian Telecommunications Carriers and Their Suppliers. Ottawa, June 1974; Cited in: Chudy. Ibid.

[405] Hunter, Lawson. Interview with author. January 30, 1990.

[406] Ibid.; and, Peter Connolly. Interview with author. October 12, 1989.

[407] See: The Canadian Press. "Bell-Northern Telecom link will be probed by Ottawa." *The Globe and Mail, Report on Business.* February 17, 1977.

[408] Hunter, Lawson. Interview with author. January 30, 1990.

[409] Ibid.

[410] Walter, Alan. Interview with author. June 12, 1990.

[411] Henderson, Gordon. Interview with author. December 5, 1990.

[412] Cited in: Dow Jones. "Bell Chairman Says Allegations Short on Facts." *The Globe and Mail, Report on Business.* December 23, 1977.

[413] Canadian Press. "Bell Cleared of Tampering in Inquiry." *The Globe and Mail, Report on Business.* November 24, 1978.

[414] Henderson, Gordon. Interview with author. December 5, 1990.

[415] See: RTPC. *Phase I Report,* pp. 1, 232-71.

[416] Hunter, Lawson. Interview with author. January 30, 1990.

[417] de Grandpré, A. Jean. Interview with author. February 20, 1990.

[418] Bell Canada. "Submission in respect of the Subject Matter of Bill C-42. An Act to Amend the Combines Investigation Act." To the Standing Committee on Finance, Trade and Economic Affairs. May 1977. Appendix FTE-52. Printed in: Minutes of Proceedings and Evidence, p. 67A:120.

[419] Henderson, Gordon. Interview with author. December 5, 1990.

[420] Canada. CRTC. *Transcript of Hearings.* Vol. 1, p. 229. Cited in: Canada. CRTC. *Telecom Decision CRTC 77-16.* Challenge Communications v. Bell Canada. Ottawa (December 23, 1977), p. 8.

[421] Kaiser, Gordon. "Competition in Telecommunications: Refusal to Supply Facilities by Regulated Common Carriers." *Ottawa Law Review 13.* No. 1 (1981), p. 97. Cited in: Babe. *Telecommunications in Canada,* p. 148.

[422] For a summary of Bill S-2 and Bill C-1001, and their legislative histories, see: Canada. Parliament. House of Commons. Debates. 13th Parliament, 3rd Session (1977). 26 ER II. Vol. 1 (November 3, 1977). (Ottawa: Queen's Printer, 1977), pp. 617-19.

[423] See: The Canadian Press. "Bell Amendments Hailed as Victory by NDP." *The Globe and Mail.* March 17, 1978.

[424] See: TransCanada Telephone System. Connecting Agreement made as of the 31st day of December, 1976. Memorandum

of Agreement Between Telesat Canada and TCTS. Schedule A, p. 8, clause 17(a).

[425] Canada. Letter from Jeanne Sauvé, minister of communications, to Harry Boyle, chairman, Canadian Radio-television and Telecommunications Commission (December 14, 1976), p. 2.

[426] Telesat Canada. Letter from D. A. Golden, president, Telesat Canada, to T. K. Shoyama, deputy minister, Department of Finance. (May 12, 1976).

[427] Canadian National and Canadian Pacific Telecommunications. Letter from A. J. Kuhr, CN, and R. T. Riley, CP, to David Golden, president, Telesat Canada, and Eldon Thompson, president, TCTS. August 14, 1976; and Memo to File, from David Golden, president, Telesat Canada. August 18, 1976.

[428] TransCanada Telephone System. Letter from Eldon Thompson, president, to Kuhr and Riley. September 1, 1976.

[429] A complete chronology of the connecting agreement is summarized in: Canada. CRTC. *Telecom Decision CRTC 77-10.* "Telesat Canada. Proposed Agreement with TransCanada Telephone System" (Ottawa, August 24, 1977), pp. 13-16.

[430] Canadian Pacific Ltd. "Statement by Board of Directors." Montreal (December 8, 1976), pp. 1-2. Mimeo.

[431] CRTC. *Decision 77-10,* pp. 1-3.

[432] Dalfen, Charles. Interview with author. December 15, 1989.

[433] CRTC. *Decision 77-10,* p. 3.

[434] See: Vandervort, Lucinda. *Political Control of Independent Administrative Agencies,* pp. 95-96.

[435] Canada. Privy Council Office. Order-in-Council P.C. 1977-3152 (November 3, 1977). Mimeo.

[436] Canada. Department of Communications. "Statement by the minister, Hon. Jeanne Sauvé." Ottawa, November 3, 1977.

[437] Cited in: Cruise and Griffiths. *Lords of the Line,* p. 425.

[438] Supreme Court of Ontario. Canadian Pacific Ltd. v. Telesat Canada, 18934/77. Statement of Claim, November 10, 1977. Appeal Book. Vol. 1, pp. 4-6.

[439] Henderson, Gordon. Interview with author. December 5, 1990.

[440] Supreme Court of Canada. *"Judgment."* Mr. Justice DuPont. March 9, 1981.

[441] See: Dow Jones. "Investors Assured by Bell." *The Montreal Gazette.* December 11, 1976.

[442] de Grandpré, A. Jean. Remarks to Luncheon of Toronto Society of Financial Analysts. Toronto, December 2, 1976. Cited in: Jane Chudy. *The Globe and Mail, Report on Business.* December 3, 1976.

[443] Lévesque, René. "For an Independent Québec." *Foreign Affairs.* (Fall 1976), p. 742.

[444] de Grandpré, A. Jean. "Where are we going?" Notes for a Talk to the Chamber of Commerce of the District of Montreal. February 15, 1977.

[445] Laurin, Camille. Minister of state for cultural development. *Quebec's*

Policy on the French Language. Quebec City, March 1977.

446 de Grandpré, A. Jean. Interview with author. February 20, 1990.

447 Quebec. Assemblée nationale. Commission permanente de l'education des affaires culturelles et des communications, 31e Législature, Deuxiéme session. Journal des Débats. No. 128. "Audition des mémoires sur le projet de loi no. 1: Charte de la langue française au Québec" (Quebec, June 21, 1977), pp. CLF 544-45.

448 Ibid., p. CLF-548.

449 Houde, Monic. Interview with author. December 13, 1989.

450 Clark, Gerald. *Montreal: The New Cité,* pp. 25-26.

451 Bandeen, Robert. Interview with author. January 8, 1990.

452 Cited in: Anderson, Ian. "Film: Cast Your Votes and Keep Your Cool." *The Montreal Gazette.* November 14, 1977.

453 de Grandpré, A. Jean. Interview with author. February 20, 1990.

454 Ibid.

455 de Grandpré, A. Jean. Remarks to Bell Canada Annual Shareholders' Meeting. Montreal, April 18, 1979; and, "Change the Language Laws Cut Executives' Taxes: Bell." *The Montreal Gazette.* April 19, 1979.

456 Bryan, Jan. "Higher Bell Rates Seen Under Provincial Regulation." *The Montreal Gazette.* January 26, 1981.

457 CRTC. Canadian Pacific Ltd. and Bell Canada. *Transcript of Hearing.* Pre-Hearing Conference. Vol. 1 (January 10, 1978), p. 20.

458 de Grandpré, A. Jean. Interview with author. April 10, 1990.

459 Canada. CRTC. Canadian Pacific Ltd. and Bell Canada. *Transcript of Hearing.* Vol. 21 (April 20, 1978), p. 3516.

460 Ibid., p. 3553.

461 Ibid., p. 3554.

462 Ibid., p. 3552.

463 Dalfen, Charles. Interview with author. December 15, 1989.

464 Ibid.

465 CRTC. *Transcript of Hearing,* pp. 3532-34.

466 Henderson, Gordon. Interview with author. December 5, 1990.

467 O'Brien, Allan. R. Interview with author. December 5, 1990.

468 CRTC. *Transcript of Hearing.* Vol. 2 (January 11, 1978), p. 1.

469 CRTC. *Transcript of Hearing.* Vol. 12 (March 15, 1978), pp. 2009-2087; and, Charles Dalfen. Interview with author. December 15, 1989.

470 CRTC. *Transcript.* Ibid., p. 2022.

471 Ibid., p. 2023.

472 See: The Canadian Press. "Judge Considers Order Request Against Bell Canada." *The Globe and Mail, Report on Business.* September 10, 1975.

473 Canada. Supreme Court. Reports, Part 3 (1979). Vol. 1. *Bell Telephone Co. of Canada and Harding Communications Ltd. et al* (Ottawa: Supply and Services, 1979), pp. 395-404.

474 Bell Canada. "Memorandum of Evidence of Eugene V. Rostow." Bell Exhibit (March 1978), p. 9.

[475] Canada. Canadian Radio-television and Telecommunications Commission. *Telecom Decision* CRTC 79-11. "CNCP Telecommunications, Interconnection with Bell Canada" (Ottawa, May 17, 1979), p. 199.

[476] de Grandpré, A. Jean. Interview with author. April 10, 1990.

[477] Dalfen, Charles. Interview with author. December 15, 1989.

CHAPTER 5

[1] de Grandpré, A. Jean. Speech. "Notes for an Address to the Annual General Meeting of the Shareholders of Bell Canada" (London, Ont., April 20, 1982), pp. 3-4. Bell Canada Transcript.

[2] de Grandpré, A. Jean. Interview with author. March 12, 1986.

[3] Ibid.

[4] Kyles, James. Interview with author. (February 18, 1990); and biographical sketch of Andrew Kovats. Unpublished mimeo.

[5] See: Raggett, R. J. "Emin Baser Outlines an Ambitious Program for the Turkish PTT." *Telephony.* (July 28, 1986), p. 39.

[6] See: "Canadian Deal with Turkey Near Signing." *The Globe and Mail, Report on Business.* November 25, 1966.

[7] Eadie, Kenneth. Interview with author. February 13, 1990.

[8] Ibid.

[9] Belford, Terrence. *The Globe and Mail, Report on Business.* July 23, 1969.

[10] Cited in: Raggett. *Telephony,* p. 42.

[11] See: The Canadian Press. In: *The Globe and Mail.* November 14, 1978.

[12] Eadie, Kenneth. Interview with author. February 13, 1990.

[13] Kyles, Jim. Interview with author. February 18, 1990.

[14] See: Leenders, Michiel. Pricing Study. In: Bell Canada. *Evidence*, p. 71.

[15] de Grandpré, A. Jean. Interview with author. February 20, 1990.

[16] See: *The Globe and Mail, Report on Business.* February 6, 1976.

[17] Leenders, Michiel. Pricing Study. Bell Canada. *Evidence,* p. 71.

[18] de Grandpré, A. Jean. Interview with author. February 20, 1990; and, Bell Canada International. "Agreement between Bell Canada and Bell Canada International Management, Research and Consulting Ltd. (December 18, 1978). Schedule A: Bell Canada International 1978 Contracts, p. 1 of 2. Exhibit to Bell Canada Annual Report on Form 10-K. Filed with: U. S. Securities and Exchange Commission. Washington.

[19] North Atlantic Treaty Organization. NATO Handbook (Brussels: NATO Information Service, 1974), p. 54.

[20] Kyles, Jim. Interview with author. February 18, 1990.

[21] Ibid.

[22] Ibid.

[23] See: Bell Canada International Inc. 1984 Annual Review (Ottawa: 1985), p. 11.

[24] Kyles, Jim. Interview with author. February 18, 1990.

[25] See: Robinson, Jeffrey. *Yamani: The Inside Story* (London: Simon & Schuster, 1988), p.73.

[26] Buchan, James. "Getting the Phones to Work: The Enormous Effort

Behind Your Telephone." Saudi Business. Supplement to *Arab News*. Riyadh, Saudi Arabia. Vol. III: No. 18 (July 12-18, 1979; corresponding Islamic Date: Shaban 18-24, 1399 AH).

27 Cited in: Powell, William. *Saudi Arabia and Its Royal Family*. (Secaucus, N.J.: Lyle Stuart Inc., 1982), p.9.

28 Background from: Bell Canada International Inc. "Your Visit to Saudi Arabia" (December 12, 1983), p. 18.

29 For details on Saudi Arabia's Second Five-Year Plan, see: Fouad Al-Farsy. *Saudi Arabia: A Case Study in Development* (London: Kegan Paul International, 1982), pp. 144-51.

30 See: Al-Farsy. *Saudi Arabia*. Ibid., p. 74.

31 Kyles, Jim. Interview with author. February 18, 1990.

32 Ibid.

33 de Grandpré, A. Jean. Interview with author. February 20, 1990.

34 BCI. "Your Visit," p. 18.

35 de Grandpré, A. Jean. Interview with author. February 20, 1990.

36 See: Holden, David, and Richard Johns. *The House of Saud* (London: Pan Books, 1981), p. 412.

37 See: BCI. Ibid., p. 18; and, R. G. Gibbens. "Bell Beats World: $3 Billion Contract in Saudi Arabia." *Montreal Star*. December 15, 1977.

38 de Grandpré, A. Jean. Interview with author. February 20, 1990.

39 BCI. "Your Visit," pp. 18-19.

40 de Grandpré, A. Jean. Interview with author. February 20, 1990.

41 Bell Canada. News Release. "Bell Canada Bidding in Saudi Arabia." July 27, 1977.

42 See: Bell Canada International. "Agreement." Schedule A, p. 1. ; and, Pierre Dupont. Interview with author. July 5, 1985.

43 Kyles, Jim. Interview with author. February 18, 1990.

44 de Grandpré, A. Jean. Interview with author. February 20, 1990.

45 Ibid.

46 Thackray, James. Interview with author. February 23, 1990.

47 de Grandpré, A. Jean. Interview with author. February 20, 1990.

48 Kyles, Jim. Interview with author. February 18, 1990.

49 Thackray, James. Interview with author. February 23, 1990.

50 Pochna, Michael. Interview with author. January 10, 1984.

51 Kyles, Jim. Interview with author. February 18, 1990.

52 See: BCI. "Your Visit," p. 19.

53 Thackray, James. Interview with author. February 23, 1990.

54 Robertson, H. Rocke, Dr. Interview with author. January 23, 1990.

55 de Grandpré, A. Jean. Interview with author. February 20, 1990.

56 Kyles, Jim. Interview with author. February 18, 1990.

57 See: Reuters News Agency. "Vance Gets Assurance of Backing From Saudis in Efforts for Peace." *The Globe and Mail*. December 15, 1977, p. 13; and, Robert Lacey. *The Kingdom: Arabia and the House of Sa'ud* (New York: Harcourt Brace Jovanovich, 1981), p. 443.

58 Bell Canada. "Contract for the Management, Operation and Maintenance of the Telephone Network in the Kingdom of Saudi Arabia" (January 25, 1978), p. 2.

Confidential. [Hereafter cited as "ATP Contract, Part 3."]

[59] BCI. "Your Visit," p. 19.

[60] Bell Canada. News Release. Montreal. January 25, 1978. Mimeo.

[61] Saudi Arabia. Council of Ministers. Royal Decree M/2. "Regulation: The Relationship Between the Foreign Contractor and his Saudi Agent" (December 31, 1977; corresponding Islamic date, 21/1/1398); translation of full text published in: Deutsche Bank. *Saudi Arabia* (Frankfurt am Main: 1983), p. 98.

[62] Cited in: Sampson, Anthony. *The Arms Bazaar: The Companies, The Dealers, The Bribes: From Vickers to Lockheed* (Toronto: Hodder and Stoughton, 1978), p. 189.

[63] See: Canada. House of Commons. Debates. 13th Parliament, 3rd Session, 27 ER II (1978). Vol. VI (June 26, 1978). (Ottawa: Supply and Services, 1978), pp. 6741-42.

[64] de Grandpré, A. Jean. Interview with author. April 10, 1990.

[65] See: Newman, Peter. "Business Watch: Engineering a Unique Success." *Macleans Magazine.* October 1986, p. 50.

[66] Bell Canada. Annual Report on Form 10-K for the year ended December 31, 1977; and Exhibit 1-F. Filed on Form 8, with U. S. Securities and Exchange Commission (Washington, D.C.).

[67] See: Bell Canada. Annual Report on Form 10-K for the fiscal year ended December 31, 1981. Filed with the U. S. Securities and Exchange Commission, p. 10.; and, Bell Canada. Prospectus. "$1.96 Cumulative Redeemable

Convertible Voting Preferred Shares of the par value of $25 each" (Montreal: April 12, 1978), p. 8.

[68] Kyles, Jim. Interview with author. February 18, 1990.

[69] Field, Michael. *The Merchants: The Big Business Families of Saudi Arabia and the Gulf States* (Woodstock, N.Y.: The Overlook Press, 1985), pp. 21, 35.

[70] Bell Canada. Letter and Agreement with Binladen Telecommunications Company (September 20, 1977). Signed by J. C. Thackray. Bell Canada and F. O. Hunnewell. Binladen Telecommunications Co.

[71] Kyles, Jim. Interview with author. February 18, 1990.

[72] Bell Canada. Letter and Agreement with Binladen.

[73] See: Lansdowne Ltd. "Certificate of Board Resolution" (January 8, 1976). Nassau, Bahamas. Filed in Quebec Superior Court. District of Montreal. *Lansdowne Financial Services Ltd. et al. vs. Binladen Telecommunications Company Ltd. and Bell Canada.* File No. 500-05-010 219-812.

[74] Thackray, James. Interview with author. February 23, 1990.

[75] Binladen Telecommunications Company Ltd. "Memorandum for Consideration by the Board of Directors." December 11, 1980; and, "Articles of Association of Binladen Telecommunications Co." (January 1976). Filed in: Quebec Superior Court. *Lansdowne et al. v. Binladen.*

[76] Field, Michael. *The Merchants*, pp. 105-106.

[77] Affadavit of Francis O. Hunnewell (January 19, 1983). Filed in Quebec

Superior Court. *Lansdowne vs. Binladen.*

[78] Bell Canada. Letter and agreement, p. 1.

[79] Ibid., pp. 2-3.

[80] Bell Canada. *ATP Contract.* Part 3. Article 23.

[81] Thackray, James. Interview with author. February 23, 1990.

[82] Bell Canada. Letter and Agreement.

[83] See: Amended Declaration. Filed by Lansdowne Financial Services Ltd. (May 20, 1982), pp. 3-6. Filed in: Quebec Superior Court. *Lansdowne et al. vs. Binladen.*

[84] See: Examination on Affidavit. "Deposition of Tahir Mohamed Bawazir" (January 15, 1982), pp. 4, 51; filed in Quebec Superior Court. *Lansdowne et al. vs. Binladen.*

[85] See: Surtees, Lawrence. "Judge Upholds Order Against Saudi Agency." *The Globe and Mail, Report on Business.* January 11, 1984, B7.

[86] Thackray, James. Interview with author. February 23, 1990.

[87] Kyles, Jim. Interview with author. February 18, 1990.

[88] Ibid.

[89] See: Dunne, Dominick. "Khashoggi's Fall: A Crash in the Limo Lane." *Vanity Fair.* September 1989, p. 305.

[90] See: McCartney, Laton. *Friends in High Places: The Bechtel Story; The Most Secret Corporation and How It Engineered the World* (New York: Simon and Schuster, 1988), pp. 209-10.

[91] See: *The New York Times.* April 16, 1980.

[92] See: Lacey, Robert. *The Kingdom,* p. 305.

[93] Decree M/2.

[94] See: Field, Michael. *The Merchants,* p. 116.

[95] Thackray, James. Interview with author. February 23, 1990.

[96] Ibid.

[97] Bell Canada Enterprises Inc. "Subsidiary and Associated Companies of Bell Canada International Inc. As at March 31, 1987" (excludes inactive companies), Chart 5, p. 56 of Briefing Document prepared for federal minister of communications, obtained by author under federal Freedom of Information Act.

[98] See: Gray, Malcolm. *The Globe and Mail, Report on Business.* March 15, 1978.

[99] Kyles, Jim. Interview with author. February 18, 1990.

[100] Bell Canada. *Bell Canada in Saudi Arabia,* p. 19. (Montreal, n.d.) Brochure.

[101] Kyles, Jim. Interview with author. February 18, 1990.

[102] Bell Canada. "Agreement between Bell Canada and Bell Canada International Management, Research and Consulting Ltd." (December 18, 1978). Filed as Exhibit to Annual Report to Form 10-K with U. S. Securities and Exchange Commission. Washington, D. C.

[103] Kyles, Jim. Interview with author. February 18, 1990.

[104] de Grandpré, A. Jean. Interview with author. April 10, 1990.

[105] See: Bell Canada. News Release. Montreal. February 24, 1978.

[106] See: Saudi Telephone. Memo to Minister of PTT, Dr. Alawi Kayal, from D. W. Delaney, general manager. "Saudi Telephone

Monthly Progress Report."
F/4019-78 (18 July 1978), pp. 2-3.
Bell Canada Archives.

[107] Thackray, James. Interview with author. February 23, 1990.

[108] Robertson, H. Rocke, Dr. Interview with author. January 23, 1990.

[109] See: Bell Canada. "Addresses 'Invented'" Bell News. Ontario Region. Special Supplement: Update Saudi Arabia. June 9, 1980, p. 2.

[110] Bell Canada. "Communications."

[111] Saudi Telephone. Report to Saudi Ministry of PTT, from Brian Tickle, general manager. "Year One in Review" (July 1979), p. 15.

[112] Ibid., p. 8.

[113] See: Bell Canada. "A 'Mini Telecom' Fit for a King." Bell/Saudi Retrospective. Bell News. Ontario Region. March 20, 1989, p. 8.

[114] Gander, Barry. Interview with author.

[115] See: Bell Canada. News Release. February 14, 1979.

[116] Saudi Telephone. "Year One," pp. 9-11; and, Saudi Telephone. Management Information. No. 1 (April 1979).

[117] See: Buchan, James. Arab News.

[118] Robertson, H. Rocke, Dr. Interview with author. January 23, 1990.

[119] Thackray, James. Interview with author. February 23, 1990.

[120] de Grandpré, A. Jean. Interview with author. February 20, 1990.

[121] Thackray, James. Interview with author. February 23, 1990.

[122] Canada. Parliament. House of Commons. Debates. 13th Parliament, 3rd Session, 27 ER II (1978). Vol. V (April 24, 1978).

(Ottawa: Queen's Printer, 1978), p. 4780.

[123] Ibid. May 15, 1978, p. 5444.

[124] See: Ontario. Legislature of Ontario. Debates. 31st Parliament, 2nd Session. No. 67 (May 23, 1978). (Toronto: Queen's Printer, 1978), p. 3698.

[125] Bell Canada. "President talks about company's Saudi Arabian contract." Bell News. Ontario Region. January 29, 1979.

[126] Cited in: Bell News. "President talks."

[127] de Grandpré, A. Jean. Interview with author. April 10, 1990.

[128] Ibid.

[129] Canadian Human Rights Commission. Cited in: Lawrence Surtees, "Rights Commission Drops Complaint Over Bell's Hiring." The Globe and Mail, Report on Business. March 23, 1984.

[130] CRTC. Transcript of Hearing. Bell Canada Increase in Rates 1978. Vol. 3 (Ottawa, May 2, 1978), pp. 351-52.

[131] Bell Canada. Memo to Guy Houle, general counsel, from S. I. Ezer. "Re: Saudi Arabia - ATP - s. 320(11) of the Railway Act" (December 20, 1977).

[132] CRTC. Transcript of Hearing. Bell Canada Increase in Rates 1978. Vol. 4 (Ottawa, May 3, 1978), pp. 645-47.

[133] Ibid., p. 675.

[134] CRTC. Transcript of Hearing. Bell Canada Increase in Rates 1978. Vol. 8 (May 10, 1978), pp. 1357-64.

[135] Kane, T. Gregory. Interview with author. October 14, 1989.

[136] Includes preferred shareholders, as of December 31, 1978. Figure from: Bell Canada. Annual Report

1981. "Selected Financial and Other Data," p. 51.

[137] Testimony of James Thackray. Cited in: CRTC. *Transcript* (1978). Vol. 4, p. 644.

[138] CRTC. *Transcript of Hearing.* Bell Canada Increase in Rates 1977. Vol. 7 (Ottawa, March 11, 1977), p. 1370.

[139] CRTC. *Transcript* (1978). Vol. 4, p. 685.

[140] Ibid., pp. 669-70.

[141] Ibid., p. 670.

[142] Bell Canada. "Notes for an address by A. Jean de Grandpré, chairman and chief executive officer, Bell Canada, to the Annual General Meeting of the shareholders of Bell Canada" (Hamilton, Ont., April 18, 1978), p. 6.

[143] CRTC. *Transcript of Hearing.* Bell Canada Increase in Rates 1978. Vol. 5 (Ottawa, May 4, 1978), pp. 911-12.

[144] Canada. Canadian Radio-television and Telecommunications Commission. *Telecom Decision CRTC 78-7.* Bell Canada Increase in Rates 1978 (Ottawa, August 10, 1978), p. 64.

[145] Ibid., p. 65.

[146] Ibid., p. 63.

[147] Ibid., p. 65.

[148] Ibid., pp. 65-66.

[149] Ibid., pp. 66-67.

[150] Dalfen, Charles. Interview with author. December 15, 1989.

[151] de Grandpré, A. Jean. Interview with author. April 10, 1990.

[152] Cited in: Canada. Canadian Radio-television and Telecommunications Commission. *Telecom Decision CRTC 79-1.* Bell Canada request to review that part of Telecom Decision CRTC 78-7 of August 10, 1978, dealing with the Saudi Arabia Telephone Project (Ottawa, February 2, 1979).

[153] Ibid.

[154] Ibid.

[155] de Grandpré, A. Jean. Cited in: Bell Canada. News Release. Montreal. February 2, 1979.

[156] Cited in: Vandervort, Lucinda. *Political Control of Administrative Agencies,* p. 100.

[157] See: Simpson, Jeffrey. "Talks with Premier and Opposition Leader; Clark Makes Pitch in Jerusalem for the Jewish Votes Back Home." *The Globe and Mail.* January 16, 1979, p. 16.

[158] Thackray, James. Interview with author. February 23, 1990.

[159] See: The Canadian Press. "PCs Would Recognize Jerusalem as Capital." *The Globe and Mail.* April 26, 1979, p. 9.

[160] See: Dow Jones. "Saudis Tell Bell Embassy Move Threatens Pact." *The Globe and Mail, Report on Business.* June 22, 1979, p. B14.

[161] Thackray, James. Interview with author. February 23, 1990.

[162] de Grandpré, A. Jean. Interview with author. April 10, 1990.

[163] Off-the-record interview.

[164] de Grandpré, A. Jean. Interview with author. April 10, 1990.

[165] Canada. Department of Industry, Trade and Commerce. Memorandum to M. A. Cohen, deputy minister (August 7, 1979).

[166] Cited in: Canada. Parliament. House of Commons. Debates. 31st Parliament, 1st Session, 29 ER II

(1979). Vol. 1 (October 12, 1979). (Ottawa: Supply and Services, 1979), p. 123.

167 de Grandpré, A. Jean. Interview with author. April 10, 1990.

168 Thackray, James. Interview with author. February 23, 1990.

169 Cited in: Webster, Norman. "Billions May be Riding on Trudeau's Arabian Tour." *The Globe and Mail.* November 17, 1980, p. A1.

170 See: Webster, Norman. "Saudi Oasis Jumps as Trudeau Dances." *The Globe and Mail.* November 20, 1980, p. A1.

171 de Grandpré, A. Jean. Interview with author. April 10, 1990.

172 See: Webster, Norman. "Trudeau is Welcomed by Riyadh Expatriates." *The Globe and Mail.* November 17, 1980, p. A13.

173 de Grandpré, A. Jean. Interview with author. April 10, 1990.

174 Bell Canada. "The Saudi Arabian Telephone Contract: Financial and Regulatory Aspects" (1983), p.3. Mimeo.

175 Davies, J. Derek. Interview with author. July 6, 1990.

176 Benger, Walter. Interview with author. March 20, 1990.

177 See: "Telecom gets new chairman." *The Montreal Gazette.* November 30, 1979.

178 de Grandpré, A. Jean. Interview with author. February 20, 1990.

179 Fitzgerald, Edmund. Interview with author. July 6, 1990.

180 See: "Cutler-Hammer shares sought by Eaton Corp." *The Globe and Mail, Report on Business.* June 27, 1978, B2.

181 Fitzgerald, Edmund. Interview with author. July 6, 1990.

182 Ibid.

183 Light, Walter. Interview with author. March 20, 1990.

184 See: Inco Limited. Appointment Notice. In: *The Globe and Mail, Report on Business.* February 12, 1980, B6.

185 Fitzgerald, Edmund. Interview with author. July 6, 1990.

186 Ibid.

187 de Grandpré, A. Jean. Interview with author. February 20, 1990.

188 Fitzgerald, Edmund. Interview with author. July 6, 1990.

189 Ibid.

190 Ibid.

191 Financial data from: Northern Telecom. *Annual Report 1980* (Mississauga, Ont., 1981).

192 Ibid.

193 See: Northern Telecom. *Annual Report 1984* (Mississauga, Ont., 1985), p. 10.

194 See: Northern Telecom. "Submission to RTPC," pp. 106-7.

195 Figures from: Northern Telecom. *Annual Report 1980.*

196 Benger, Walter. Interview with author. March 20, 1990.

197 Fitzgerald, Edmund. Interview with author. July 16, 1985. Unpublished quote.

198 *Annual Report 1980*, p. 24.

199 Fitzgerald, Edmund. Interview with author. July 6, 1990.

200 Ibid.

201 CRTC. *Transcript of Hearing.* Bell Canada General Increase in Rates 1980. Vol. 20 (Ottawa, June 26, 1980), pp. 3522-23.

202 CRTC. Bell Canada General Increase in Rates. Proceeding 80-14. Bell Canada Exhibit 80-626. Letter from Francis Fox, minister

of communications, to A. Jean de Grandpré, chairman, Bell Canada. May 26, 1980; and letter from A. Jean de Grandpré to Francis Fox, June 5, 1980.

[203] Ibid. Fox to de Grandpré.

[204] CRTC. *Transcript of Hearing* (1980). Vol. 17 (June 20, 1980), p. 2886-87.

[205] Ibid., p. 2889.

[206] Canada. CRTC. *Telecom Decision CRTC 80-14.* Bell Canada General Increase in Rates (Ottawa, August 12, 1980), p. 85.

[207] Ibid., pp. 86-87.

[208] Thackray, James. Interview with author. February 23, 1990.

[209] CRTC. Decision 80-14, p. 61.

[210] Ibid., p. 60.

[211] Thackray, James. Interview with author. February 23, 1990.

[212] Meisel, John, Dr. Interview with author. November 4, 1989.

[213] Ibid.

[214] Canada. Department of Communications. Statement by Francis Fox, minister of communications. "In respect of Petitions to the Governor-in-Council to vary Telecom Decision 80-14" (Ottawa, February 3, 1981).

[215] Fox, Francis, Hon. Interview with author. February 8, 1990.

[216] Lawrence, John. Interview with author. December 12, 1989.

[217] Fox, Francis, Hon. Interview with author. February 8, 1990.

CHAPTER 6

[1] Canada. Canadian Radio-television and Telecommunications Commission. *Telecom Decision CRTC 81-15.* "Bell Canada General Increase in Rates" (Ottawa, September 28, 1981), pp. 30-31.

[2] de Grandpré, A. Jean. Interview with author. March 12, 1986. Quoted in: Lawrence Surtees. "Profile: Now It's On to the Next Billion." *The Globe and Mail, Report on Business.* July 1986, p. 52.

[3] Bell Canada. "Information requested by the Canadian Radio-television and Telecommunications Commission." Item No. 133A. Bell Canada - Proposed Corporate Reorganization (December 7, 1982).

[4] CRTC. *Decision 81-15,* p. 30.

[5] Ibid., p. 31.

[6] Lawrence, John. Interview with author. December 12, 1989.

[7] de Grandpré, A. Jean. Interview with author. April 10, 1990; and see: Thomas Hockin et al. "Bell Canada's Corporate Reorganization." In: Mark Baetz and Donald Thain, eds. *Canadian Cases in Business-Government Relations* (Toronto: Methuen Publications, 1985), p.85.

[8] de Grandpré, A. Jean. Interview with author. April 10, 1990.

[9] Howard, Kenneth. Interview with author. November 30, 1990.

[10] Campeau, Arthur. Interview with author. November 30, 1989.

[11] Bell Canada. *Annual Report 1981* (Montreal, February 1982), p. 14.

[12] Cited in: Quebec Superior Court. *Bell Canada and Bell Canada Enterprises Inc. v. Attorney-General of Canada.* Transcript (September 14, 1982), p. 33. In: File 500-05-010452-827.

[13] de Grandpré, A. Jean. Interview with author. April 10, 1990; and see: A.

Jean de Grandpré. Press Conference. Bell Canada Transcript (Montreal, June 23, 1982), p. 14.

[14] Scrivener, R. C. Interview with author. March 11, 1990.

[15] Canada. Senate. Proceedings of the Standing Committee on Transport and Communications (December 14, 1976), p. 1:10.

[16] Scrivener, R. C. Interview with author. March 11, 1990.

[17] Canada. Revised Statutes of Canada. *Canada Business Corporations Act, Chapter 33.* 30th Parliament, 1st Session, 23-24 ER II (1975), section 181.

[18] Scrivener, R. C. Interview with author. March 11, 1990.

[19] Canada. Revised Statutes of Canada. "An Act to amend the Canada Business Corporations Act," c. 9, 30th Parliament, 4th Session, 27-28 ER II (1978). Sub-section 84(3). Nicknamed "the Bell amendment." It removed the restriction contained in clause 261(6)(b)(i).

[20] A federal Access to Information request by the author in 1989 failed to find any records in the federal government's files or at National Archives on the reasons for the repeal of clause 261(6)(b), or on Bell Canada's role in obtaining that amendment. Canada. Department of Consumer and Corporate Affairs. Access to Information request ATI-89-90-25. Letter from Ken Huband, access coordinator, to author, September 14, 1989; and, National Archives of Canada. Letter from Roy Maddocks, chief of access section, to author. August 22, 1989.

[21] Thackray, James. Interview with author. February 23, 1990.

[22] de Grandpré, A. Jean. Interview with author. April 10, 1990.

[23] Howard, John. Cited in: Canada. Senate. Standing Senate Committee on Banking, Trade and Commerce. First Proceedings on Bill S-2. "An Act to amend the Canada Business Corporations Act." Issue No. 6 (November 23, 1977). (Ottawa: Queen's Printer, 1977), pp. 6-36.

[24] Cited in: Ibid., pp. 6-38.

[25] Ibid., pp. 6-38, 6-39.

[26] Ibid., pp. 6-39.

[27] Ibid., pp. 6-38.

[28] See: U.S. v. AT&T et al. Civil Action No. 74-1698. "Stipulation for Voluntary Dismissal" (January 8, 1982).

[29] de Grandpré, A. Jean. Interview with author. April 10, 1990.

[30] de Grandpré, A. Jean. Interview with author. April 10, 1990; and, Dr. H. Rocke Robertson. Interview with author. January 23, 1990.

[31] Howard, Ken. Interview with author. November 30, 1990.

[32] Bell Canada. Letter from Richard Marchand, general solicitor, to F. H. Sparling, director-corporations March 17, 1982. Confidential. [Obtained under Access to Information request ATI 89-90-25.]

[33] Bell Canada. "Articles of Continuance." Form 11. Canada Business Corporations Act. Filed with: Department of Consumer and Corporate Affairs. Corporation No. 10245.

[34] Cited in: Rose, Frederick, and Alan Freeman. "Canada Bell, by Outsmarting Government, May Sharply Reduce Regulation of

Itself." *The Wall Street Journal.*
September 1982.

[35] Ogilvy, Renault. "Response to Bell
Canada Working Paper on
Proposed Corporate
Reorganization," p. 21.
Memorandum (n.d.)

[36] Canada. Department of Consumer
and Corporate Affairs -
Corporations Branch. Memorandum
from Jean Turner, to F. H. Sparling.
"Application for Continuances.
261(1.3) and (6) Bell Canada."
March 22, 1982. [Obtained under
Access to Information request ATI-
89-90-25.]

[37] Bell Canada. Marchand to Spalding.
March 17, 1982.

[38] de Grandpré, A. Jean. Interview with
author. April 10, 1990.

[39] Canada. Consumer and Corporate
Affairs. Corporations Branch.
Letter from H. Denolf, deputy
director, to Richard Marchand, Bell
Canada. April 7, 1982. [Obtained
under Access to Information
request ATI 89-90-25.]

[40] Bell Canada. Letter from Richard
Marchand, general solicitor, to
director, corporations branch. April
16, 1982. Confidential. [Obtained
under Access to Information
request ATI-89-90-25.]

[41] See: Bell Canada. Resolution adopted
by the Board of Directors.
"Continuance of Bell Canada under
the Canada Business Corporations
Act." *Resolution No. 3* (April 20,
1982); and, Canada. Department of
Consumer and Corporate Affairs.
"Certificate of Continuance,
Canada Business Corporations
Act." Bell Telephone Company of
Canada. N. 10245 (April 21, 1982).

Both documents filed with Quebec
Superior Court (July 16,1982).

[42] de Grandpré, A. Jean. "Notes for an
address to the Annual General
Meeting of the Shareholders of Bell
Canada." London, Ont. April 20,
1982. Bell Canada Transcript, p. 11.

[43] Ibid., p. 11.

[44] See: Canada. Department of
Consumer and Corporate Affairs.
Certificate of Amendment. Canada
Business Corporations Act. Form
4. Tele-Direct Ltd. Corporation
No. 14216 (June 22, 1982).

[45] de Grandpré, A. Jean. Interview with
author. April 10, 1990.

[46] Ibid.

[47] See: Quebec Superior Court.
Transcripts, p. 32; and, The
Consumers' Gas Company.
"Notice of General Meeting of
Shareholders." March 14, 1980.

[48] Consumers' Gas. Ibid., p. 89; and, Bell
Canada. *Annual Report 1982*, p. 46.

[49] Scrivener, R. C. Interview with
author. March 11, 1990.

[50] de Grandpré, A. Jean. Interview with
author. April 10, 1990.

[51] Campeau, Arthur. Interview with
author. November 30, 1989.

[52] Fox, Francis, Hon. Interview with
author. February 8, 1990.

[53] Cruickshank, Donald. Interview
with author. January 10, 1990.

[54] Campeau, Arthur. Interview with
author. November 30, 1989.

[55] Bell Canada. "List of Investments to
be Transferred to BCE."
Information requested by CRTC.
Filed in: Bell Canada - Proposed
Corporate Reorganization
proceedings. Response to
Interrogatory Bell (CRTC)
December 7, 1982-130 BCR." Item

130." Appendix 1 (December 29, 1982).

56 Bell Canada. "Transfer of Bell Canada Investments to BCE." Background Information. August 30, 1982.

57 de Grandpré, A. Jean. Interview with author. April 10, 1990; and, Hockin et al. "Bell Canada's Corporate Reorganization," p. 89.

58 Bell Canada. Resolution adopted by the Board of Directors. _Resolution No. 7._ "Reorganization of the Bell Group of Companies" (June 23, 1982).

59 See: Hockin et al. "Bell Canada's Corporate Reorganization," p. 88.

60 Bell Canada. Transcript of Press Conference. Montreal, June 23, 1982, p. 7; and see: Bell Canada. _Bell News_ (Quebec region). Special Edition. "Bell Proposes to Set Up New Parent Company." June 28, 1982, p. 3.

61 Cited in: Canada. Canadian Radio-television and Telecommunications Commission. Bell Canada - Proposed Corporate Reorganization. _Transcript of Hearing._ Vol. 2 (Hull, Que., February 1, 1983), p. 226.

62 de Grandpré, A. Jean. Interview with author. April 10, 1990.

63 Ibid.

64 Fox, Francis, Hon. Interview with author. February 8, 1990.

65 Bell Canada. Letter from J. Albany Moore, assistant vice-president, to A. Gourd, senior assistant deputy minister, Department of Communications. June 23, 1982. [Obtained under federal Access to Information request.]

66 Meisel, John. Interview with author. November 4, 1989.

67 Campeau, Arthur. Interview with author. November 30, 1989.

68 de Grandpré, A. Jean. "Notes for introductory remarks to Bell Canada news conference." Queen Elizabeth Hotel. Montreal, June 23, 1982. Bell Canada Transcript, p. 7.

69 Fox, Francis, Hon. Interview with author. February 8, 1990.

70 Meisel, John. Interview with author. November 4, 1989.

71 See: Janisch, Hudson. "Bell Canada's Reorganization: Private Interests and Public Concerns." Canadian Law Information Reporter. _Regulatory Reporter_, 3 CRR [1982], pp. 5-46, fn. 23.

72 Bell Canada. Canada Business Corporations Act. "Form 11 - Articles of Continuance." Submitted to Department of Consumer and Corporate Affairs (March 17, 1982). Draft. [Obtained under Access to Information request ATI-89-90-25.]

73 Robertson, H. Rocke, Dr. Interview with author. January 23, 1990.

74 Vaillancourt, Louise Brais. Interview with author. February 20, 1990.

75 Campeau, Arthur. Interview with author. November 30, 1989.

76 Nicholls, John. "The Transforming Autocrat." _Management Today._ March 1988, p. 114.

77 Ibid., pp. 114-18.

78 Vaillancourt, Louise Brais. Interview with author. February 20, 1990.

79 Campeau, Arthur. Interview with author. November 30, 1989.

80 Ibid.

81 Ogilvy, Renault. "Transcript of Hearing for Order." Quebec Superior Court (June 25, 1982), p. 4. Translation.

[82] Campeau, Arthur. Interview with author. November 30, 1989.

[83] Canada. Restrictive Trade Practices Commission (RTPC). Letter from G. M. Payette, secretary, to Warren Grover, Blake, Cassels & Graydon. July 8, 1982. In: Canada. Department of Consumer and Corporate Affairs. RTPC. *Telecommunications in Canada. Part II: The Proposed Reorganization of Bell Canada*. Appendix 1 (Ottawa: July 26, 1982).

[84] Canada. Department of Consumer and Corporate Affairs. RTPC. *Telecommunications in Canada. Part I.*

[85] Benger, Walter. Interview with author. March 20, 1990.

[86] See: Chevreau, Jonathan. "RTPC Claims Bell Units Operated Beyond Firm's Statutory Power." *The Globe and Mail, Report on Business.* September 22, 1981, B4; and, "Bell is Arguing for Status Quo." *The Globe and Mail, Report on Business.* November 11, 1981, B2.

[87] Henderson, Gordon. Interview with author. December 5, 1990.

[88] Canada. Department of Consumer and Corporate Affairs. RTPC. *Telecommunications in Canada. Part III.*

[89] Blake, Cassels & Graydon. Letter from Warren Grover to G. M. Payette, secretary, Restrictive Trade Practices Commission. July 21, 1982. In: RTPC. *Telecommunications in Canada, Part II.*

[90] Ibid., p. 17.

[91] Blake, Cassels & Graydon. Letter from A. J. MacIntosh to Ernest E. Saunders, vice-president law, Bell Canada. September 2, 1982; and,

Blake, Cassels & Graydon. Memorandum to W. M. Grover from Andrea Vabalis. "Re: The Legal Status of the Report of the RTPC on the Proposed Reorganization of Bell Canada." September 2, 1982. Unpublished. [In author's files.]

[92] Canada. Department of Communications. Briefing Document for the Minister. "Bell Canada Reorganization." July 16, 1982. [Obtained under Access to Information request.]

[93] Ibid., p. 5.

[94] Ibid., pp. 5-6.

[95] Ibid., p. 6; and, Bell Canada. News Release. "Bell Seeking 15 Per Cent Increase in 1983 Revenue." June 28, 1982.

[96] Canada. Canadian Radio-television and Telecommunications Commission. *CRTC Telecom Public Notice 1982-31.* "Bell Canada: Articles of Continuance and Proposed Corporate Reorganization" (Ottawa, August 12, 1982), p. 22.

[97] Bell Canada. News Release. "Bell Canada Studies CRTC Concerns About Proposed Corporate Reorganization." August 12, 1982.

[98] Bell Canada. "Notes on intervention and comments of A. J. de Grandpré to E. E. Saunders." August 20, 1982. Memo to File. [Copy in author's files.]

[99] Ogilvy, Renault. "Corporate Reorganization: Memorandum to Legal File," p. 8. (n.d.)

[100] See: Bell Canada. Management Information. "Bell Shareholders Authorize Reorganization." August 19, 1982; and, Exhibit I

to Affadavit No. 3 of Guy Houle, August 19, 1982. Quebec Superior Court.

101 de Grandpré, A. Jean. "Notes for an Address at Special Meeting of Bell Canada Shareholders." Queen Elizabeth Hotel. Montreal, August 18, 1982. Bell Canada Transcript, p. 6.

102 Campeau, Arthur. Interview with author. November 30, 1989.

103 Quebec Superior Court. "Procès-verbal des procédures." August 23, 1982. File 05-010452-827.

104 Hunter, Lawson. Interview with author. January 30, 1990.

105 Campeau, Arthur. Interview with author. November 30, 1989; and Quebec Superior Court. *Transcript of Hearing* August 23, 1982, pp. 95-96.

106 Campeau, Arthur. "Briefing notes for A. J. de Grandpré." (n.d.)

107 See: Quebec Superior Court. "Memorandum of Law, CRTC Approval is not required." Filed by Bell Canada, September 14, 1982; and, Bell Canada. Memo to R. K. Shapiro from Michel Racicot. "Re: Proposed Stock Split of Bell Canada Common Shares." June 8, 1978.

108 Bell Canada. *Bell Canada Act of Incorporation.* Office Consolidation. (May 1979), section 18, p. 12.

109 Campeau, Arthur. Interview with author. November 30, 1989; and see: Chapter 4, p. 119.

110 Quebec Superior Court. "Memorandum that Prior CRTC Approval Is Not Required." Filed by Bell Canada. September 14, 1982, p. 3.

111 Quebec Superior Court. *Order.* Mr. Justice Charles Gonthier (September 24, 1982), p. 24.

112 Lawrence, John. Interview with author. December 12, 1989.

113 See: Canada. Department of Finance. *The Budget. Speech to the House of Commons.* By Allan J. MacEachen, deputy prime minister and finance minister (Ottawa: June 28, 1982).

114 Canada. Parliament. House of Commons. Debates. 32nd Parliament, 1st Session. Vol. 124, no. 374 (July 6, 1982), p. 19039.

115 Canada. Parliament. House of Commons. Debates. 32nd Parliament, 1st Session. Vol. 124, no. 376 (July 8, 1982), p. 19126.

116 Meisel, John. Interview with author. November 4, 1989.

117 Canada. Office of the Prime Minister. Letter from P. E. Trudeau. July 27, 1982.

118 Bell Canada. News Release. "Bell Canada is Reassessing the Situation. July 22, 1982.

119 Canada. Privy Council Office. Order-in-Council P.C. 1982-2350 (August 5, 1982). Mimeo.

120 Cited in: Canada. Department of Communications. News Release. "Bell Canada Rate Increases Limited to 6% and 5% for the Next Two Years." Ottawa: August 5, 1982.

121 Bell Canada. News Release. "Bell Canada Will Comply with Federal Budget Price Guidelines." August 5, 1982.

122 Fox, Francis, Hon. Interview with author. February 8, 1990.

123 Canada. Department of Communications. News Release. "Government Directs

CRTC to Examine the Bell Canada Reorganization" (October 25, 1982).

124 Canada. Privy Council Office. Order-in-Council P.C. 1982-3253 (October 22, 1982). Mimeo.

125 Quebec. Minister of Communications. "Summary" (Quebec City, October 7, 1982).

126 Campeau, Arthur. Interview with author. November 30, 1989.

127 Cruickshank, Donald. Interview with author. January 10, 1990.

128 de Grandpré, A. Jean. Interview with author. April 10, 1990.

129 Canada. Department of Communications. "Briefing Note to the Minister on the Proposed Bell Canada Reorganization" (December 29, 1982), p. 2. [Obtained under Access to Information request, October 1989, 5210-3 (648).]

130 Ibid., p. 3.

131 Meisel, John. Interview with author. November 4, 1989.

132 Cited in: Canada. DOC. News Release. "Government Directs CRTC to Examine the Bell Canada Reorganization" (Ottawa, October 25, 1982).

133 Off-the-record interview.

134 Fox, Francis, Hon. Interview with author. February 8, 1990.

135 Canada. DOC. Memorandum to the Minister, from Alain Gourd. "Bell Reorganization - Tax Status" (October 21, 1982), p. 2. [Obtained under Access to Information request 5210-3(648).]

136 Fox, Francis, Hon. Interview with author. February 8, 1990.

137 Meisel, John. Interview with author. November 4, 1989.

138 de Grandpré, A. Jean. Interview with author. April 10, 1990.

139 Lawrence, John. Interview with author. December 12, 1989.

140 CRTC. Telecom Public Notice 1982-49. "Bell Canada - Proposed Corporate Reorganization" (Ottawa, November 2, 1982).

141 Canada. DOC. Letter from Alain Gourd, senior assistant deputy minister, to the Minister (January 24, 1983); and, "Background Briefing Paper on the Proposed Bell Canada Reorganization." Prepared for government caucus Members of Parliament, p. 3. [Obtained under Access to Information request.]

142 Registered holders of common shares as of January 27, 1983. See: Bell Canada. *Annual Report 1982* (Montreal: 1983), p. 44.

143 Bell Canada. "Letter from the Chairman." In: "Shareholder Newsletter," 1982 Third Quarter Report (November 17, 1982), pp. 1-4.

144 de Grandpré, A. Jean. Interview with author. April 10, 1990.

145 de Grandpré, A. Jean. Cited in: Bell Canada. Transcript. "Questions and Answers, Special Meetings of Shareholders" (August 18, 1982), p. 12.

146 de Grandpré, A. Jean. Interview with author. April 10, 1990.

147 DOC. Gourd Briefing Paper, p. 3.

148 Campeau, Arthur. Interview with author. November 30, 1989.

149 de Grandpré, A. Jean. Interview with author. April 10, 1990.

150 Campeau, Arthur. Interview with author. November 30, 1989.

[151] Bell Canada. Standby Information. "Motion for a Preferential Hearing" (December 2, 1982).

[152] See: Quebec. Quebec Superior Court. Court of Appeal - District of Montreal. "In the Matter of a Proposed Arrangement Concerning Bell Canada and certain of its Shareholders under section 185.1 of the Canada Business Corporations Act as amended." *The Attorney-General of Canada v. Bell Canada et al.* "Motion for Preferential Hearing (Article 523 CCP)" (November 26, 1982). File: 500-09-001 410-828.

[153] Bell Canada. Management Information. "Hearing on the Attorney-General's Appeal is Scheduled to take Place the Week of Feb. 21" (December 7, 1982).

[154] Lawrence, John. Interview with author. December 12, 1989.

[155] Cruickshank, Donald. Interview with author. January 10, 1990.

[156] Meisel, John. Interview with author. November 4, 1989.

[157] Ibid.

[158] Ibid.

[159] Canada. Canadian Radio-television and Telecommunications Commission. *Transcript of Public Hearing.* Bell Canada - Proposed Corporate Reorganization. Vol. 2 (Hull, Que., February 1, 1983), p.138.

[160] Ibid., p. 140.

[161] Bell Canada. CRTC Telecom Public Notice 1982-49. Bell Canada Proposed Corporate Reorganization. *Evidence of Bell Canada* (November 15, 1982), p. 4.

[162] Campeau, Arthur. Memo to file, p. 7.

[163] Bell Canada. *Evidence,* p. 6.

[164] Ibid., p. 8.

[165] Bell Canada. CRTC Telecom Public Notice 1982-49. Bell Canada - Proposed Corporate Reorganization. *Final Argument* (February 21, 1983), p. 5.

[166] National Anti-Poverty Organization. CRTC Telecom Public Notice 1982-49. Bell Canada - Proposed Corporate Reorganization. *Reply Argument* (February 28, 1983), p. 2.

[167] Canada. Canadian Radio-television and Telecommunications Commission. *Transcript of Public Hearing.* Bell Canada - Proposed Corporate Reorganization. Vol. 4 (Hull, Que. February 3, 1983), p. 513.

[168] Bell Canada. "Analysis of the Impact of the Reorganization on Subscribers in 1982" (August 1982). Unpublished memo.

[169] See: Bell Canada. Notices of Special Meetings of Shareholders; Information Circular; Notice of Motion (Montreal, June 25, 1982), p. 7; and: *Evidence,* p. 5.

[170] Consumers' Association of Canada. CRTC Telecom Public Notice 1982-49. Bell Canada - Proposed Corporate Reorganization. *Evidence.* "Statement to the CRTC by Myron Gordon" (September 7, 1982), p. 17.

[171] Bell Canada. CRTC Telecom Public Notice 1982-49. Bell Canada - Proposed Corporate Reorganization. *Reply Argument* (February 28, 1983), p. 26.

[172] Ibid., p. 22.

[173] Consumers' Association of Canada. CRTC Telecom Public Notice 1982-49. Bell Canada - Proposed Corporate Reorganization. *Supplementary Argument* (March 16, 1983).

[174] CRTC. Transcript. Vol. 4, pp. 514-15.

[175] Canada. *RTPC Report, Part II*, pp. 10-11; cited in: Canada. Department of Consumer and Corporate Affairs. Director of Investigation and Research. CRTC. "Bell Canada - Articles of Continuance and Proposed Corporate Reorganization." *Comments* (September 13, 1982), p. 15.

[176] Bell Canada. Comments of E. E. Saunders, vice-president law, to J. G. Patenaude, secretary-general, CRTC. Re: CRTC Telecom Public Notice 1982-31 (September 27, 1982), p. 11.

[177] See: DIR. Comments, p. 70, and: Canada. Canadian Radio-television and Telecommunications Commission. *Transcript of Public Hearing.* Bell Canada - Proposed Corporate Reorganization. Vol. 7 (February 8, 1983), p. 1055.

[178] DIR. Comments, p. 74.

[179] Cited in: CRTC. Transcript. Vol. 3 (February 2, 1983), p. 365.

[180] Ibid, p. 366.

[181] Ibid., p. 360.

[182] Ibid., pp. 361-62.

[183] National Anti-Poverty Organization. CRTC Telecom Public Notice 1982-49. Bell Canada - Proposed Corporate Reorganization. *Evidence.* "Submission by Peter Oliphant" (January 20, 1983), p. 4.

[184] National Anti-Poverty Organization. CRTC Telecom Public Notice 1982-49. Bell Canada - Proposed Corporate Reorganization. *Final Argument* (February 21, 1983).

[185] Bell Canada. *Reply Argument,* pp. 29-30.

[186] Campeau, Arthur. Interview with author. November 30, 1989.

[187] Quebec. Court of Appeal. District of Montreal. *Attorney-General of Canada v. Bell Canada et al.* File 500-09-001410-828. "Opinion of Montgomery J. A." (March 24, 1983), p. 7.

[188] Ibid., p. 6.

[189] Canada. Department of Communications. Memorandum from Vince Hill, director-general Telecommunications Branch, to Joe Thornley, executive assistant to Francis Fox. "Re: Caucus' meetings with A. Jean de Grandpré." File 5608-4 (February 16, 1983). [Obtained under Access to Information request 5210-3(648).]

[190] Fox, Francis, Hon. Interview with author. February 8, 1990.

[191] Off-the-record interview.

[192] See: Canada. Department of Communications. Briefing Document to the Minister, from Alain Gourd, SADM. "Bell Canada Reorganization" (February 15, 1983), p. 4. [Obtained under Access to Information request 5210-3(648).]

[193] Canada. Department of Communications. News Release. "Deadline for CRTC Report on Bell Reorganization Extended to April 19, 1983 (March 25, 1983); and, Privy Council Office. Order-in-Council, P.C. 1983-862 (March 24, 1983).

194 Intven, Hank. Interview with author. August 8, 1991.

195 Canada. Canadian Radio-television and Telecommunications Commission. *Report of the Canadian Radio-television and Telecommunications Commission on the Proposed Reorganization of Bell Canada* (Ottawa: April 18, 1983), pp. 66-70.

196 Ibid., p. 66; and see: Canadian Industrial Communications Assembly (CICA) et al. Bell Canada Proposed Corporate Reorganization. *Reply Argument* (February 28, 1983), pp. 10-15.

197 Lawrence, John. Interview with author. December 12, 1989.

198 CRTC. Transcript. Vol. 3, p. 319.

199 Lawrence, John. Interview with author. December 12, 1989.

200 CRTC. *Report,* p. 27.

201 Bell Canada. *Reply Argument,* p. 75.

202 CRTC. *Report,* p. 57.

203 Hunter, Lawson. Interview with author. January 30, 1990.

204 Lawrence, John. Interview with author. December 12, 1989; and, CRTC. *Report,* pp. 32-35; 42-44.

205 Lawrence, John. Ibid.

206 Canada. Department of Communications. Briefing Paper for the Minister, from Robert Rabinovitch, deputy minister. "CRTC Report on the Bell Canada Reorganization Inquiry" (April 18, 1983), p. 9. [Obtained under Access to Information request 5210-3(648).]

207 See: Chevreau, Jonathan. "CRTC's Green Light on Bell Inspires a Mixed Reaction." *The Globe and Mail, Report on Business.* April 19, 1983, B1.

208 de Grandpré, A. Jean. Interview with author. April 10, 1990.

209 Off-the-record interview with author.

210 DOC. News Release. April 21, 1983.

211 Chevreau, Jonathan. Interview with author.

212 de Grandpré, A. Jean. Interview with author. April 10, 1990.

213 Department of Communications. "Bell Canada Reorganization." Briefing Note for the Minister for use during Question Period. September 6, 1983. [Obtained under Access to Information request, 5210-3(648).]

214 Lawrence, John. Interview with author. December 12, 1989.

215 Meisel, John. Interview with author. November 4, 1989.

216 Off-the-record interview with author.

217 Fox, Francis, Hon. Interview with author. February 8, 1990.

218 Roman, Andrew. Interview with author. January 9, 1990.

CHAPTER 7

1 Newman, Peter. *The Canadian Establishment,* p. 214.

2 Sawatsky, John. *The Insiders: Government, Business, and the Lobbyists* (Toronto: McClelland & Stewart, 1987).

3 Sawatsky, John. "Mulroney's First Term: Rating the Powers that Were." *Vista,* April 1989, p. 93.

4 Bryan, Jay. "Bell Canada Usually Gets What it Wants." *The Montreal Gazette,* February 2, 1978.

5 de Grandpré, A. Jean. Interview with author. February 7, 1990.

6 de Grandpré, A. Jean. interview with author. April 10, 1990.

7 Ibid.

8 de Grandpré, A. Jean. Interview with author. February 7, 1990.

9 Fox, Francis, Hon. Interview with author. February 8, 1990.

10 Armstrong, John. Interview with author. October 5, 1989.

11 See: Canada. Revised Statutes, 35-36-37 ER II, c. 53. "An Act Respecting the Registration of Lobbyists" (Ottawa: Queen's Printer, 1988), sec.5; and, Canada. Consumer and Corporate Affairs. Lobbyists Registration Branch. "Information on the Lobbyists Registration Act and Regulations" (Ottawa: 1988). Booklet.

12 de Grandpré, A. Jean. Interview with author. February 7, 1990.

13 Scrivener, R. C. Interview with author. March 11, 1990.

14 Ibid.

15 Fox, Francis, Hon. Interview with author. February 8, 1990.

16 Canada. Department of Consumer and Corporate Affairs. Lobbyists Registration Branch. "Subject Matters Report (Communications)." August 15, 1990.

17 Fox, Francis, Hon. Interview with author. Februry 8, 1990.

18 Langlois, Raynold. Interview with author. January 9, 1990.

19 Fox, Francis, Hon. Interview with author. February 8, 1990.

20 Ibid.

21 Hunter, Lawson. Interview with author. January 30, 1990.

22 Northern Electric Co. Ltd. Submission to the Royal Commission on Corporate Concentration (October 31, 1975), p. 2.

23 Gotlieb, Allan. Interview with author. January 12, 1990.

24 Meisel, John. Interview with author. November 4, 1989.

25 Lawrence, John. Interview with author. December 12, 1989.

26 Ibid.

27 Canada. Canadian Radio-television and Telecommunications Commission. Office of the Secretary-General to the Commission. Letter from Alain F. Desfossés to Monty Richardson, executive-director, Communications Competition Coalition. September 4, 1990.

28 See: Canada. Parliament. House of Commons. Special Commons Committee on the Meech Lake Accord (Ottawa: Queen's Printer, May 17, 1990).

29 de Grandpré, A. Jean. Interview with author. June 14, 1990.

30 de Grandpré, Louis-Philippe. Interview with author. February 7, 1990.

31 Canada. Minister of Communications. Letter from Flora MacDonald to A. Jean de Grandpré, chairman, Bell Canada Enterprises. Filed by Bell Canada with the CRTC as Exhibit B-87-254 RR88 in Evidence for 1987 Revenue Requirement Hearing. July 14, 1987.

32 See: Consumers' Association of Canada. Press Release. "Consumer and Business Groups Accuse Communications Minister of Dictating to CRTC." September 8, 1987.

33 Canada. Canadian Radio-television and Telecommunications Commission. *Telecom Decision* CRTC 88-4. "Bell Canada - 1988

[34] Canada. Privy Council Office. Order-in-Council P.C., 1988-762 (April 22, 1988).

[35] Canada. Federal Court. *Federal Court Reports* [1989] 3 C.F. Part 4: "Attorney General of Canada v. National Anti-Poverty Organization and Arthur Milner and Bell Canada International Inc. and BCE Inc." A-798-88, pp. 684-709.

[36] Canada. Minister of Communications. Letter from Flora MacDonald to Janet Yale, general counsel, Consumers' Association of Canada. September 9, 1987.

[37] National Anti-Poverty Organization. "Backgrounder for press release" (September 8, 1987), p. 3.

[38] Canada. Minister of Communications. Letter from Flora MacDonald to André Bureau, chairman, CRTC. October 9, 1987.

[39] NAPO. Backgrounder, p. 4.

[40] de Grandpré, A. Jean. Interview with author. April 10, 1990.

[41] Ibid.

[42] Off-the-record interview with author. April 22, 1991.

[43] Lucien Bouchard was appointed Minister of Environment after he won a federal by-election in the riding of Lac-Saint-Jean, Quebec, in June 1988. He defected from the Tory ranks following the defeat of the Meech Lake Accord in the summer of 1990 and now sits as a member representing the separatist Bloc Québécois party he co-founded.

[44] "Canada Elections Act: Registered Part Fiscal Period Return, 1988, on behalf of the Progressive Conservative Party of Canada" (Ottawa: July 1989), p. 130.

[45] "Canada Elections Act: Registered Part Fiscal Period Return, 1988, on behalf of the Liberal Party of Canada," p. 69.

[46] Figures from: Canada. Elections Canada. Office of the Chief Electoral Officer. Returns for 1984-87.

[47] Scrivener, R. C. Interview with author. March 11, 1990.

[48] Canada. Elections Canada. Returns for 1988, p. 20.

[49] See: Sawatsky, John. *The Insiders*, pp. 68-69.

[50] See: Canada. Department of Consumer and Corporate Affairs. Lobbyists' Registration Branch. "Subject Matters Report (Communications)" (January 9, 1990), p. 1.

[51] Comeau, Bill. Interview with author. June 1987.

[52] Fitzgerald, Edmund. Interview with author. July 6, 1990.

[53] Ibid.

[54] United States. Congress. Senate. 97th Congress, 1st Session. *S.898.* "Telecommunications Competition and Deregulation Act of 1981" (October 20, 1981).

[55] United States. Congress. Senate. 97th Congress, 1st Session. *Congressional Record,* S11215 (October 7, 1981).

[56] United States. Congress. House of Representatives. 97th Congress, 1st Session. H. R. 5158. "Telecommunications Act of 1981" (December 10, 1981), S. 267, pp. 62-65.

[57] See: Dunlop, John. "U.S. Plans Telecom Import Barrier."

Canadian Electronics Engineering (January 1982), pp. 20-21.

[58] See: Hockin, Thomas. "Northern Telecom and the American Political Process." In: Baetz and Thain. *Canadian Cases in Business-Government Relations*, p. 397.

[59] Cited in: Ibid., p. 397.

[60] Fitzgerald, Edmund. Cited in: Lawrence Surtees. "Nortel Objects to U.S. Proposal." *The Globe and Mail, Report on Business.* September 15, 1984, B4.

[61] Little, Arthur D. and Associates Inc. International Trade Policies: Implications for U.S. Business. Boston, July 1985. Cited in: J. D. Davies. *Background Paper*, p. 20. Unpublished.

[62] Fitzgerald, Edmund. Interview with author. July 6, 1990.

[63] Cited in: Berry, John. "Timothy E. Wirth: Now, FCC and Market Forces." *Financier: The Journal of Private Sector Policy.* Vol. IX: No. 6 (June 1985), p. 10.

[64] Fitzgerald, Edmund. Interview with author. July 6, 1990.

[65] Surtees, Lawrence. "Nortel Switch to be Used in White House." *The Globe and Mail, Report on Business.* December 4, 1985, B1.

[66] Bamford, James. "Carlucci and the N.S.C." *The New York Times Magazine.* January 18, 1987, p. 16.

[67] Northern Telecom. News Release. "Frank Carlucci Named to Northern Telecom Board of Directors." October 17, 1989.

[68] Fitzgerald, Edmund. Interview with author. July 6, 1990.

[69] See: International Telecommunication Union. Report on the Activities of the International Telecommunication Union in 1976 (Geneva: ITU, 1977), pp. 46-47.

[70] Hébert, Michel. "Panaftel West: Bringing Microwave to Darkest Africa." *Telephony.* November 24, 1986, p. 50.

[71] See: Canadian International Development Agency (CIDA). Project Approval Memorandum, 83-84/41. "Développement des télécoms en Afrique de l'Ouest." Project No. 784/08203 (August 1983).

[72] Hébert. "Panaftel West," p. 50.

[73] CIDA. Consultant and Industrial Relations Directorate. Business Cooperation Branch. "Service Contracts: Continent and Country" (January 5, 1987), p. 25. Public version, published in: Canada. CIDA. The Business of Development: Active Contracts (Ottawa: Winter 1987).

[74] Hébert. "Panaftel West," p. 52.

[75] Canada. The Treasury Board. "Decision of the Treasury Board - Meeting of October 20, 1983." T. B. Minute 790590.

[76] See: Canada. CIDA. "Articles de convention services de consultants et de professionnels." Contract C4137. Project 10373 (January 4, 1985).

[77] Canada. Department of Consumer and Corporate Affairs. Canada Business Corporations Act. "Certificate of Incorporation." Douserv-BCI Inc. No. 176877-8 (October 2, 1984); and, "Form 6." Notice of Directors (September 6, 1984).

[78] See: Canada. Receiver-General for Canada. *Public Accounts of*

Canada, 1984-1985; 1985-1986; 1986-1987; Vol. II: Part II, Details of Expenditures (Ottawa: Supply and Services, 1985, 1986, 1987); and, notes in author's files provided by Stevie Cameron.

[79] *CIDA. The Business of Development;* and, Monique Landry. "Minister's Message." In: CIDA. *The Business of Development.* Consultant Selection: Criteria and Procedures (Ottawa, 1987).

[80] Canada. CIDA. "Service Contracts for Continent and Country" (protected, July 7, 1986).

[81] Canada. CIDA. Letter from Richard Beattie to anonymous. Re: Devtelao (July 31, 1984). [Original in French; copy provided to author by Andrew MacIntosh.]

[82] Canada. The Treasury Board Secretariat. Decision of the Treasury Board - Meeting of January 24, 1985." T.B. Minute 796505. To: Margaret Catley-Carlson, president, CIDA. [Provided to author by Andrew MacIntosh.]

[83] See: Belford, Terrence. "Bell President on Pay Review for Civil Staff." *The Globe and Mail.* June 1, 1973.

[84] See: The Canadian Press. "Ten Executives to Advise Clark Before Summit." *The Montreal Gazette.* June 21, 1979.

[85] Canada. Privy Council Office. Order-in-Council P.C. 1987-2707. "Advisory Council on Adjustment Order."

[86] Canada. Parliament. House of Commons. Debates. 34th Parliament, 1st Session. Vol. 130. No. 3 (December 14, 1988). (Ottawa: Queen's Printer, 1988), p. 51.

[87] Ibid., p. 48.

[88] The Canadian Press. "Open up Markets: Bell Head." *The Montreal Gazette.* October 10, 1984.

[89] See: Bell Canada. Positions on Canada-U.S. Trade Negotiations and Related Issues (July 1986).

[90] Canada. Advisory Council on Adjustment. *Report: Adjusting to Win* (Ottawa: Supply and Services, March 28, 1989).

[91] Gessell, Paul. The Canadian Press. "Powis, de Grandpré Denounce Government's Economic Policies." *The Montreal Gazette.* April 5, 1976.

[92] La Chambre de Commerce de la province de Québec. "La bosse des affaires: Monsieur Jean de Grandpré." *Action Chambre de Commerce.* No. 2 (Septembre 1977).

[93] Bandeen, Robert. Interview with author. January 8, 1990.

[94] See: Ravensbergen, Jan. "An Inactive Retirement Just Wouldn't Feel Right." *The Montreal Gazette.* September 9, 1989.

[95] Scott, Bruce. Interview with author. January 18, 1990.

[96] Salter, Michael. "The Power and the Profit." *The Globe and Mail, Report on Business,* p. 30.

[97] de Grandpré, A. Jean. Chairman. Northern Telecom Ltd. "Remarks to Annual General Meeting of Shareholders" (Ottawa, April 23, 1981).

[98] de Grandpré, A. Jean. Interview with author. December 3, 1990.

[99] Bell Canada. Letter from J. V. Raymond Cyr, chairman and chief executive officer, to Rt. Hon. Brian Mulroney, Prime Minister of Canada. September 23, 1985. [Copy in author's files.]

[100] de Grandpré, A. Jean. Speech to Investment Institute of Canada (Montreal, March 25, 1983).

[101] de Grandpré, A. Jean. "Canada in Perspective." Speech to the 68th National Convention of the Canadian Construction Association (Montreal, February 20, 1986).

[102] de Grandpré, A. Jean. Interview with author. June 14, 1990.

[103] Stanbury, W. T. "Corporate Power and Political Influence." In: R. S. Khemani et al., eds. *Mergers, Corporate Concentration and Power in Canada* (Halifax: The Institute for Research on Public Policy, 1988), p. 408.

[104] Beck, Stanley M. "Corporate Power and Public Policy." In: *Consumer Protection and Environmental Law and Corporate Power* (Toronto: University of Toronto Press, 1985), p. 201. Cited in: Stanbury, p. 408.

[105] de Grandpré, A. Jean. Interview with author. June 14, 1990.

[106] Ibid.; and, Business Council on National Issues. News Release. "Business leaders endorse Goods and Services Tax initiative" (Ottawa, September 20, 1989).

[107] de Grandpré, A. Jean. Interview with author. June 14, 1990.

[108] de Grandpré, A. Jean. Interview with author. June 14, 1990.

[109] See: Newman, Peter. *The Canadian Establishment,* pp. 422-24.

[110] Ibid., p. 426.

[111] de Grandpré, A. Jean. Interview with author. June 14, 1990.

[112] See: Newman. *The Canadian Establishment.* "List of Canada's Top One Hundred Bank Directors and their Corporate Connections" [as of 1972], p. 111.

[113] de Grandpré, A. Jean. Interview with author. June 14, 1990.

[114] See: Mimick, Richard. "The 'New Age' board of directors." *Business Quarterly.* Winter 1985, pp. 48-55.

[115] Fitzgerald, Edmund. Interview with author. July 6, 1990.

[116] de Grandpré, A. Jean. Interview with author. June 14, 1990.

[117] BCE's audit committee is fully described in the introduction to the consolidated financial statements in: BCE Inc. *Annual Report 1990,* p. 36.

[118] de Grandpré, A. Jean. Interview with author. June 14, 1990.

[119] See: Fruhan, William E., Jr. "Corporate Raiders: Head 'em off at Value Gap." *Harvard Business Review.* 88:4 (July/August 1988), pp. 63-68.

[120] See: Farrell, Christopher. "LBOs: The stars, the strugglers, the flops." *Business Week.* January 15, 1990, pp. 58-62.

[121] The Ross Johnson and RJR Nabisco saga is chronicled in: Bryan Burrough and John Helyar. *Barbarians at the Gate: The Fall of RJR Nabisco* (Toronto: Harper Collins, 1990).

[122] Conference Board of Canada, cited in: Bud Jorgensen. "Street Talk: Stock Bonuses for Executives Don't Seem to Guarantee Loyalty." *The Globe and Mail, Report on Business.* May 11, 1990, B7.

[123] From: BCE Inc. "Notice of Annual Meeting 1991 and Information Circular," 1991, pp. 12, 15, 16.

124 BCE, Notice of Annual Meetings, 1988, 1989, 1990.

125 BCE, Notice 1991, p. 15.

126 See: Tauber, Yale. "Trends in Compensation for Outside Directors." *Compensation and Benefits Review.* January/February 1986, pp. 43-52.

127 See: Southerst, John. "How to Attract Right Directors." *The Financial Post.* September 28, 1987; Stratford Sherman. "Pushing Corporate Boards to be Better." *Fortune.* July 18, 1988, pp. 58-67; and, Judith Dobrzynski. "Taking Charge: Corporate directors start to flex their muscle." *Business Week.* July 3, 1989, pp. 66-71.

128 For a recent example, see: Bell Canada. "Strategic Thrusts For 1990 and Beyond" (February 9, 1990), p. 2. Restricted. Filed at: CRTC. Interexchange Competition 1990, Unitel Exhibit 143.

129 Robertson, H. Rocke, Dr. Interview with author. January 23, 1990.

130 de Grandpré, A. Jean. Interview with author. June 14, 1990.

131 Dimma, William. "Viewpoint: The Role of the Board of Directors." *Business Quarterly. Summer 1989,* pp. 28-33.

132 Robertson, H. Rocke, Dr. Interview with author. January 23, 1990.

133 Vaillancourt, Louise Brais. Interview with author. February 20, 1990.

134 Robertson, H. Rocke, Dr. Interview with author. January 23, 1990.

135 See: Schoenberg. Geneen, p. 155.

136 de Grandpré, A. Jean. Interview with author. June 14, 1990.

137 Mimick, Richard. 'New Age' Directors. *Business Quarterly,* p. 54.

138 de Grandpré, A. Jean. Interview with author. June 14, 1990.

139 Ibid.

140 For a general discussion of the duties of chartered bank directors and the function of the board, see: Toronto-Dominion Bank. Brief to the Royal Commission on Corporate Concentration. Toronto, pp. 21-24. Mimeo. (n.d.)

141 de Grandpré, A. Jean. Interview with author. June 14, 1990.

142 de Grandpré, A. Jean. Interview with author. April 10, 1990.

143 See: Canada. Royal Commission on Corporate Concentration. Study No. 3: A Corporate Background Report by Ira Gluskin. "Cadillac Fairview Corporation Limited." January 1976, p. 51.

144 de Grandpré, A. Jean. Interview with author. June 14, 1990.

145 de Grandpré, A. Jean. Interview with author. June 14, 1990.

146 Ibid.

147 Ibid.

148 See: Iacocca, Lee. *Iacocca: An Autobiography* (Toronto: Bantam Books, 1984), pp. 153-54.

149 Ibid., p. 154.

150 de Grandpré, A. Jean. Interview with author. June 14, 1990.

151 *The New York Times Service.* "House approves $1.5 billion bill to save Chrysler." In: *The Globe and Mail.* December 19, 1979, p. 11.

152 Reuters. "Chrysler Clears Loans Seven Years Ahead of Schedule." *The Globe and Mail, Report on Business.* August 13, 1983, B5.

153 See: Iacocca, *Autobiography,* p. 297.

154 de Grandpré, A. Jean. Interview with author. June 14, 1990.

[155] See: Surtees, Lawrence. "Workers are Warily Adapting to Robotics on Assembly Line." *The Globe and Mail, Report on Business.* March 16, 1984, B13.

[156] See: Surtees. "Bell Feeling Affects of Lost Monopoly." *The Globe and Mail, Report on Business.* July 20, 1984, B1.

[157] de Grandpré, A. Jean. Interview with author. June 14, 1990.

[158] Ibid.

[159] See: "Iacocca Talks on What Ails Detroit." *Fortune.* February 12, 1990, pp. 68-72; and, James Treece and David Woodruff. "Crunch Time Again for Chrysler." *Business Week.* March 25, 1991, pp. 92-94.

[160] See: Saunders, John. "Rise in Share Price Forces Directors to Meet at Night." *The Montreal Gazette.* March 28, 1981.

[161] Slocum, Dennis. "Du Pont Bid for Conoco May Have Thwarted Seagram." *The Globe and Mail, Report on Business.* July 7, 1981, B1.

[162] See: Hunter, Nicholas. "Doubts Continue on Seagram's Costly Du Pont Buy-in." *The Globe and Mail, Report on Business.* August 2, 1982, B5.

[163] de Grandpré, A. Jean. Interview with author. June 14, 1990.

[164] See: Rothman, Andrea, with Mark Maremont. "The Maverick Boss at Seagram: Edgar Bronfman Jr. is Reshaping the Company for a Tougher World." *Business Week.* December 18, 1989, pp. 90-98.

[165] de Grandpré, A. Jean. Interview with author. June 14, 1990.

[166] Ibid.

[167] Ibid.

[168] Ibid.

[169] See: "Hail - But Not Farewell." *McGill Reporter.* November 30, 1983.

CHAPTER 8

[1] See: Javetski, Bill. "Bell Canada. 'They're Going to be a Great Big Octopus.'" *Business Week.* March 18, 1985, pp. 57-58.

[2] Cited in: Won, Shirley. "BCE Aims to Expand Around the World." *The Montreal Gazette.* May 29, 1985.

[3] de Grandpré, A. Jean. Interview with author. March 13, 1986; and, Lawrence Surtees. "BCE Head May be Around When Profit Hits $2 Billion." *The Globe and Mail, Report on Business.* March 14, 1986, B1.

[4] Cited in: Surtees, Lawrence. "Bell Canada's Structure is Model for U. S. West." *The Globe and Mail, Report on Business.* January 20, 1984, B17.

[5] See: Schlesinger et al. *Chronicles of Corporate Change*, p. 199.

[6] See: Mittelstaedt, Martin. "How George Mann Stalked Union Gas." *The Globe and Mail, Report on Business.* July/August 1985, pp.54-57.

[7] See: Martin, Douglas. "Rapid Growth at Bell Canada." *The New York Times.* April 17, 1986, p. 31.

[8] de Grandpré, Louis-Philippe. Interview with author. February 7, 1990.

[9] Cunningham, Douglas. "Bell Canada." Equity Research. McLeod Young Weir. April 19, 1983, p. 2.

[10] de Grandpré, A. Jean. Interview with author. March 13, 1986.

[11] de Grandpré, A. Jean. Interview with author. June 14, 1990.

[12] See: Enchin, Harvey, and Virginia Galt. "BCE to Buy Montreal Trustco From Power." *The Globe and Mail, Report on Business.* March 9, 1989, B1.

[13] de Grandpré, A. Jean. Interview with author. December 3, 1990.

[14] Figures from: Compagnie Financière Paribas. *Annual Report 1988.* Paris; and, see: Darrell Delamaide. "Who's Minding the Bank?" *Euromoney.* May 1990, pp. 49-55.

[15] See: Hockin, Thomas. New Directions for the Financial Sector. Ottawa: Department of Finance. December 18, 1986; and, Thomas Courchene. "Re-regulating the Canadian Financial Sector: The Ownership Controversy." In: Khemani et al. *Mergers, Corporate Concentration,* pp. 521-82.

[16] See: Enchin, Harvey. "Cabinet Remark May Jeopardize BCE Offer for Montreal Trustco." *The Globe and Mail, Report on Business.* March 25, 1989, B4.

[17] See: BCE. *Annual Report 1989,* p. 44.

[18] See: Enchin, Harvey. "Giants' Common Goal Led to Trust Firm Sale." *The Globe and Mail, Report on Business.* March 10, 1989, B2.

[19] See: Philp, Margaret. "Chairman Gratton Leaves Montreal in a Power Grab." *The Globe and Mail, Report on Business.* July 27, 1989.

[20] de Grandpré, A. Jean. Interview with author. December 3, 1990.

[21] See: Surtees, Lawrence, and Brian Milner. "BCE Appoints Chief."

The Globe and Mail, Report on Business. October 31, 1990, B1; and, Brian Milner and Lawrence Surtees. "Succession Shuffle; He Came, He Saw, He Left." *The Globe and Mail, Report on Business.* November 1, 1990, B7.

[22] de Grandpré, A. Jean. Interview with author. June 14, 1990.

[23] Mandelman, Avner. "The Belle of Bay Street: Canada's Telephone Giant Turns in a Ringing Performance." *Barron's.* October 8, 1984, pp. 13, 26. Mandelman was then executive vice-president of Brown, Baldwin, Nisker Ltd.

[24] Mandelman. Ibid., p. 26.

[25] See: Surtees, Lawrence. "TCPL Stake Could Give Bell Access to Rights of Way." *The Globe and Mail, Report on Business.* July 16, 1984, B1.

[26] See: Houle, Bernard. "Against the Odds." *Telephony.* June 6, 1988.

[27] Cited in: Surtees, Lawrence. "BCE Boss Gives Law Firms Ultimatum." *The Globe and Mail, Report on Business.* January 24, 1987, p. B5.

[28] See: Partridge, John. "'Corporate-Style Differences' Led to Departure From TCPL." *The Globe and Mail, Report on Business.* June 20, 1985, B4.

[29] Kyles, Jim. Interview with author. February 18, 1990.

[30] de Grandpré, A. Jean. Interview with author. December 3, 1990.

[31] See: TransCanada Pipelines Ltd. Notice of Annual General Meeting of Shareholders, 1986. Calgary, March 1986; and, Dan Westell. "Parachute Softens Fall for ex-TCPL President." *The Globe and Mail, Report on Business.* April 1, 1986, B3.

[32] Westell, Dan. "TCPL Has Prospects in the Pipeline." *The Globe and Mail, Report on Business.* July 15, 1985, B5.

[33] de Grandpré, A. Jean, to author. June 25, 1985.

[34] de Grandpré, A. Jean. Remarks to BCE press conference, Toronto, May 1, 1986.

[35] de Grandpré, A. Jean. Interview with author. June 14, 1990.

[36] See: TransCanada Pipelines Ltd. "TransCanada Pipelines and Encor Energy Announce Corporate Restructuring." January 5, 1989. News Release; and, TRAP. "Material Change Report." January 5, 1989.

[37] See: Fagan, Drew. "NEB Head Steps Down From TCPL Hearing." *The Globe and Mail, Report on Business.* March 27, 1990, B1.

[38] BCE Inc. June 27, 1990. News Release.

[39] BCE Inc. September 10, 1990.The deal involved the sale of units consisting of one TRAP share, paid for in two installments, and one warrant to buy another TRAP share from BCE until December 15, 1992. The first transaction would give BCE $637.5 million in two installments ($328.1 million in October 1990 and $309.4 million payable in October 1991). Exercise of all warrants would give BCE another $656.3 million. News Release.

[40] de Grandpré, A. Jean. Interview with author. December 3, 1990.

[41] de Grandpré, A. Jean. Interview with author. June 14, 1990.

[42] Cunningham, Douglas. Interview with author. September 10, 1990; in: Lawrence Surtees and

Christopher Donville. "TCPL Deal Ends BCE Sell-off." *The Globe and Mail, Report on Business.* September 11, 1990, B1.

[43] See: Simon, Bernard. "U. S. Go-Ahead for C$3bn Canadian Gas Pipeline." *Financial Times* (London). November 15, 1990, p. 38.

[44] Maier, Gerald. Cited in: Surtees and Donville.

[45] BCE. News Release.

[46] de Grandpré, A Jean. Interview with author. December 3, 1990.

[47] Cruickshank, Donald. Interview with author. January 10, 1990.

[48] Richardson, Robert J. Interview with author. February 25, 1985.

[49] BCE. "BCE President to Retire." November 27, 1986. News Release.

[50] Bell Canada Enterprises Inc. *Notice of Annual Meeting 1986;* and Information Circular. March 6, 1986, pp. 12-13.

[51] de Grandpré, A. Jean. Interview with author. December 3, 1990.

[52] Bell Canada. *Bell News.* "Inquiry: Is the World Really Flat? And - Now that You Mention It - Why Doesn't BCE Keep Us Better Informed About Why It's Doing the Things It's Doing?" January 23, 1984, p. 3.

[53] Bell Canada Enterprises Inc. Letter from Marc Ryan, assistant general counsel, to Ontario Securities Commission. March 9, 1984.

[54] Cyr, J. V. Raymond, chairman and CEO, Bell Canada. Speech to the South-Shore (Longeuil) Chamber of Commerce. February 4, 1986; Reprinted in: *Bell News.* March 3, 1986, p. 8.

[55] See: Bell Canada Enterprises Inc. *Annual Report 1984,* p. 14.

[56] See: Surtees, Lawrence. "BCE Unit Buys U. S. Printing Firm." *The Globe and Mail, Report on Business.* March 17, 1984, B6.

[57] de Grandpré, A. Jean. Interview with author. December 3, 1990.

[58] See: Partridge, John. "BCE Deals For Stake in Quebecor." *The Globe and Mail, Report on Business.* June 23, 1988, B1.

[59] de Grandpré, A. Jean. Interview with author. December 3, 1990.

[60] See: Enchin, Harvey. "Quebecor Gets Licence to Print Money." *The Globe and Mail, Report on Business.* October 6, 1988, B14.

[61] de Grandpré, A. Jean. Interview with author. December 3, 1990.

[62] Fox, Francis, Hon. Interview with author. February 8, 1990.

[63] Surtees, Lawrence. "Ottawa Wants Marketplace to Set Customer Rates for Cellular Service." *The Globe and Mail, Report on Business.* March 30, 1984, B1.

[64] See: Minister of Communications. Letter from Hon. Francis Fox to J. V. Raymond Cyr. March 14, 1984; and, Lawrence Surtees. "Tensions High Over Cellular Radio Delay." *The Globe and Mail, Report on Business.* March 16, 1984, B13.

[65] Fox, Francis, Hon. Interview with author. March 29, 1984; cited in: Surtees, Lawrence. "Ottawa Wants Marketplace."

[66] Fox. Ibid.

[67] de Grandpré, A. Jean. Remarks at Inaugural ceremonies of Bell Cellular Mobile Telephone Service. Toronto, June 25, 1985.

[68] Bell Canada. "Bell Creates Private Mobile Telecommunications Subsidiary." NR-86-16 (September 2, 1986). News Release.

[69] Bell Canada. Quarterly Overview: 3rd Quarter 1986. "National Mobile: Combining Know-how with Entrepreneurship." November 24, 1986, p. 1.

[70] Bell Canada. "Bell Canada Reacts to CRTC Rate of Return Decision." News Release. October 15, 1986.

[71] Cyr, J. V. Raymond. Interview with author. October 15, 1986. Unpublished.

[72] Canada. Canadian Radio-television and Telecommunications Commission. CRTC *Telecom Decision* 86-17. Bell Canada - General Increase in Rates and Revenue Requirement. Ottawa, October 14, 1986.

[73] Bell Canada. Management Information. "Bell Canada Sells Most of its BCMC Investments to Bell Canada Enterprises." MI No. 2-87 (January 14, 1987).

[74] Montambault, Léonce. Interview with author. January 20, 1987.

[75] BCE Inc. *Annual Report 1987,* p. 17.

[76] BCE Inc. *Annual Report 1988,* p. 18.

[77] See: Enchin, Harvey. "Newsmaker: BCE Mobile Chief Casts Worldwide Net." *The Globe and Mail, Report on Business.* May 23, 1990, B5.

[78] de Grandpré, A. Jean. Interview with author. December 3, 1990.

[79] Northern Telecom Ltd. Telex from Walter Light, president, to Toji Tashiro, Nippon Telegraph & Telephone. June 24, 1976; in: Northern Telecom Ltd. J. D. Davies. *Nature of the Telecommunications Equipment Industry Throughout the World:* Reference Binder 1. Evidence

with Restrictive Trade Practises Commission. Exhibit T-1566A (May 1980).

[80] See: Smith, Charles. "Canada 'Barred' From Winning Telecommunications Contracts." *Financial Times* (London). July 21, 1976.

[81] Electronic Industries Association. Memo to Gene Lotochinski, Northern Telecom Ltd. March 16, 1976. Position paper from Liberal Democratic Party mission on NTT procurement attached. Reproduced in: Davies. Reference Binder.

[82] See: *Nihon Keizai Shimbun (Japan Economic Journal)*. "Halfway Compromise is not Good: NTT Head Comments on Procurement from U.S." March 20, 1979.

[83] Fitzgerald, Edmund. Interview with author. July 6, 1990.

[84] Ibid.

[85] Hamilton, Hugh. Interview with author. July 23, 1985.

[86] Fitzgerald, Edmund. Interview with author. July 6, 1990.

[87] de Grandpré, A. Jean. Interview with author. December 3, 1990.

[88] Fitzgerald, Edmund. Interview with author. July 6, 1990.

[89] Benger, Walter. Interview with author. March 20, 1990.

[90] Light, Walter. Interview with author. March 20, 1990.

[91] Fitzgerald, Edmund. Interview with author. July 6, 1990.

[92] See: Northern Telecom Ltd. *Annual Report 1988*, p. 7.

[93] de Grandpré, A. Jean. Interview with author. December 3, 1990.

[94] Northern Telecom Ltd. June 1, 1990. News Release.

[95] Fitzgerald, Edmund. Interview with author. July 6, 1990.

[96] Stern Paul, and Tom Shachtman. *Straight to the Top* (New York: Warner Books Inc., 1990).

[97] See: Surtees, Lawrence. "Merger Proposal Frightens Investors." *The Globe and Mail, Report on Business.* November 9, 1990, B1.

[98] Stern, Paul. Interview with author. April 25, 1990.

[99] Canada. Treasury Board of Canada. Secretariat. *Crown Corporations and Other Canadian Government Corporate Interests.* December 1982. (Ottawa: Supply and Services, 1983), p. 3.

[100] Canada Development Corp. *Annual Report 1982.* "CDC Relationship with Federal Government." Text of letters between H. A. Hampson and Jack Austin (Vancouver: 1983), pp. 49-51.

[101] Canada. Privy Council. Order-in-Council P.C. 1982-3575 and P.C. 1982-3576. November 23, 1982. Mimeos.

[102] Canada. Minister of State for Social Development. "Statement by Senator Jack Austin" (Ottawa: May 25, 1983). Mimeo.

[103] See: Dow Jones. "Bell Searching For Acquisitions in Canada, U.S., Chairman Says." *Montreal Gazette.* October 30, 1984.

[104] See: Surtees, Lawrence. "Telephone Firms Eye Teleglobe, Telesat." *The Globe and Mail, Report on Business.* January 11, 1985, B12.

[105] Surtees, Lawrence. "CN, CP Join in Making Bid To Buy Teleglobe from CDIC." *The Globe and Mail, Report on Business.* May 18, 1985, B4.

[106] de Grandpré, A. Jean. Speech to the Canadian Club. "Canada Has a

Global Future." Toronto, December 9, 1985. BCE Transcript, p. 20.

[107] de Grandpré, A. Jean. Remarks to Press Conference. Toronto, May 1, 1986. [Notes in author's files.]

[108] de Grandpré, A. Jean. Remarks to Shareholders. BCE Annual Shareholders' Meeting. Toronto, May 1, 1986.

[109] de Grandpré, A. Jean. Press Conference. May 1, 1986.

[110] See: Canada. Minister of State for Privatization and Minister Responsible for Regulatory Affairs. "Teleglobe Canada to be Sold to Memotec Data Inc." February 11, 1987. News Release.

[111] BCE Inc. May 7, 1987. News Release.

[112] de Grandpré, A. Jean. Interview with author. December 3, 1990.

[113] Gordon Capital Corporation. "Information Circular in Support of the Election of Three New Directors of Memotec Data Inc." Toronto, May 8, 1991.

[114] See: Canada. CRTC. Telecom Decision 1991-19. Teleglobe Canada Inc. Cash Management System (Ottawa: April 19, 1991); and, Lawrence Surtees. "CRTC Slams Memotec: High-Risk Cash Management Called Into Question." *The Globe and Mail, Report on Business.* May 10, 1991, B6.

[115] Cyr, J. V. Raymond. Interview with author. May 16, 1991; and, Lawrence Surtees. "BCE Mulls Next Move On Memotec: Cyr Holds Out Prospect of Forcing Holders' Vote on Management, Restructuring." *The Globe and Mail, Report on Business.* May 17, 1991, B4.

[116] Cyr, J. V. Raymond. Interview with author. May 16, 1991.

[117] Memotec Data Inc. " Message to Shareholders from William McKenzie." Advertisement in: *The Globe and Mail, Report on Business.* May 13, 1991, B14.

[118] Monty, Jean. Interview with author. May 10, 1991; and, see: Lawrence Surtees. "BCE Seeks More of Teleglobe." *The Globe and Mail, Report on Business.* May 11, 1991, B1.

[119] Cyr, J. V. Raymond. Interview with author. May 16, 1991.

[120] Bell Canada. "Supplemental Final Argument." Submitted to CRTC in Teleglobe Canada Inc., 1991 Revenue Requirement Hearing (Ottawa, August 28, 1991). Mimeo.

[121] de Grandpré, A. Jean. Interview with author. December 3, 1990.

[122] Ibid.

[123] Cruickshank, Donald. Interview with author. January 8, 1990.

[124] Lawrence, John. Interview with author. December 12, 1990.

[125] de Grandpré, A. Jean. Interview with author. December 3, 1990.

[126] Canada. Department of Communications. Department of Communications Act. Canada Gazette Notice No. DGTP-09-89. "Local Distribution Telecommunication Networks." Ottawa, September 2, 1989.

[127] Canada. Department of Communications. Notes for a Speech by Marcel Masse concerning broadcasting and technical convergence at the

Annual Convention of the Canadian Cable Television Association. Toronto, May 8, 1989, p. 11.

[128] See: Lush, Patricia. "BCE Chief Assails Curbs; Plans Thrust Into Cable TV." *The Globe and Mail, Report on Business.* May 3, 1989, B1.

[129] BCE Inc. September 27, 1989. News Release.

[130] Connolly, Gerald, to author. January 25, 1991.

[131] Cyr, J. V. Raymond. Cited in: Lawrence Surtees. "BCE Chairman Weathers Anger of Shareholders." *The Globe and Mail, Report on Business.* May 4, 1990, B1.

[132] Ibid.

[133] de Grandpré, A. Jean. Interview with author. December 3, 1990.

[134] Scrivener, R. C. Interview with author. March 11, 1990.

[135] Olympia & York Developments Ltd. Letter from Paul Reichmann to J. J. Hillel, Bell Canada. November 23, 1979. Filed with the CRTC 1980 Rate Application. Exhibit B-80-624.

[136] See: Bryan, Jan. "Trizec Holds Trump Card in Bell Project." *The Montreal Gazette.* December 9, 1978.

[137] Ibid.

[138] de Grandpré, A. Jean. Press Conference. Cited in: Fran Halter. "Reorganized Bell Looks Back on 'Exciting' First Year." *The Montreal Gazette.* April 26, 1984.

[139] BCE Inc. *Annual Report 1984,* p. 15; and, "Bell, Oxford Plan Project in Toronto." *The Globe and Mail, Report on Business.* July 11, 1984, B9.

[140] BCE. Ibid.; and, "Company News: Canada Trust," *The Globe and Mail, Report on Business.* Oct. 2, 1984, B6.

[141] Cyr, J. V. Raymond. Chamber of Commerce speech, p. 8.

[142] See: Daon Development Corp. Information Circular. February 15, 1985, p. 4.

[143] See: Johnson, Arthur. "Pension Fund Group Acquires Big Stake in Park Place Tower." *The Globe and Mail, Report on Business.* June 19, 1985, B3.

[144] Ibid., and, de Grandpré, A. Jean. Interview with author. December 3, 1990.

[145] See: Supreme Court of Ontario. Selim Anter v. Bimcor Inc. Bell Canada Enterprises Inc., 149177 Canada Ltd. and BCE Realty Inc. "Amended Reply." File 12308/86, p. 1.

[146] BCE Inc. *Annual Report* on Form 10-K, For the Fiscal Year ended December 31, 1985, p. I-14.

[147] Lush, Patricia. "BCE to Make Offer to Daon Holders." *The Globe and Mail, Report on Business.* October 17, 1985, B8.

[148] BCE Inc. October 24, 1985. News Release.

[149] BCE Inc. October 28, 1985. News Release.

[150] Daon Development Corp. November 29, 1985. News Release.

[151] Cited in: Surtees, Lawrence. "Daon Creditors Support Marriage to BCE." *The Globe and Mail, Report on Business.* November 30, 1985, B1.

[152] de Grandpré, A. Jean. Interview with author. December 3, 1990.

[153] Ibid.

[154] See: Surtees, Lawrence. "Oxford Sells Top Holdings to BCE Unit." *The Globe and Mail, Report on Business.* March 13, 1986, B1.

[155] Ibid.

[156] BCE Development Corp. BCED Capital I Corp. Prospectus (Initial public offering). February 12, 1987, p. 17.

[157] Supreme Court of Ontario. "Statement of Claim." June 20, 1986, p. 5.

[158] Supreme Court of Ontario. Maron Land Development Inc. v. Bell Canada Enterprises Inc. et al. "Statement of Claim." File 9666/86 (February 24, 1986).

[159] Surtees, Lawrence, and Geoffrey York. "Blocked $63 Million Quick Flip Shakes up Subsidiary of Bell." *The Globe and Mail.* March 12, 1986, A15.

[160] de Grandpré, A. Jean. Interview with author. March 12, 1986.

[161] Metropolitan Toronto Police. Interview with author. March 14, 1986.

[162] Smith, Garry. Interview with author. March 11, 1986; cited in: Surtees and York.

[163] Supreme Court. Anter v. Bimcor. "Amended Statement of Defence" (August 29, 1986), pp. 11-21.

[164] Supreme Court. Maron v. BCE. "Affidavit of Documents" (September 30, 1986), Schedule B, p. 8.

[165] de Grandpré, A. Jean. Interview with author. December 3, 1990.

[166] Ibid.

[167] BCED. *Annual Report 1988.* "Report to Shareholders, John W. Poole." March 17, 1989, p. 3.

[168] de Grandpré, A. Jean. Interview with author. December 3, 1990.

[169] BCE Inc. *Annual Report 1988.* p. 55; and, Dow Jones Service. "O & Y in Talks to Acquire 49.9 Per Cent Stake in BCE Unit." *The Globe and Mail, Report on Business,* B3.

[170] BCE Inc. Joint announcement from BCE Inc., BCE Development and Olympia & York. March 17, 1989. News Release; and, see: Harvey Enchin. "Deal Dies for O & Y to Pump Cash Into BCED." *The Globe and Mail, Report on Business.* March 18, 1989, B1.

[171] See: Lush, Patricia. "Poole Plans to Step Down as Focus Shifts for BCED." *The Globe and Mail, Report on Business.* May 2, 1989, B1.

[172] See: Stinson, Marian. "Olympia & York Offers $557 Million for BCED." *The Globe and Mail, Report on Business.* June 28, 1989, B1.

[173] Olympia & York Developments Ltd. August 15, 1989. News Release; and, Margaret Philp. "O & Y Fails to Get Enough of BCED." *The Globe and Mail, Report on Business.* August 16, 1989, B1.

[174] See: Stackhouse, John. "The Mighty Crash of BCED." *The Globe and Mail, Report on Business,* 1990, p. 54.

[175] de Grandpré, A. Jean. Interview with author. December 3, 1990.

[176] Naiman, Robert. Interview with author. September 15, 1989; and, BCE Development Corp. Sept. 15, 1989. News Release.

[177] BCED Corp. "BCE Development announces line of credit

arrangement." September 27, 1989. News Release.

[178] BCED Corp. October 6, 1989. News Release.

[179] de Grandpré, A. Jean. Interview with author. December 3, 1990.

[180] Brookfield Development Corp. *Annual Report 1989*, p. 2.

[181] L'Heureux, Willard. Remarks to Debenture Holders. January 25, 1991. [Notes in author's file.]

[182] Denton, J. A. "BCED Default." Mimeo. (n.d.)

[183] de Grandpré, A. Jean. Interview with author. December 3, 1990.

[184] See: Hatter, David. "BCE Phones Home." *Canadian Business.* August 1990, p. 32.

[185] BCE Inc. December 19, 1990. News Release.

[186] de Grandpré, A. Jean. Interview with author. December 3, 1990.

[187] See: Cyr, J. V. Raymond. Chamber of Commerce Speech; and, Karen Howlett. "BCE Buys Campeau Stake in Computer Retailer." *The Globe and Mail, Report on Business.* July 25, 1985, B1.

[188] Bryden, Roderick. Interview with author. August 9, 1989.

[189] SHL Systemhouse Inc. September 12, 1989. News Release.

[190] de Grandpré, A. Jean. Interview with author. December 3, 1990.

EPILOGUE

[1] Ravensbergen, Jan. "The de Grandpré Legacy." *The Montreal Gazette.* September 9, 1989, C1.

[2] Adams, Walter. "Corporate Concentration and Power." In: R. S. Khemani et al., ed. *Mergers, Corporate Concentration and Power in Canada* (Halifax: The Institute for Research on Public Policy, 1988), p. 253.

[3] Laurin, Pierre. Interview with author. January 16, 1990.

[4] Cyr, J. V. Raymond. In BCE *Annual Report 1989*, p. 5.

[5] de Grandpré, A. Jean. Interview with author. December 3, 1990.

[6] Figures from: BCE *Annual Report 1990*, pp. 66-67.

[7] de Grandpré, A. Jean. Interview with author. December 3, 1990.

[8] Scrivener, R. C. Interview with author. March 11, 1990.

[9] Fitzgerald, Edmund. Interview with author. July 6, 1990.

[10] Dalfen, Charles. Interview with author. December 15, 1989.

[11] Scrivener, R. C. Interview with author. March 11, 1990.

[12] Vaillancourt, Louise Brais. Interview with author. February 20, 1990.

[13] de Grandpré, A. Jean. Interview with author. December 3, 1990.